Wang, Chang, Hsu

Molding Simulation:
Theory and Practice

Maw-Ling Wang
Rong-Yeu Chang
Chia-Hsiang (David) Hsu

Molding Simulation: Theory and Practice

2nd Edition

HANSER
Hanser Publishers, Munich

The Authors:

Maw-Ling Wang, Chupei City, Hsinchu County 302, Taiwan
Rong-Yeu Chang, Chupei City, Hsinchu County 302, Taiwan
Chia-Hsiang (David) Hsu, Chupei City, Hsinchu County 302, Taiwan

MIX
Paper from
responsible sources
FSC® C083411

Distributed by:
Carl Hanser Verlag
Postfach 86 04 20, 81631 Munich, Germany
Fax: +49 (89) 98 48 09
www.hanserpublications.com
www.hanser-fachbuch.de

Editor: Mark Smith
Production Management: Der Buchmacher, Arthur Lenner, Windach
Cover concept: Marc Müller-Bremer, www.rebranding.de, Munich
Cover image: © CoreTech System Co., Ltd. (Moldex3D)
Cover design: Max Kostopoulos
Typesetting: Eberl & Koesel Studio, Altusried-Krugzell, Germany
Printed and bound by CPI books GmbH, Leck
Printed in Germany

ISBN: 978-1-56990-884-6
E-Book ISBN: 978-1-56990-885-3

Acknowledgments

We would like to thank and acknowledge Beaumont Technologies, Inc. for the flow imbalance case in Section 7.1.4 and Section 7.2.1, OPM Laboratory Co., Ltd. for the conventional and conformal cooling design cases in Section 8.2.1, Associate Prof. Chao-Tsai Huang for the coupling effects between flow and fibers study of ASTM D638 standard test pieces in Section 10.2.3, Ann Tong Industrial Co., Ltd. For the single-gate hot runner system case in Section 11.2.1, Associate Prof. Chao-Tsai Huang for co-injection molding of standard tensile bar test piece in Section 12.2.3, Prof. Shi-Chang Tseng and Prof. Shia-Chung Chen for the gas-assisted injection molding cases in Section 13.2.1 and Section 13.2.3, respectively, Prof. Shih-Jung Liu for the water-assisted injection molding case in Section 13.2.2, Trexel Inc. for the MuCell® case in Section 14.2.1, Proplast, Inc. for the U-shape model of microcellular injection molding in Section 14.2.2, Prof. Shyh-Shin Hwang for the chemical foaming injection molding case in Section 14.2.3, Prof. Shun-Tian Lin for the metal injection molding case in Section 15.2.1, Atech Composites Co., Ltd. for resin transfer molding cases in Section 16.3.3 and Section 16.3.5, Associate Prof. Yuan Yao for the resin transfer molding case in Section 16.3.4, and Amkor Technology Korea Inc. for the IC packaging cases in Section 17.2.1 and Section 17.2.2. These practical cases are quite valuable and helpful to illustrate how to co-develop innovative molding technologies and solve molding issues with the CAE tool.

We would also like to thank the following for their contributions: Dr. Che-Ping (Barton) Lin on Chapter 1, Dr. Chen-Chieh (Jye) Wang and Dr. Chih-Wei (Joe) Wang on Chapter 2, Tsai-Hsin (Sam) Hsieh, Tsai-Heng (Paul) Tsai, Dr. Ying-Mei (May) Tsai, Dr. Yao-Chen (Cloud) Tsai, and Yao-Wei (Willie) Chuang on Chapter 3, Tsai-Hsin (Sam) Hsieh, Wen-Bing (Webin) Liu, Dr. Ying-Mei (May) Tsai, and Dr. Chuan-Wei (Arvid) Chang on Chapter 4, Hsien-Sen (Ethan) Chiu and Dr. Ying-Mei (May) Tsai on Chapter 5, Yu-Chih (Goran) Liu, Wen-Bing (Webin) Liu, and Wen-Hsin (Debbie) Weng on Chapter 6, Dr. Che-Ping (Barton) Lin, Dr. Sung-Wei (Franz) Huang, Dr. Yao-Chen (Cloud) Tsai, and Yao-Wei (Willie) Chuang on Chapter 7, Dr. Chih-Wei (Joe) Wang, Dr. Sung-Wei (Franz) Huang, and Hung-Chou (Kent) Wang on Chapter 8, Dr. Shih-Po (Tober) Sun and Wen-Hsin (Debbie) Weng on Chapter 9, Dr. Huan-Chang (Ivor) Tseng on Chapter 10 and Chapter 15, Tsai-Hsin (Sam) Hsieh on Chapter 11, Dr. Chih-Chung (Jim) Hsu and Yu-Sheng (Tim) Chou on Chapter 12 and Chapter 13, Yuan-Jung (Dan) Chang, Li-Yang (Robert) Chang and Chih-Wei (Joe) Wang on Chapter 14, Hsun (Fred) Yang on Chapter 16, and Chih-Chung (Jim) Hsu, Chia-Peng (Victor) Sun, Chen-An (Jennan) Wang, and Yu-En (Joseph) Liang on Chapter 17. They dedicated their wisdom and skills, and a great deal of time, to complete this wonderful book.

Moreover, a very special thanks to Chia-Lin (Carol) Li for redrawing figures, Pao-Hui (Ryan) Wan for his assistive editing, and Dr. Ying-Mei (May) Tsai and Dr. Che-Ping (Barton) Lin for their executive editing.

Preface

Injection molding techniques have been developed over decades and well-applied in automotive, 3C (Computer, Communication, and Consumer electronics), optics, medical products, and in daily necessities, among other areas. Due to this long-term development and widely ranging applications, the individual molding criteria have been specialized in several industries to fit various product specifications and innovative materials.

The increasing requirements and diversity of plastic products demand a shorter time to market. However, much time can be spent in developing the procedures for some products, from concept generation, design drawing, mold tooling and assembling, and trial-molding through to mass production. "How can the procedures be shortened using CAE (Computer Aided Engineering) tools?" then becomes a key question for industry. The idea is to predict potential molding problems and defects by CAE during the design stage, modify the design according to these results, and then re-analyze until the best design is obtained. This book is written to provide practical and user-friendly guidance in this area. Since the 1970s, virtual trial moldings have been implemented by computer using injection molding simulation CAE tools to check whether the molding parameters are good enough for manufacture. These parameters are part design, gate design, runner layout, cooling layout, molding materials, process conditions, and so on. From CAE, the optimized parameters can be estimated efficiently and provided as the initial-guess settings for the real molding to cost down in time, manpower, material, and energy. To summarize, CAE is a decades-proven design-verification tool for real applications of the injection molding process.

In addition to conventional injection molding, there are many innovative molding processes that have appeared. Molding issues become more challenging and complicated with innovations in processes and materials, which can lead to a longer time and higher costs in conditions optimization. In this 2nd edition, the material measurement data are elaborated in Chapter 2 for deeper understanding in polymer processing property effects, including the viscosity comparison between different types or grades of polymer; measurement of viscoelasticity and method to obtain the master curve; data interpretation of PVT, thermal conductivity, and dynamic modulus of solid-like viscoelastic; and comparison of curing kinetics and reactive viscosity between epoxy molding compounds (EMC). The new concept of connecting Smart Design and Smart Manufacturing is conveyed in Section 4.6 with the methods of Machine Characterization, combination of CAE and injection machine on site, and a practice case.

For further understanding the product quality affected by processing, the flow-fiber coupling model is expressed in Section 10.1.2 to show the flow-induced fiber orientation effect; iARD-RPR models applied with GNF-decoupling and IISO coupling are compared in Section 10.2.2; a study on the tensile strength of test pieces with different gating systems is demonstrated in Section 10.2.3; a microcellular injection part is molded and the sink marks, warpage, and micro characteristics of cell size and cell density are validated by CAE in Section 14.2.2; PU reactive foaming is also addressed in Section 14.2.4 with the material characterization method and a practice case; the experiment with a professional instrument of EASYPERM is illustrated in Section 16.2.3 to obtain the more accurate permeabilities for RTM (Resin Transfer Molding) simulation. And for IC packaging, the warpage of a bi-material component model and a bi-material strip are expounded in Section 17.2.5 and Section 17.2.6, and the effect of dispensing control and creeping behaviors on underfill process is discussed in Section 17.2.7.

From decades of experience in CAE assistance in molding troubleshooting, we have found that processing knowledge is as important as software operation to CAE users. To make a high-quality molded product, the total effects of part design, mold design and manufacture, machine capability, and material properties must all be taken into account and then integrated into the CAE tool to implement design verification and conditions optimization wisely. Each of these definitely involves a deep knowledge, whether in theory and/or empirical formula. When talking about molding issues, plastics rheology and the designs of part and mold are especially the key criteria since their interactions will dominate the material property variations inside the mold.

At Moldex3D, as worldwide leaders in molding simulation software, we are not just continuously enhancing CAE capability but also intend to help industry people improve their molding-related abilities. The importance of training and instruction has become strongly apparent to us. As a result, this book consists of plastics molding theory, practical applications, and case studies intended to elaborate the molding system and melt flowing behaviors in an easy-to-understand way. The practical examples show how to use CAE to achieve design verification and process innovation in conventional injection molding, G/WAIM, co-/bi-injection, foam injection molding, PIM, RTM, and IC packaging. With this book, readers can effectively learn molding simulation applications and its importance in molding industries.

The CAE case study exercises found in the book for execution in the Moldex3D software can be downloaded from the Website: *https://moldex3d.box.com/s/zr6fvc1vlhbi4ocx111jwd3wmxt4ooif*, for which the QR code is as follows:

Maw-Ling Wang March 2022

Rong-Yeu Chang

Chia-Hsiang (David) Hsu

Contents

1 Overview of Plastics Molding

Che-Ping (Barton) Lin

The context of plastics molding will be briefly introduced in this chapter using the most popular method of injection molding. Two major topics are included in this chapter:

1. Introduction to Injection Molding: The systems of injection molding and the defects of injection molded products are described.
2. Core Values of Molding Simulation: The core values of simulation in injection molding will be introduced at the end of this chapter.

1.1 Introduction to Injection Molding

Plastics can be shaped because of their ductility and plasticity. Therefore, plastics have been widely used in daily life and become a necessary part of the current world. The source of plastic products comes from customers' "needs", which are then developed to "design concepts". Such design concepts will be delivered to product designers for product design before being handed over to mold designers for mold design and development, as shown in Figure 1.1.

There are four stages from product development to mass production:

1. Design product drawings according to its functions, appearance, material, and processes, and hand over to mold factories for the design and manufacturing of molds.
2. Mold designers undertake discussion, drawing design, machining, mold-closing, and other procedures upon receiving product drawings, samples, or relevant specifications regarding material, weight, color, etc., which are used to manufacture the molds based on the conclusions made in the mold manufacturability meeting. The manufactured molds will be delivered to molding factories for mold test, modification, and detection.

3. Molding engineers execute tests to obtain better molding conditions for smooth production during the mold test stage, and provide feedback comments for mold modification regarding the difficult points for molding. The mold test will be executed repeatedly after the mold modification until the product quality achieves the specification of the mold test.
4. The production yield is improved via small-scale production and quality certification before the mass production stage begins. To obtain high-quality plastic products, one must first understand the relevant principles and mechanisms of the molding process. This section will briefly introduce the most common components of injection molding systems.

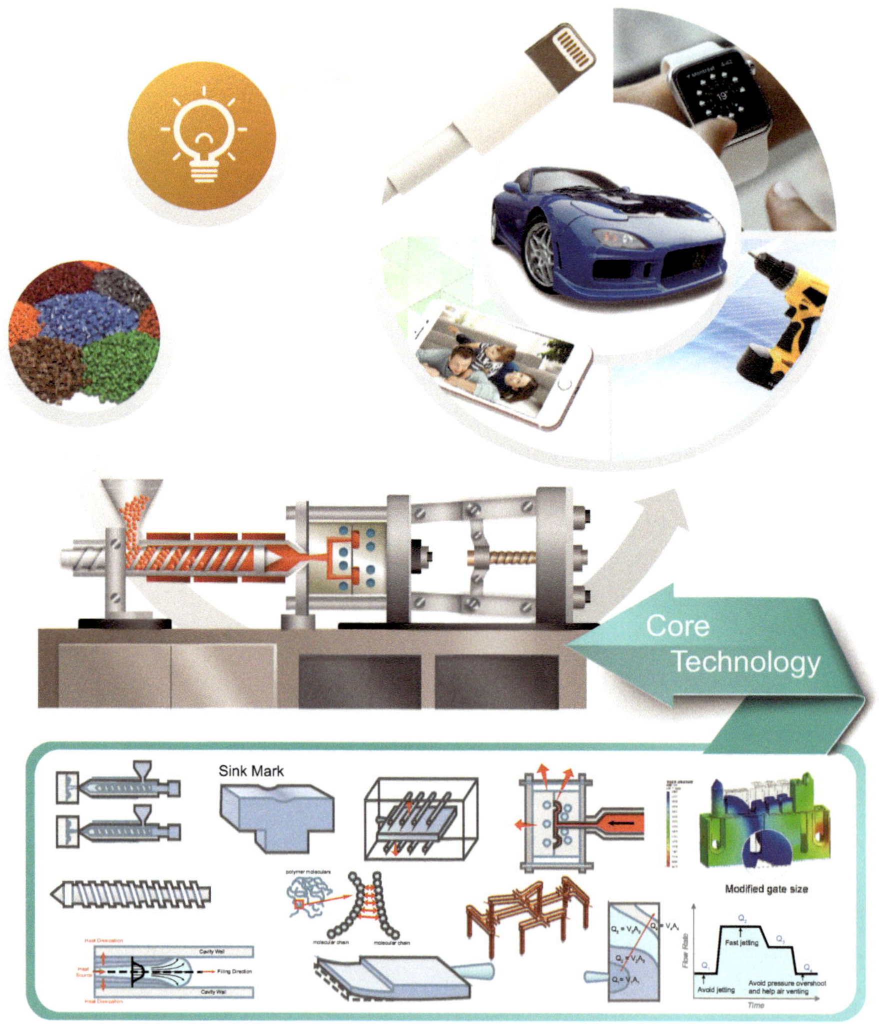

Figure 1.1 Development of workflow of mold products

1.1.1 The Systems of Injection Molding

1.1.1.1 The Cycle of Injection Molding

First, what is injection molding? Simply speaking, it is a process of making a product by injecting plastic material of liquid state into a mold cavity via the help of injection molding machines. When the plastic material enters the injection molding machine through a hopper, it is turned into a melted state after being squeezed by the screw from which a large amount of heat is generated through friction. The melted plastic accumulates in the front of the cylinder and is constantly heated in order to maintain the temperature for injection. The process mentioned above is called plastification, as shown in Figure 1.2.

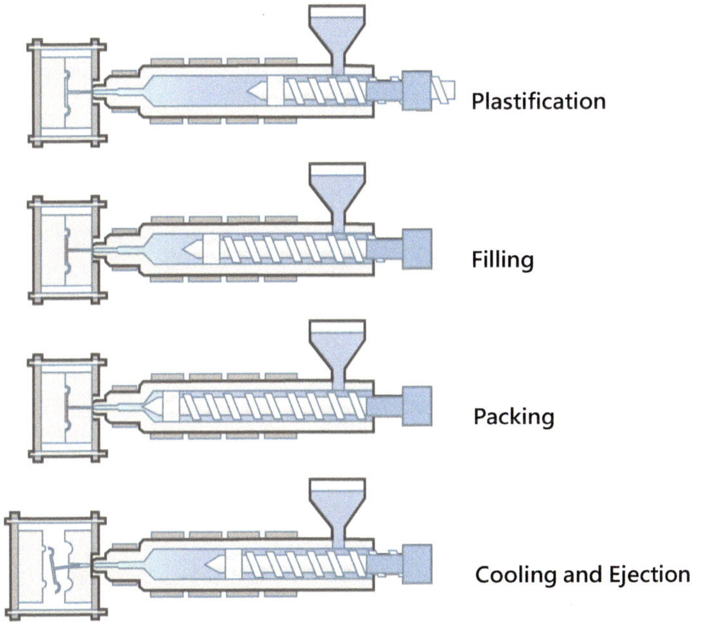

Figure 1.2 Cycle of injection molding

Then, the melted plastic will be pushed forward into the closed mold cavity by the screw, a process that is called injection. After the initial injection is completed, when the high molecular weight melted plastic has fully filled the mold cavity, more melted plastic is injected under high pressure in order to compensate for the decrease in the volume of the plastic due to cooling as well as to make sure the mold cavity is perfectly filled until the sprue is solidified, a process that is called packing. Finally, the movable side moves back until the ejection pin reaches the rear platen to eject the molded product, runner system, and waste. This cycle is known as the molding cycle of injection molding.

1.1.1.2 Injection Machine

Figure 1.3 shows a basic injection molding machine (injection machine), which is suitable for manufacturing products of different shapes from thermoplastic or thermosetting plastics. There are two basic functions: 1) heating of the plastics to a melted state, and 2) application of high pressure to inject the melted plastic to fill the mold cavity completely.

Injection machine systems often have different components due to different types and uses, but generally comprise injection unit, clamping unit, mold unit, plastification unit, feed unit, and control unit.

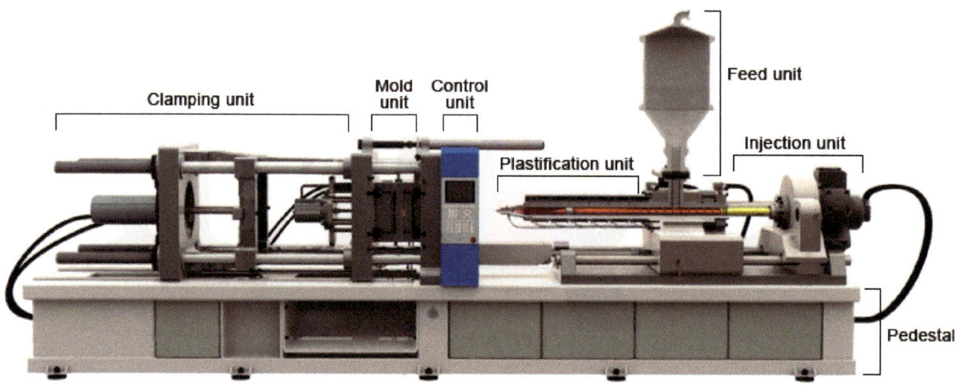

Figure 1.3 Injection machine

In industrial plastics processing methods, whether extrusion, injection, calendering, blow molding, film blowing, or spinning, a huge quantity of additional auxiliary equipment is always required to complete each processing step. The optimization, automation, and rationalization of auxiliary equipment for plastics processing play a role in determining product quality and the economic viability of the process.

Next, we will introduce the equipment required for injection molding in detail. This includes the feed, control, plastification, injection, and mold systems.

1.1.1.2.1 Feed System

Generally, there are two kinds of feed systems, namely independent and central. Independent feed equipment is usually set above the feeding port of the injection machine, i.e. the hopper with inverted cone structure, as shown in Figure 1.4(a). Some are supplemented by a heating or drying device, and also with a metering device to quantify the plastics, as shown in Figure 1.4(b).

The other kind of central feed system (also called automatic feed system or centralized feed system), including suction machine, central material loader, and conveying control system, has the advantages of continuous feeding, centralized automatic control, uniform mixing, and flexibility to change colors.

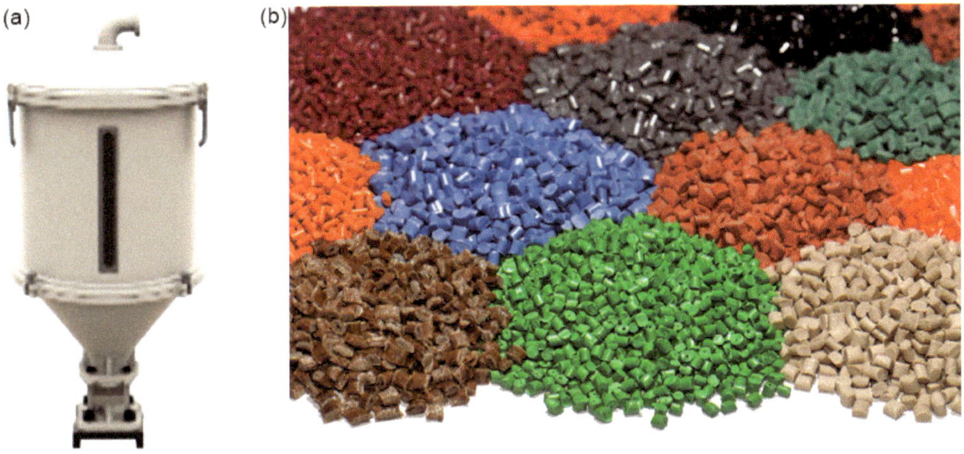

Figure 1.4 (a) Hopper and (b) plastic pellets

1.1.1.2.2 Control System

The control system (Figure 1.5) can be said to be the brain and nerve center of the entire injection machine. The pressure, temperature, speed and time of the machine are controlled via the set values input by the operator that will directly affect the molding cycle and product quality.

The key physical parameters of injection molding processing are temperature, pressure, speed, time, and position, each of which includes the following:

1. Temperature includes resin drying temperature, barrel temperature, melt temperature, mold temperature, machine oil temperature, and working environment temperature.
2. Pressure includes filling pressure, holding pressure, back pressure, ejection pressure, mold opening pressure, and mold clamping pressure.
3. Speed includes injection speed, screw rotation speed, mold opening and closing speeds, and ejection speed.

   Time includes filling time, pressure holding time, cooling time, and resin drying time.
4. Position covers measuring position, VP switch position, eject position, and switch mode position.

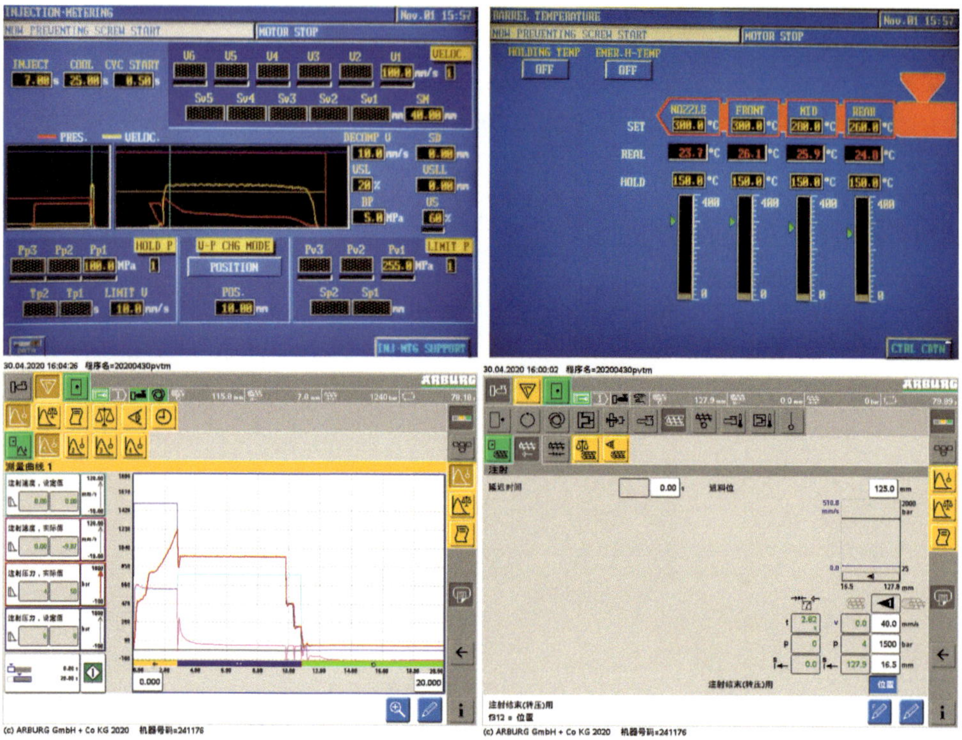

Figure 1.5 Control panels from Nissei and Arburg

One of the biggest problems encountered in common development and quality control of molded products during the production process is ineffective quality control due to using a one-way, indirect method for operating injection machines and system control devices. Therefore, it is important to analyze the characteristics of the machines so as to understand the different ways in which they can be used to exert good control over the process conditions.

1.1.1.2.3 Plastification System

Plastification is a process that uses the mechanical energy of the screw and the thermal energy of the heater to melt the incoming solid plastic, which is then applied with high pressure to be ready for injection. The plastic is turned into a melted state after being squeezed by the screw from which a large amount of heat is generated through friction. The melted plastic accumulates in the front of the cylinder and is constantly heated in order to maintain the temperature.

As shown in Figure 1.6 and Figure 1.7, the solid plastic enters the screw channel via the inlet hopper. With the high rotation speed of the screw that generates a shear stress effect with the barrel, the plastic is mixed and transferred along the screw channel. As the solid plastic is heated by the electric heaters outside the barrel and due to the shear stress effect, it turns into a melted state as the temperature rises inside the barrel.

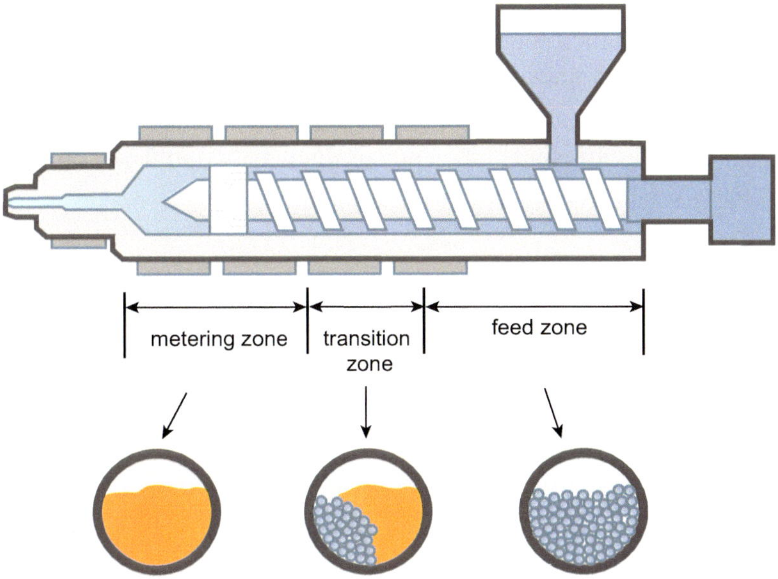

Figure 1.6 Barrel and plastification

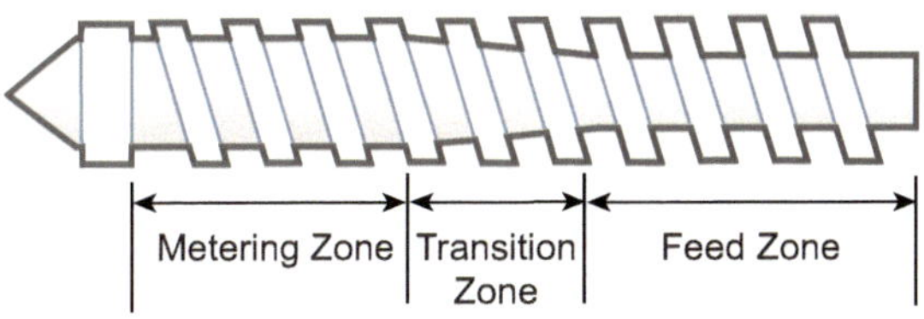

Figure 1.7 Screw

The screw is divided into three zones (Figure 1.7):

1. Feed zone: The fixed feed depth of the screw channel is for pre-heating, transferring, and pushing the plastic granules, which start to melt at the end of the feed zone.
2. Transition zone: The thread depth is gradually decreased in the transition zone, the purpose of which is melting, mixing, shearing, and compressing of the plastic, and pressurized venting. The plastic must be completely melted in this zone. The volume of the melted plastic will be reduced and should be compressed accordingly to avoid incomplete compression and poor venting.
3. Metering zone: The fixed metering depth of the screw channel is for transfer and metering of the melted plastic, as well as providing sufficient pressure to

maintain a uniform temperature and stabilize the flow of the melted plastic. The length of this zone greatly affects the quality of the melt injected into the mold cavity. The longer the metering zone, the better is the chain mixing result. But a zone that is too long will cause a long dwell time of the melt in the barrel and will induce thermal decomposition. In contrast, a zone that is too short will lead to an uneven melt temperature.

The length and geometry of the zones in the screw directly affect the extent of plastification and the transport efficiency of the plastic. Judicious screw selection is very important for injection molding.

1.1.1.2.4 Injection System

The injection system is mainly responsible for filling and packing (Figure 1.8). For the filling stage, the screw moves forward to inject the melted plastics into the closed mold cavity through the nozzle to finish the filling process. When the melted plastic enters the cavity, the air is expelled from the ejection pin, parting line, and vent holes. Under-injection would occur if the liquidity is poor or the injection pressure is insufficient; in contrast, if the liquidity is too high, flash (see Section 1.1.2.3) would occur on the parting facet of the plastic part.

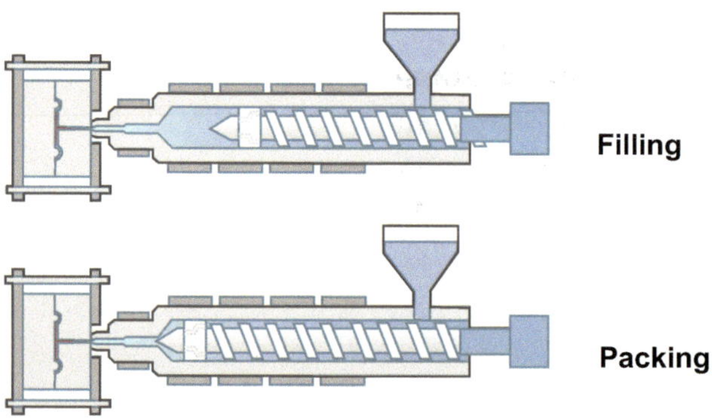

Figure 1.8 Injection system

After the mold cavity has been almost filled by the melt in the filling stage, the machine continues to apply high pressure to inject more melt into the mold cavity, to pre-compensate for the plastic volume shrinkage caused by cooling and to ensure that the melt completely fills the mold in the packing stage. Packing continues until the gate is sealed or the packing time ends.

1.1.1.2.5 Mold System

The mold system generally includes a fixed side and a movable side platen. The space inside the platens is the mold cavity, and is where the plastic is molded into shape. The mold also consists of ejection mechanisms such as ejector pins or ejector platens for ejecting the solidified product from the cavity. A typical structure is shown in Figure 1.9.

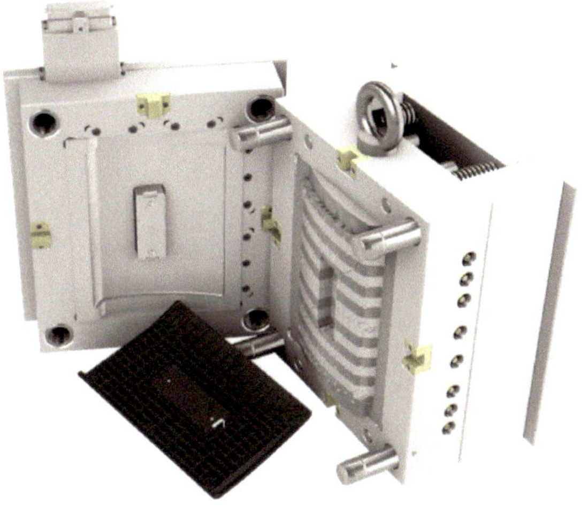

Figure 1.9 Open mold

Molds are important in injection molding. The basic structure of a mold is generally divided into three types: two-platen mold, three-platen mold, and hot runner. The decision for a particular mold structure is generally made by customers or according to products. A plastic mold consists of seven major systems: guiding, support, molding component, pouring, cooling, ejection, and venting systems. Using a sliding block is a way to handle undercuts, but normally the mechanism of mold opening and closing is sufficient for lateral parting, core extracting, and position reset.

A complete cycle time of plastics injection molding consists of filling, packing, cooling, and mold opening times, among which the cooling time has the highest proportion at about 70–80%. Therefore, the design of cooling system is a critical step which directly affects the length of cycle time, production efficiency, and cost. The following chapters will introduce the characteristics of plastic processing and the key points of mold design in detail.

1.1.2 Defects in Injection Molded Products

Product defects present another concern when plastic is turned into the final product through the foregoing processes. The plastic takes its shape gradually through the cooling during the process of injection molding, and usually it has the form of a near-finished product when it leaves the mold. If there are defects in the plastic product, it is necessary to analyze and understand the factors that are the cause.

The common injection molding defects are briefly described as follows:

1.1.2.1 Short Shot

The phenomenon called the "short shot", shown in Figure 1.10, gives a defective appearance to the final plastic product, and is caused by under-filling of the mold cavity. It is most apparent at in thinner zones or at the end of runners, and is mainly caused by insufficient plastic supply or poor liquidity of the plastic itself, so that the liquid state halts prematurely during the filling process. Therefore, any factor that affects the smooth flowing of the melted plastic is likely to cause short shot defects, such as insufficient amount of plastic injected, too high flow resistance, or insufficient liquidity.

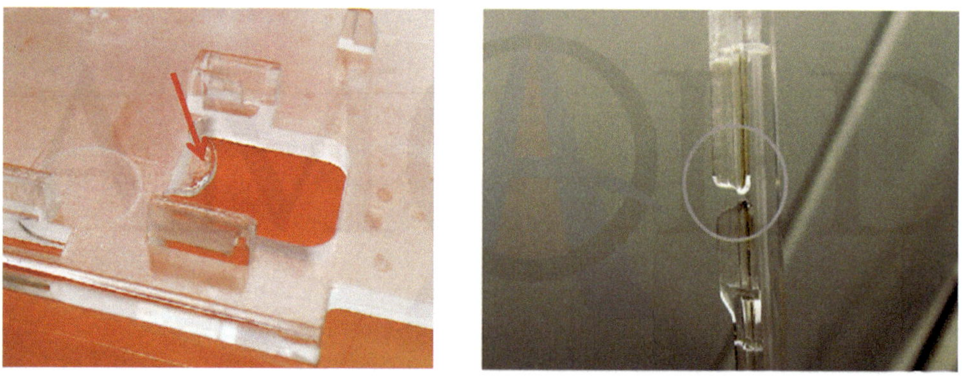

Figure 1.10 Short shot

In addition to a low temperature of melted plastic and mold wall, a thin part geometry or improper sprue location or length can also generate short shot defects as the mold cavity cannot be filled up. Improper configuration of the vent hole is also likely to cause a short shot.

It should be confirmed that the hopper has enough plastic if a short shot is observed. Then, the cylinder should be checked for blockage and the back-pressure valve checked for failure, which can result in a low injection pressure or material leakage. However, a long injection time can also cause a short shot.

1.1.2.2 Warp

Warp denotes the distortion or deformation of a product after injection molding. It is the defect type most commonly seen in injection molded products. Figure 1.11 shows a warped finished product which has two parts that cannot be assembled together. However, even if the product is not a combinative part but a single product, warp can also give rise to customer complaints and product returns. Therefore, product warp should be strictly controlled to within the tolerance.

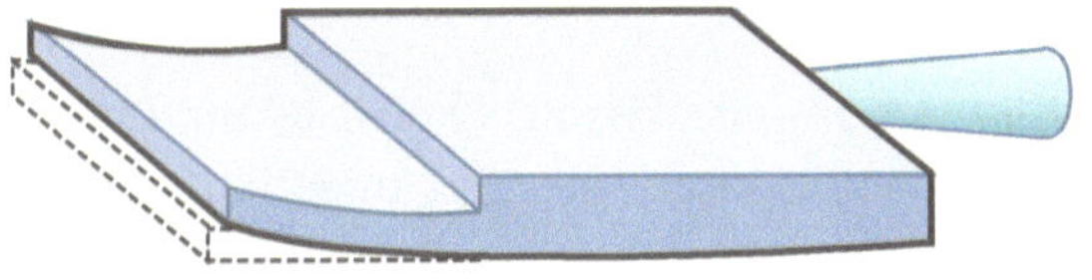

Figure 1.11 Warp resulting in a molded product that cannot be fitted

Thermal expansion and contraction are also seen in plastics. The melted plastic starts to cool down and solidify as it enters the mold cavity, and it contracts during the process of cooling and solidification. If the contraction rate is evenly distributed across the product, warp would not be seen and only shrinkage would result. However, with the interaction between the external factors, e.g. molding conditions, mold cooling design, product appearance design, and the plastics characteristics, e.g. molecular chain and fiber orientation, it is very difficult for plastic finished products to contract evenly or with low contraction rate.

It can be seen from the above that the causes of warp are various, complicated, and interdependent. Ways of reducing warp are a very important concern for producers.

1.1.2.3 Flash

Flash is generated as there is a gap existing at the split plane via which the melted plastics spills outside the mold cavity, as shown in Figure 1.12.

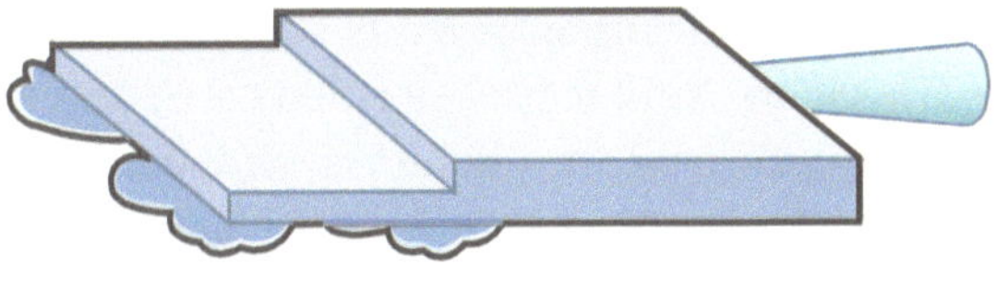

Figure 1.12 Flash

The main causes of such formation are as follows:

1. Mold clamping force is too small: A pushing force is applied to the mold by the melted plastics during the injection molding process, especially if the central area of the mold cavity is subjected to excessive high pressure the mold will separate from the parting plane.
2. Mold gap: The causes that the moving and stationary side of the mold cannot contact completely consist of a) a parting plane that is defective and not parallel between each side, and b) impurities on the parting plane that create gaps on the parting plane.
3. Improper molding conditions: Wrong choice of molding machine, over-temperature of melted plastics, and excessive injection pressure are all causes of flash generation.
4. Improper venting system: Flash will be generated if the venting is insufficient or the venting groove is too deep.

1.1.2.4 Sink Mark and Void

The formation of sink marks and voids is a phenomenon observed in thick areas where there is not enough plastic supplied during the cooling process, as shown in Figure 1.13.

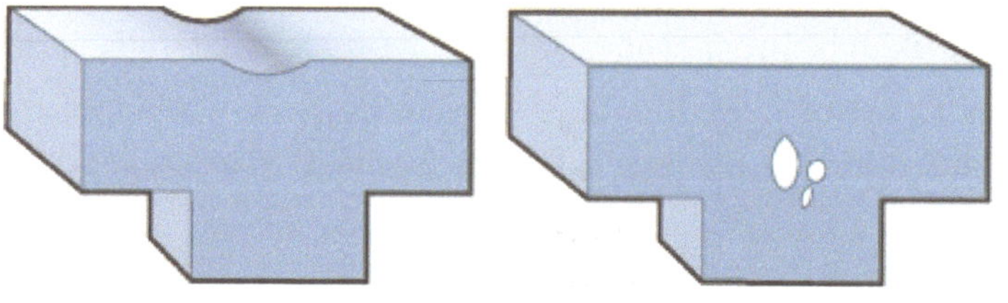

Figure 1.13 Sink mark/void

Sink marks are generated as the plastic contacting the mold walls cools and hardens before the inner plastic starts to cool down, and hence the surface is pulled inward by contraction. If the surface strength is sufficient, voids are generated instead of sink marks. Therefore, sink marks and voids are often seen at the rib parts or the backside of a convex surface. In conclusion, sink marks and voids are generated easily if the contraction is uneven between the inner and outer part in some areas.

1.1.2.5 Air Trap

Air trap denotes a condition in which the melt front of the melted plastic traps the air inside the mold cavity, so that the air cannot escape from the venting holes or the gaps in the mounting parts. A possible consequence is shown in Figure 1.14.

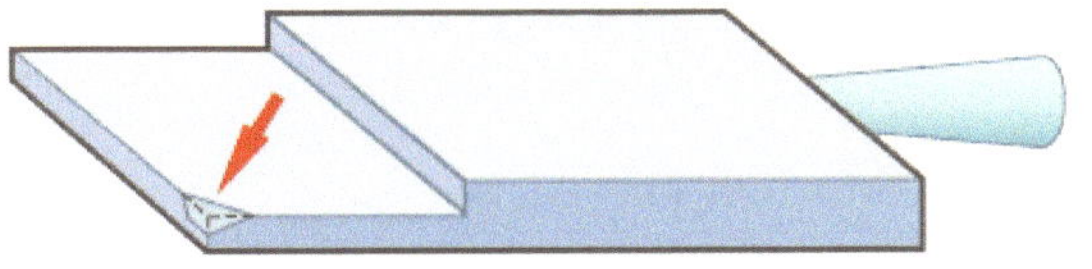

Figure 1.14 Air trap

Generally speaking, air trap mostly occurs in the area filled last, where there is no venting hole or the venting holes are too small, with the result that bubbles, short shot, or surface defects are generated inside the plastic part. In addition, any large difference in the thickness of the parts will lead the melt flows toward the thicker area instead of the thinner area, giving rise to the race-tracking effect, which is also one of the reasons for air trap.

1.1.2.6 Burn Mark

The causes of burn marks are very similar to those of air traps. The major cause is that the air trapped inside the mold cavity is overheated by compression and creates dark marks on the plastic surface as shown in Figure 1.15. When the air inside the mold cavity is compressed, the pressure and temperature rise so rapidly that the surface of the plastic part at the end of the flowing path is decomposed and thereby burn marks are generated. Generally, burn marks usually occur in tandem with air traps.

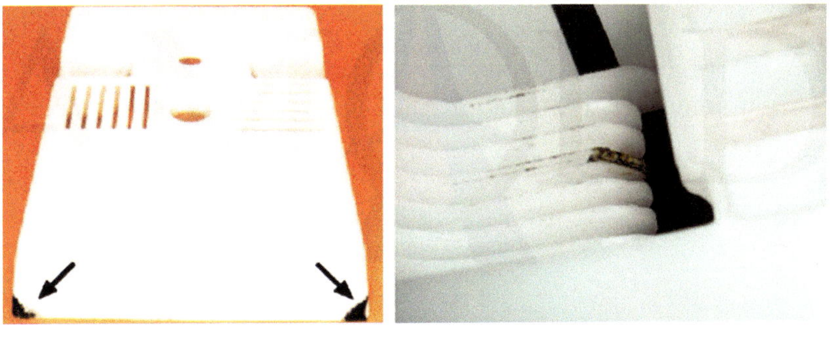

Figure 1.15 Burn mark

1.1.2.7 Delamination

The main cause for delamination (layer separation) is due to the mixing of two incompatible materials or material types used in the molding process that are too dissimilar, as shown in Figure 1.16. In addition, delamination is also possible to occur if the temperature of the melted plastic is too low, the humidity of the material is too high, or the runner and sprue are not smooth.

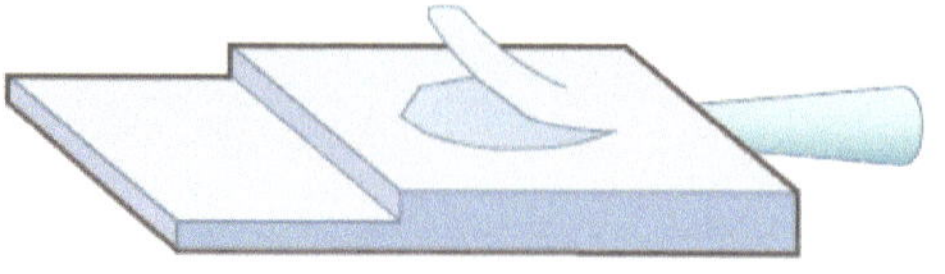

Figure 1.16 Delamination

1.1.2.8 Fish Eye

Fish eyes, as shown in Figure 1.17, are usually caused by unmelted plastic because of insufficient cylinder temperature and screw rotating speed, and low backside pressure. The phenomenon can also result from using too much recycled material or contaminated plastic. Therefore, if the above-mentioned situation can be avoided, the probability of fish-eye occurrence can be effectively reduced.

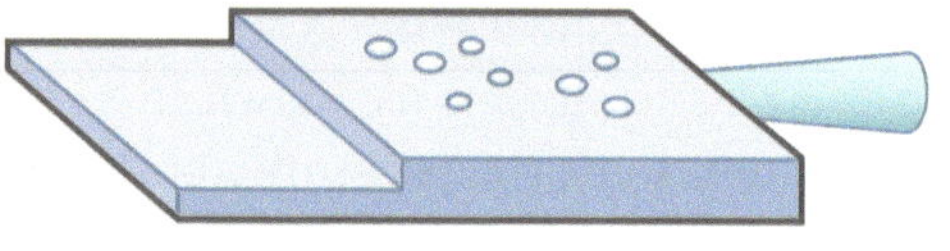

Figure 1.17 Fish eye

1.1.2.9 Flow Mark

Flow marks, as shown in Figure 1.18, are generated mainly if the temperature of the melted plastic is not evenly distributed or the viscosity of the melted plastic is excessively high.

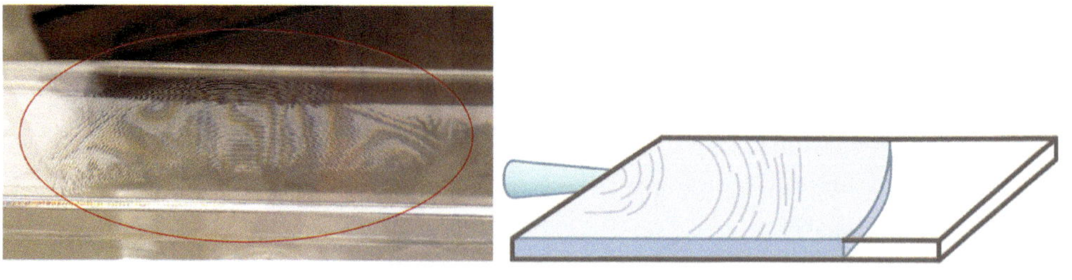

Figure 1.18 Flow mark

An excessively low temperature causes friction and pushing between the plastic and the mold cavity, and results in the plastic's hardening too quickly and leaving a flow mark, as shown in Figure 1.19. In addition, flow marks may occur if the melt near the gate solidifies due to unexpected cooling, or the melt is unable to provide compensation in the packing stage.

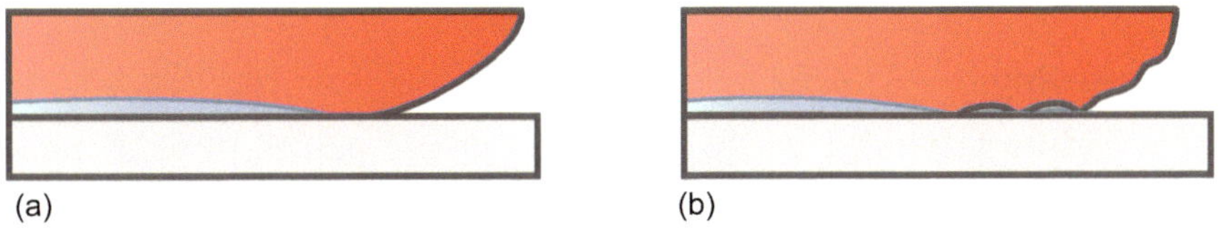

Figure 1.19 (a) General fountain flow and (b) over-rapid cooling conditions that give rise to uneven local temperatures will readily cause flow marks

1.1.2.10 Stress Mark

On the surface of the injection molded part, a shiny white mark occasionally appears and is called a stress mark. Generally speaking, a stress mark is induced by internal stress. If the thickness of a plastic part varies a lot, the cooling speed at a thinner area differs significantly from that of other areas. Then, the uncooled melted plastic will apply stress on the cooled plastics, which generates a stress mark due to inner stress, as shown in Figure 1.20.

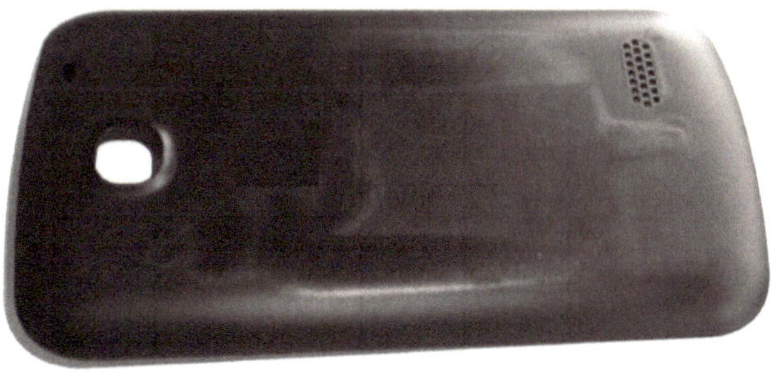

Figure 1.20 Stress mark

1.1.2.11 Hesitation

When the mold cavity is being filled, the melted plastic tends to move to a thicker and low-flow-resistance area, which it fills up first before filling the thinner area. Therefore, the melted plastic usually hardens at a stagnation point as the flow stops at the thinner area. It is highly possible for the solidified plastic to be pushed to the surface of the plastic part and create a hesitation mark when the following melted plastic starts to move to such a thinner area, as shown in Figure 1.21. In addition, hesitation often leads to short shots because the plastic usually completely solidifies at the stagnation point.

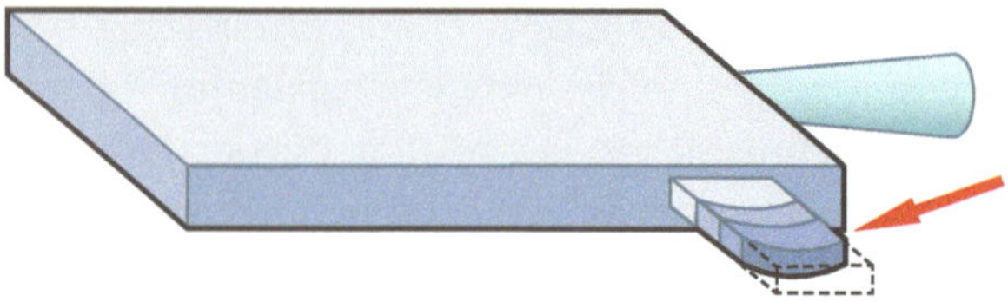

Figure 1.21 Hesitation

1.1.2.12 Jetting

Jetting is usually generated when the melted plastic passes through a narrow sprue or runner into the mold cavity with a high speed, as shown in Figure 1.22. Jetting normally causes contact between cold materials as the temperature of the plastic strips injected into the mold cavity drops and the strips make contact with each other afterward. The strength of the plastic part will be reduced when jetting happens, which is more likely to cause surface and internal defects at the same time. Thus, jetting should be avoided as much as possible in order to assure production quality.

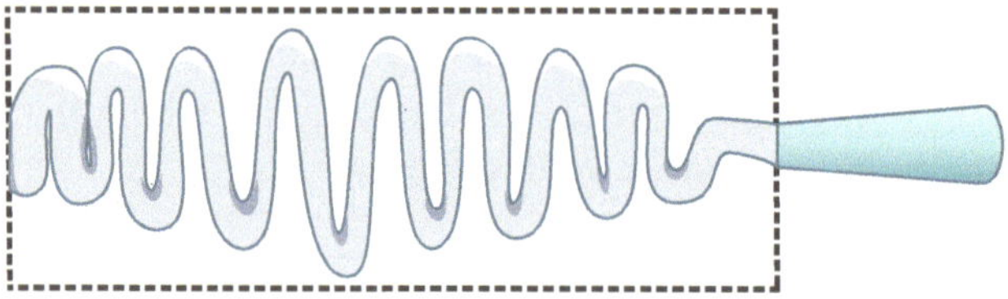

Figure 1.22 Jetting

1.1.2.13 Splay

Splays, as shown in Figure 1.23, are caused by the incomplete release of humidity or other volatile gases mixed in the plastic melt during the plastification stage, or by gases generated from slight decomposition of the plastic due to an excessively high pre-heating temperature.

If the plastic is not properly dried before manufacturing, the moisture inside will evaporate into steam during injection filling, which causes bubbles to flow along with the melted plastics inside the mold cavity and thus silver stripes (splays) will occur along the flowing direction. If the bubbles cannot be expelled entirely upon completion of filling, splays will also occur on the surface of the plastic part.

Figure 1.23 Splay

1.1.2.14 Weld Line

When two or more plastic flow fronts merge together during the injection process, incomplete fusion can occur as the melt front is of lower temperature and hardens first. Hence, weld lines, as shown in Figure 1.24, are generated. Such defects are usually seen around the holes or the merging boundaries of the finished products. Therefore, when race-tracking effects take place, they are usually accompanied by weld lines. To avoid the generation of weld lines, extreme care should be taken regarding conditions such as significant thickness changes or multiple sprues in the mold.

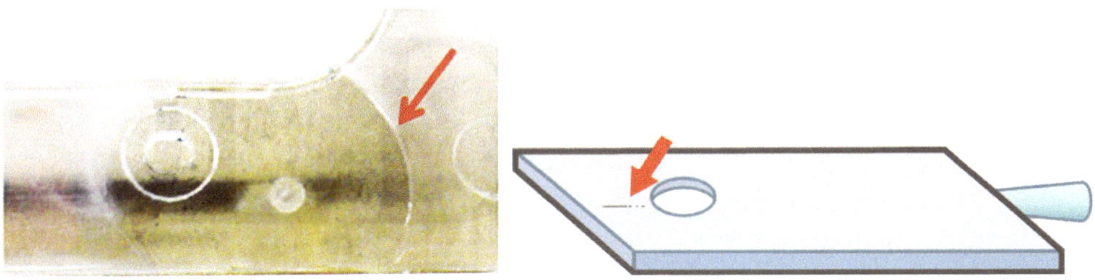

Figure 1.24 Weld line

The plastic melts cannot be completely fused at the weld line, as a result of which the strength of the plastic part is lowered. Therefore, if weld lines cannot be avoided, the position and size of the gate should be adjusted to make the weld lines occur in as inconspicuous a low-stress area as possible. Generally, melt fronts converging from two directions at less than 135° will give rise to weld lines, as shown in Figure 1.25.

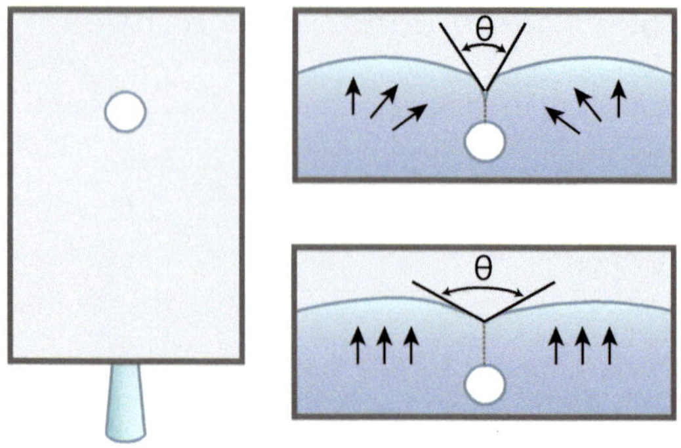

Figure 1.25 The converging of melt fronts

■ 1.2 Core Values of Molding Simulation

From the foregoing introduction to plastics injection molding, it can be seen that the types of defects are of various kinds, which can be categorized as appearance or dimensional issues, as shown in Figure 1.26.

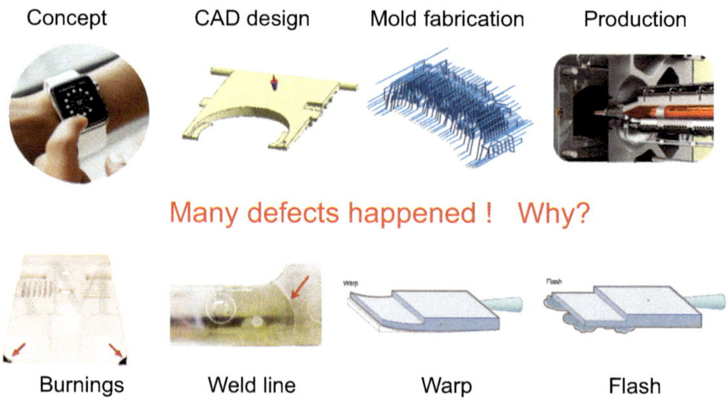

Figure 1.26 Molding challenges: design quality

When facing these defect issues, the traditional solution is the trial-and-error method, i. e. to repeat mold test and mold modification until the product specifications are met, which usually takes gigantic amount of time, labor, and other costs. However, with the assistance of CAE (Computer Aided Engineering) tools, the production process becomes more efficient and the product quality gets better, while waste is further reduced, with environmental benefits. An introduction to CAE, including its application possibilities in injection molding, is given below.

1.2.1 Application of CAE Technology in Injection Molding

CAE is a kind of computer aided engineering software and technology that uses computer simulation and analysis to assist the diagnosis and development of complicated injection molding processes. CAE is able to integrate the complicated rheological, thermal, and mechanical properties of a material, and enables designers and developers to do qualitative and quantitative analysis and diagnosis for mold design as well as analysis and diagnosis for existing molds and operating conditions.

From the CAE analysis result, developers can explore the causes of problems that occur and test different design changes to find out the most appropriate solution, which is not achievable by the traditional trial-and-error method. Furthermore, if the design change involves modifications of products or molds, the cost of time, labor, machine, material, and energy by repeated mold tests and modifications is even beyond estimation. Hence, it is common to implement CAE for design verification during the development process.

We know that the injection process is a major factor that determines product quality. Going through solid, melted, and back to solid states in a short time involves rearrangement of plastic molecules. If we can effectively control the transition of plastic properties in the process, the structural strength of the product can be assured.

Where is CAE used?

CAE is not always suitable to provide assistance for injection molding, as shown in Figure 1.27.

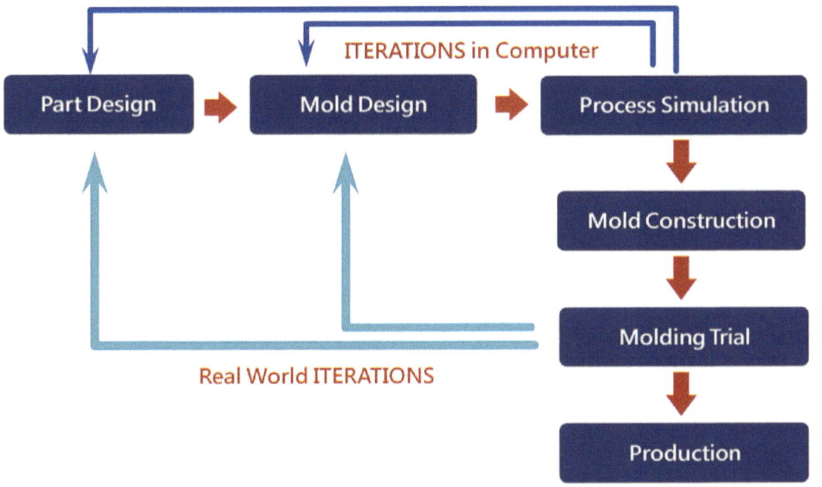

Figure 1.27 New concepts in product development of injection molding processes

The best opportunities for using CAE in the injection molding process are:

1. At the product design stage and before mold construction: Predict and amend possible defects in product design and reduce the cost for mold opening.
2. After mold construction but before mass production: If the product is defective, but the causes are difficult to solve via onsite mold tests, this can be analyzed to determine the causes for defect generation through regenerating the actual defects and issues via simulation.
3. After mold construction and during mass production: Find out any possibility for yield improvement and molding cycle time shortening via simulation to further increase the productivity.
4. Create an in-house database: Summarize the problem-solving knowledge and create problem-determination SOPs (standard operation procedures) by constantly accumulating project counts.

Hereunder is an example of a cap for spraying deodorant in mass production. How do we utilize CAE to help the manufacture? This situation meets the above-mentioned CAE application opportunity item 3. For the product in mass production, using trial-and-error to further improve productivity or yield is time- and labor-consuming, which makes CAE a pretty good choice in this application.

As shown in Figure 1.28(a), with CAE analysis we can see that the temperature difference between inside and outside of the cap is 45 °C in EOC (end of cooling), which leads to thermal stress and causes quality issues. The high product temperature also affects the ejection time (problem discovery).

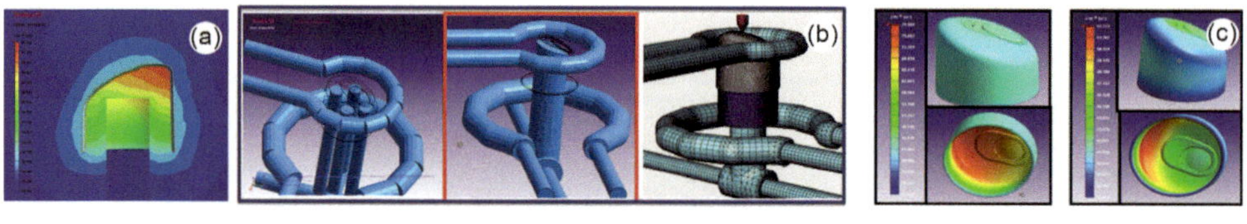

Figure 1.28 CAE example: spray deodorant cap

As shown in Figure 1.28(b), originally, the cooling channel design was a big pipe; however, by considering abnormal cooling channel design and comparing three different models, the one in the middle shows the best result. This is a good example of using CAE to save time and labor without redesigning the mold.

As shown in Figure 1.28(c), the new design reduces the temperature difference by 15 °C.

Final result: The productivity is increased by 25% as the cycle time is shortened. For an annual production of 4 million pieces, 4 million seconds (46 days) are saved therefrom. In addition, the quality is improved simultaneously as the temperature difference is reduced.

This example clearly tells us that, with proper use of CAE tools, we can improve existing processes without affecting current production lines and with only little cost.

The following chapters will introduce the principles, models, and usage of CAE in detail, from theory to practice, as well as its applications in various industries and molding technologies, so that readers can better understand and learn how to use CAE to solve problems.

2 Material Properties of Plastics

Chen-Chieh (Jye) Wang

Generally, the plastic material for injection molding is in a pellet form, which is heated and melted during the molding process, and is given shape and function before cooling down to become solid again. Therefore, an understanding of different materials regarding their rheological, thermal, thermodynamic, kinetic, chemical, and mechanical properties is essential to obtain the desired application characteristics, functions, molding parameters, product quality, and precision of CAE simulation. Material properties are explored individually in the following sections.

2.1 Overview

Polymer materials are everywhere, from our daily lives to high-tech fields (such as electro-optical, biomedical, and high performance applications). Polymers consist of many chemical units (monomers) after polymerization, such as polyethylene (PE), whose monomer is ethylene. Polymers have tremendous molecular weights (MW) ranging from tens of thousands to millions and thus are called "macromolecules".

Figure 2.1(a) shows that the bonding of monomers can be linear or branched. For example, in branched PE, the molecules cannot pack closely together due to steric hindrance of the branches, and so its density is lower; it is therefore called low-density polyethylene (LDPE). In contrast, linear PE is more regular and the polymer chains can pack together more closely and crystallize efficiently; it is therefore called high-density polyethylene (HDPE).

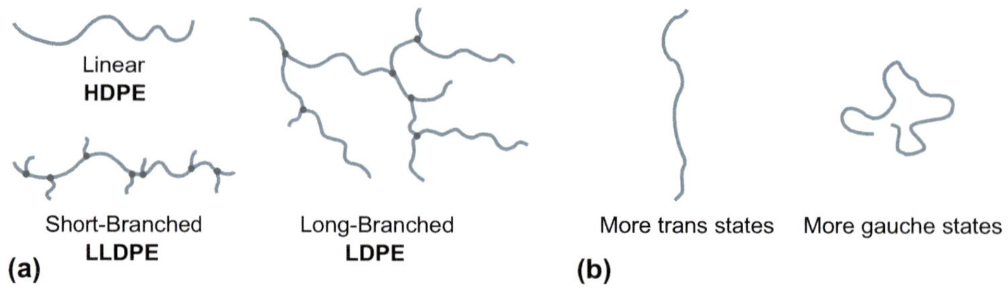

Figure 2.1 (a) Molecular configurations of polyethylene and (b) conformations of polymer chains

In the polymer chain, the atoms are connected by covalent bonding forces. The bond energy of the C-C covalent bond is around 80 kcal/mol. The rotation of a single covalent bond gives rise to the bond "conformation", i.e., cis, trans, and gauche states. Since there are many bonds in a polymer chain, a single polymer chain can have tremendously accessible chain conformations. The different chain conformations have different shapes, a fact which also implies a different volume of occupied space, as shown in Figure 2.1(b). Additionally, intermolecular interaction (van der Waals forces) occurs among molecular chains, inducing atoms to pack more closely together, preventing atoms from chain crossover and leading to chain entanglement, as shown in Figure 2.2(a). The entanglement of chains plays an important role in the viscoelastic behavior of materials.

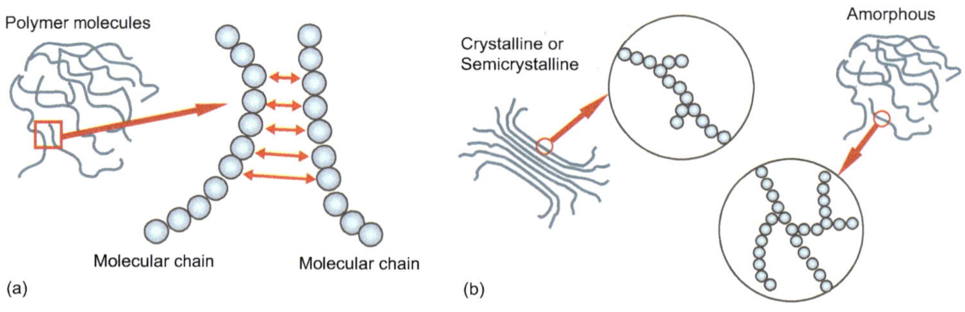

Figure 2.2 (a) Interactions among macromolecules and (b) classification of thermoplastic polymers

When the polymer material is heated, the thermal energy imparted to the atoms is sufficient to overcome the energy for performing large-scale molecular movements, and the material softens. When the temperature is high enough for the polymer chain to move freely, the material flows like a liquid. A plastic that can be melted by heat and reverts to the solid state after cooling is generally called "thermoplastic". If the temperature is further increased, the thermal energy may be high enough to break the covalent bonds. The molecular chain may then break into fragments in a process called "degradation". In summary, the polymer behaves like a solid when the temperature is lower than the glass transition temperature at which the molecules freeze. The material behaves like a rubber when molecules are able to perform large-scale movements and like a liquid when it can move freely.

In accordance with their molecular morphology in the solid state, thermoplastic polymers can be classified into two categories: crystalline and amorphous. The amorphous morphology has no long-range order structure. In other words, the chains pack randomly throughout the material, e.g. as in PMMA, a-PS, and ABS. In contrast, crystalline polymers (such as i-PP, PE, nylon) can form orderly stacks of folded chains called lamellae, which are like crystals with an orderly arrangement of atoms.

In contrast to thermoplastics, thermoset polymers cure irreversibly. After heating, they cannot return to the original state, which means that they cannot be processed again. Examples are phenolic resins, melamine, UP, PU, and epoxy.

When a thermoset material is heated, its molecular chains will cross-link with each other to form a three-dimensional network structure due to chemical reaction. After the cross-linking reaction, the material cannot flow or be re-formed under the constraint of the cross-linked structure, and exhibits solid behavior, as shown in Figure 2.3. The higher the cross-linking density, the more constrained it becomes and therefore the material becomes harder.

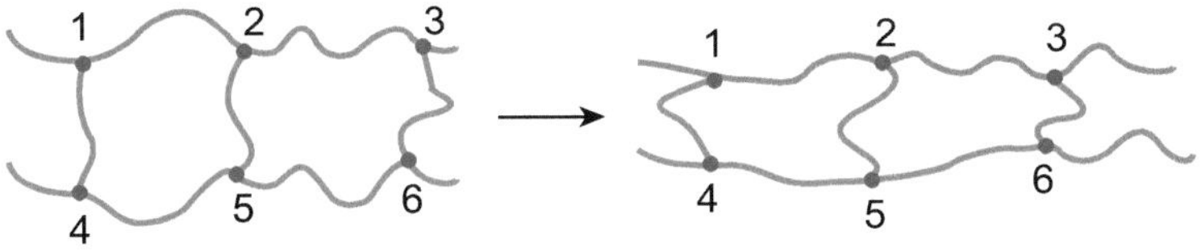

Figure 2.3 Network structure of thermoset materials

■ 2.2 Rheological Properties

The flow behavior of polymers is within the realm of rheology, which is the study of the deformation and flow of materials. Polymer rheology covers experimental research on the measurement of fluid viscosity, the relationship between molecular structure, pressure, and temperature, and the effect on rheological properties of solvents and plasticizers, as well as Newtonian and non-Newtonian fluid mechanics, and other theoretical research such as molecular rheology, which has become a popular field in recent years. The reason why the rheological behavior of polymeric fluid is different from that of simple molecules is that a polymer's molecular weight is often as high as 10^5 to 10^8 g/mol, consisting of long-chain molecules, sometimes with branched structure. The rheological behavior of polymers therefore plays an important role in injection molding. When melted plastic is injected into the mold cavity, the flow resistance and behavior are related to its rheological properties with regard to how the mold cavity is filled, how much injection pressure is required to inject the melted plastic through sprues and runners, what the temperatures of the melt and mold should be, etc. It is also instrumental in designing the runner and sprue locations so as to avoid unnecessary packing and weld lines. The shrinkage, warp, and deformation of the product are affected by the molecular orientation, crystallinity, and residual stress after filling and cooling of the melted plastic. In summary, an understanding of rheology is beneficial to product and mold design, which can also help to solve injection molding issues, and derive the greatest benefits from CAE.

2.2.1 Viscosity

Consider an incompressible fluid located in the space between two parallel plates, as shown in Figure 2.4.

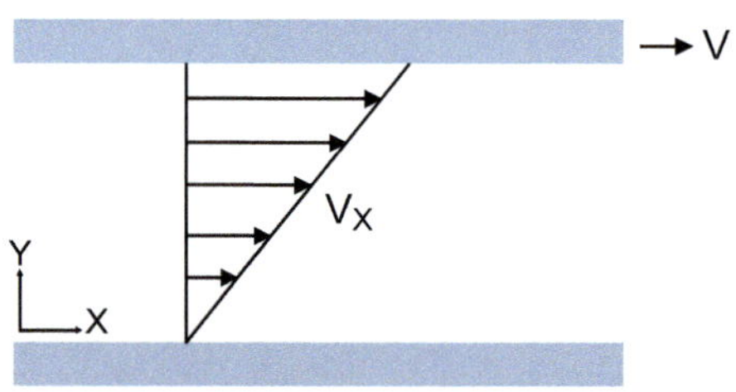

Figure 2.4 Flow between two parallel plates

Assume the areas of top and bottom plate are both A, and the two plates maintain a small distance Y between each other. If an external force F is applied to the top plate, the top plate can move at a constant velocity V, while the bottom plate remains stationary. Such flow is called simple shear flow. It was found by experiment that when a steady state is reached, (F/A) is in proportion to (V/Y), where the proportional constant η denotes the viscosity of the fluid (Equation 2.1). (F/A) is called the shear stress, τ_{xy}, and (V/Y) is called the shear rate, $\dot{\gamma}$, giving Equation 2.2, which is known as Newton's Law of Viscosity:

$$\frac{F}{A} = \eta \frac{V}{Y} \tag{2.1}$$

$$\tau_{xy} = \eta\dot{\gamma} \tag{2.2}$$

Shear rate denotes the difference in velocity of each point in the flow field, reflecting the velocity gradient in a flow.

2.2.1.1 Effects of Non-Newtonian and Molecular Conformation

A fluid is defined as being Newtonian only if the stress and shear rate are related by a constant viscosity (Equation 2.2) that does not depend on the stress or shear rate of the flow. The viscosity can be regarded as the resistance to flow and deformation. The resistance results from the collision of and friction among the molecules during flow. It can be influenced by material compositions, temperature, and pressure.

The flow resistance (viscosity) of a polymer depends on the orientation and conformation of molecular chains. Since orientation and conformation vary with the flow fields, the viscosity may change with shear rate, as shown in Figure 2.5.

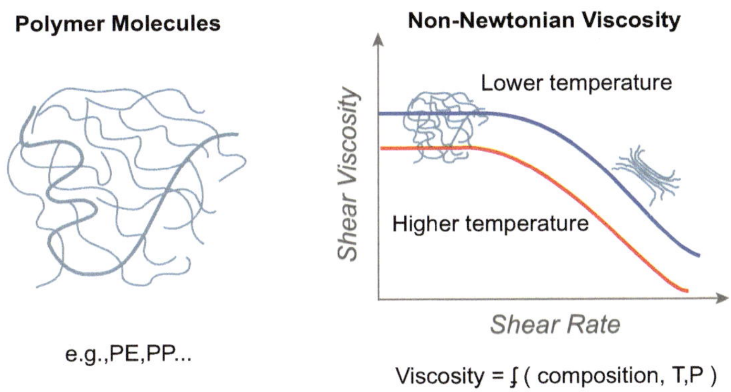

Figure 2.5 Molecular structure and viscosity of polymer molecules

For a polymer of low molecular weight, the flow resistance and viscosity are lower because its molecules are shorter, whereas a polymer with longer molecules has a higher viscosity due to the molecular size and the number of entanglements among the molecular chains. The relationship between viscosity and molecular weight is shown in Figure 2.6, where a larger slope is observed once the molecular weight exceeds a critical value, owing to the effect of entanglement.

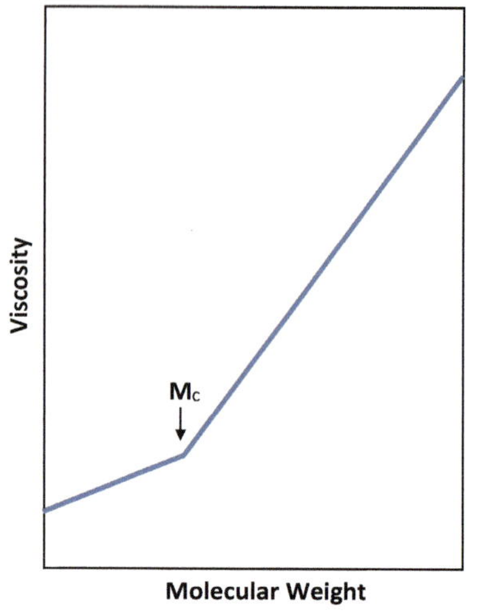

Figure 2.6 Viscosity at various molecular weights

In macromolecules, entanglements act as obstacles to the movement of a polymer chain, creating a restricted tube with surrounding chains in such a way that the chain can only wriggle along the tube, while the lateral motions of the chain are restricted. Such motion is called "reptation". Based on this model, Doi and Edwards [1] proposed a mathematical derivation that links microscopic molecular topology to macroscopic rheological properties. The results also show that the viscosity increases with about a 3.4 power of the molecular weight [2]. This conclusion is in agreement with experimental measurement for many polymers [3].

2.2.1.2 Effects of Shear Rate

Fluids that obey Newton's Law of Viscosity (Equation 2.2) are called Newtonian fluids, and can be observed in many gases, or liquids and solutions of low molecular weight. Since the viscosity of a Newtonian fluid is irrelevant to shear rate at a constant temperature, the graph showing the relationship between them is just a horizontal line (Figure 2.7(a)). There is a linear relationship between shear stress and shear rate which takes the form of a straight line passing through the origin on the graph (Figure 2.7(b)).

For polymeric fluids, the viscosity usually exhibits shear-rate-dependent properties. A fluid whose viscosity decreases with increase in shear rate (Figure 2.7(a)) is called a shear thinning fluid or pseudoplastic fluid. In contrast, a fluid whose viscosity increases with increase in shear rate is called a dilatant fluid, also known as a shear-thickening fluid.

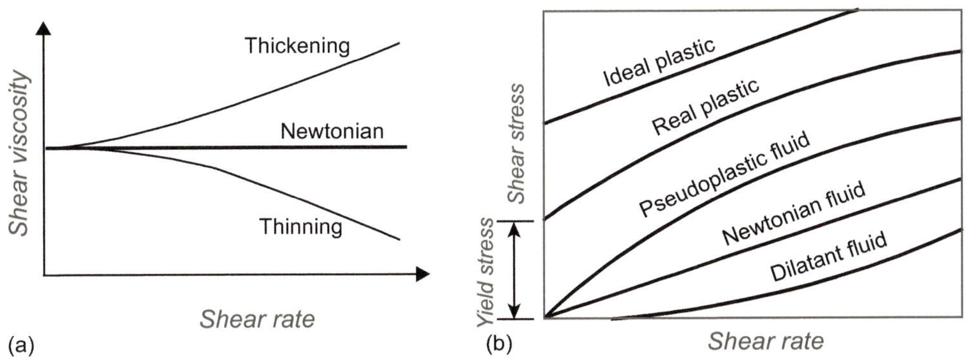

Figure 2.7 (a) Shear viscosity vs. shear rate and (b) shear stress vs. shear rate

Some fluids, such as suspensions or slurry solutions, usually do not flow under small shear stress. The flow starts when the external force is larger than the yield stress; such a fluid is called a viscoplastic fluid, also known as a Bingham fluid.

Most polymeric fluids can be classified as pseudoplastic fluids. Figure 2.8 shows the dependence of shear rate on shear stress and viscosity for a typical polymeric fluid under shear flow. With increase in shear rate, the polymer chains start to disentangle and align along the direction of flow. The alignment reduces the molecular interactions, as a result of which the viscosity continues decreasing as the shear rate increases, i. e. shear-thinning behavior occurs. When the shear rate gets extremely high, the oriented molecules do not have sufficient time to revert to the coiled state and are completely aligned under the shear stress of the fluid. The flow resistance and viscosity reach the minimum (η_∞) theoretically. However, this phenomenon is usually difficult to measure in polymer melts, because the degradation will be very acute under such high shear rates and stress.

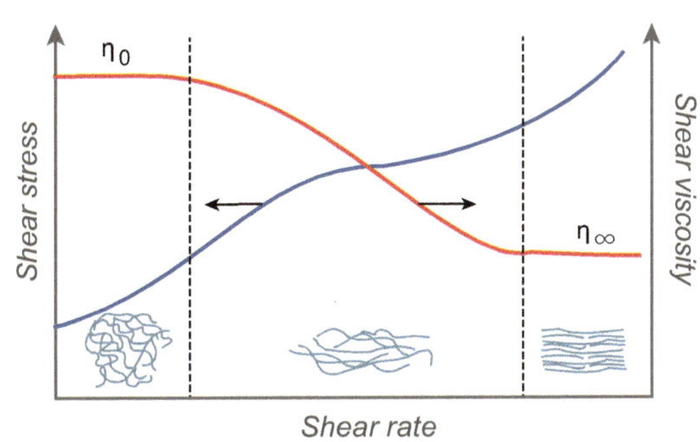

Figure 2.8 Shear viscosity and shear stress vs. shear rate

2.2.1.3 Effects of Temperature

The viscosity of polymeric fluids not only is sensitive to the shear rate, but also changes with temperature. This phenomenon of viscosity changing with temperature is also observed in common Newtonian fluids. As shown in Figure 2.9, the curve of viscosity moves toward the bottom left of the graph as the temperature rises. At higher temperatures, the polymer chains have a larger free volume and mobility and so find it much easier to overcome the interaction energy of molecules. Therefore, raising the processing temperature is commonly used to increase the flowability of polymer melts.

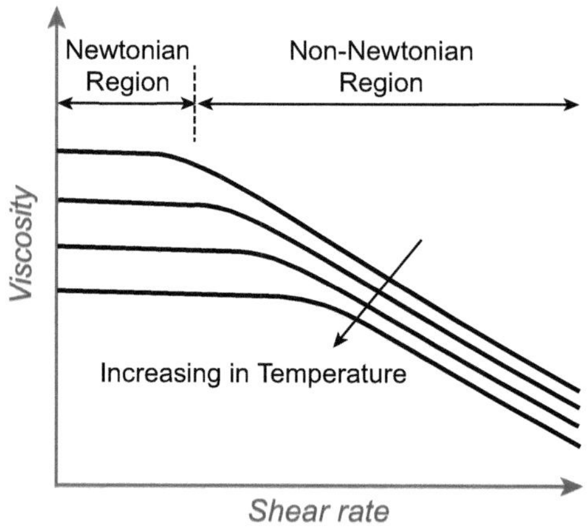

Figure 2.9 Shear viscosity of various temperatures

To describe the temperature effect on viscosity, the shift factor, a_T is defined as:

$$a_T \equiv \frac{\eta_0(T)}{\eta_0(T_0)} \tag{2.3}$$

where η_0 is the zero shear rate viscosity under temperature T and T_0, respectively.

The viscosity under various temperatures and shear rates can be expressed as the viscosity at reference temperature:

$$\eta(T, \dot{\gamma}) = a_T \eta(T_0, \dot{\gamma} \cdot a_T) \tag{2.4}$$

Commonly, there are two models to describe the relationship between a_T and temperature:

1. Arrhenius-type equation:

$$a_T = \exp\left(\frac{\Delta H}{R}\left(\frac{1}{T} - \frac{1}{T_0}\right)\right) \tag{2.5}$$

where ΔH is known as the activation energy, familiar from chemistry, and R is the gas constant. T is the absolute temperature; T_0 is the reference temperature. ΔH depends on the type of polymer. The model provides a good description of molten polymers at temperatures much higher than the glass transition temperature. It can be described alternatively as:

$$a_T = A\exp\left(\frac{T_b}{T}\right) \tag{2.6}$$

where $A = \exp\left(-\frac{\Delta H}{RT_0}\right)$, and $T_b = \frac{\Delta H}{R}$.

2. WLF equation (Williams-Landel-Ferry equation [4]):

$$\log a_T = \frac{-C_1\left(T - T_0\right)}{C_2 + \left(T - T_0\right)} \tag{2.7}$$

where C_1 and C_2 are nearly universally constant (C_1 = 17.44; C_2 = 51.6 K) for many polymers [4].

The WLF equation is derived from the free volume theory [4]. It is readily applicable within the range between T_g and T_g + 100 °C, which makes it suitable for many various types of polymeric materials.

2.2.1.4 Effects of Pressure

Although pressure can affect the viscosity of polymeric fluids, the effect may be not high enough in a common injection molding filling process. However, in the case of high-speed injection molding of thin-wall parts at pressures greater than 100 MPa, the viscosity increases that accompany the pressure increases start to become significant. The pressure effect can also be described by free volume theory, namely that the free volume decreases with increase in pressure, and the shift factor of pressure [5] a_p can be defined in a manner similar to the temperature effect. The viscosity, along with the pressure effect, can be expressed as:

$$\eta(T, P, \dot{\gamma}) = a_T a_P \eta(T_0, P_0, \dot{\gamma} a_T a_P) \tag{2.8}$$

where P_0 is the reference pressure. The model for combining both temperature and pressure effects can also be expressed by both the Arrhenius-type and WLF equations that will be discussed in the next section (Equation 2.13 and Equation 2.14).

2.2.1.5 Theoretical Models

There are many different mathematical models to represent the relationship between viscosity and shear rate, four of which are introduced as follows: the power-law, Carreau-Yasuda, modified Cross, and Herschel-Bulkley models.

1. Power-Law Model

For the shear thinning regions, the power law can be used to represent the variation of viscosity with shear rate:

$$\eta = m\dot{\gamma}^{n-1} \tag{2.9}$$

where m and n are characteristic rheological parameters, with m being the flow consistency index with unit Pa · s^n, and n being the flow behavior index or power law index (dimensionless). The parameters m and n are fitted via data analysis from experiments. If $n = 1$, then m is the viscosity of a Newtonian fluid; if $n < 1$, it reflects the shear thinning property of a pseudoplastic fluid; on the other hand, if $n > 1$, it reflects the shear thickening property of a dilatant fluid.

The power-law model is quite simple and useful, as it is not only able to represent the characteristics of polymeric fluids in shear flow within the shear thinning region by just two parameters, but can also be used to deduce many easy but useful equations and analytical solutions.

2. Carreau-Yasuda Model

The Carreau-Yasuda model uses five parameters to accurately represent the variation of viscosity with shear rate from first to second Newtonian plateau. It is very helpful for numerical calculations and is also widely applied in analytical calculations of some processing procedures. The model is given by the following equation:

$$\eta = \frac{\eta_0 - \eta_\infty}{\left(1 + \left(\frac{\eta_0}{\tau^*}\dot{\gamma}\right)^a\right)^{\frac{1-n}{a}}} + \eta_\infty \tag{2.10}$$

where η_0 denotes the viscosity of zero shear rate, while η_∞ denotes the viscosity of infinite shear rate (second Newtonian plateau). τ^* denotes the shear stress during the transition between Newtonian and shear thinning region, n is the power-law index, and a is a dimensionless value representing the transition from the zero shear rate zone to the power-law zone.

3. Modified Cross Model

Another frequently used experimental model is the modified Cross model, which is given by the following equation:

$$\eta = \frac{\eta_0}{1 + \left(\frac{\eta_0}{\tau^*}\dot{\gamma}\right)^{1-n}} \tag{2.11}$$

where τ^* denotes the critical shear stress which is the same as the one in the Carreau-Yasuda model, and also n has the same effect as in the Carreau-Yasuda model.

4. Herschel-Bulkley Model

As mentioned above, the yield stress will be observed for highly filled or complex materials with high fiber content (Bingham-type fluid), which means that the material does not flow until the stress applied is higher than the yield stress, τ_y:

$$\eta = \frac{\tau_y}{\dot{\gamma}} + \frac{\eta_0}{1 + \left(\frac{\eta_0}{\tau^*}\dot{\gamma}\right)^{1-n}} \tag{2.12}$$

The first term is the effect of yield stress on viscosity, and the second term describes the shear thinning effect of polymer media.

The temperature and pressure effects on viscosity are combined within the zero shear viscosity for both Arrhenius-type and WLF equations.

Arrhenius-type:

$$\eta_0 = A\exp\left(\frac{T_b}{T} + DP\right) \tag{2.13}$$

WLF equation:

$$\eta_0 = D_1\exp\left(\frac{A_1(T - T_c)}{A_2 + (T - T_c)}\right) \tag{2.14}$$

$$T_c = D_2 + D_3P$$

$$A_2 = \tilde{A}_2 + D_3P$$

Under processing at high speed, the viscous dissipation energy can become significant.

The energy of viscous heat generation (H_{vis}) can be determined by calculating the product of the viscosity and the shear rate squared:

$$H_{vis} = \eta\dot{\gamma}^2 \tag{2.15}$$

The shear rate has a stronger influence on heat generation. Consequently, the temperature will rise intensely with increase in shear rate, and may lead to acute degradation of material.

2.2.1.6 Viscosity Properties of Plastics

As mentioned above, the rheological properties of melted plastics are affected by chemical structure, molecular configuration, conformation, molecular weight, temperature, pressure, shear rate, and other factors. Generally, the viscosity of amorphous polymer is more sensitive to the temperature effect than is crystalline polymer. As shown in Figure 2.10, where PP (crystalline) is compared with PS (amorphous), it is clear that the curves of PS (red lines) separate more than those of PP (blue lines) between different temperatures, while in the shear-thinning region, the slope of the PP curves is larger than those of PS, indicating a stronger shearing effect in the case of PP.

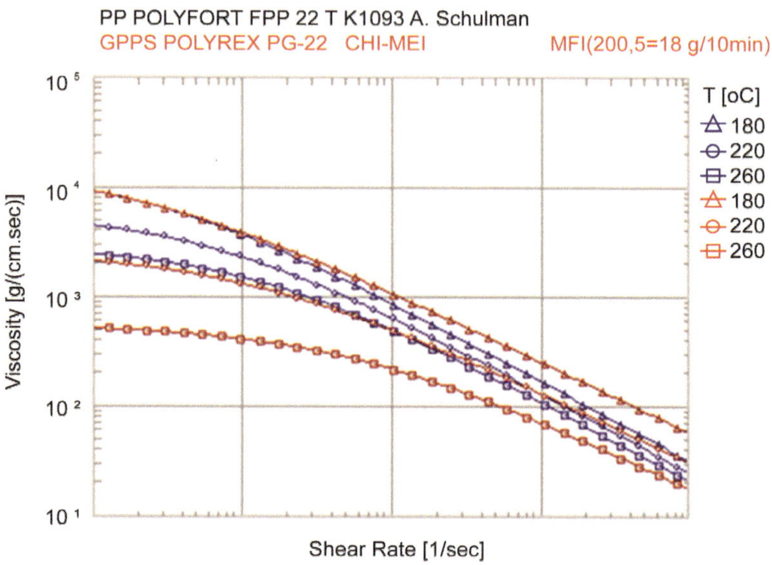

Figure 2.10 Viscosity comparison for PP and PS

The viscosity comparison of two grades of PS is shown in Figure 2.11, where the slopes in the shear-thinning region of both PS grades are the same (this may indicate that the molecular structures of the two grades are similar). The overall viscosity of POLYREX PG-80 (red lines) is significantly higher, and the critical shear rate transition from Newtonian region to shear-thinning region is also lower. Generally speaking, the reciprocal of the critical shear rate can be considered as the relaxation time of material [2]. In other words, the polymer of higher molecular weight has a longer relaxation time, and thus the transition point appears at a lower shear rate, while the viscosity in the Newtonian region is also higher (as mentioned in Section 2.2.1.1, the viscosity in Newtonian region is related to the molecular weight by the power of 3.4).

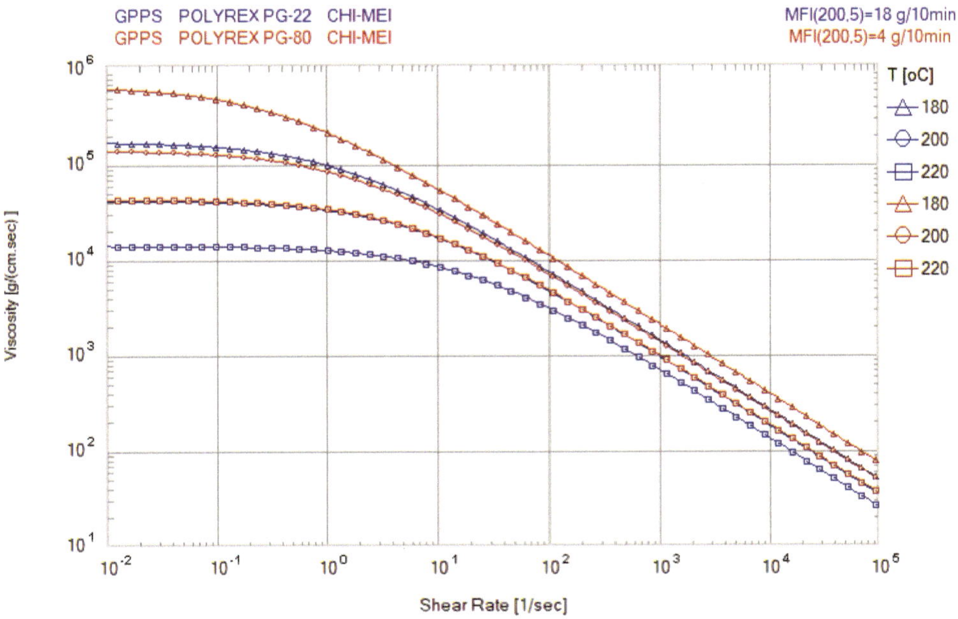

Figure 2.11 Viscosity comparison for different grades of PS

2.2.2 Viscoelastic Fluids

In many situations, polymeric fluids not only exhibit elastic behavior like that of solids, but also behave like the aforementioned viscous fluids. Such polymeric fluids are thus called viscoelastic fluids. Whether a polymer material shows more elastic or viscous behavior is related not only to the thermodynamic states (i.e. temperature and pressure) but also to the observation time. Thus, time (or history) dependence is an important feature of viscoelastic behavior.

2.2.2.1 Viscoelastic Behavior

The flow behavior of polymers that causes them to perform differently from a simple fluid stems from their viscoelasticity, such as in the following cases:

1. Rod-Climbing

When a rod rotates in a Newtonian fluid, the fluid will move away from the rod under inertia. However, in the case of a melted polymer or concentrated polymer solution, the fluid will climb the rod as shown in Figure 2.12(a) because of the normal stress difference, and this can be interpreted in such a way that the velocity on the shaft surface is higher and the molecular chains close to the shaft are directionally stretched onto the shaft. The stretched chains try to return to the coiled condition spontaneously, as this creates a pulling force on the fluid near the shaft. Such pulling force tends to reduce the diameter of the circular ring generating the stress in the centripetal direction and makes the fluid move centripetally until the centripetal force is balanced by the inertia (centrifugal force) of the fluid. The fluid pressure on the top side is lower, and the fluid is driven to move upward along the shaft until the upward force is balanced by gravity. Such a phenomenon is also called the Weissenberg or rod-climbing effect.

2. Extrudate Swell

In polymer extrusion, it can be observed that the dimension of the fluid exiting the die is usually larger than the die; this is known as die swell or extrudate swell, as shown in Figure 2.12(b). A simplified interpretation explains this as being due to the relaxation of normal stress difference. The polymer chains of the fluid orient along the streamline in the mold, and the fluid contracts axially and expands radially once it is outside the mold. In injection molding, when the melted plastic passes through a small sprue under higher shear stress, the result is a normal stress difference (due to the molecular orientation) of the melted plastic inside the runners, whereas the melt front swells up rapidly as it passes through the sprue, because the shear rate is suddenly reduced and hence the sectional area of the extrudate is larger than that of the sprue.

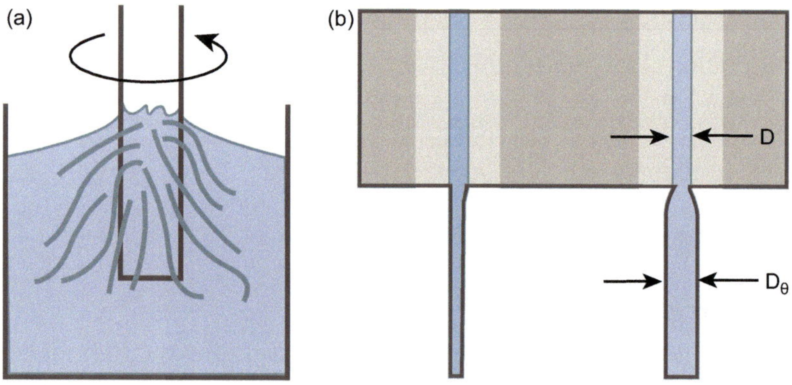

Figure 2.12 (a) Rod-climbing and (b) extrudate swell (also called "die swell")

3. Tubeless Siphon

Siphonage, a common phenomenon, works only if one end of the tube inside the beaker is lower than the liquid level of a Newtonian fluid. However, siphonage still works for polymeric fluids even if the tube is held above the liquid surface within a certain distance, as shown in Figure 2.13(a). The main reason for this phenomenon is considered to be the result of molecular chains of the polymer being directionally stretched by the fluid field.

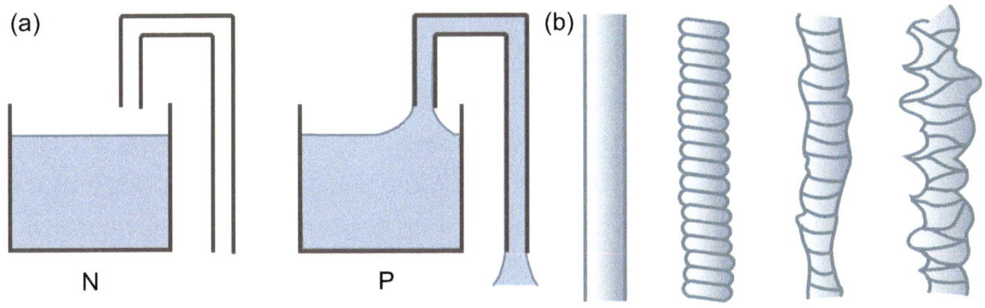

Figure 2.13 (a) Tubeless siphon and (b) extrusion instability

4. Elastic Recoil

Water and a polymeric fluid are injected into tubes respectively. If the injection is stopped after a certain time, the wavefront of water stops moving, whereas that of the polymeric fluid still moves a certain distance forward, then recoils for a short distance, indicating elastic behavior by the fluid [6].

5. Flow Instabilities

It is known that a fluid may exhibit unstable flow when inertial force dominates over viscous force (turbulent flow). Generally, inertia-driven instability seldom happens because Re is very small for a polymer melt of high viscosity. However, flow instabilities may occur when elastic force dominates over viscous forces, and this is called elastic instability. A notorious example is polymer extrusion. As shown in Figure 2.13(b), with increase in extruding speed, the corresponding shear stress exceeds a critical value, the surface of the extrudate starts to change from slick and smooth to rough like shark skin. At even higher speed, the extrudate becomes distorted and breaks; this is called melt fracture.

2.2.2.2 Theoretical Models

1. Linear Differential Model

For describing the viscoelastic behavior of a material, the Maxwell model is not only the simplest but also a linear viscoelastic type. As shown in Figure 2.14, the viscous behavior is modeled by a dashpot and an elastic spring connected in series.

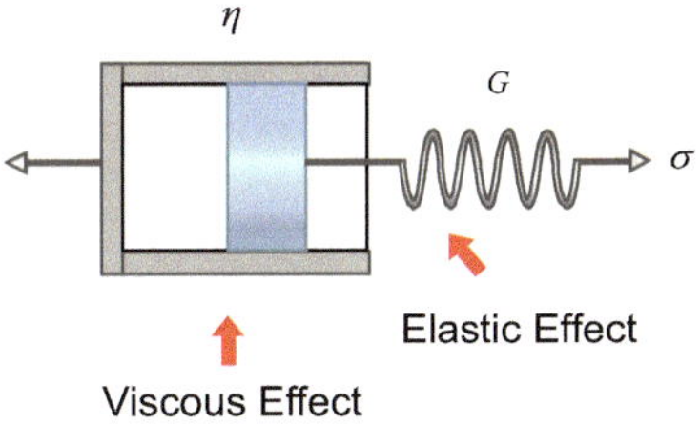

Figure 2.14 Representation of a single mode Maxwell model

Elastic behavior is represented by Hooke's law, whose corresponding mechanical model is a spring that presents a linear relationship between stress and strain, whereas viscous behavior is represented by Newton's law of viscosity, whose corresponding model is a dashpot that presents a linear relationship between stress and strain rate. By combining both, the Maxwell model uses a differential equation to represent the relationship between stress and strain (or strain rate):

$$\sigma + \lambda\dot{\sigma} = \eta\dot{\gamma} \qquad (2.16)$$

$$\lambda = \frac{\eta}{G}$$

where σ is the applied stress, η is the viscosity of the dashpot, G is the modulus of the spring, and λ is a time constant often called the "relaxation time". A higher relaxation time indicates more memory (elastic) effects. In this model, the stress, strain, and their rates of change with respect to time are all related to the time-dependent stress.

(a) *Upper convected Maxwell model (UCM)*

To apply the model in a multi-dimensional system, it is necessary to find a frame-invariant way of expressing time-changing stress and deformations. Therefore, the contravariant (upper convected) time derivative is used instead of the partial time derivative. The upper convected Maxwell model is written as:

$$\boldsymbol{\tau} + \lambda\overset{\nabla}{\boldsymbol{\tau}} = \eta\left(\nabla\mathbf{V} + \nabla\mathbf{V}^{\mathrm{T}}\right) \qquad (2.17)$$

$$\overset{\nabla}{\boldsymbol{\tau}} = \frac{\partial\boldsymbol{\tau}}{\partial t} + \mathbf{V}\cdot\nabla\boldsymbol{\tau} - \nabla\mathbf{V}^{\mathrm{T}}\cdot\boldsymbol{\tau} - \boldsymbol{\tau}\cdot\nabla\mathbf{V}$$

For steady-state motions and extremely small relaxation times, this equation simplifies to the Newtonian fluid with viscosity η. For sudden changes in stress, the time derivative term gives the Hookean solid with elastic modulus $G(\eta/\lambda)$.

(b) *Convected Jeffrey model (Oldroyd-B)*

This model adds an additional Newtonian component to UCM model:

$$\boldsymbol{\tau}_1 + \lambda_0\left(\frac{\partial\boldsymbol{\tau}_1}{\partial t} + \mathbf{V}\cdot\nabla\boldsymbol{\tau}_1 - \nabla\mathbf{V}^{\mathrm{T}}\cdot\boldsymbol{\tau}_1 - \boldsymbol{\tau}_1\cdot\nabla\mathbf{V}\right) = \eta_1\left(\nabla\mathbf{V} + \nabla\mathbf{V}^{\mathrm{T}}\right) \qquad (2.18)$$

$$\boldsymbol{\tau}_2 = \eta_2\left(\nabla\mathbf{V} + \nabla\mathbf{V}^{\mathrm{T}}\right)$$

$$\boldsymbol{\tau} = \boldsymbol{\tau}_1 + \boldsymbol{\tau}_2$$

$$\eta_2 = r\eta_0$$

If $r = 0$, the model reduces to UCM model. The model predicts a constant viscosity and a quadratic first normal difference. It should be further noted that the linear (or quasi-linear) model has been widely used for viscoelastic flow calculations for its simplicity, but it is not suitable for injection molding process because of the lack of shear thinning characteristics. It is suggested to use those models in polymeric solutions for research purposes.

2. Nonlinear Differential Models

As a linear model cannot provide a good portrayal of nonlinear rheological properties, such as shear thinning and non-quadratic first normal difference observed in typical melts, it needs to be extended to cover nonlinear behavior. The following are the common nonlinear Maxwell-type models that are easier to implement numerically:

(a) *White-Metzner model*

This model is modified from the UCM model. It can illustrate a reasonable curve of shear rate-dependent shear viscosity because the curve is equal to the Cross model of steady shear flow. However, unrealistic behavior may be predicted for elongational flow. The model is suitable for fast time-dependent motions.

$$\boldsymbol{\tau}+\lambda(T,\dot{\gamma})\left(\frac{\partial\boldsymbol{\tau}}{\partial t}+\mathbf{V}\cdot\nabla\boldsymbol{\tau}-\nabla\mathbf{V}^{\mathrm{T}}\cdot\boldsymbol{\tau}-\boldsymbol{\tau}\cdot\nabla\mathbf{V}\right)=\eta(T,\dot{\gamma})\left(\nabla\mathbf{V}+\nabla\mathbf{V}^{\mathrm{T}}\right) \tag{2.19}$$

$$\lambda(T,\dot{\gamma})=\frac{\eta(T,\dot{\gamma})}{G}$$

(b) *Modified White-Metzner model*

This model is modified from the White-Metzner model. The variation of relaxation time is independent of the viscosity and can be fitted to the Cross model.

$$\boldsymbol{\tau}+\lambda(T,\dot{\gamma})\left(\frac{\partial\boldsymbol{\tau}}{\partial t}+\mathbf{V}\cdot\nabla\boldsymbol{\tau}-\nabla\mathbf{V}^{\mathrm{T}}\cdot\boldsymbol{\tau}-\boldsymbol{\tau}\cdot\nabla\mathbf{V}\right)=\eta(T,\dot{\gamma})\left(\nabla\mathbf{V}+\nabla\mathbf{V}^{\mathrm{T}}\right) \tag{2.20}$$

$$\lambda(T,\dot{\gamma})=\frac{\lambda_0 a_T}{1+\left(\dfrac{\lambda_0 a_T}{k}\dot{\gamma}\right)^{1-m}}$$

where λ is the relaxation time, λ_0 is the relaxation time at zero shear rate, and a_T is the shift factor. The model provides flexibility by allowing different variation profiles for the shear rate dependence of viscosity and relaxation time.

(c) *Giesekus model*

This model describes the rheological properties with nonlinear stress terms as follows:

$$\boldsymbol{\tau}+\frac{\alpha\lambda_0}{\eta}\boldsymbol{\tau}\cdot\boldsymbol{\tau}+\lambda_0\left(\frac{\partial\boldsymbol{\tau}}{\partial t}+\mathbf{V}\cdot\nabla\boldsymbol{\tau}-\nabla\mathbf{V}^{\mathrm{T}}\cdot\boldsymbol{\tau}-\boldsymbol{\tau}\cdot\nabla\mathbf{V}\right)=\eta_0\left(\nabla\mathbf{V}+\nabla\mathbf{V}^{\mathrm{T}}\right) \tag{2.21}$$

where λ_0 is the relaxation time at zero shear rate, η_0 is the shear viscosity at zero shear rate, and α (0~1) is the dimensionless mobility factor. The origin of the term involving α can be associated with anisotropic Brownian motion or anisotropic hydrodynamic drag on the constituent polymer molecules. Large decreases in viscosity and normal stress coefficients with increasing shear rate are possible, which is much more realistic than in the linear models.

(d) *PTT (Phan-Tien and Tanner) model*

This model is a simplified form of the PTT model, which describes the rheological properties with nonlinear stress terms as follows:

$$\left(1+\frac{\varepsilon\lambda_0}{\eta}\mathrm{tr}(\boldsymbol{\tau})\right)\boldsymbol{\tau}+\lambda_0\left(\frac{\partial\boldsymbol{\tau}}{\partial t}+\mathbf{V}\cdot\nabla\boldsymbol{\tau}-\nabla\mathbf{V}^{\mathrm{T}}\cdot\boldsymbol{\tau}-\boldsymbol{\tau}\cdot\nabla\mathbf{V}\right)=\eta_0\left(\nabla\mathbf{V}+\nabla\mathbf{V}^{\mathrm{T}}\right) \tag{2.22}$$

where λ_0 is the relaxation time at zero shear rate, η_0 is the shear viscosity at zero shear rate, and ε is the material property that controls nonlinear behavior.

The model exhibits shear thinning and non-quadratic first normal stress difference, which is similar to the predictions of the Giesekus equation, except for some minor viscometric features.

(e) *Multimode model*

In commercial polymer material, chain relaxations happen over a period of time, instead of at a single point in time, due to the contribution of chain lengths or to the various characteristic lengths in a polymer chain. The molecular segments of different lengths give rise to various relaxation times, called a relaxation spectrum. Multiple elements are assembled in parallel as Maxwell elements, as shown in Figure 2.15.

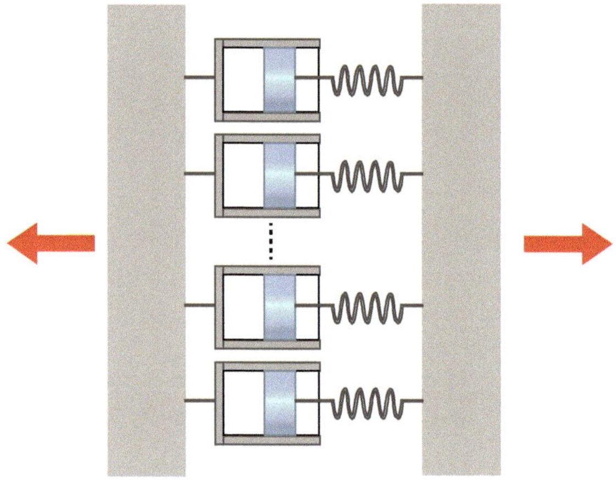

Figure 2.15 Representation of multimode model

To take the distribution of various relaxations into account, the total stress $\boldsymbol{\tau}$ is the sum of all elements $\boldsymbol{\tau}_i$ in the following equation, where $\boldsymbol{\tau}_i$ could be expressed as any constitutive model mentioned above.

$$\boldsymbol{\tau} = \sum_{i-1}^{n} \boldsymbol{\tau}_i \tag{2.23}$$

2.2.2.3 Measurement of Viscoelasticity

Rotational rheometers are able to measure the viscoelastic properties of liquids. Step strain is commonly employed and consists in exerting a sudden strain on the material at time = 0 and then measuring the stress variation over time. The obtained relaxation modulus $G(t)$ [7] varies with time, as shown in Figure 2.16, where the modulus can be divided into a few regions of physical implications:

1. Glassy region: The chains are frozen, with the segments vibrating only around these fixed positions on the sites of a disordered quasi-lattice; the material is hard and brittle, and the modulus is 10^9–10^{10} Pa.
2. Glassy transition: Short-range motions begin that make the chains change to a resilient leather-like state from the brittle glass-like state.
3. Rubbery region: After the modulus has decreased by several orders of magnitude in the glass transition, a plateau region, called the rubbery plateau, is reached again. The modulus is 10^5–10^6 Pa.
4. Viscoelastic flow: The material starts to flow and becomes a fluid possessing elastic behavior.
5. Viscous flow: The elasticity almost disappears and thus the modulus is immeasurable.

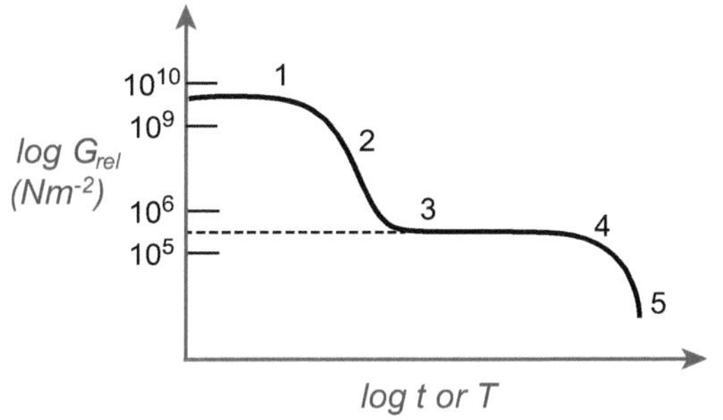

Figure 2.16 Relaxation modulus of a typical amorphous polymer

A similar curve can also be obtained for modulus as a function of temperature at constant time. Generally speaking, the rotational rheometer is mainly employed for measuring materials in the liquid state (regions 3 to 5). On the other hand, the torque needed for measuring from solid will exceed the instrument's limitations, with the result that the fixture of common parallel plates or cone plates cannot grasp the material firmly. Therefore, DMA (dynamic mechanical analysis) is recommended for the solid-like state of regions 1 to 3.

Oscillatory Strain (Dynamic Measurement)

Another experiment commonly used to obtain the viscoelasticity is called dynamic measurement. That is, if the strain applied to the material is sinusoidal, the strain rate (or shear rate) is a cosine wave, and the stress measured also presents as a periodic wave. The waveform of the stress has to be in phase with the strain for an ideal elastomer material, and the waveform of stress has to be in phase with the strain rate for a purely viscous fluid, while the phase of a viscoelastic fluid has to lie between these two.

Sinusoidal strain:

$$\gamma(t) = \gamma^0 \sin(\omega t) \tag{2.24}$$

where ω is the angular frequency, t is time, γ^0 is the amplitude of strain, and shear rate can be deduced as:

$$\dot{\gamma}(t) = \gamma^0 \omega \cos(\omega t) \tag{2.25}$$

The measured stress response can be represented by the stress amplitude $A(\omega)$ and the phase angle δ, or by moduli G' and G''.

$$\sigma(t) = -A(\omega)\gamma^0 \sin(\omega t + \delta) = \gamma^0 \left(G'(\omega)\sin(\omega t) + G''(\omega)\cos(\omega t)\right) \tag{2.26}$$

where G' is the storage modulus representing the elastic behavior and G'' is the loss modulus representing viscous behavior. $G^*(\omega) = G(\omega) + iG''(\omega)$ is the complex modulus and can be derived from $G(t)$ through a Fourier transform. $G'(\omega)$, $G''(\omega)$, relaxation modulus and $G(t)$ contain the same viscoelastic information.

Figure 2.17 shows the results for storage and loss modulus measurements made on polystyrene at different temperatures with a parallel plate rheometer, with higher modulus appearing at lower temperatures. Since the parallel plate rheometer is mainly used for measuring the rheological properties of fluids, the measuring limitation with respect to modulus is about 10^6 Pa.

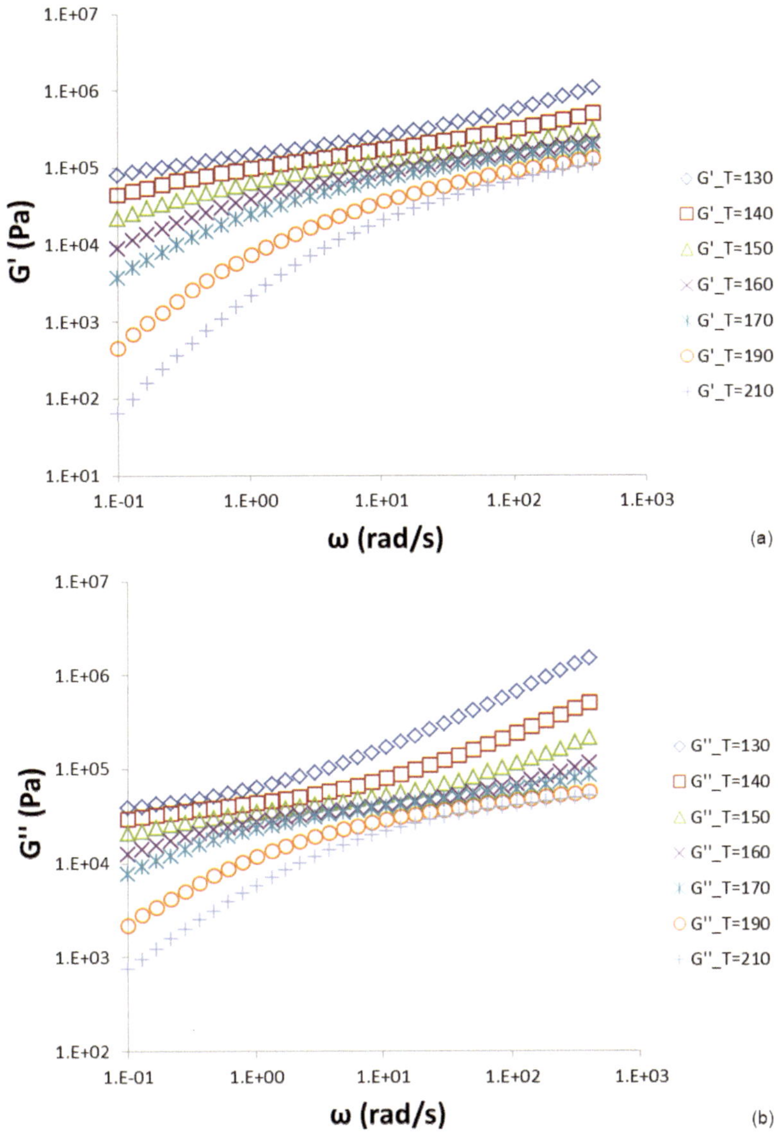

Figure 2.17 Modulus of polystyrene measured by parallel plate rheometer: (a) storage modulus and (b) loss modulus

The frequency range of a common rotational rheometer is 0.01 to 100 Hz, while the magnitude of polymeric relaxation behavior often covers 10 orders of magnitude, with the result that data cannot be directly measured under all frequencies at a constant temperature. The time-temperature superposition principle can thus be used to extend the frequency range by superposing the curves of other temperatures, i.e. the curve obtained at lower temperature corresponds to that obtained under higher frequency. As shown in Figure 2.18, a reference temperature (T_2) is chosen and the curves for the remaining temperatures are shifted horizontally to construct a master curve, which is a complete modulus-time relationship at a constant temperature. This procedure makes it possible to extend the data to a wide range of frequency (or time) from the data obtained at a constant temperature. This also implies that a master curve obtained at a given temperature can serve as a reference for predicting curves at various temperatures through the application of a shift operation.

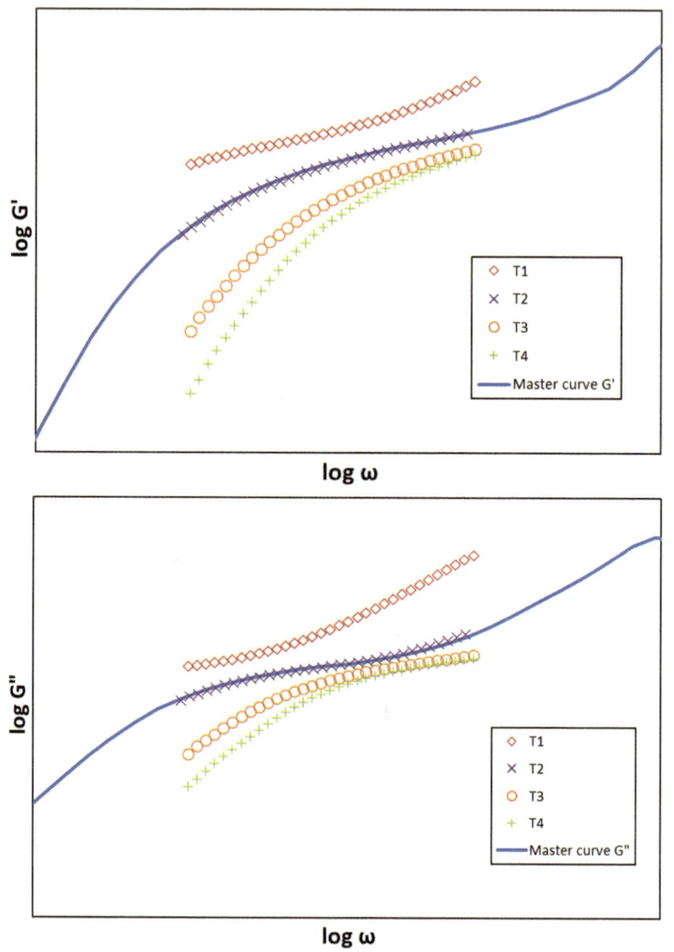

Figure 2.18 Moduli of various temperatures and the master curve

The shifting quantity a_T for moving each temperature to the master curve is called the shift factor and represents the effect of temperature on rheological properties, as shown in Figure 2.19. This shift factor has the same physical implications as viscosity, and so it can be represented by the same mathematical model, such as a WLF or an Arrhenius-type equation.

Figure 2.20 shows the results after shifting the curve into the master curve by the time-temperature superposition principle using the data in Figure 2.17, where the data were measured with a parallel plate rheometer from the lowest viscous region to the highest rubbery region.

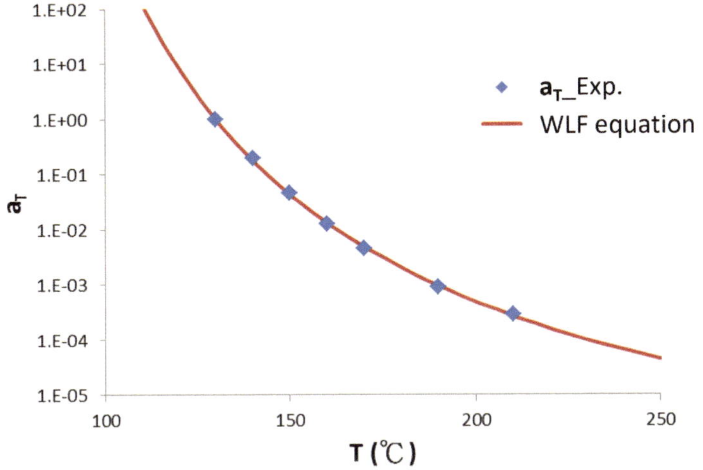

Figure 2.19 Shift factors for polystyrene fitted by an WLF equation with reference temperature of 130 °C

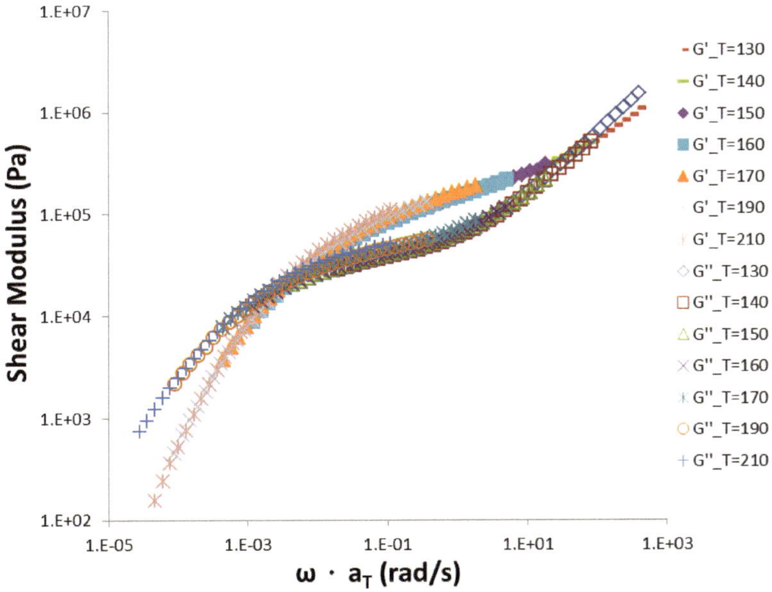

Figure 2.20 Master curve for polystyrene at a reference temperature of 130 °C

2.3 Thermodynamic and Thermal Properties

During the molding process, the material is heated to the melt state, then cooled back to the solid state. Part temperature will not be uniform since heat dissipation is governed by conservation of energy including thermal conduction, convection, and heat generation etc. In the injection molding process, the injection temperature should be higher than the melting temperature T_m, so that the material is of low-enough viscosity to be properly injected into the mold cavity whereas, before ejection, the material should be cooled until its temperature becomes lower than the glass transition temperature T_g, so that solidification of the material and the mechanical properties can be assured.

Thermal conductivity (k) and specific heat capacity (C_P) also play important roles in injection molding. Because the material is constantly chilled and cooled in the mold, thermal conductivity and specific heat capacity are the material characteristics for determining how easily the material can be cooled.

The density (or specific volume) of the material changes as temperature and pressure vary. The relationship between temperature, pressure, and volume is called the P-V-T relation.

2.3.1 Specific Heat Capacity

Specific heat capacity is defined as the heat required to produce a unit change in temperature. The specific heat capacity is usually stated in two ways, depending on which measuring method is used: constant pressure or constant volume.

$$C_p = \left(\frac{dQ}{dT}\right)_P \tag{2.27}$$

$$C_V = \left(\frac{dQ}{dT}\right)_V \tag{2.28}$$

Table 2.1 shows the specific heat capacity of different materials, from which we can see that the specific heat capacity of common plastic materials is about 3–5 times that of metals.

Table 2.1 Specific Heat Capacity of Polymers and Steel

Material	Specific Heat Capacity [J/(kg K)]
ABS	1250–1700
POM-Acetal	1500
PA66	1700

Material	Specific Heat Capacity [J/(kg K)]
PC	1300
PE	2300
PP	1900
PS	1300
PVC	800-1200
Steel (AISI 1020)	460
Steel (AISI P20)	460

From the data obtained by DSC (differential scanning calorimetry), a peak value of C_P can be observed within the temperature range of the phase change, as shown in Figure 2.21, which is a phenomenon caused by the latent heat of the phase transition that occurs during melting or crystallization of the material. The area under the peak line denotes the latent heat of the material.

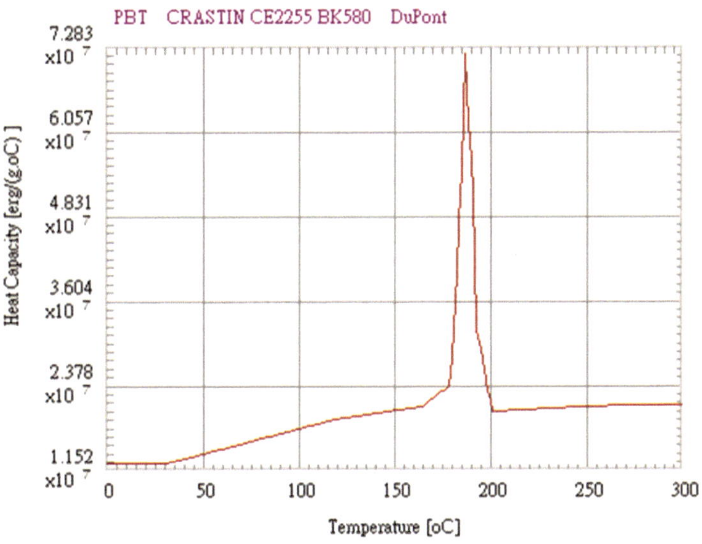

Figure 2.21 Specific heat capacity vs. temperature

Theoretical Models

1. Constant C_P:

It is generally a convenient approximation to assume the specific heat capacity to be a constant:

$$C_P = C_{P_0} \tag{2.29}$$

where C_P is the specific heat capacity and C_{P0} is its given value.

2. Linear model:

The linear interpolation function is another common approximation of the dependence of the specific heat capacity on temperature. Given specific heat capacities, C_{PL} and C_{PS}, at two different temperatures, T_L and T_S, we can obtain the following linear equation:

$$C_P = mT + b \tag{2.30}$$

$$m = \frac{C_{PL} - C_{PS}}{T_L - T_S}$$

$$b = \frac{T_L C_{PS} - T_S C_{PL}}{T_L - T_S}$$

3. Three-stage linear interpolation model:

Another three-stage linear interpolation function of C_P is usually used for CAE simulation. The C_P can be interpolated from four different values. Typically, two of them, C_{PS1} and C_{PS2}, are taken in the solid state, and the other two, C_{PL1} and C_{PL2}, are taken in the liquid state.

$$\text{If } T_{S1} < T < T_{S2},\ C_P = \frac{C_{PS2} - C_{PS1}}{T_{S2} - T_{S1}} T + \frac{T_{S2} C_{PS1} - T_{S1} C_{PS2}}{T_{S2} - T_{S1}} \tag{2.31}$$

$$\text{If } T_{S2} \le T < T_{L1},\ C_P = \frac{C_{PL1} - C_{PS2}}{T_{L1} - T_{S2}} T + \frac{T_{L2} C_{PS2} - T_{S2} C_{PL1}}{T_{L1} - T_{S2}}$$

$$\text{If } T_{L1} < T < T_{L2},\ C_P = \frac{C_{PL2} - C_{PL1}}{T_{L2} - T_{L1}} T + \frac{T_{L2} C_{PL1} - T_{L1} C_{PL2}}{T_{L2} - T_{L1}}$$

4. Tabulated:

As for the specific heat capacity between two given temperatures, a standard linear interpolation of the specific heat capacity is employed.

$$\text{If } T_i < T < T_{i+1},\ C_P = C_{P,i} + (T - T_i)\left(\frac{C_{P,i+1} - C_{P,i}}{T_{i+1} - T_i}\right) \tag{2.32}$$

2.3.2 Melting Point and Glass Transition Temperatures

When the material undergoes a phase change, some enthalpy signals might be observed due to latent heat. With crystalline materials, as shown in Figure 2.22, an endothermic peak at melting point temperature can be observed during scanning, and its area is the enthalpy of melting. In contrast, a similar measurement on cooling yields the crystallization temperature and the enthalpy of crystallization. No crystallization peak or significant melting peak would be observed for non-crystalline (amorphous) materials, but a step change in heat flow would indicate the glass transition temperature.

For amorphous polymers, the conformation of polymer chains in the matrix is random and the chains only exhibit a very small amount of short-range order, whereas considerable three-dimensional ordered structures and long-range order can be found in crystallized polymers.

If the temperature is below T_g, the material is in the glassy state, which is hard and brittle. In contrast, the material becomes soft when the temperature rises above T_g, because the molecules have more thermal energy to perform large-scale movement. As shown in Figure 2.22, the melting temperature (T_m), crystallization temperature (T_c), and glass transition temperature (T_g) happen over a range instead of at constant values.

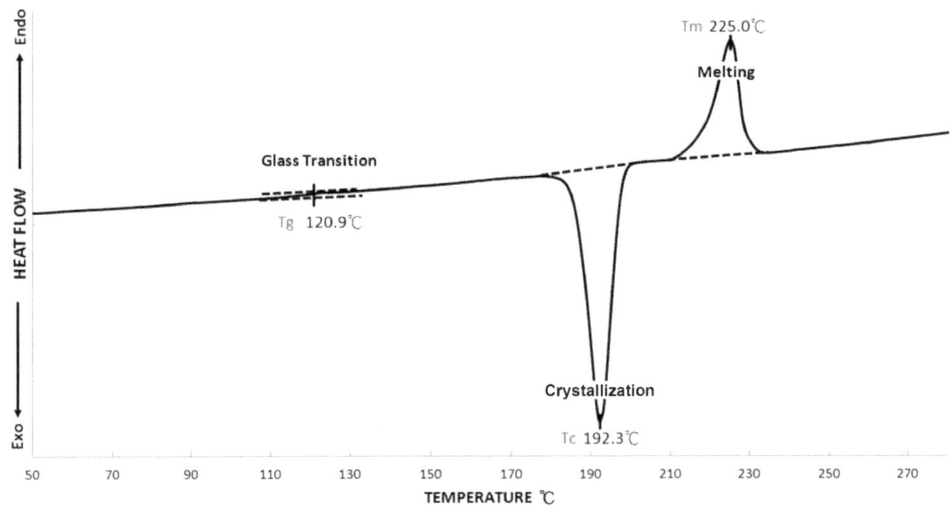

Figure 2.22 DSC result for a crystalline polymer

2.3.3 PVT Equation of State

2.3.3.1 Definition

Like other common materials, plastics also have a tendency to undergo thermal expansion, and their density (or specific volume) also changes if compressed by pressure. The relationship for volume changes with temperature and pressure is called the P-V-T relation.

When the material is at a phase change point, such as the glass transition temperature or melting point, the behavior of volume change is also different. In most cases, the relationship between P-V-T can be considered as a state function; that is, by fixing two, the other is therefore determined. Hence, under the assumption of a state function, the P-V-T relationship could be expressed as:

$$f(p,\hat{V},T)=0 \tag{2.33}$$

The state function relates to three important process variables: pressure (P), temperature (T), and specific volume (V). A change in one variable affects the other two. Given any two variables, the third one can be determined:

$$\hat{V}=f(p,T) \tag{2.34}$$

where *f* is some function determined experimentally.

During the molding process, the temperature difference could be over 350 °C since the plastic is cooled down from a very high temperature, possibly 400 °C, to room temperature. And the pressure may range from 350 MPa to atmospheric, which may cause the maximum change in specific volume of over 10%. The change in specific volume will interfere with the part quality significantly.

2.3.3.2 Data Interpretation

The PVT of amorphous materials is shown in Figure 2.23(a), where a significant transition point occurs on the slope of volume against temperature; this transition point is the T_g. No large-scale movement occurs, as molecular chains are frozen at temperatures below the T_g and thus the rate of volume change accompanying a temperature change is smaller; large-scale movements occur at temperatures higher than T_g as the molecular chains can move freely, which results in a larger free volume for each molecule and hence a steeper slope. For semi-crystalline materials, as the molecules are more tightly packed due to crystallization, a tremendous change in volume around the crystalline temperature may be observed, as shown in Figure 2.23(b).

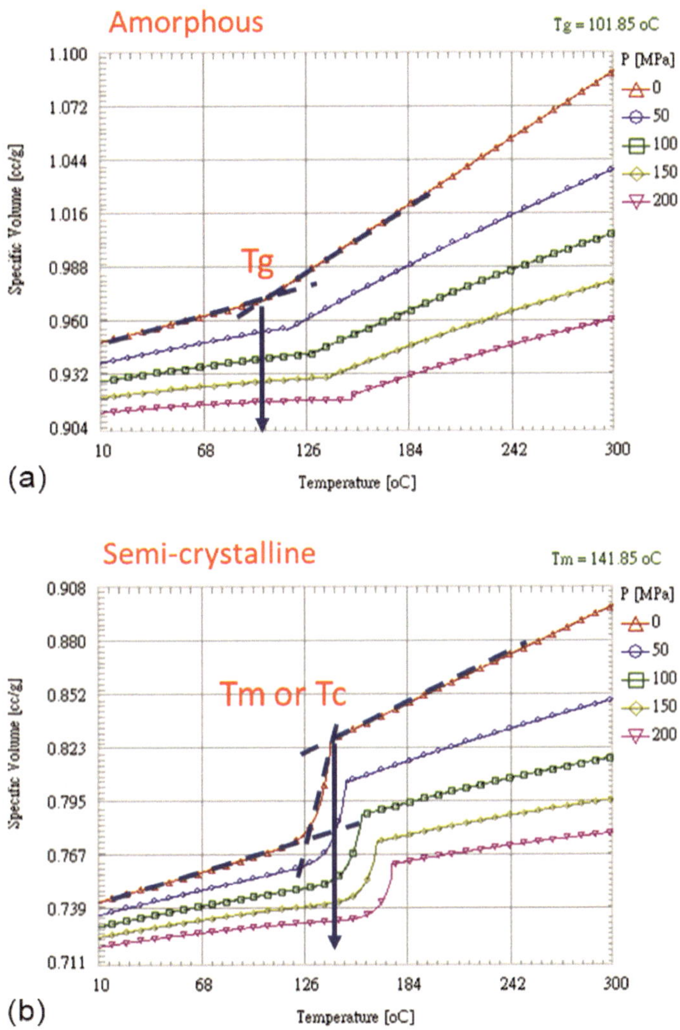

Figure 2.23 Specific volume of (a) amorphous polymer and (b) semi-crystalline polymer

2.3.3.3 Theoretical Models

A good PVT model should characterize the dependence of specific volume on temperature and pressure, and also the different types of materials.

1. Constant Specific Volume

The constant specific volume model assumes that the specific volume is independent of pressure and temperature. It corresponds to incompressible materials.

$$\hat{V} = \hat{V}_0 \tag{2.35}$$

where $\hat{V}$ and $\hat{V}_0$ are the specific volume of the material.

2. Spencer-Gilmore Model

$$\hat{V} = \hat{V}_0 + \frac{R'T}{P + P_0} \tag{2.36}$$

where $\hat{V}$ is the specific volume of the material and $\hat{V}_0$ is its reference specific volume in given conditions. There are three parameters to be determined, $\hat{V}_0$, R', and P_0. T is the material temperature in K.

3. Tait Model

$$\hat{V} = \hat{V}_0 \left[1 - C \ln\left(1 + P / B\right)\right] \tag{2.37}$$

$$\hat{V}_0 = b_1 \exp\left(-b_2 T\right)$$

$$B = b_3 \exp\left(-b_4 T\right)$$

This is the original version of the Tait model. There are five parameters to be determined, b_1, b_2, b_3, b_4, and C. The Tait model was derived from the empirical equation of compressibility [8] and showed that C is almost a universal constant ($C = 0.0894$) for many hydrocarbon liquids [9].

4. Modified Tait Model (1)

$$\hat{V} = \hat{V}_0 \left[1 - C \ln\left(1 + P / B\right)\right] \tag{2.38}$$

$$\hat{V}_0 = b_1 + b_2 \overline{T}$$

$$B = b_3 \exp\left(-b_4 \overline{T}\right)$$

$$\overline{T} = T - T_t$$

$$T_t = b_5 + b_6 P$$

where b_1, b_2, b_3, b_4, b_5, b_6, and C are the fitting parameters. T_t is the pressure-dependent transition temperature of the material. The equation reflects the fact that the transition temperature is generally higher if the pressure is higher.

5. Modified Tait Model (2)

The modified Tait model (1) cannot deal with the abrupt volumetric change of semi-crystalline material, so the following modification was proposed and has been widely employed in all major CAE calculations:

$$\hat{V} = \hat{V}_0 \left[1 - C \ln\left(1 + P / B\right)\right] + \hat{V}_t \tag{2.39}$$

$$\hat{V}_0 = \begin{cases} b_{1S} + b_{2S}\overline{T}, & \text{if } T \le T_t \\ b_{1L} + b_{2L}\overline{T}, & \text{if } T > T_t \end{cases}$$

$$B = \begin{cases} b_{3S}\exp\left(-b_{4S}\overline{T}\right), & \text{if } T \le T_t \\ b_{3L}\exp\left(-b_{4L}\overline{T}\right), & \text{if } T > T_t \end{cases}$$

$$\hat{V}_t = \begin{cases} b_7\exp\left(b_8\overline{T} - b_9 P\right), & \text{if } T \le T_t \\ 0, & \text{if } T > T_t \end{cases}$$

$$\overline{T} = T - b_5$$

$$T_t = b_5 + b_6 P$$

$$C = 0.0894$$

where $\hat{V}_t$ characterizes the abrupt volumetric change undergone by a semi-crystalline material around its melting point; b_7, b_8, and b_9 are zero for amorphous materials. T_t is the phase transition temperature corresponding to the glass transition temperature of amorphous polymer and to the crystalline temperature of semi-crystalline polymer.

The cure-induced volumetric shrinkage usually has been neglected in the warpage simulation. But recently there is more and more evidence showing that warpage and residual stress are underestimated if only the PVT relationship is included, especially when the length to thickness ratio of the part is larger.

6. Spencer-Gilmore-C Model

$$\hat{V} = \hat{V}_0 + \frac{R'T}{P + P_0}(1 - \zeta C) \tag{2.40}$$

Compared to the original Spencer-Gilmore model, there is an additional parameter, ζ, to be determined to evaluate the effect of cure (C).

7. Tait-C Model

$$\hat{V} = \hat{V}_0\left[1 - C\ln\left(1 + P/B\right)\right]\left(1 - \zeta C'\right) \tag{2.41}$$

$$\hat{V}_0 = b_1 + b_2 T$$

$$B = b_3\exp\left(-b_4 T\right)$$

Compared to original Tait model, there is one additional parameter, ζ, to be determined to evaluate the effect of cure (C').

The curing effect in the Tait model can readily describe some thermoset materials, such as rubber or liquid silicone rubber. However, it cannot describe the phase transition behavior, i.e. the change of slope across T_g. For high-T_g material, such as epoxy, a two-domain-modified Tait model was proposed [3], and can be expanded by the Modified Tait (2) model, which takes the glass transition effect into account.

8. Two domain modified Tait Model

$$\frac{1}{V}=\frac{1-C'}{V_{uncured}}+\frac{C'}{V_{cured}} \tag{2.42}$$

$$V_{cured}=V_{0_{cured}}\left[1-C\ln\left(1+\frac{P}{B_{cured}}\right)\right]$$

$$V_{uncured}=V_{0_{uncured}}\left[1-C\ln\left(1+\frac{P}{B_{uncured}}\right)\right]$$

where $V_{uncured}$ and V_{cured} are the specific volume of uncured and cured sample, respectively.

2.3.3.4 Effects of Non-Equilibrium State on PVT

The long-chain structure of a polymer leads to a large number of chain conformations, which become more and more numerous as the molecules get larger. The stretched molecule has lower energy (enthalpy) and lower entropy. According to the second law of thermodynamics, materials with lower energy but higher entropy are more stable. At a higher temperature, the contribution of entropy becomes more significant and so the molecular chain tends to form a random coil easily, whereas, at a lower temperature, the enthalpy becomes dominant and molecules tend to be packed regularly (crystallization) thermodynamically.

A semi-crystalline polymer may be completely amorphous in the solid state, which means that the chains in the specimen are oriented in a totally random fashion. With temperature decrease, the volume change in amorphous polymers follows the curve A–D in Figure 2.24. In the region D to C, the polymer is a glass, but as the sample is heated, it passes through a temperature T_g, called the glass transition temperature, beyond which it softens and becomes rubber-like. This is an important temperature because it marks the point where significant changes in material properties occur, i.e., the material may be more easily deformed or become ductile above T_g. A continuing increase in temperature will increase the free volume along C–B–A, leading to a change of the rubbery polymer to a viscous liquid.

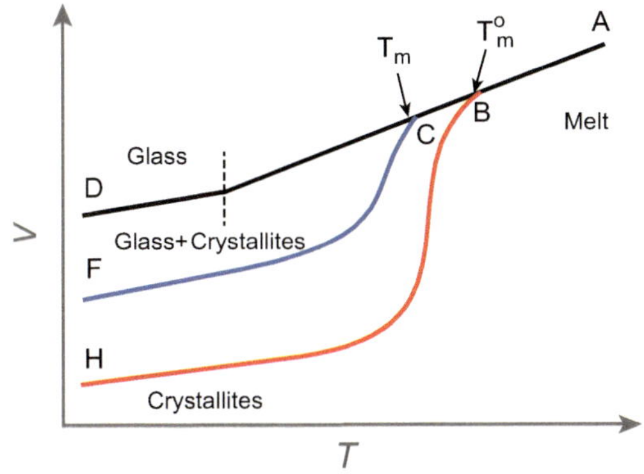

Figure 2.24 Schematic representation of the change of specific volume of a polymer with temperature *T* for (i) a completely amorphous sample (A–C–D), (ii) a semi-crystalline sample (A–C–F), and (iii) a perfectly crystalline material (A–B–H)

In reality, perfectly crystalline polymers do not exist because it would take infinite time for the thermodynamically reversible process of crystallizing all the polymer chains to occur. Thus, polymers will contain varying portions of crystallized and amorphous regions. These semi-crystalline polymers usually possess both T_g and T_m, corresponding to crystallization and amorphous portions, and follow a pathway similar to F–C–B–A. As T_m^0 represents the melting temperature of a perfectly crystalline polymer in a state of thermodynamic equilibrium, T_m is lower and more often represents a melting range, because the semi-crystalline polymer contains a spectrum of chain lengths and crystallites of various sizes with many defects.

As polymers usually exist in non-equilibrium states, the measured crystallization and melting points are therefore different under different thermodynamic processes. As mentioned in Section 2.3.2, what is observed from the phase change in crystalline materials during heating are the melting behavior and the melting temperature, whereas it is the crystallization temperature that is observed when the material is cooled down from high to low temperatures and there is a significant difference between the two temperatures. For example, in the DSC measurement of polypropylene, the melting observed temperature is 170–180 °C at a temperature ramping rate of 5 °C/min; at the same cooling rate, the crystallization temperature is 120–130 °C. The outcome of measuring heating and cooling at the same rate using a PVT instrument is shown in Figure 2.25. We can see that the transition point of the curve is at around 180 °C (melting point) in a temperature-ramping measurement, while the transition point appears at 130 °C (crystallization point) in a cooling measurement. Since injection molding is a cooling process, it is more realistic in reality to use the cooling curve as a simplified state function to represent PVT variation during injection molding. Moreover, crystallization is a kinetic process which depends heavily on the cooling rate. The fact that the temperature of crystallization varies with the cooling rate indicates that the PVT curves will vary with different cooling rates. This phenomenon will be further discussed in Section 2.5.3.

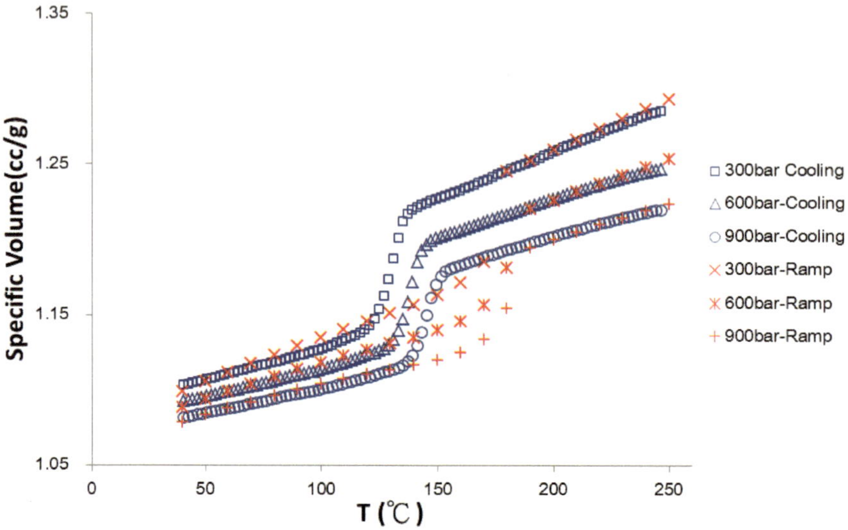

Figure 2.25 PVT curves of polypropylene at both cooling and ramping rates of 5 °C/min

2.3.4 Thermal Conductivity and Heat Transfer Coefficient

2.3.4.1 Definition

Fourier's law of heat conduction is shown in Equation 2.43, where the thermal conductivity k is a proportional constant of heat flux and thermal gradient and denotes the ability of a material to conduct heat. Plastic materials, having a thermal conductivity around 1000 times lower than that of metals, can be deemed heat insulators.

$$\left(\frac{dQ}{dt}\right) = -kA\left(\frac{dT}{dx}\right) \tag{2.43}$$

where dQ/dt is the heat flux, dT/dx is the temperature gradient, and A is the surface area.

2.3.4.2 Theoretical Models

1. Constant *k* Model

The simplest model is the constant k model. It assumes that the thermal conductivity is independent of temperature:

$$k = k_0 \tag{2.44}$$

where k is the thermal conductivity and k_0 is its specified value.

2. Linear Model

The linear interpolation function is another common approximation of the dependence of thermal conductivity on temperature. Given thermal conductivities, k_L and k_S, at two different temperatures, T_L and T_S, then the following linear equation can be obtained:

$$k = mT + b \tag{2.45}$$

$$m = \frac{k_L - k_S}{T_L - T_S}$$

$$b = \frac{T_L k_S - T_S k_L}{T_L - T_S}$$

3. Tabulated Data

The input of discrete thermal conductivity data at different temperatures is one of the most-used models. It is therefore very flexible to describe the thermal conductivity variation accurately over a broad range of temperatures.

As for the thermal conductivity between two given temperatures, a standard linear interpolation of the thermal conductivity is employed as shown in Equation 2.46.

$$k = k_i + (T - T_i)\left(\frac{k_{i+1} - k_i}{T_{i+1} - T_i}\right) \quad \text{for } T_i < T < T_{i+1} \tag{2.46}$$

2.3.4.3 Data Interpretation

For polymer materials, there is usually only a small correlation between thermal conductivity and temperature. As the temperature rises, the kinetic energy of the molecules increases accordingly and thus the thermal conductivity rises somewhat. For crystalline materials, the more densely packed crystalline structure exhibits a higher thermal conductivity, and therefore thermal conductivity can be improved with increase in material crystallinity. Figure 2.26 shows the thermal conductivity of two materials, PS and PP, at different temperatures. It can be seen that the values of PP at the temperatures below 120 °C are significantly higher than those above 150 °C, whereas PS exhibits a weak temperature dependence. The reason for this is the crystalline structure, which facilitates thermal conductivity, as mentioned above.

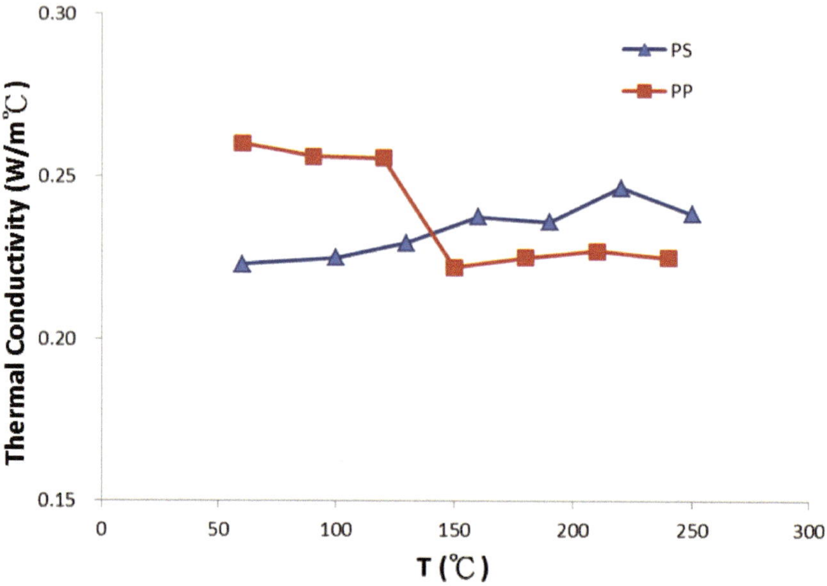

Figure 2.26 Thermal conductivity of PS (polystyrene) and PP (polypropylene)

2.3.4.4 Mold-Melt Contact and Heat Transfer Coefficient (HTC)

During mold filling in injection molding, thermal transfer between the mold and melt may not be ideal owing to thermal contact resistance arising from imperfect contact. The mold-melt interface temperature (T_{mt}) is usually higher than the set mold temperature (T_{mb}), as shown in Figure 2.27. The difference between T_{mb} and T_{mt} depends on the thermal properties of mold metal and melt, the surface roughness of the mold and the process conditions.

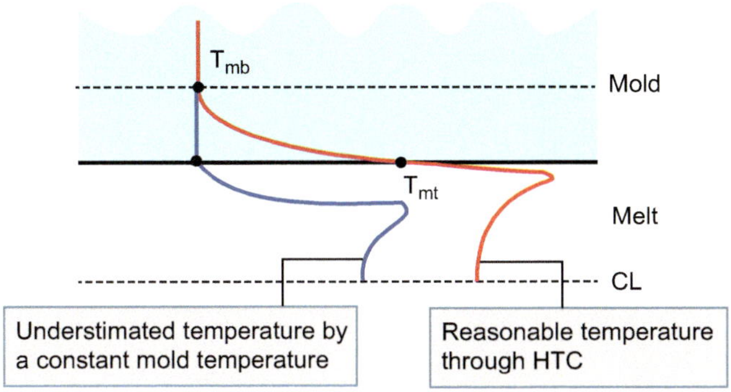

Figure 2.27 Schematic demonstration of mold-melt contact thermal heat transfer effect

It is often convenient to work with the HTC, which is defined by Newton's Law of Cooling:

$$q = h_c\left(T_{mt} - T_{mb}\right) \tag{2.47}$$

where the heat transfer coefficient (h_c) between the polymer and mold base depends on materials, temperature, and process conditions. Generally, the range of h_c is from 1000 to 25,000 W/m^{2}K for injection molding.

When using a higher heat transfer coefficient, the heat transfer flux q is larger. The mold-melt interface temperature T_{mt} will be closer to the mold base temperature T_{mb}. In contrast, when a lower heat transfer coefficient is used, the temperature T_{mt} will be much higher than the mold base temperature T_{mb}.

■ 2.4 Mechanical Properties

2.4.1 Stress and Strain of Plastics

For common plastic materials, the relationship between stress and strain is shown in Figure 2.28, which is linear for small strain, and the slope is called Young's modulus or tensile modulus. As the strain becomes larger, nonlinear behavior starts to appear and plastic deformation may occur. When a material undergoes plastic deformation, it cannot return to its original shape. Different behaviors in the nonlinear region are observed for different materials. Generally speaking, in injection molding, plastic materials shrink and warp when they are solidified in molds; however, if the deformation is not large, using the linear elasticity model is an easy and efficient method.

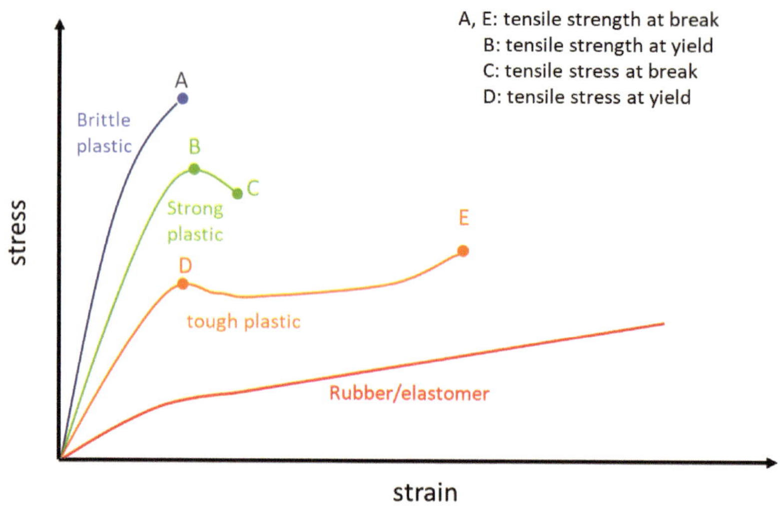

Figure 2.28 Stress vs. strain of plastics

2.4.2 Solid-Like Viscoelasticity

For tensile relaxation experiments, the behaviors of a typical amorphous polymer are shown in Figure 2.29. At low temperatures, in the glassy region (1), the modulus is high, and the polymer is hard and brittle. With increase in temperature to region (2), the polymer is at the glass transition temperature and behaves like a resilient leather. As the temperature is increased further, the modulus reaches a plateau region again, i.e. the rubbery plateau (3). The longer the chain (the larger the molecular weight), the longer the rubber plateau will be. As the temperature is further increased to region (4), the modulus drops, resulting in a considerable amount of flow. If the temperature is increased further, the polymer will become a viscous liquid in region (5).

As far as the injection molding process is concerned, the melt temperature is usually higher than the nozzle temperature due to viscous heating in the filling stage. The temperature falls in the viscous flow region, indicating that inelastic (generalized Newtonian) models can be employed. In the packing stage, the melt temperature starts to cool down and it changes from viscoelastic (4) to rubbery (3), which indicates that elastic behavior is starting to play a crucial role. After packing, the polymer passes through glass transition in the cooling state, and reaches the glassy region in the ejection stage.

The viscoelasticity modeling is helpful for simulating polymer viscoelastic mechanisms. Therefore, the flow-induced residual stress in fill/pack and cool stages can be predicted. Moreover, the viscoelastic effect can also be considered while simulating warpage, annealing, and optics. This will make the prediction more reasonable and accurate.

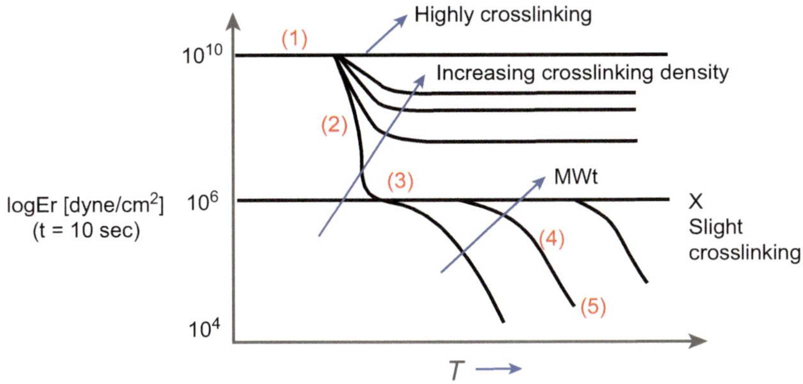

Figure 2.29 Modulus and viscosity of amorphous polymer PS

The cooling and ejecting stages of injection molding are mainly located in regions (1) and (2), whereas flow-induced residual stress or molecular orientation take place in the filling and packing stages in regions (3) to (5).

2.4.3 Theoretical Model

The relaxation behavior of polymers can be described by the relaxation modulus, $G(t)$. The stress responses of a solidified polymer also can be modeled by the generalized Maxwell model, as mentioned above (Figure 2.14). The Maxwell model is the simplest model for describing the behavior of viscoelastic liquids in linear region (small strain or low strain rate). As can be figured from this setting, the dashpot allows for indefinite deformation; for a viscoelastic fluid, no limiting cross-links exist, unlike in viscoelastic solids. The generalized Maxwell model consisting of a series of n Maxwell elements (Figure 2.15) describes the relaxation time spectrum of a polymer; each length of a molecule has a characteristic relaxation time and each element represents one of them.

For the generalized Maxwell model, the elastic modulus during stress relaxation can be expressed as a function of time:

$$G(t) = G_\infty + \sum_{i=1}^{n} G_i \exp\left(-\frac{t}{\lambda_i}\right), \lambda_i \equiv \frac{\eta_i}{G_i} \tag{2.48}$$

where G_∞ is the long-term modulus once the material is totally relaxed.

As mentioned in Section 2.2.1.3, the temperature effect on relaxation time can be modeled by the WLF or Arrhenius equation.

2.4.4 Data Interpretation

In injection molding, usually the material is mostly solid-like in the cooling, demolding or annealing stages. If it is located in the glass-transition-to-glassy regions, it behaves as a viscous solid. In order to measure the viscoelastic property of a solid-like material in such a temperature range, DMA (dynamic mechanical analysis) instruments are most frequently used to measure the viscoelastic property "solid-like state", mainly by the flexure or torsion methods.

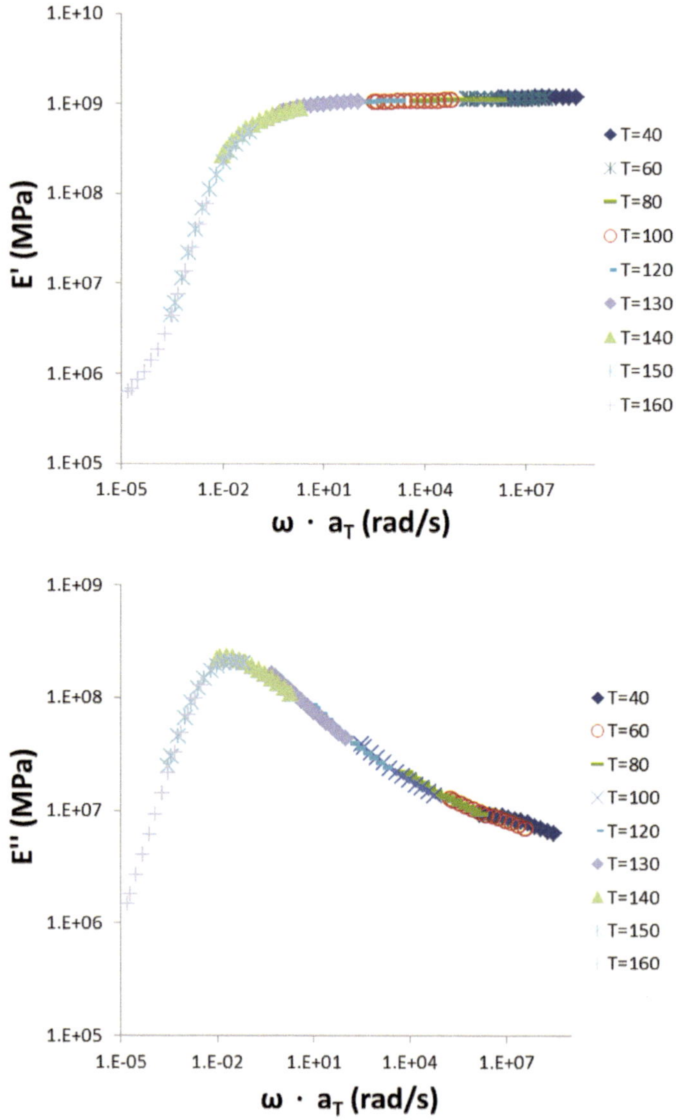

Figure 2.30 Dynamic modulus of solid-like viscoelastic properties obtained from DMA measurements (polycarbonate)

The master curve of a polycarbonate, shown in Figure 2.30, was measured by the flexure method at the reference temperature of 130 °C via time-temperature superposition (as mentioned in Section 2.2.2.3). E' on the master curve shows a nearly constant value (10^9 Pa) when the frequency is above 1.0 rad/s, which is the glassy region, while it starts to drop when the frequency goes below 0.1 rad/s, which is the glass transition. The E' measured at the lowest frequency (highest temperature) tends to plateau in the rubbery region. If the temperature goes even higher, the test sample will turn softer and undergo plastic deformation. Also, reliable data might not be obtained as the measurement is being conducted close to the instrument's limits.

The relationship between shift factor and temperature is shown in Figure 2.31. As mentioned in Section Section 2.2.1.3, modeling can also be done using WLF or Arrhenius-type equations. However, it is obvious from the experimental result that there are two stages in the curve which cannot be characterized by WLF only. It is commonly understood that the WLF equation is applicable in the range from T_g to T_g + 100 °C, and thus that other equations are required to characterize the data at the temperatures below T_g. Since the experimental data cover T_g and the range is greater than 100 °C, a two-part model to describe the shift factor is needed, as set out below:

$$a_T = \begin{cases} \exp\left(\dfrac{-A_1(T-T_f)}{A_2+(T-T_f)}\right), & T \geq T^* \\ \exp\left(\dfrac{\Delta H_T}{R}\left(\dfrac{1}{T}-\dfrac{1}{T_0}\right)\right), & T < T^* \end{cases} \tag{2.49}$$

where the WLF equation is used to describe the data for the temperature above T_g, and the Arrhenius-type is used for that below T_g.

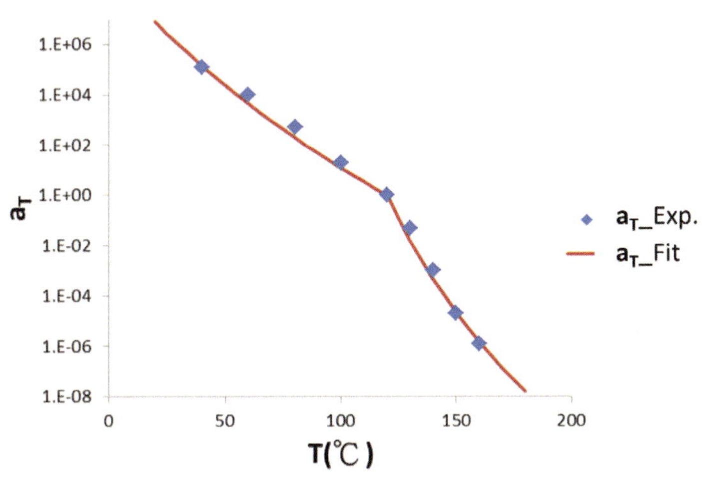

Figure 2.31 Shift factor for polycarbonate

2.5 Kinetic Properties

2.5.1 Crystalline

Normally the alignment of long-chain polymeric crystals is stacked by folding into a plate, called a lamellae structure. As a sphere is the form expressing the lowest surface energy, and many plate crystals will be twisted to form spherical crystals when being stacked, which requires a nucleation site to be generated first, i.e., a few molecules must be aligned first. Such a nuclear generation process is called nucleation, which can be divided into homogeneous and heterogeneous nucleation.

Generally, polymers do not crystallize immediately upon reaching their crystallization temperature. It takes time for the molecules to rearrange their conformation to favor the formation of crystals, called the induction time, as shown in Figure 2.32. The crystallization process starts slowly, and then the crystals grow rapidly. The time at which the crystallinity reaches half of the maximum crystallinity is used to determine the crystallization rate of polymers.

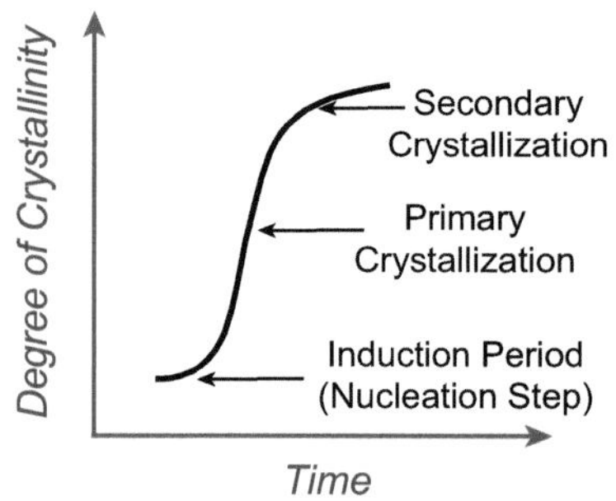

Figure 2.32 A schematic diagram of degree of crystallinity versus time

2.5.2 Theoretical Models

Crystallization kinetics is generally described by the Avrami equation:

$$\theta(t) = \frac{X(t)}{X_\infty} = 1 - \exp\left(-kt^n\right) \tag{2.50}$$

where $\theta(t)$ is the relative crystallinity at time t, $X(t)$ is the absolute crystallinity at time t, X_∞ is the ultimate crystallinity, n is the Avrami index, and k is the Avrami crystallization rate constant. There is no clear physical interpretation of the Avrami constants k and n. Originally, n was held to have an integer value between 1 and 4 which reflected the nature of the transformation in question. The induction time t_i can be described by an empirical relation [10]:

$$t_i = t_m \left(T_m - T\right)^{-a} \tag{2.51}$$

where t_m and a are material constants, T is the crystallization temperature, t_i is the induction time at temperature T, and T_m is the equilibrium melting temperature.

The Nakamura equation developed from the Avrami theory is used to describe the crystallization kinetics. The following equations show the model:

$$\frac{D\theta}{Dt} = nK(t)(1-\theta)\left[-\ln(1-\theta)\right]^{\frac{n-1}{n}} \tag{2.52}$$

$$K(T) = \ln(2)^{1/n}\left(\frac{1}{t_{1/2}}\right)_0 \exp\left(-\frac{U^* / R}{T - T_\infty}\right)\exp\left(-\frac{K_g}{T\Delta Tf}\right)$$

where $K(T)$ is the non-isothermal crystallization rate constant, $t_{1/2}$ is the crystallization half-time, T is the crystallization temperature, R is the universal gas constant, $\Delta T = T_m - T$ is the supercooling, $f = 2T/(T + T_m)$ is a correction factor, accounting for the reduction in the latent heat of fusion as the temperature is decreased, U^* is the activation energy for the crystallizing units to transport across the phase boundary, and T_∞ is the temperature that crystallization ceases.

Stress-induced crystallization is considered by assuming that shear stress only affects the equilibrium melting point. The following equations show the relationship between the equilibrium melting temperature shift and the shear stress:

$$T_{shift} = C_1 \exp\left(-\frac{C_2}{\tau}\right) \tag{2.53}$$

$$T_m = T_m^0 + T_{shift}$$

2.5.3 Effects of Cooling Rate on Crystallization

In conclusion, the crystallization of a polymer does not occur instantly once the crystallization temperature is reached as time is required for molecules to align, and therefore the degree of crystallinity and rate of crystallization are kinetic behaviors. As shown in Figure 2.33, different cooling rates under a constant pressure will result in different curves in volume measurement.

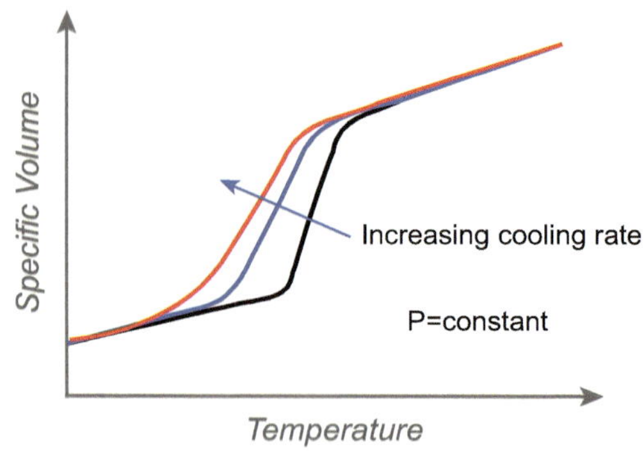

Figure 2.33 A schematic diagram of PVT curve under various cooling rates

■ 2.6 Curing Kinetics

2.6.1 Curing Phenomenon

Another important kinetic behavior in injection molding is the chemical reaction of a thermoset, called curing kinetics. The raw material of a thermoset is a pre-polymer of lower molecular weight, which is cross-linked through chemical reaction and finally becomes a network structure, as shown in Figure 2.34. The pre-polymer has a low viscosity initially, but as the molecular weight increases by cross-linking, the viscosity becomes higher. Entanglement between molecules become significant when the molecular weight becomes large enough. When the cross-linking density reaches a certain value that a network structure is formed, the molecular weight is nearly infinite due to the cross-linking reaction. Therefore, molecular chains are not able to move freely, so the material cannot flow and solidifies.

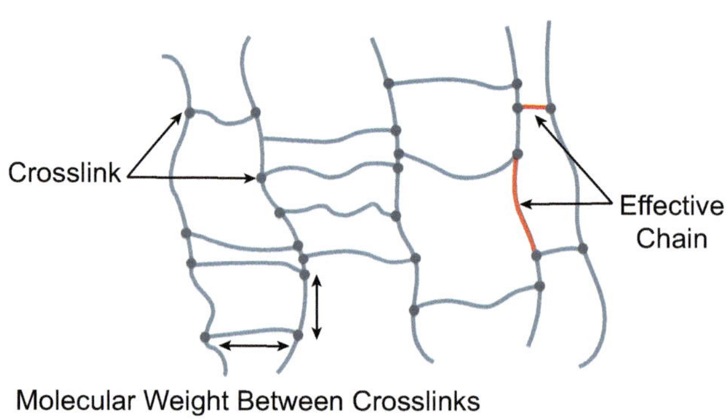

Figure 2.34 Chemical structure of curing reaction

The same trend is also observed from measured modulus [11]. At the beginning of a reaction when the molecular weight is low, the relaxation time could be short enough to show the elastic behavior and the molecules are able to move freely. As cross-linking develops the relaxation time increases with the increase of molecular weight. At a certain degree of cross-linking the modulus cannot be relaxed completely and molecular chains are restrained by cross-linked points. Hence, a rubbery plateau is formed due to the build-up of a network structure.

Figure 2.35 shows the relaxation modulus of epoxy measured under different degrees of cure. It can be found that as the degree of cure gets higher (0.6–1.0), the rubbery plateau becomes higher, and the curves move rightward gradually, which indicates a growth on relaxation time.

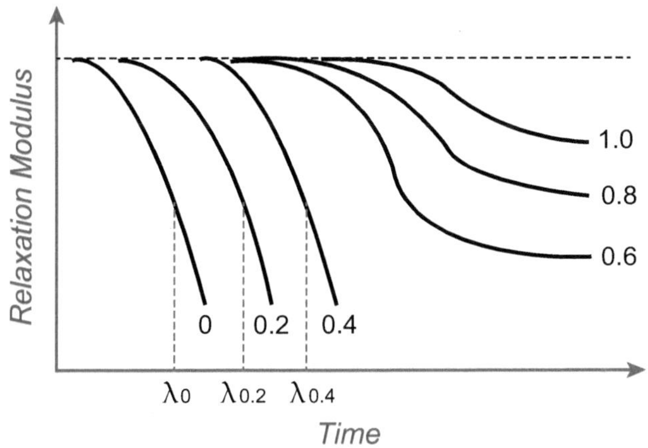

Figure 2.35 A schematic diagram of relaxation modulus of epoxy for various degrees of cure

2.6.2 Theoretical Models

The n-th order reaction is mostly seen in reaction kinetics, i. e., the reaction rate is proportional to the n-th order of the concentration of reactant.

1. n-th order kinetic model:

$$\frac{d\alpha}{dt} = k(1-\alpha)^n \tag{2.54}$$

where α is the conversion of chemical reaction (or degree of cure).

2. Autocatalytic kinetic model:

$$\frac{d\alpha}{dt} = k\alpha^m(1-\alpha)^n \tag{2.55}$$

3. Combined model:

Usually the formula of a material is hard to know in practical applications, and therefore the best method that is often used is the combination of both, which is called the combined model. It is applicable to both reaction types.

$$\frac{d\alpha}{dt} = \left(k_1 + k_2\alpha^m\right)(1-\alpha)^n \tag{2.56}$$

2.6.3 Curing Effect on Viscosity

The most important material information in flow simulation is viscosity. As mentioned above, the viscosity of a melted polymer relates to temperature and shear rate. It is obvious that by measuring the viscosity of each instant moment at a certain heating rate under a constant shear rate, the curves similar to Figure 2.36 can be obtained. Because there is almost no chemical reaction at low temperatures, the viscosity decreases as the temperature rises. When the temperature is high enough, the cross-linking effect starts to occur, and thus the viscosity starts to increase as well.

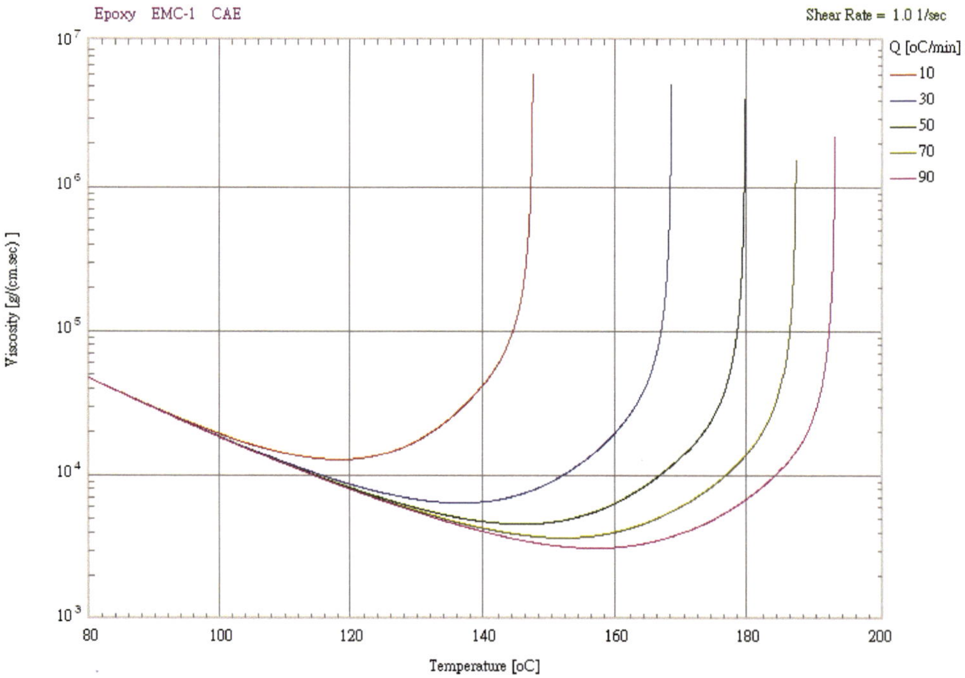

Figure 2.36 Viscosity of epoxy for various ramping rates

Chemorheology Model

As mentioned above, a typical U-shape viscosity curve can be observed when heating a thermoset. The viscosity drops at first due to the temperature effect. With the increase of cross-linking, the molecular weight of thermosets increases. Therefore, the viscosity increases accordingly. After reaching a minimum, the viscosity soon increases dramatically due to the build-up of a molecular network structure. To model such phenomena the following models could be used:

1. Castro-Macosko model

This model assumes that the viscosity depends on temperature and degree of cure, α:

$$\eta = \eta_0 \left(\frac{\alpha_g}{\alpha_g - \alpha} \right)^{c_1 + c_2 \alpha} \tag{2.57}$$

$$\eta_0 = A \exp\left(\frac{T_b}{T} \right)$$

The dependence on degree-of-cure is described by three parameters (α_g, c_1, and c_2). The parameter α_g correlates with the gel conversion, at which point the viscosity of the material grows dramatically. The temperature dependence is exponential and there is no dependence on shear rate.

2. Power-law Castro-Macosko model

This model describes the curing effect nonlinearly on η_0, T_b, and n, with power-law-type shear rate dependence:

$$\eta = \eta_0 \exp\left(\frac{T_b}{T} \right) \dot{\gamma}^{n-1} \tag{2.58}$$

$$\ln \eta_0 = a_0 + a_1 \alpha + a_2 d$$

$$T_b = b_0 + b_1 \alpha + b_2 d$$

$$n = c_0 \exp(c_1 \alpha + c_2 d)$$

$$d = \tan\left[\frac{\pi}{2} \left(\frac{\alpha - \alpha_i}{\alpha_g} \right) \right]$$

From the viewpoint of kinetics, there are basically two major kinds of curing reactions: n-th order and autocatalytic reaction. The rate of an n-th order reaction depends on the concentrations of reactants only, while the rate of an autocatalytic reaction depends on the concentration of both reactants and products. In other words, the products in the autocatalytic reaction help to accelerate the reaction itself. Most thermosets show a combination of both curing kinetics. The combined model is therefore widely employed in modeling the kinetics. The following equations show the combined model:

$$\frac{d\alpha}{dt} = \left(k_a + k_b\alpha^m\right)(1-\alpha)^n \tag{2.59}$$

$$k_a = A\exp\left(\frac{-T_A}{T}\right)$$

$$k_b = B\exp\left(\frac{-T_B}{T}\right)$$

where α represents the degree of cure of the reaction. $\alpha = 0$ implies that no reactants reacted at all, while $\alpha = 1$ implies the curing reaction complete. α is defined as the ratio of the heat released in reaction versus the total reaction heat of the curing reaction.

The model combines the n-th order contribution and the autocatalytic one linearly. The equation can be reduced to an autocatalytic equation ($k_a = 0$) and an n-th order equation ($k_b = 0$).

The heat of chemical reaction might affect the temperature distribution significantly. To model such an effect, another source term can be added into the energy equation, as shown below.

$$\rho C_p\left(\frac{\partial T}{\partial t} + V\cdot\nabla T\right) = \nabla\left(k\nabla T\right) + \eta\dot{\gamma}^2 + \rho\left(dH\frac{\partial\alpha}{\partial t}\right) \tag{2.60}$$

where dH is the amount of heat per unit mass of reactive material during the curing process; thus, the temperature of the material will increase as the reaction occurs.

2.6.4 Data Interpretation

Of the thermosets, epoxy is a widely used material, especially for IC packages. Commercially available EMC (epoxy molding compound) can modify different formulas to meet the demands of actual processes or products, such as mechanical strength, thermal expansion coefficient, fluidity, and reaction rate. The reaction curves for all three formulas do not vary a lot, as shown in Figure 2.37(a). However, the reactive viscosity varies much more, relatively. For instance, although the viscosity of Hitachi CEL-9000 XU (LF) is close to that of Sumikon EME-6300 at low temperatures, as the temperature rises, the slope during the decrease in viscosity displays different behavior, and the difference in temperature (up to 20 °C), along with curing and viscosity increase, is extremely large. Although the viscosity curves of Hitachi CEL-9000 XU (LF) and Panasonic CV8710TU differ by a factor of 10, the dependence on temperature change is similar, i. e. the slope of temperature change and the temperature of curing and viscosity increase.

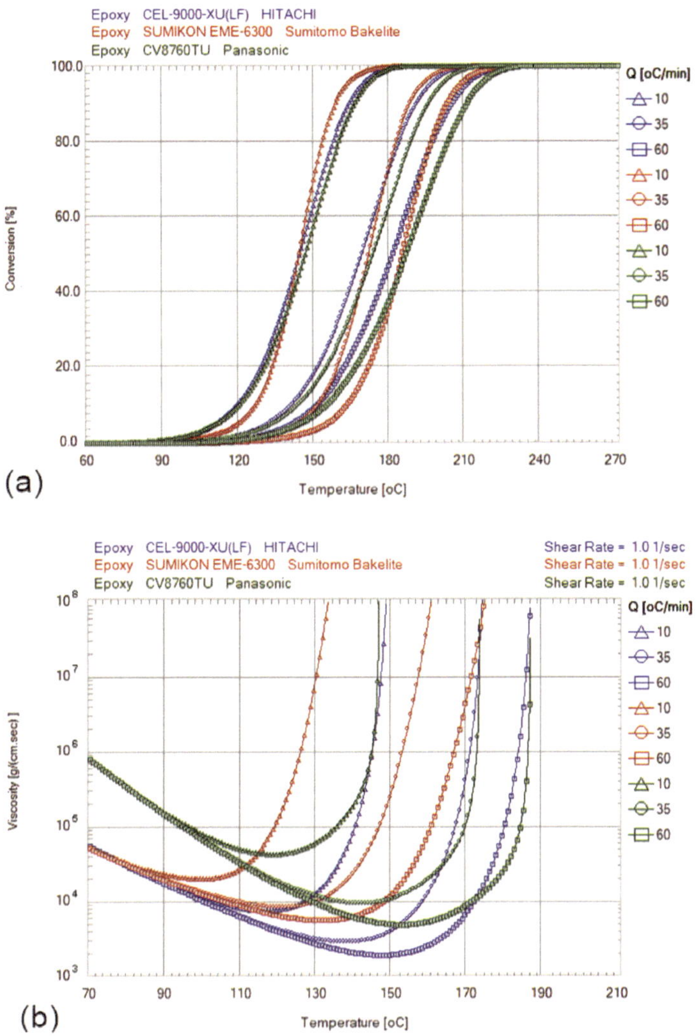

Figure 2.37 Comparison of (a) curing kinetics and (b) reactive viscosity for various EMCs

As mentioned above, in order to satisfy the constantly evolving process and demands imposed on product characteristics, the EMC formula has been extensively modified, e. g. with regard to the chemical composition of epoxy, filler content, size, shape, distribution, etc., all of which exert an influence on the physical and chemical properties and make material property control all the more challenging.

References

[1] Doi, M.; Edwards, S.: *The Theory of Polymer Dynamics*. Clarendon Press 1988

[2] Doi, M.: Explanation for the 3.4 power law of viscosity of polymeric liquids on the basis of the tube model. *Journal of Polymer Science, Part C: Polymer Letters*, 19, 1981, p. 265

[3] Berry, G.; Fox, T.: The viscosity of polymers and their concentrated solutions. *Advances in Polymer Science*, 5, 1968, p. 261

[4] Williams, M.; Landel, R.; Ferry, J.: The Temperature Dependence of Relaxation Mechanisms in Amorphous Polymers and Other Glass-forming Liquids. *Journal of the American Chemical Society*, 77, 1955, p. 3701

[5] John, D. F.: *Viscoelastic Properties of Polymers*. New York: John Wiley & Sons 1980

[6] Bird, R.; Armstrong, R.; Hassager, O.: *Dynamics of Polymeric Liquids*, 2nd ed., Vol. 1. Wiley-Interscience 1987

[7] Shaw, M. T.; MacKnight, W. J.: *Introduction to Polymer Viscoelasticity*. John Wiley & Sons 2005

[8] Tait, P. G.: Report on some of the physical properties of fresh water and of sea water. Report on the scientific results of the voyage of H. M.S. Challenger during the years 1873–76. *Physics and chemistry*, 2, 1889

[9] Dymond, J.; Malhotra, R.: The Tait equation: 100 years on. *International Journal of Thermophysics*, 9, 1988, p. 941

[10] Godovsky, Y.; Slonimsky, G.: Kinetics of polymer crystallization from the melt (calorimetric approach). *Journal of Polymer Science: Polymer Physics Edition*, 12, 1974, p. 1053

[11] O'Brien, D.; Mather, P.; White, S.: Viscoelastic properties of an epoxy resin during cure. *Journal of Composite Materials*, 35, 2001, p. 883

3 Part and Mold Design

Tsai-Hsin (Sam) Hsieh, Yao-Chen (Cloud) Tsai,
Yao-Wei (Willie) Chuang

In this chapter, we introduce the second of the three major factors in molding: part and mold design. Nowadays, the machining process of the mold is mostly emphasized [1, 2], but few consider the integration of the design guidelines with molding analysis. The main consideration in part design is to comply with market demands, such as aesthetics, function, laws, and regulations. Mold design considers the manufacturing cost of each component such as sliding blocks, ejection systems, two- or three-plate mold, etc. Therefore, in order to design good parts and molds, many detailed and complicated theories and mechanisms must be considered.

3.1 Part Design

There are five principles in part design:

1. Golden rule for part design, that is, ensure even distribution of wall thickness.
2. Relationship between part thickness and flowing length.
3. Dimension specification of radius, fillets, and chamfer angles.
4. Design principle for ribs and bosses.
5. Relationship between draft allowance and mold opening.

3.1.1 Golden Rule: Uniform Wall Thickness

Plastics are poor conductors of heat. Taking a plate as an example, then the cooling time is in proportion to the square of the product thickness (Equation 3.1).

$$t_c = \frac{t^2}{\pi^2 \alpha} \ln\left(\frac{8}{\pi^2} \frac{T_{melt} - T_{mold}}{T_E - T_{mold}}\right) \tag{3.1}$$

where t_c is the cooling time, indicating the time required for cooling from T_{melt} to T_{mold}, T_{melt} and T_{mold} are the melt processing temperature and mold temperature, respectively, t is the part thickness, T_E is the ejection temperature, and α is the thermal diffusivity of the plastic.

The wall thickness not only determines the cooling efficiency but also affects the melting behavior during filling and packing stages. For a part with a thickness distribution as shown in Figure 3.1, a drastic change in the wall thickness is the worst design, because it does not possess the transition region for shrinkage during plastics molding.

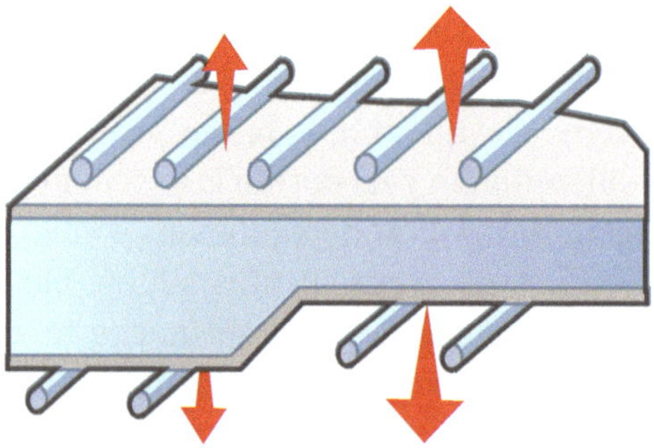

Figure 3.1 Different flow behaviors and heat dissipation at sites with different thickness

If such a thickness difference is necessary, a design of gradual variation in thickness as shown in the middle one in Figure 3.2 is recommended, but to keep a uniform thickness is of course the best situation.

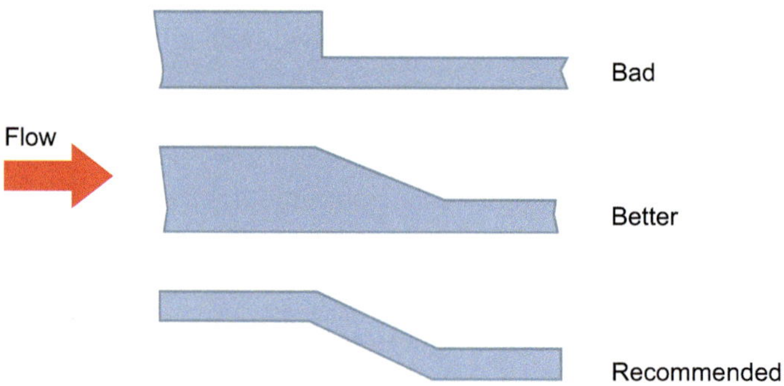

Figure 3.2 Recommended design of the part thickness

For the transition region, the gradient change as shown in Figure 3.3 is recommended, that is, for a difference t in thickness, the transition region shall be $3t$ in length. And for an optimum design, the start and endpoint of the thickness difference shall be rounded, because for plastics a bigger molding stress would possibly be accumulated in the sharp corners.

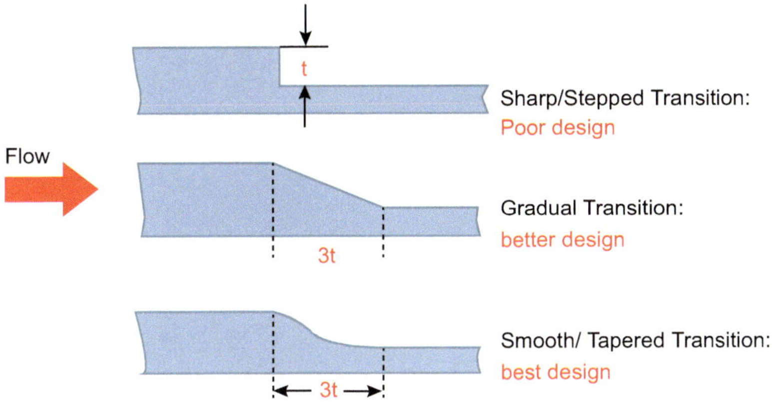

Figure 3.3 The design of thickness transition region

For the original design as shown in Figure 3.4, filling is done by the central gating method. Because the melt front has a circular flow pattern and the distance along the vertical direction (top to bottom) is shorter than along the horizontal direction (left to right), the top and bottom edges will be filled first. Overpacking occurs at those first-completely-filled regions, which eventually causes the part to warp and deform.

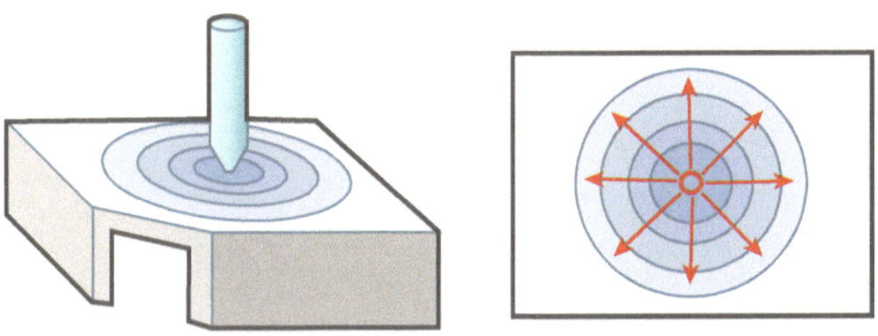

Figure 3.4 Overpacking happens due to the uniform thickness design

To facilitate the flow toward the edges, the flow guidance design shown in Figure 3.5, thickening the wall thickness diagonally on the part (colored areas), improves the flowing, which makes it more uniform and fills out the whole mold. Such a design is suitable for plastic parts with thin walls.

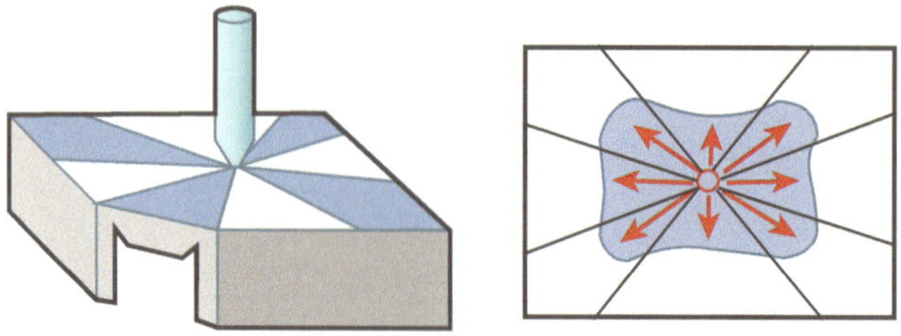

Figure 3.5 Guide flow design

Although the golden rule of part design requires uniform wall thickness, it is basically difficult to fulfill, because requirements such as aesthetics, strength, assembly, cost, etc., have to be considered as well.

3.1.2 Wall Thickness versus Flow Length

In injection molding, the length that the plastic flows is related to its flowability and wall thickness design. The simplest way is to check whether the flow length/wall thickness ratio of the part is within a reasonable range. Take the model in Figure 3.6 as an example: a good design allows melted plastic to flow smoothly from the gate to the end. The smoothness of melt filling is related to the resistance encountered along the flow path.

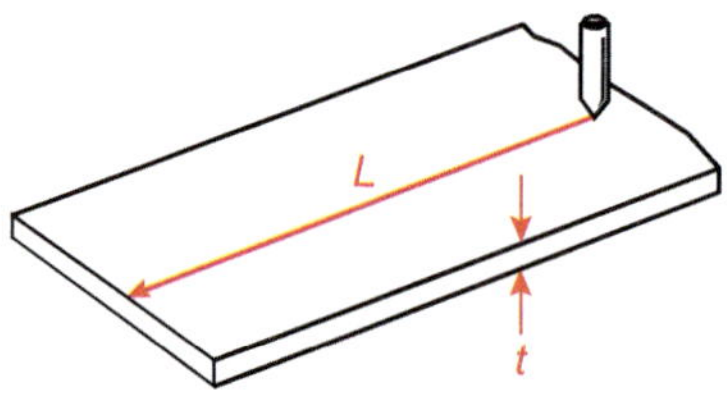

Figure 3.6 Flow length/wall thickness ratio (L/t ratio)

Because the part thickness and flow length affect the flowability, the flow length to wall thickness ratio (Equation 3.2) is commonly used in product design to check whether a design is good or not, and it is defined as

$$\frac{L}{t} = \frac{\text{Flow length from gate to end}}{\text{Average wall thickness}} \tag{3.2}$$

where L denotes the filling distance and t the average part thickness.

As shown in Figure 3.7, the flow length to wall thickness ratio can help to predict the difficulty of molding. If the flow length to wall thickness ratio is smaller than 100, filling in such designs is very easy for melts. The flow length to wall thickness ratio of most parts is between 100 and 200, but for thin-wall parts, it would go over 200 and the design should be reconsidered and modified to avoid the occurrence of short shot or other defects. If the flow length to wall thickness ratio goes beyond 300, the design is very difficult for filling and requires plastics with good flowability, possibly even the development of an exclusive plastic material with high flowability for this part, or high-speed injection molding machines for manufacturing.

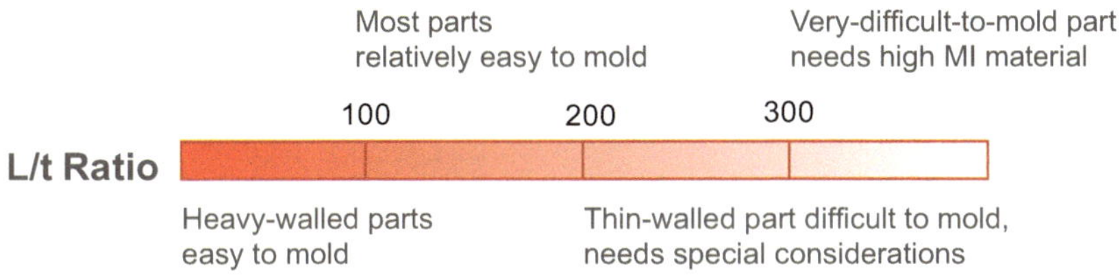

Figure 3.7 Scale of flow length to wall thickness ratio

Table 3.1 shows the recommended ranges of flow length to wall thickness ratios of different plastic materials.

Table 3.1 Maximum Flow Length of Plastics Melt [1]

Material	Flow length to wall thickness ratio	Material	Flow length to wall thickness ratio
ABS	100–200	PET	200–350
ASA	180–230	PMMA	110–170
HDPE	200–270	POM	100–250
HIPS	250–340	PP	230–340
LDPE	200–300	PPO	100–200
LLDPE	180–250	PPS	120–185
PA6	160–300	PS	150–200
PA66	180–300	PSU	60–120
PBT	140–220	SAN	170–200
PC	30–110		

Therefore, a good product design in the beginning can not only reduce molding difficulties, but also manufacturing cost.

3.1.3 Radius/Fillets and Chamfer Angle

In product design, radius, fillets, and chamfer angle are also important. The boundary of geometric change, such as the vertical and horizontal boundary shown in Figure 3.8, should have radius (rounded) design. It is known that the smaller the radius the higher the concentration of stress will be. The reason is the same as mentioned before.

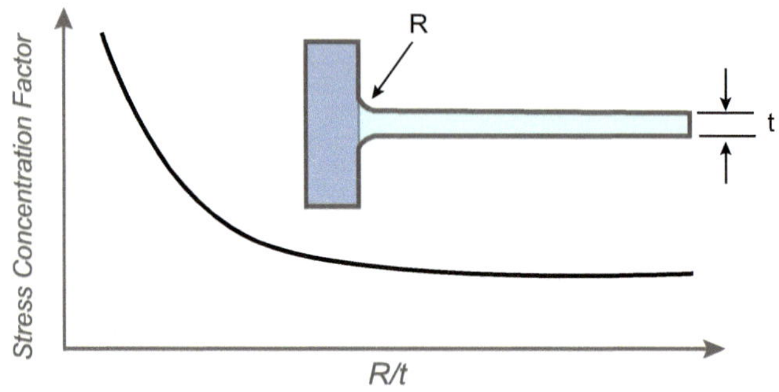

Figure 3.8 Fillet radius on stress concentration

CAE analysis can depict the situation of volumetric shrinkage and stress concentration. In Figure 3.9(a), we can see that the top case displays a bigger volumetric shrinkage at the mismatched area as there is no chamfer angle. However, with larger chamfer angles, the variation gradient of volumetric shrinkage becomes smaller, which reduces the warp caused by the difference of volumetric shrinkage.

In Figure 3.9(b), a stress concentration phenomenon by shear stress is seen without chamfer angle design, which is a structural weak point of the product and breakage is likely to occur from here. As the chamfer angle becomes bigger, the stress value decreases.

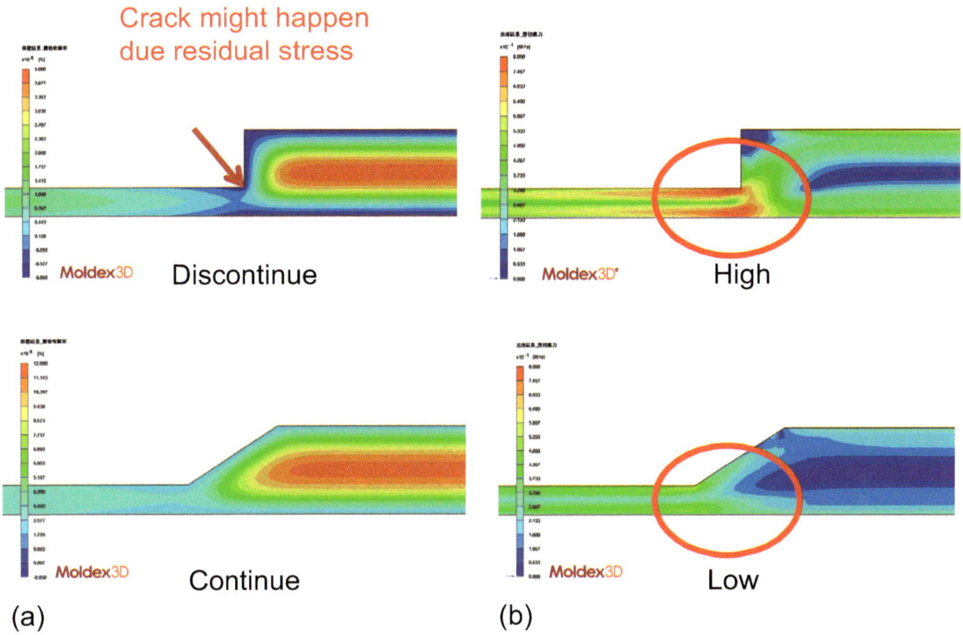

Figure 3.9 (a) Volumetric shrinkage and (b) shear stress of different radius designs

3.1.4 Rib and Boss

For ribs, the most common molding defect is the sink mark at the back of a rib, as shown in Figure 3.10. This happens because the heat dissipation at the relatively thicker ribs is slower and therefore the shrinkage is still going on while the shrinkage of thin-wall areas is finished already.

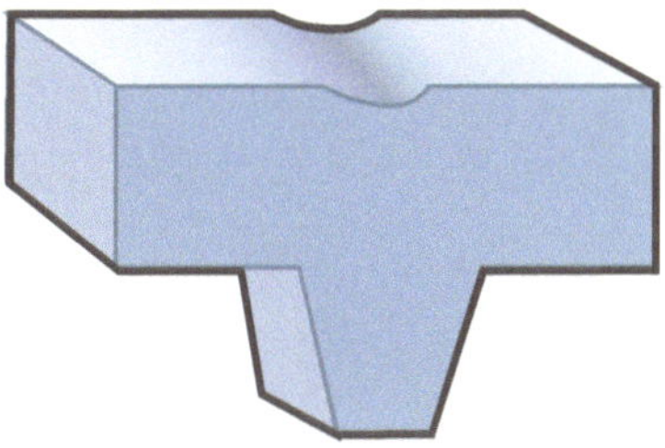

Figure 3.10 Sink mark in the opposite of a rib

Special attention concerning the size of a rib is required to avoid the occurrence of sink marks. Generally, it is recommended that the thickness of a rib be less than half of that of the wall of a part, a draft angle of at least 0.5° be reserved, and a radius fillet design be applied at the bottom of a rib, whose radius should be 0.125 times the wall thickness, as shown in Figure 3.11.

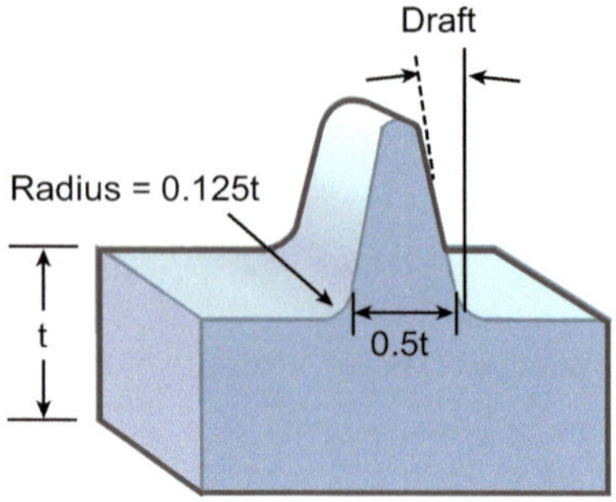

Figure 3.11 Recommended dimensions of a rib

Recommended rib dimensions of some plastic materials are shown in Table 3.2, among which the rib thickness allowances of PA and PBT are much stricter and should be less than 40% of the wall thickness of a part. If the part surface is required to be of high gloss, the thickness of a rib should be much thinner, which would cause filling difficulties. Therefore, a compromise between part quality and moldability has to be made.

Table 3.2 Recommended Thickness of a Rib [2]

Material	Minimal Sink [%]	Slight Sink [%]
PC	50 (40 if high gloss)	66
ABS	40	60
PC/ABS	50	66
PA (unfilled)	30	40
PA (glass-filled)	33	50
PBT (unfilled)	30	40
PBT (filled)	33	50

As shown in Figure 3.12(a), a poor design with the rib of the same thickness as the part will result in serious sink marks on the surface, whereas by increasing the rib length and decreasing the rib thickness the equivalent strength can be achieved. However, the rib design shown in Figure 3.12(b) is too tall after thinning, which may cause filling problems, so using two ribs with original height is a way to solve the problem (Figure 3.12(c)).

Also, the design of thin and thick sections together should be avoided, which not only causes sink marks on the outside, but also generates voids on the inside, as shown in Figure 3.13. What should be done is to replace the thick sections with multiple ribs, whose thickness should be well-controlled as well.

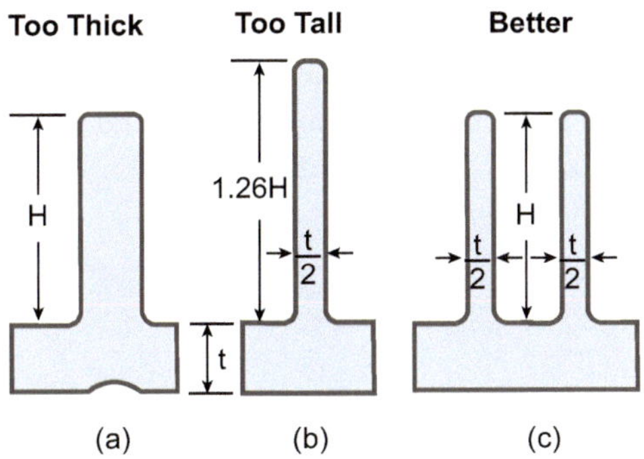

Figure 3.12 Equivalent designs of a rib

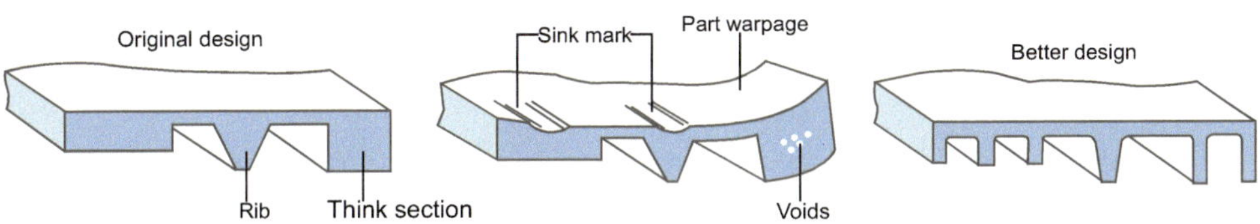

Figure 3.13 Better design of a rib

In conclusion, guidelines for designing rib dimensions are shown in Figure 3.14: a given part thickness t limits the bottom thickness of a rib to be less than $0.5t$, the height of a rib to be less than $3t$, the radius of fillet at the bottom to be $0.25–0.4t$, the draft angle to be 0.5° or more, and the span between ribs to be at least $2t$. In addition, by comparing the volume of two spheres a and b as indicated, if the volume ratio is larger than 1.34, that is $4/3\pi(r''/r')^3 > 1.34$, the design is more likely to generate voids.

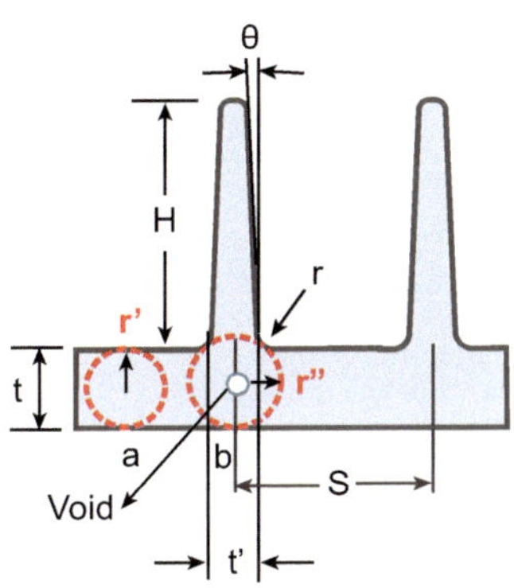

Figure 3.14 Design criteria of a rib

Because the thickness of a rib is smaller than that of the part, the allocation of ribs should be specially considered to avoid hesitations. As shown in Figure 3.15, the melt at the gate is free to move to the main plane or the ribs.

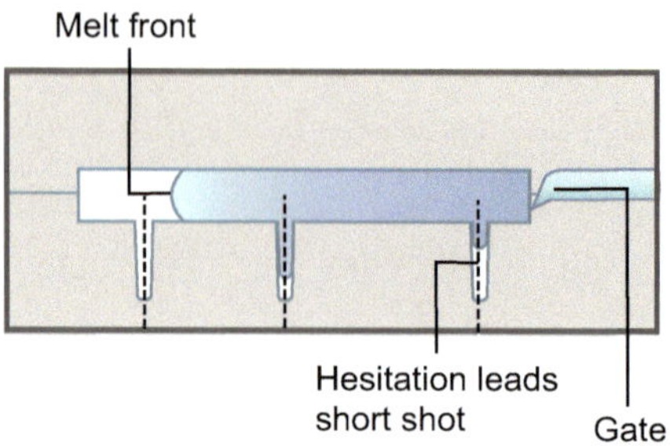

Figure 3.15 Hesitation at ribs

Incorporating a boss is another common design for fixing other parts. Similar to ribs, the common defect is the sink mark on the backside of the boss. Figure 3.16 shows the recommended dimensions of a boss design [4]. The depth of a boss should be larger than the external height by less than 30% of the part thickness *t*. Furthermore, the wall thickness at the top of a boss should correspond to the diameter of a screw head. The outer diameter at the top is recommended to be 2 to 2.4 times of that of the screw head.

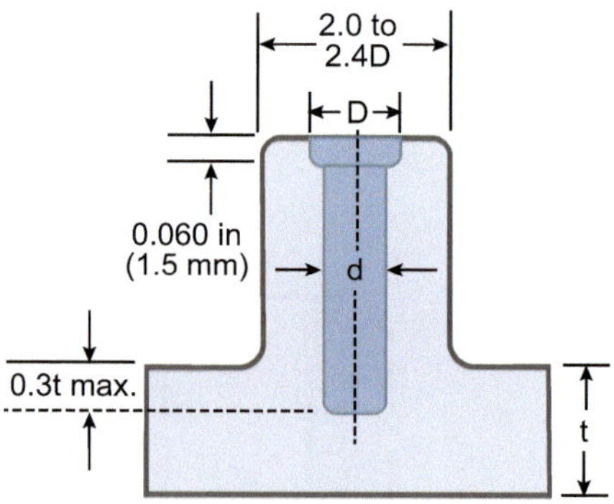

Figure 3.16 Recommended dimensions of a boss [4]

The part thickness should be reduced at the base of a boss or otherwise serious recessions might easily occur. The depth is also recommended to be 30% of the part thickness and the slope to be 30°, as shown in Figure 3.17.

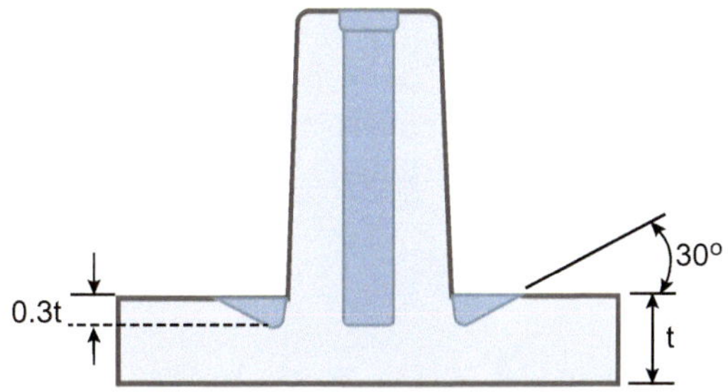

Figure 3.17 Part thickness should be reduced at the base of a boss

Ribs and bosses have different functions but at the same time have similar design principles. One is to pay attention to the ratio between wall thickness and product thickness to reduce sink marks. CAE analysis can be used to check the severity of sink marks, as shown in Figure 3.18, and depicts sink mark conditions at the backside of ribs and bosses.

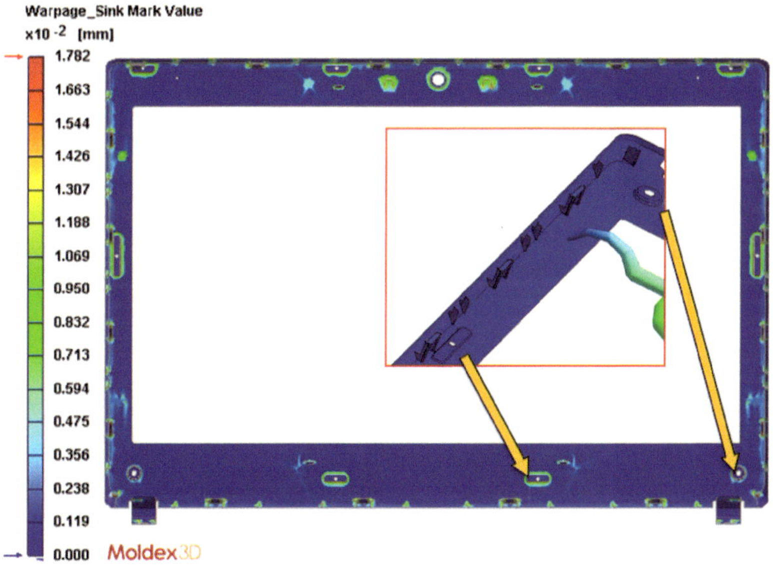

Figure 3.18 CAE result of a sink mark

Short shot defects are another problem caused by insufficient wall thickness, especially for reinforcing ribs around a boss, as shown in Figure 3.19, which are not only at the end of filling but their thin walls also makes them more difficult to fill. Also, venting design in mold should be enhanced in order to prevent burn marks.

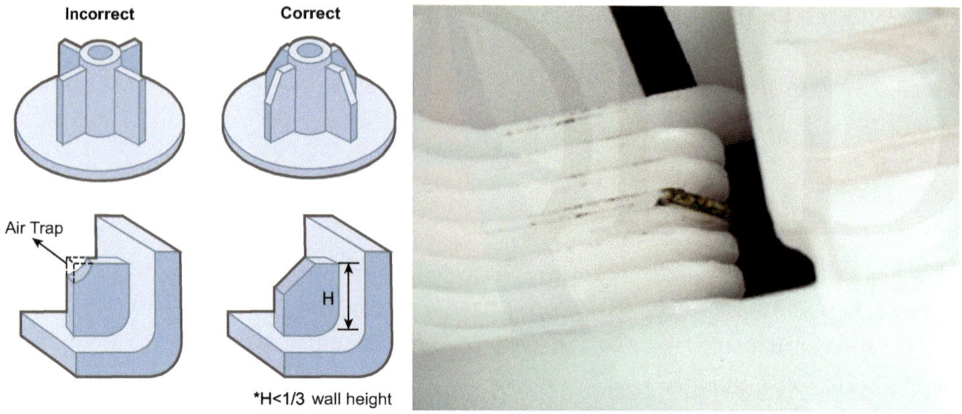

Figure 3.19 Thin wall leads short shot and burn mark defects

3.1.5 Draft Angle

Draft angles are necessary for ribs or bosses, as shown in Figure 3.20, as they reduce the friction between parts and molds at mold opening. The larger the angle the smaller is the resistance encountered while demolding. Therefore, any contact sites of parts and molds should have a sufficient draft angle, which is recommended to be 0.5° as the minimum.

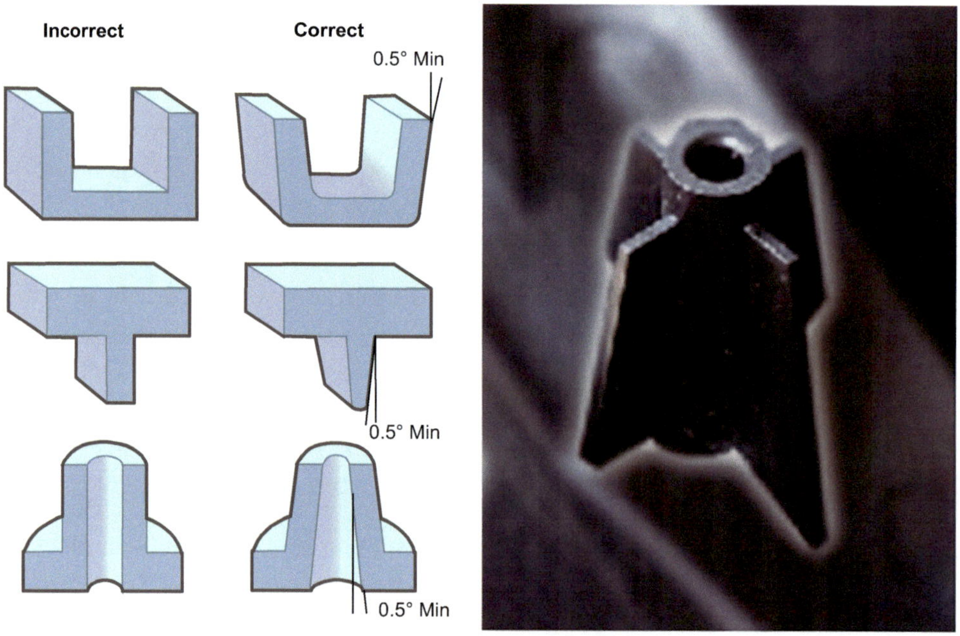

Figure 3.20 Draft angle design (where Min means minimum)

Increasing the draft angle is necessary if the part surface is surface-treated or the part surface would be scratched during demolding. It is recommended to increase the draft angle by 1–1.5° for each increase of depth for surface treatment by 0.001 inch [3].

3.1.6 Design for Manufacturing (DFM)

Nowadays, mechanical engineers have to review the manufacturability with tool makers after finishing the 3D CAD model, because many designers do not know the constraints in mold design as well as in manufacturing. Some young designers even take for granted that once the drawing is finished, the tool maker can definitely make the mold without consideration of the manufacturing. The following characteristics should be reviewed:

1. Wall thickness:

   The wall thickness of non-reinforced resins is around 0.5 to 5.0 mm, whereas that of the reinforced resins is around 0.7 to 3.0 mm. The wall thickness should be kept uniform as much as possible.

2. Location of the parting line:

   The parting line is where the top and bottom half of a mold close with each other. Usually, a flash on the molded part would occur here as well after injection. Flashes normally cause only cosmetic issues, but a few are related to functionality. Therefore, it is better to keep flashes away from where there is concern of functionality and cosmetic issues, such as at the inside of the part.

3. Avoid undercuts:

   If the molds have undercuts, the plastic parts would be hooked up and not able to come off the molds. Therefore, no plastic should be in the way of the demolding direction.

4. Draft angle:

   Once the plastic parts have been molded, a certain draft angle is needed for the demolding mechanism going deep into the mold to take it off smoothly.

Users can take CAE analysis results, for example those of the common vehicle interior accessory shown in Figure 3.21, and observe the flow rate analysis of each gate. From these results it is apparent which area is of poor flow efficiency, so that the location of the gate can be adjusted accordingly to optimize the flow equilibrium.

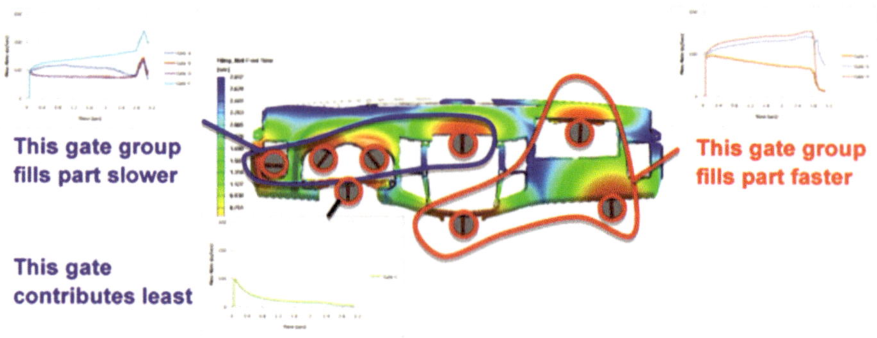

Figure 3.21 Flow rate curves of each gate

3.1.7 Summary

Using CAE analysis in the design phase is very efficient. It helps designers to identify the areas with potential high shrinkage ratio, as shown in the red areas in Figure 3.22, which can be focused on for product design improvement.

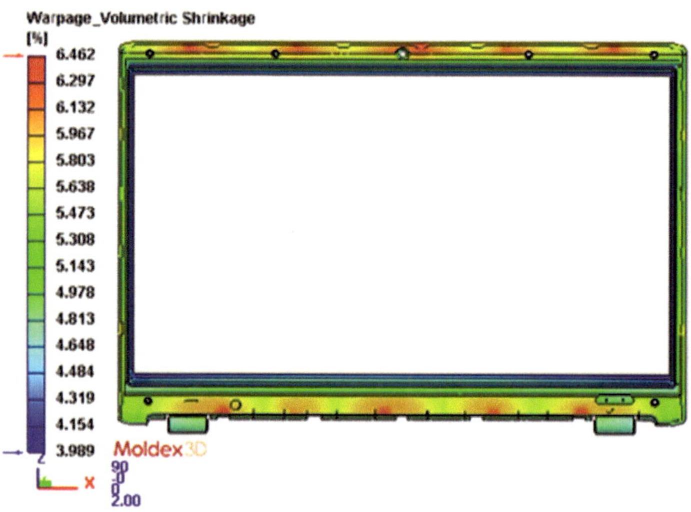

Figure 3.22 CAE prediction of shrinkage

■ 3.2 Mold Design

After kickoff and the design for manufacturing has been checked, the workflow of mold design requires the following operation steps, as shown in Figure 3.23:

1. Drawing confirm: Review drawing, feedback information and receive customer's reply.
2. Tooling process design: Make detailed drawings and design the process.
3. Tooling process confirms: Analyze and discuss the process, and confirm the process.
4. Tooling review: Review the mechanism design drawing and order materials.
5. Tooling: Make the mold based on the final drawing and material selected.
6. Molding trial: Assemble the mold, take test injections and record problems.
7. Record: Summarize the defects.
8. Database: Back up and store related data.

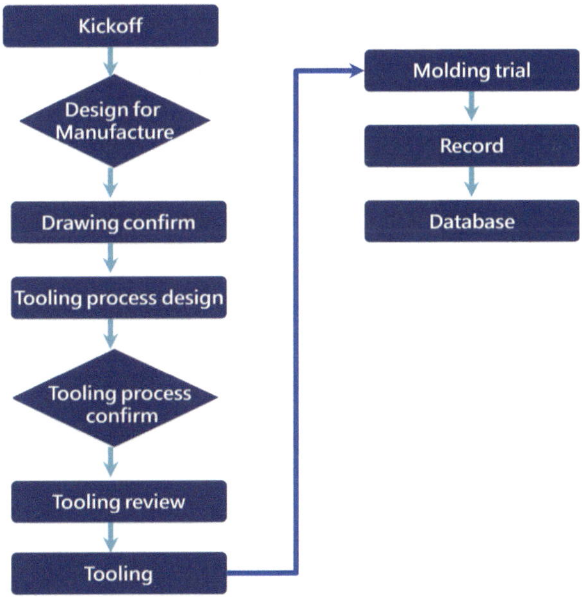

Figure 3.23 Workflow of mold design

A complete mold design consists of many systems, involving slides, guiding pins, ejection device, and other mechanical parts. However, only basic principles are introduced in this book, among which the part related to molding analysis is focused upon, including introduction of the basic concept, gate, runner, and cooling system design principles, and other systems that directly affect product quality.

3.2.1 Basics

Upon receipt of a part design drawing, the mold designer first confirms that ribs and bosses are in line with the mold opening direction, as shown in Figure 3.24, as well as that their draft angles are sufficient.

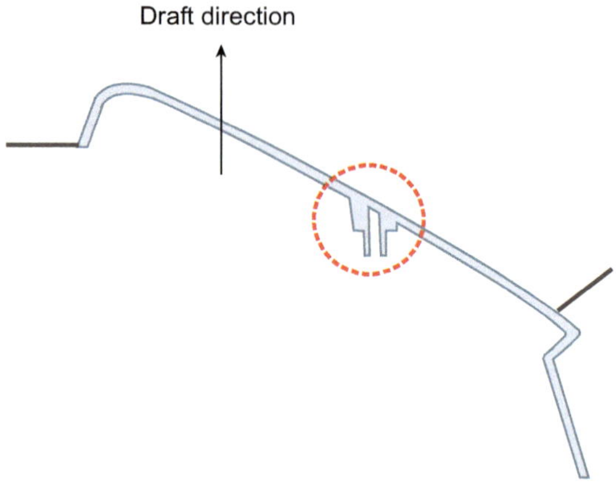

Figure 3.24 Check the draft direction

If there is an undercut design in the part, a slide should be properly allocated by referencing the parting line, as shown in Figure 3.25. Potential interferences between the moving path of these slides and the mold should be confirmed one by one as well.

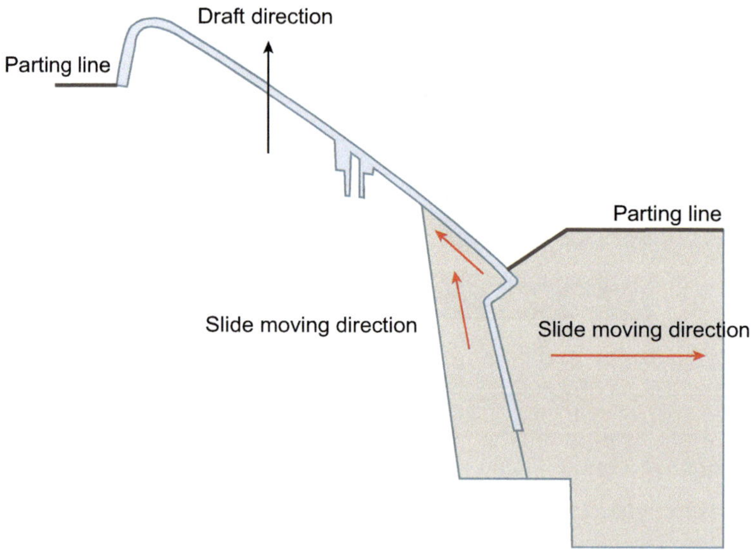

Figure 3.25 Define parting line and slide, and then check the interference

The definition of the parting line location depends very much on experience, but the rough rule is to use the outline of the maximum projection area of the product, which is better for the ejection mechanism, as shown in Figure 3.26. Another consideration is to define the location without affecting the appearance of the part and the convenience for mold construction.

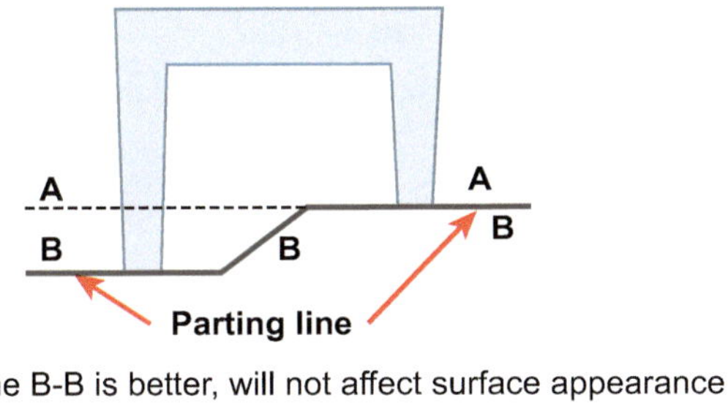

Figure 3.26 General design guidelines for the parting line

If there is an undercut design in the part as shown in Figure 3.27, the parting line should be shifted from the end upward to the outermost location of the projection of a part. Also, an arrangement of the parting line in a cramped space must be avoided or part breakage may occur in demolding due to overstress.

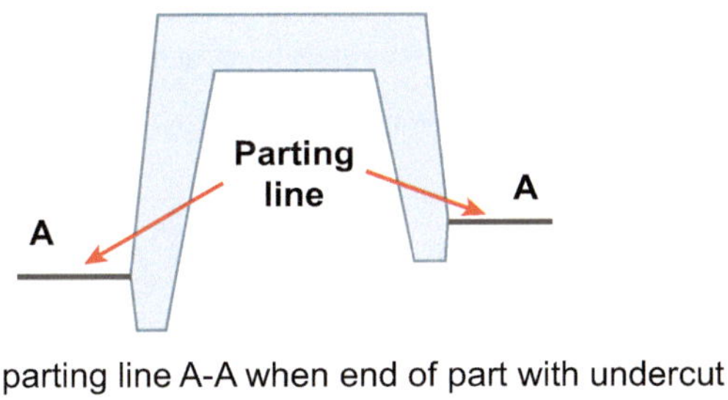

Figure 3.27 General design guidelines for the parting line for parts with undercuts

If the part is designed with zero draft angle, as shown in Figure 3.28 where the outward walls of the part are in parallel to the demolding direction, the mold design requires slides to reduce the resistance during demolding.

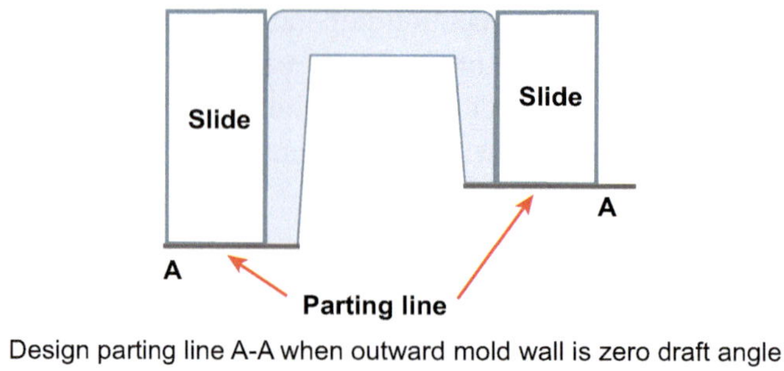

Figure 3.28 General design guidelines of the parting line for parts with zero draft angles

Sometimes the designed parting line may encounter an open area, as shown in Figure 3.29(a). This example shows a discontinuous and unconnected segment on the part, which requires the mold designer to properly define the location of the parting lines. Directly extending the original two parting lines is not a solution here (as shown by the dotted lines in the figure), whereas going along the edges of the part (as shown with the solid lines) is preferred so that the core underneath can be manufactured. Remember also to design in the draft angle where the parting line is located to have a smooth demolding, as shown in Figure 3.29(b).

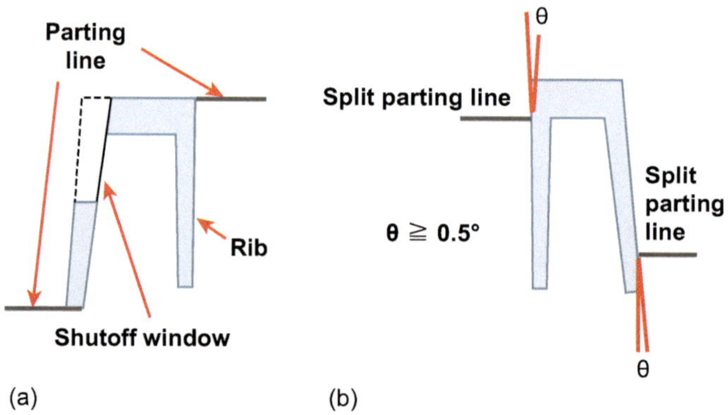

Figure 3.29 (a) Connection and (b) split of the parting lines

The actual mold base components are quite complex. When using CAE in the development and design stage, the mold base system is often simplified in order to quickly evaluate the advantages and disadvantages of the design. The function to generate simple mold bases is provided in Moldbase Wizard of Moldex3D, i. e., the settings are completed after the size of the overall mold base and the relative positions between the product and the mold base are given, as shown in Figure 3.30. The processing of the parting surface, the sliding block, and other mechanisms can be omitted whether for a single or a multi-cavity mold, so that the interaction between the product, the runner, and the cooling system can be quickly analyzed.

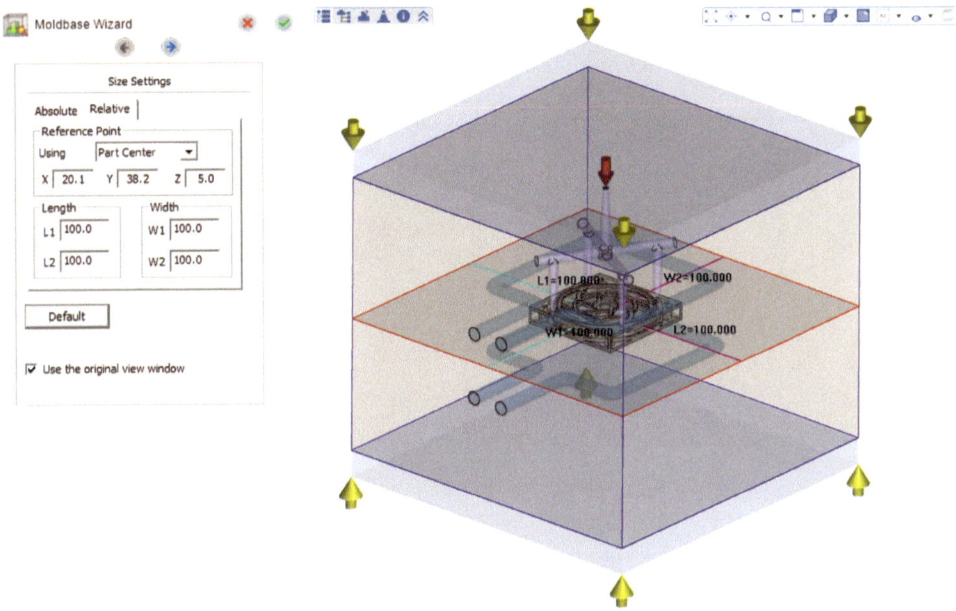

Figure 3.30 Moldbase Wizard in Moldex3D Studio

3.2.2 Gate Design

As a small cross-section of a gate leads to high injection pressure, gates should be designed at thick-wall areas to avoid huge pressure drops for better pressure maintenance. Other considerations are as follows:

1. Flow direction: Avoid melts generating molecular orientation that causes severe warpage.
2. Weld line: Keep the flow lengths inside mold cavities to equal distance to reduce air traps and weld lines.
3. Filling: With proper locations, more disturbance can be induced to decrease the occurrence of jetting and flow marks.

3.2.2.1 Gate Number

The following example shows a rough rule to determine the gate number. As shown in Figure 3.31(a), a single gate has to transfer melts up to 250 mm with a wall thickness of 1 mm, resulting in a flow length to wall thickness ratio (L/t ratio) of 250. In Figure 3.31(b), two gates are used and the L/t ratio is reduced to 180; however, weld lines might occur. As already mentioned in Section 3.1.2, the *L*/*t* ratio is very critical and a sufficient number of gates should be included according to the flowing properties of the used plastics, as shown in Table 3.1. Also, CAE can be used to verify the flow balance, pressure loss, and location of filling ends and weld lines under different gate numbers.

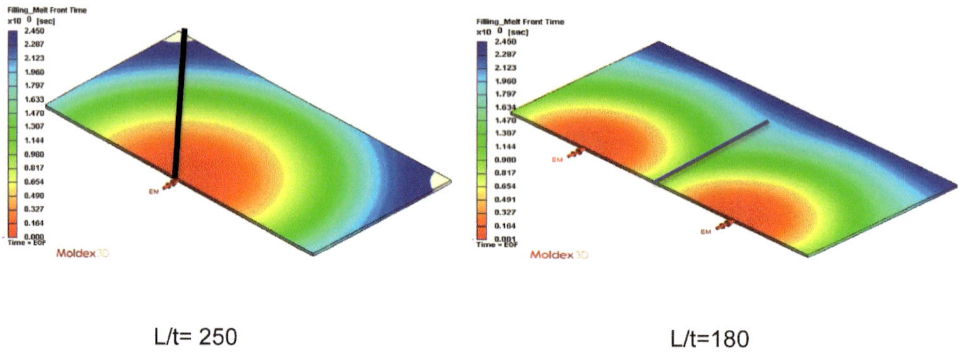

Figure 3.31 Flow length for different gate numbers: (a) one gate and (b) two gates

3.2.2.2 Gate Location

The gate location determines the packing conditions of each area. As shown in Figure 3.32, the area close to the gate has better packing and less shrinkage, and vice versa.

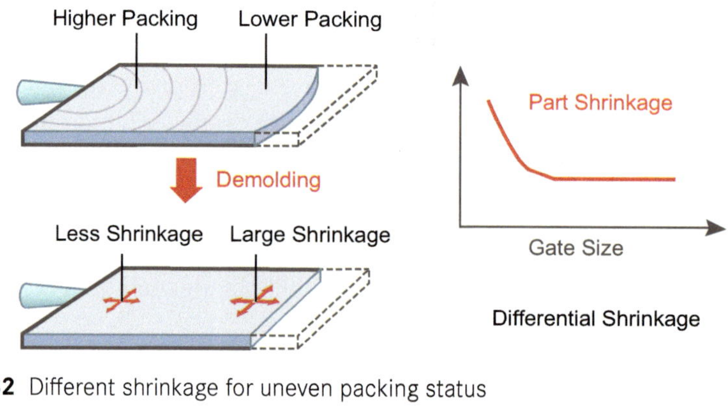

Figure 3.32 Different shrinkage for uneven packing status

Basically, the location of a gate should follow part specifications and not be designed on the appearance of a part, and neither should weld lines occur on the outside because of melt convergence. As shown in Figure 3.33, the CAE analysis indicates that three weld lines would possibly occur as displayed, caused by three gates. The location of these gates should be shifted to move the location of weld lines away if this area is critical for appearance.

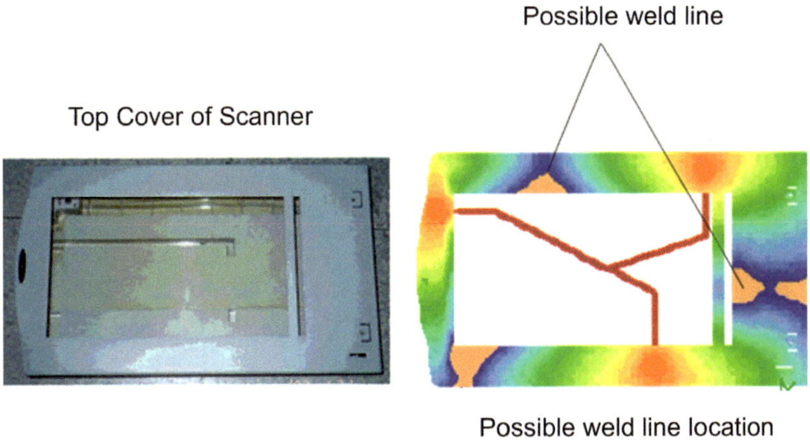

Figure 3.33 CAE prediction of weld lines

The design of the gate location should also avoid the occurrence of weld lines at structural weak points. As shown in Figure 3.34, the weld line can be moved from the short side to the corner by changing the gate location.

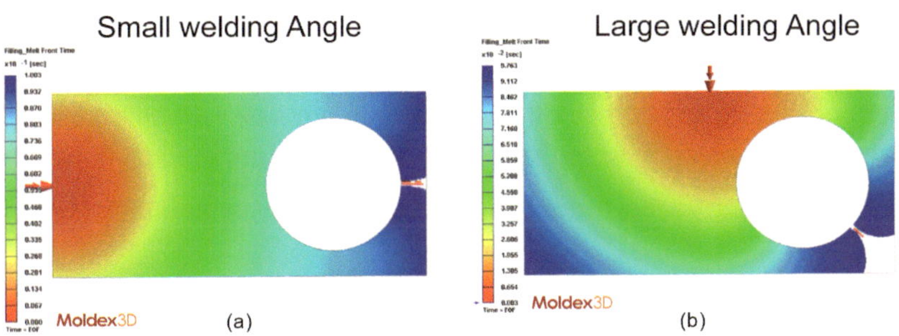

Figure 3.34 Welding angle for different gate locations: (a) gate at the short side and (b) gate at the long side

Generally, big parts need more gates. An example of a mold with two cavities is shown in Figure 3.35. The original design (a) has two gates in each cavity and the predicted weld lines would occur in the middle of the part, which is more visible as the flow length is long and results in a higher temperature drop. A modified three-gate design is shown in Figure 3.35(b). A new weld line occurs but it is less visible because the weld lines in this design are formed at a higher temperature. Therefore, a better welding quality can be achieved by adjusting proper gate location through CAE analysis.

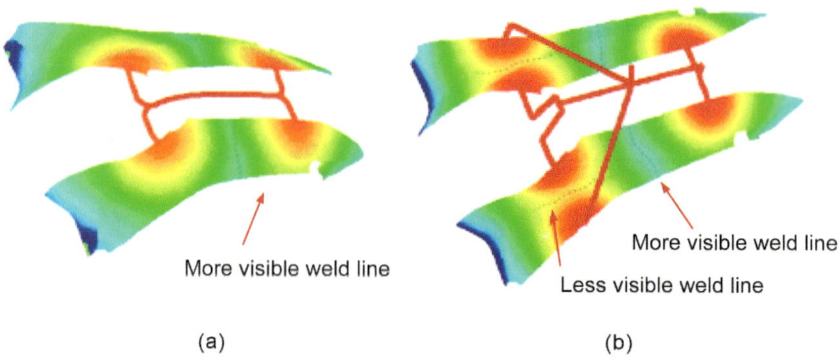

Figure 3.35 Weld line quality estimated by CAE analysis: (a) two gates for each part and (b) three gates for each part

The appearance quality is the first priority of gate location design, followed by smoothness of flowing. In the example shown in Figure 3.36, there are five gating points A, B, C, D, and E, among which there is a thin tab design near gate A. By injection through gate A, hesitation occurs at the tab due to a race-tracking effect, because melts tend to flow to thick-wall areas when reaching a branch, and so a selective flowing is seen; however, by injection through gate D, this phenomenon, which is in principle the same for ribs as already shown in Figure 3.15, can be eliminated and the filling can be finished properly. Therefore, using CAE to verify the adequacy of gate location change through simple simulations rapidly provides references for the selection of multiple preliminary design scenarios.

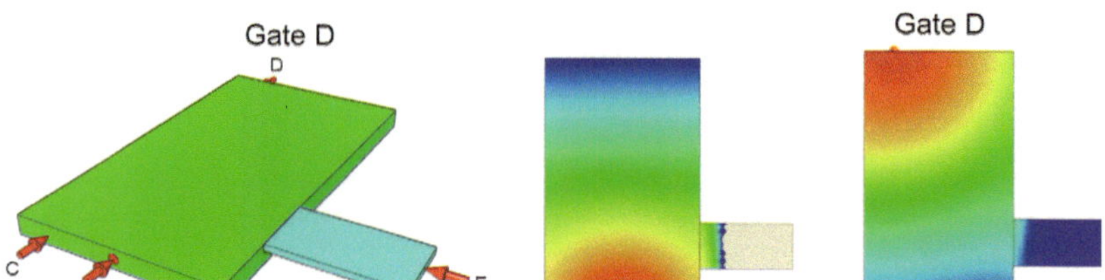

Figure 3.36 CAE filling results of the different gate locations

Another consideration in gate location is the instant switching behavior of a melt after passing through a narrow gate showing the elasticity effect. If the melt is not confined along the flow direction in the cavity, jetting occurs and thus a change of location and type of the gate is required, for example, by use of the overlapping gate to replace the edge gate as shown in Figure 3.37.

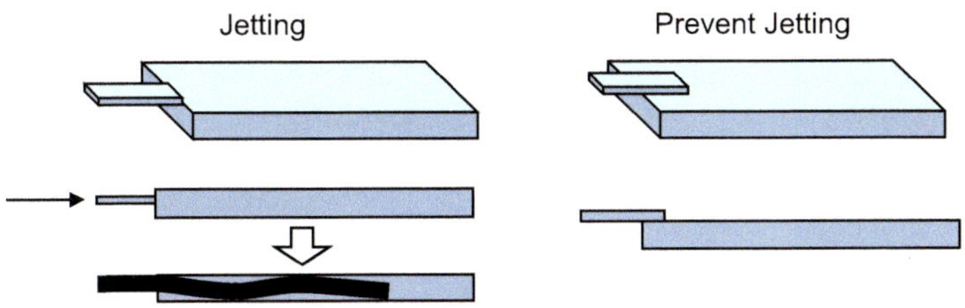

Figure 3.37 Gate design to prevent jetting

If fibers are used to strengthen the plastics, the gate location becomes more important because the way the melt flows affects fiber distribution and orientation, so ultimately affects the structural strength of the part, as shown in Figure 3.38. When fibers are oriented along a certain direction, the strength along this direction is enhanced; however, the strength along the perpendicular direction is relatively weakened, so cracking along the fiber direction is likely under an external force along the perpendicular direction.

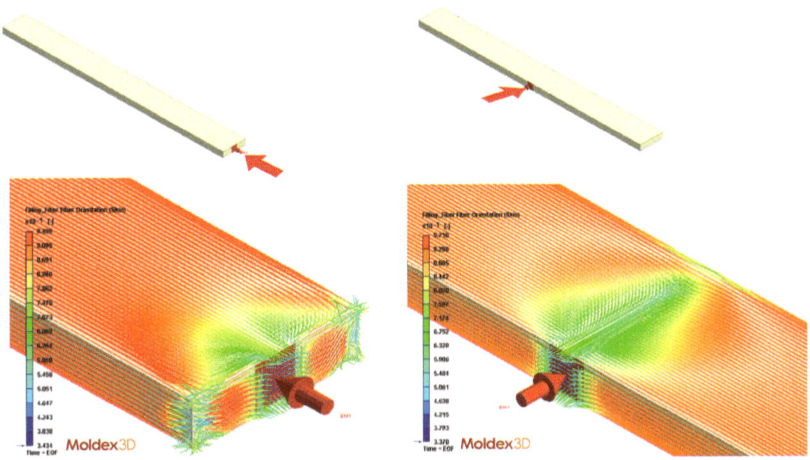

Figure 3.38 Fiber orientation for different gate locations

3.2.2.3 Gate Types

There are many gate types and the type to be used is determined by part appearance requirements, gate removal method, part geometry, mold manufacturing, plastic characteristics, etc. As shown in Figure 3.39, the types include the cashew gate or cashew gate with ejector pin, the fan gate, the overlapping gate, the tunnel gate, and the pin gate. The application scenario for each gate is briefly introduced in Table 3.3. A more detailed description is given in Section 7.1.2.

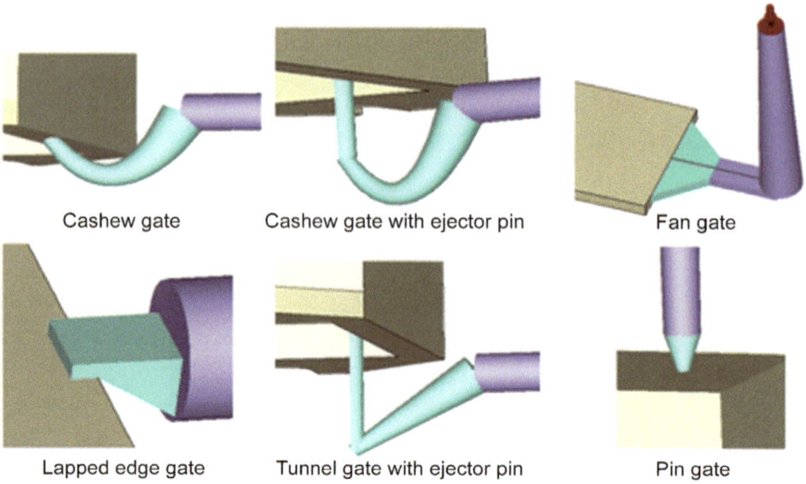

Figure 3.39 General gate types

Table 3.3 Common Gate Types

Gate Type	Application Timing
Sprue gate	For single-cavity molding and allows gate mark to form in the center of finished part.
Edge gate	Often used in flat or thin plastic parts, for which it is easier for processing.
Fan gate	Suitable for fast filling but hard to cut off the gate without leaving marks.
Pin gate	Runner can be separated from part automatically. Often used in three-plate molds with multiple cavities, but with higher pressure drop and fast solidification.
Tunnel gate	Runner can be separated from part automatically. Often used in two-plate molds with small cavities.
Cashew gate	Allowed only for gating from the interior. Higher mold cost.
Hot runner valve gate	The gate can be opened or closed on demand. Often used in big parts, but with higher mold cost.

In addition to locations and types, the size of a gate is also critical in gate design, which is mainly determined by timeliness of filling and packing, bearable shear rate of the plastic itself, and the method of gate removal (Figure 3.40).

The shear rate of a gate can be calculated according to Equation 3.3:

$$\dot{\gamma} = \frac{4Q}{\pi r^3} \text{ for round gates} \tag{3.3}$$

$$\dot{\gamma} = 6Q / w t_g^2 \text{ for rectangular gates}$$

where Q is the flow rate, r is the gate radius, w is the gate width, and t_g is the gate thickness.

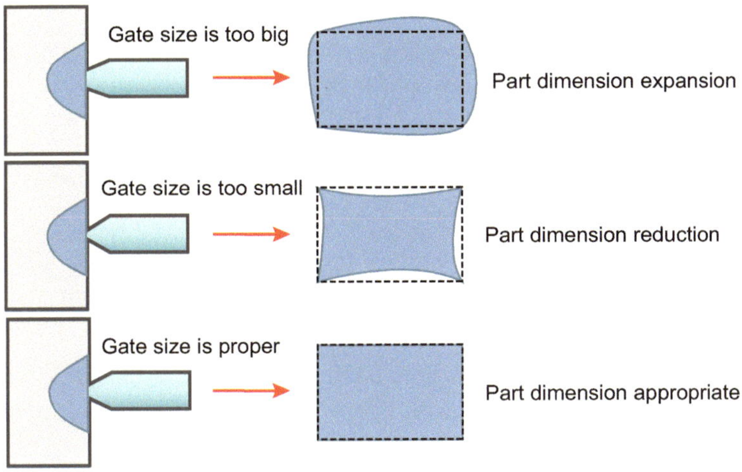

Figure 3.40 Effects of gate size on part dimension

3.2.3 Runner Design

As shown in Figure 3.41, a plastic injection system includes a runner, a sprue, and a nozzle. The main function of a runner is to deliver the melt evenly from nozzle to cavity, which directly affects filling path, cycle time, and, ultimately, the product quality.

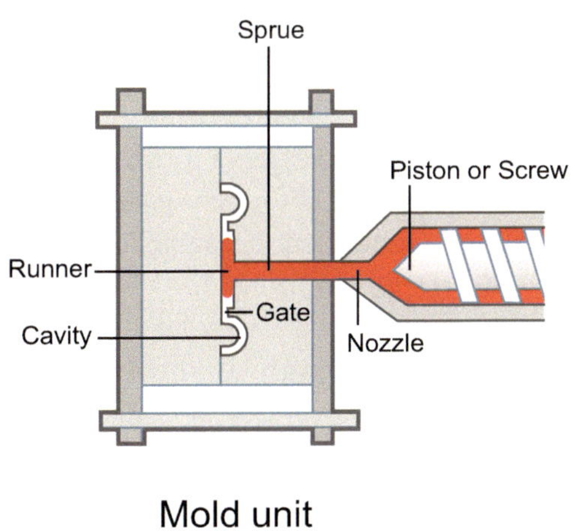

Figure 3.41 Plastics melt delivery system

Four major aspects are considered in runner design:

1. Runner shape: A circular, parabolic or trapezoidal cross-section determines pressure loss and cost of mold manufacturing, and influences considerations in demolding.
2. Runner dimension: Determines the melt flow rate and affects pressure loss and demolding mechanism.
3. Mold cavity number: Should consider the balance of geometric allocation in a multi-cavity system, as well as the volume containable by each mold, productivity, and cost.
4. Overall runner layout: Affects flow balance, demolding design, and manufacturing cost. Commonly seen layout types of runner design are series layout, H-style branching layout, circular layout, and X-style layout, as shown in Figure 3.42.

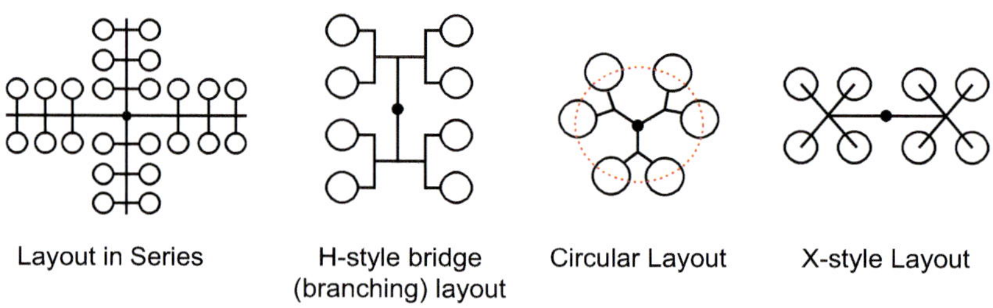

Figure 3.42 General layouts of runners

3.2.3.1 Runner Shape and Dimension

Regarding the shape and dimension design of a runner, the circular, trapezoidal, and parabolic cross-sections are commonly seen types, as shown in Figure 3.43. In fluid dynamics, the circular shape is best because it has optimum flow efficiency and low pressure loss, but processing on molds at both sides is a drawback that leads to higher cost. The processing of a trapezoidal runner is easy and thus it is frequently used, whereas the parabolic runner works better with parting lines, but has a higher cost than the trapezoidal runner.

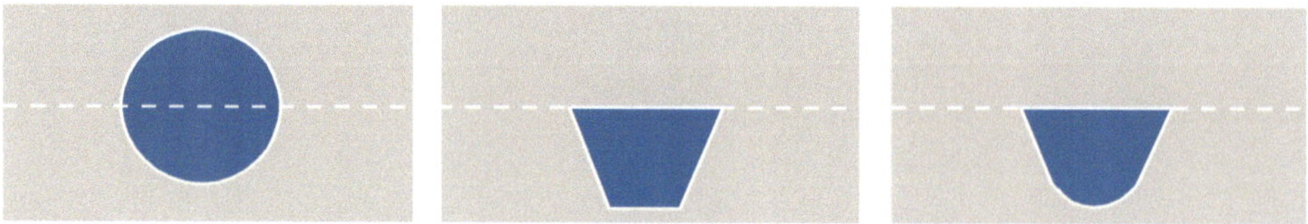

Figure 3.43 Cross-sections of general runner types

Runners are responsible for delivering melts from nozzles to sprues smoothly. Because materials have different viscous characteristics, plastics with good flowability can be used in narrow flowing paths. Therefore, the flowing properties of a material should be considered in runner dimension design as well. Recommended runner diameters of common materials are listed in Table 3.4.

For a hot runner system, the common runner diameter is 8 to 20 mm depending on the vendor. More details for runner design are given in Section 7.1.1.

Table 3.4 Recommended Runner Diameters

Plastic material	Recommended Runner Diameter	Plastic Material	Recommended Runner Diameter
ABS, SAN	0.187-0.375" (4.7-9.5 mm)	PC	0.187-0.375" (4.7-9.5 mm)
Acetal	0.125-0.375" (3.1-9.5 mm)	Polyester	0.187-0.375" (4.7-9.5 mm)
Acrylic	0.312-0.375" (7.5-9.5 mm)	PE	0.062-0.375" (1.5-9.5 mm)
Butyrate	0.187-0.375" (4.7-9.5 mm)	PP	0.187-0.375" (4.7-9.5 mm)
Cellulosics	0.187-0.375" (4.7-9.5 mm)	PPO	0.250-0.375" (6.3-9.5 mm)
Fluoro-carbon	0.187-0.375" (4.7-9.5 mm)	Poly-sulfone	0.250-0.375" (6.3-9.5 mm)
Ionomer	0.093-0.375" (2.3-9.5 mm)	PS	0.125-0.375" (3.1-9.5 mm)
PA	0.062-0.375" (1.5-9.5 mm)	PU	0.250-0.313" (6.4-8.0 mm)
Polyamide	0.187-0.375" (4.7-9.5 mm)	PVC	0.125-0.375" (3.1-9.5 mm)

3.2.3.2 Multi-Cavity Runner Design

For multi-cavity molding, the imbalance issues between cavities still frequently occur even if the runner system is designed geometrically symmetrical. This is related to viscosity differences resulting from shear heating of the melt in runners, as shown in Figure 3.44. This phenomenon is described in more detail in Section 6.3 of reference [5]. The flow behavior of a melt in the cavity is described in Section 6.1.1, and relevant case studies are explored in detail in Section 7.2.1.

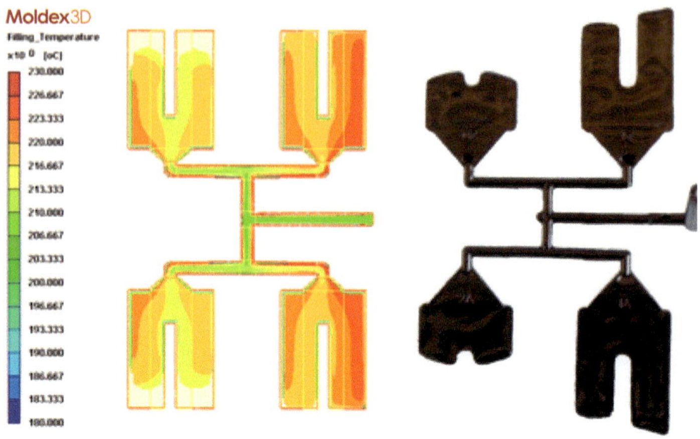

Figure 3.44 Flow imbalance in a runner system with symmetrical geometry [6]

3.2.4 Cooling Design

The connection of the cooling channel also affects heat dissipation. Cascade (serial) and parallel cooling system designs are the two types most commonly seen, as shown in Figure 3.45. Other kinds of connections, for example a design that uses one-to-many branching, may cause selective flowing and generate dead corners. The cooling channel should be divided into multiple segments as shown in Figure 3.46 (from one-in/one-out to multi-in/multi-out) in the case of a long cascade with a temperature difference of the coolant over 3 °C between the inlet and outlet.

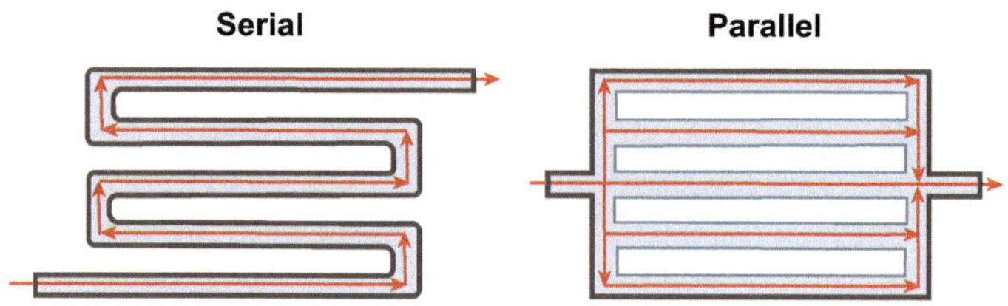

Figure 3.45 Serial and parallel cooling systems

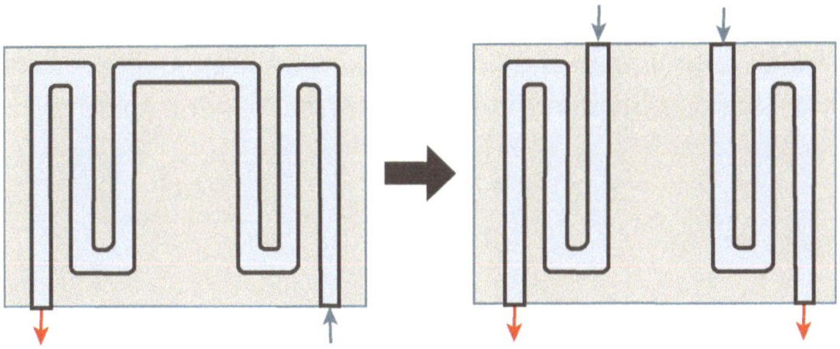

Figure 3.46 Multi-inlets and outlets of the cooling system is better

If the area cannot be reached by a regular cooling channel, traditional auxiliary cooling channel designs can be implemented, such as baffle (Figure 3.47(a)) and bubbler (Figure 3.47(b)), or heat pipes, or even the mold insert material can be replaced with a beryllium copper alloy of high heat dissipation coefficient. Also, the new conformal cooling techniques shown in Figure 3.47(c) enhance the cooling efficiency in these areas. By comparing the temperature distribution of these three designs, the conformal cooling is able to take away the internal heat of the part, which makes it the best cooling design among all three.

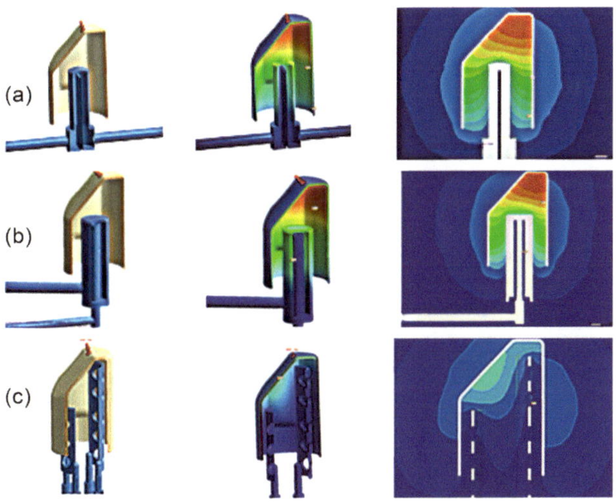

Figure 3.47 Enhanced cooling designs: (a) baffle, (b) bubbler, and (c) conformal cooling

In addition to a good cooling channel layout, a sufficient flow rate of the coolant is important because turbulence can take away heat three to five times faster than with a steady laminar flow. A Reynolds number larger than 6000 is recommended to maintain the turbulent condition of the coolant. A complete introduction on the cooling system is given in Chapter 8, including the theoretical heat dissipation mechanism of injection molding, golden rules for uniform mold temperature, quantitative index of cooling design, cooling time estimation, interpretation of CAE cooling analysis, and application cases of conformal cooling channels.

3.2.5 Others

Besides products, runners, and cooling systems, there are other systems that directly affect the quality of the finished part as well.

3.2.5.1 Ejector System

Take an ejector system as an example: If the ejector pin is designed at the corner with poor cooling efficiency as shown in Figure 3.48, ejector damage (Figure 3.48(a)) may occur because this area of the part is ejected at too high a temperature. Therefore, a better design is to incorporate a bubbler to improve the cooling efficiency in the corners (Figure 3.48(b)), by which the molding cycle time can be reduced, or to design an ejector rail of bigger area for ejection to reduce the risk of ejection deformation.

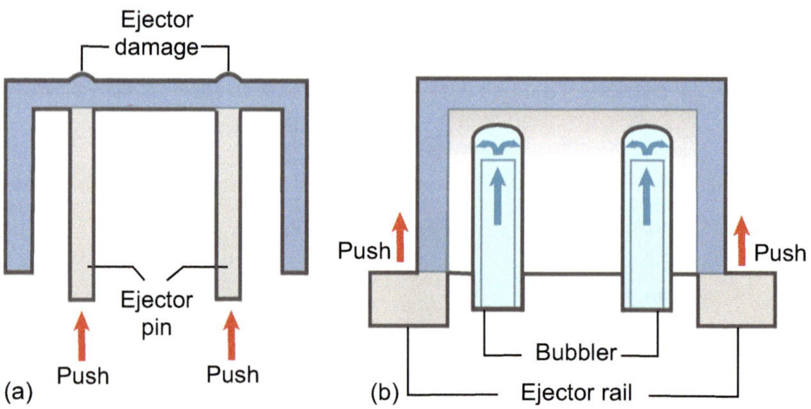

Figure 3.48 Ejector design at the corner

3.2.5.2 Venting Design

Usually, a gap of 0.02 to 0.03 mm is kept at the ejector pin for air to escape and venting gaps are also designed at filling ends as well.

CAE tools become significantly important for the arrangement of venting locations. As shown in Figure 3.49, the simulation result precisely predicts the final filling position, and thus the venting locations can be arranged properly on the mold in advance, which shows how CAE can help in the design process.

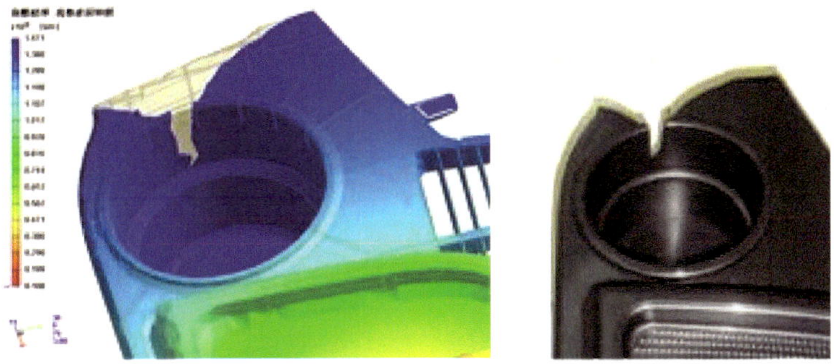

Figure 3.49 CAE prediction of filling end

The size of venting gaps is also related to the used plastics. As shown in Table 3.5, the basic rule is to design a smaller gap for materials with good flowability, such as PA and PBT, whereas for materials with higher viscosity a bigger gap is required.

Table 3.5 Venting Gaps for Different Plastics

Material	Venting Gap [mm]	Material	Venting Gap [mm]
PE	0.02	POM	0.01~0.03
PP	0.01~0.02	PMMA	0.03
PS	0.02	PA	0.005~0.015
ABS	0.03	PPS	0.01~0.03
SAN	0.03	PC	0.01~0.03
PPO	0.03	PBT	0.005~0.015

In conclusion, product and mold design is not easy because different parts require different design rules. 3C products (computer, communication, and consumer electronics), for example, need to be light weighted and durable for collision and wear in daily use, whereas for optical products, the requirements on precision and surface quality are extremely strict, and therefore an exclusive mold material and processing methodology have to be developed, such as steel molds with special heat treatment or a mirror-grade surface polishing technique.

References

[1] John, G.: Investigating Melt Flow, *Injection World*, no. 2, 2011, p. 39

[2] Bayer MaterialScience LLC: Part and Mold Design, Pittsburgh, 2000

[3] Campbell, P. D. Q.: *Plastic Component Design*, New York: Industrial Press Inc., 1996, p. 214

[4] Moldex3D: "Shear Stress and Shear Rate Information", [Online]. *http://support.moldex3d.com/r16/en/sync/sync-for-solidworks/shearstressandshearrateinformation.html*

[5] Mennig, G.; Stoeckhert, K.: *Mold-Making Handbook*, 3rd ed., Cincinnati: Carl Hanser Verlag, 2013

[6] Menges, G.; Michaeli, W.; Mohren, P.: *How to Make Injection Molds*, 3rd ed., Munich: Carl Hanser Verlag, 2001

4 Process Conditions

Chuan-Wei (Arvid) Chang

This chapter is divided into five sections. The first section (Section 4.1) gives the main steps and items of a molding cycle, and from Section 4.2 to Section 4.5 the stages of a molding cycle are introduced in order, which can be roughly categorized as the plasticizing stage that turns plastics from solid state to melt, the filling stage when plastics enter the mold system through a nozzle, the packing stage that compensates shrinkage, and finally the cooling stage which takes most of the time in a cycle.

4.1 Introduction to the Injection Molding Cycle

4.1.1 Brief Introduction to Injection Molding Machine Units

The main components of an injection molding machine are the feed, injection, mold, mold clamping, driving, and control units, as shown in Figure 4.1.

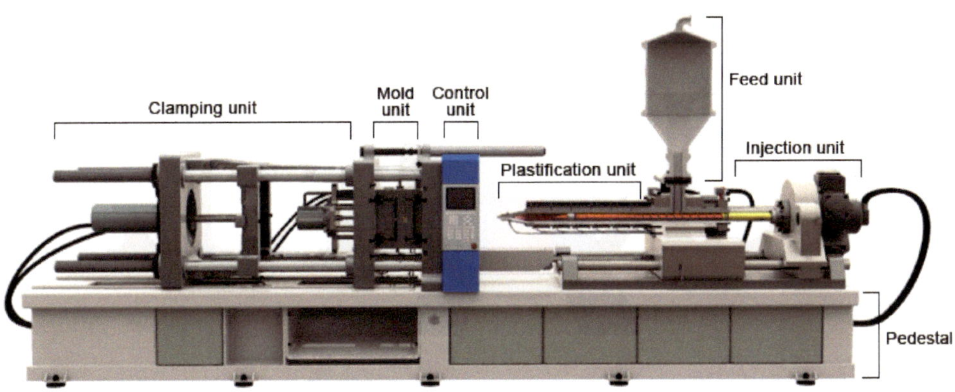

Figure 4.1 Injection molding machine

1. Feed Unit

The function of a feed unit is mainly to allow the plastic to enter the screw smoothly. Usually, plastics are preprocessed by drying under a certain temperature and time before entering the feed.

2. Injection Unit

As shown in Figure 4.2, inside the screw there is a feed section (L_f), a compression section (L_C), and a metering section (L_M). L is the effective length of the screw, D is the external screw diameter, P is the pitch of the screw, N is the non-return valve, and h_M and h_F are the flight depths in metering and feed sections, respectively. The screw moves the solid plastic toward the nozzle through constant rotating, during which the plastic is turned from the solid state to melt and accumulates in front of the screw ready to be injected.

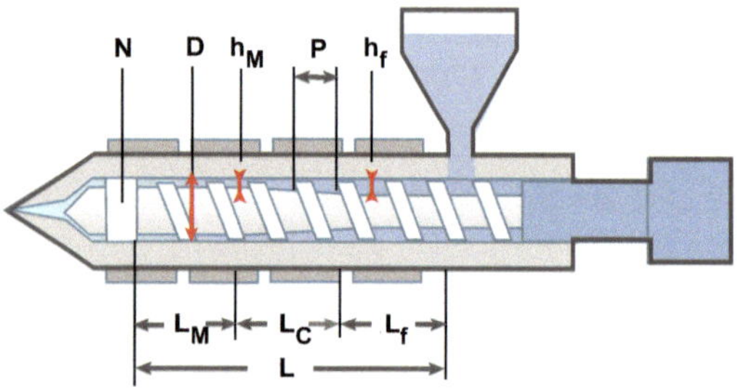

D	External screw diameter	
L	Effective length of screw	20-23 D
L_F	Length of feed section	0.5-0.55 L
L_C	Length of compression section	0.25-0.3 L
L_M	Length of metering section	0.2 L
h_M	Fight depth in the metering section	
h_F	Fight depth in the feed section	
P	Pitch of screw	0.8-1.0 D
N	Nonreturn valve	

Figure 4.2 Screw specifications

3. Driving Unit

The injection process includes filling and packing, and the types of driving systems are hydraulic, electric servo motor, and hydraulic-electric hybrid.

4. Mold Unit

Melt enters the mold system from nozzles and passes through the sprue, runners, and gates to form a part. Cooling channels that solidify the plastics for follow-up ejection are also included in the mold unit.

5. Mold Clamping Unit

The mold opening and closing in the mold cycle is controlled by a mold clamping unit, which also represents a specification of a machine. Usually, injection machines can be distinguished by maximum clamping force (tons).

6. Control Unit

Molding conditions are entered into a machine through a control unit that defines all machine actions, for example, the time for mold opening/closing, clamping force setting, injection speed of screw, packing pressure, etc., which will be introduced in this chapter.

4.1.2 Injection Molding Cycle

Plastics injection molding is a process that continuously repeats injection cycles. Figure 4.3 shows a molding process from the perspective of the mold. The stages are as follows:

1. Mold closing stage:

   After closing, the mold is clamped with high pressure and ready for the high-temperature and high-pressure conditions in the following melt injection stage.

2. Injection stage:

   Generally, the injection stage is defined as filling and packing, which is controlled by the speed and pressure of the screw for injecting melt into the mold.

3. Cooling stage:

   During cooling, the screw is preparing for the next injection by executing plasticizing actions, including feed, compression, and metering. Therefore, cooling and plasticizing metering proceed concurrently.

4. Mold opening stage:

   After cooling, the temperature and pressure of the melt drop down. Pressure is released prior to mold opening and removal of the finished part.

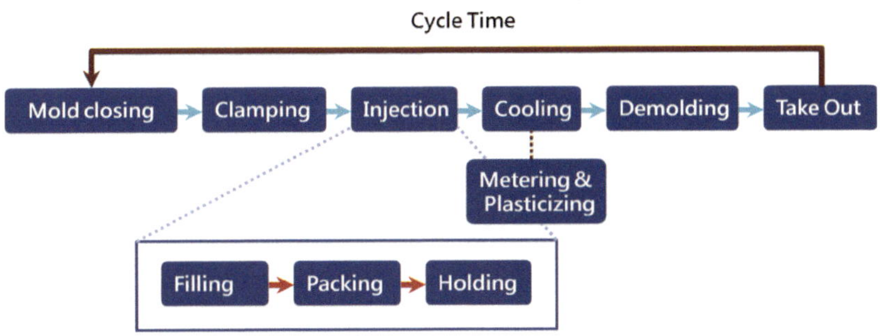

Figure 4.3 Injection molding cycle

Figure 4.4 shows the molding process from the perspective of the product. The injection includes filling, packing, and holding stages. Packing is to push more melt into the cavity through the gate, so the melt density near the gate is higher than that at the filling end. Holding is to make the melt density a more uniform distribution inside the cavity by application of moderate pressure provided by the injection machine.

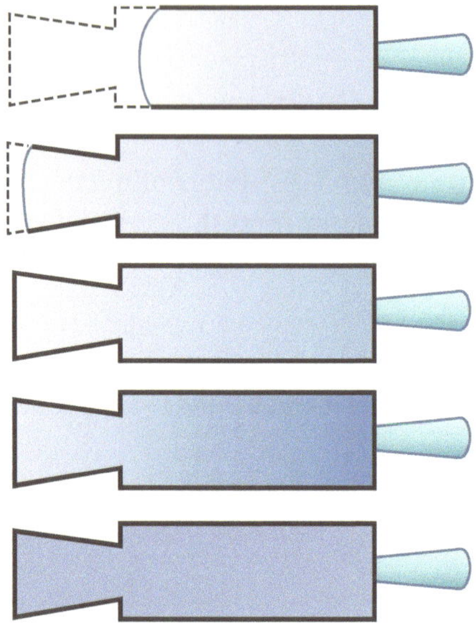

Figure 4.4 Part filled status at each stage

Figure 4.5 shows the time of each stage in a mold cycle for a general part with 2 to 3 mm wall thickness. The whole cycle takes about 30 to 40 seconds, which consists of 1 to 2 s for mold opening, 1 to 5 s for filling, 8 to 10 s for packing (and holding), 10 to 20 s for cooling, 2 to 5 s for screw plasticizing, 1 to 3 s for mold opening, and 1 to 3 s for ejection, among which the cooling stage is the longest step.

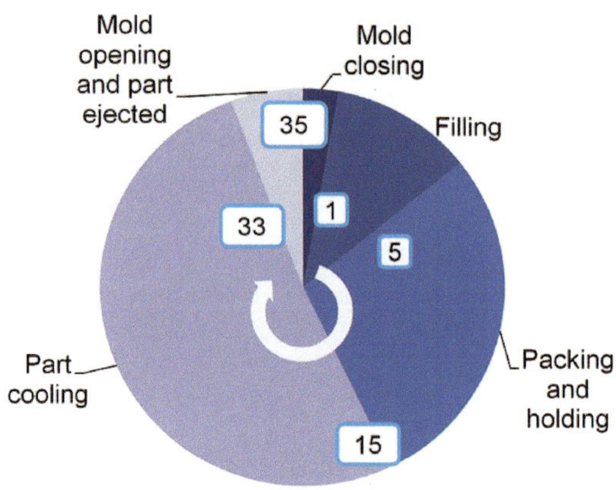

Figure 4.5 Duration of the stages in an injection molding cycle

Figure 4.6 shows the volumetric or weight change of the melt filled into the mold with the screw stroke. The volumetric filling (red line) can be deemed theoretically as the product of the forward distance and the cross-sectional area of the screw, which can be used as the volumetric filling for simulations. The filling starts from zero until it reaches the velocity/pressure (VP) switch point where the volumetric filling is nearly 100%.

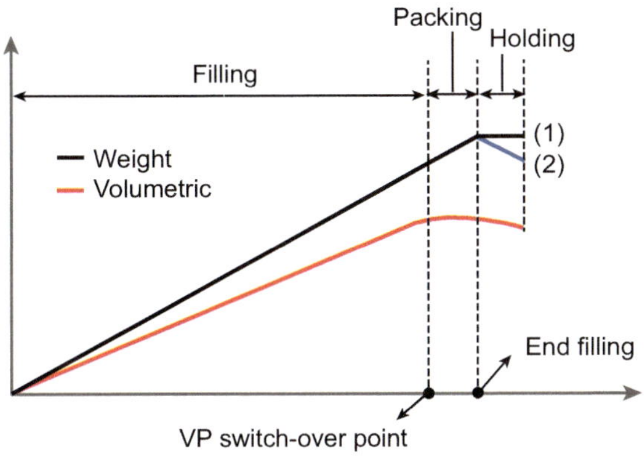

Figure 4.6 Volumetric and weight filling during injection procedure

Theoretically, the weight would not change at the holding stage, and the melt inside the cavity keeps relaxing to reduce residual stress and to reach a more uniform density distribution, as shown in line (1) in Figure 4.6. However, if the pressure drops too soon, melt back flow from the gate would occur, which results in a decreased weight as shown in line (2) in Figure 4.6. It may increase the risk of shrinkage.

Figure 4.7 shows the changes from the perspective of plastic during injection molding, which include heating of the barrel heater, shear grinding and melting of plastic pellets in the barrel, pressurized melt flowing, filling and packing completion, cooling within molds, solidification, and ejection.

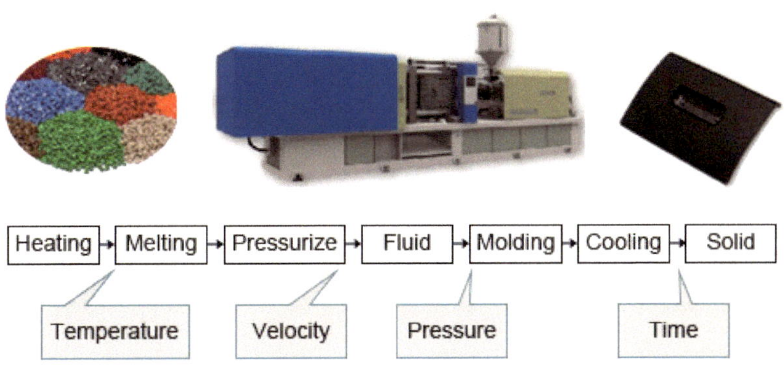

Figure 4.7 From raw material to product

4.1.3 Molding Window

Because plastic properties vary differently, including specific volume change under different temperatures and pressures, viscosity change, heat capacity, thermal conductivity, etc., their behavior is affected in a molding cycle. From a graph of plastics temperature and pressure toward molding defects as shown in Figure 4.8, it can be seen that the risk of thermal degradation gets higher as the temperature rises, whereas short shot defects appear at low temperature or under insufficient pressure, suggesting that by reducing the process temperature the pressure should be raised to transfer melt to mold cavities, but flashes would occur if the pressure goes too high.

Table 4.1 shows the flow properties and processing temperature ranges of common plastics, including melt temperature, mold temperature, and ejection temperature. For example, the polyamide (PA) series has a very high melt flow rate (MFR) and good flowability, which enables the material to flow at a rate of 100 grams in 10 minutes by applying a loading weight of just 5 kg. However, the mixture of polycarbonate/polybutylene terephthalate (PC/PBT) can flow at only 46 grams per 10 minutes, less than half of the amount under the same temperature and loading weight conditions. The table clearly identifies the recommended upper and lower limits of processing temperatures.

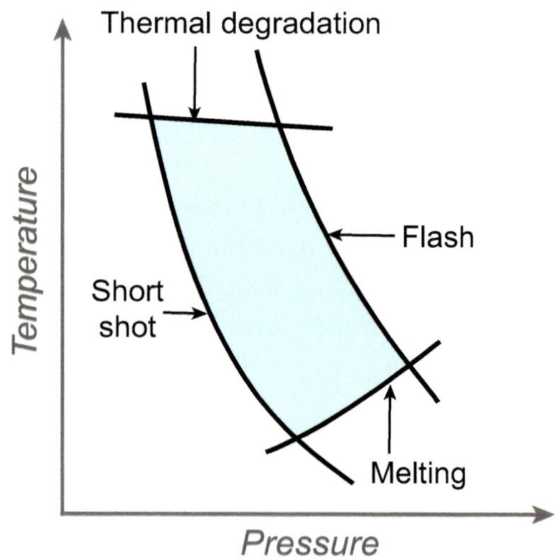

Figure 4.8 Molding window

Table 4.1 Recommended Process Conditions of General Plastics

Material	Flow Properties			Recommended Process Conditions		
	MFR [g/10 min]	Loading [kg]	Temperature [°C]	Melt Temperature [°C]	Mold Temperature [°C]	Ejection Temperature [°C]
ABS	35	10	220	200-280	25-80	88
PA6	110	5	275	230-300	70-110	133
PA66	100	5	275	260-320	70-110	158
PBT	35	2.16	250	220-280	15-80	125
PC	20	1.2	300	260-340	70-120	127
PC/ABS	12	5	240	230-300	50-100	117
PC/PBT	46	5	275	250-280	40-85	125
HDPE	15	2.16	190	180-280	20-95	100
LDPE	10	2.16	190	180-280	20-70	80
PET	27	5	290	265-290	80-120	150
PMMA	10	3.8	230	240-280	35-80	85
POM	20	2.16	190	180-235	50-105	118
PP	20	2.16	230	200-280	20-80	93
PS	15	5	200	180-280	20-70	80

4.1.4 PVT Variations during Injection Stages

Pressure-volume-temperature (PVT) variations during the plasticizing, filling, and cooling stages and after demolding are discussed in this section.

1. Plasticizing Stage

The plasticizing stage of solid pellets is carried out in the barrel as shown in Figure 4.9. The original plastic status is marked as (1) in the figure, and the status of the melt accumulated in front of the screw after metering is marked as (2). To keep the plastic at a certain dense condition in the barrel, a back pressure should be applied at the back end of the screw to hold the melt until the melt pressure is higher than the back pressure, enough to push back the screw, by which an accurate metering stroke can be reached. In a hydraulic machine as example, the pressure of the melt in front of the screw is about 10 times that at the back end of the hydraulic cylinder.

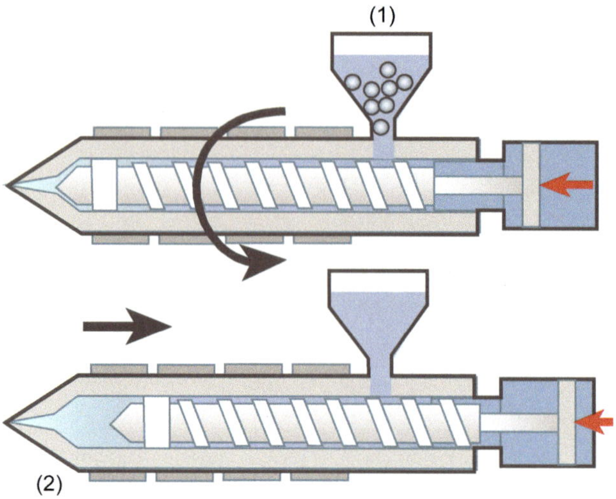

Figure 4.9 Plasticizing stage

The overall melt state is not uniform throughout the whole injection process, that is, temperature, pressure, and density of the melt are not at fixed values in a molding system but are distributed within certain ranges. A practical situation is shown in Figure 4.10, where the time points 1 to 8 are as follows:

1. Beginning of injection.
2. Melt reaches location gate (G).
3. Melt reaches location filling end (E), filling completed, and then packing to compensate melt in the part begins.
4. Melt pressure at gate (G) rises to the maximum value.
5. Machine pressure is kept at a middle packing pressure through time points 4 to 5, while the melt pressure at the filling end (E) gradually decreases due to the decreasing temperature. Then the machine pressure is set as a stepwise reduction to hold the cavity pressure.
6. Melt pressure at filling end (E) drops to zero.
7. Gate is frozen so that packing (including holding) is finished.
8. Melt pressure at gate (G) drops to zero, and the part starts to shrink.

The injection pressure, that is, the maximum pressure borne by the melt at the nozzle, decreases as the melt flows further. The pressure at the gate (G) is lower than that at the nozzle, while the pressure at the filling end (E) is lower than that at the gate, and drops quickly to zero in the course of time.

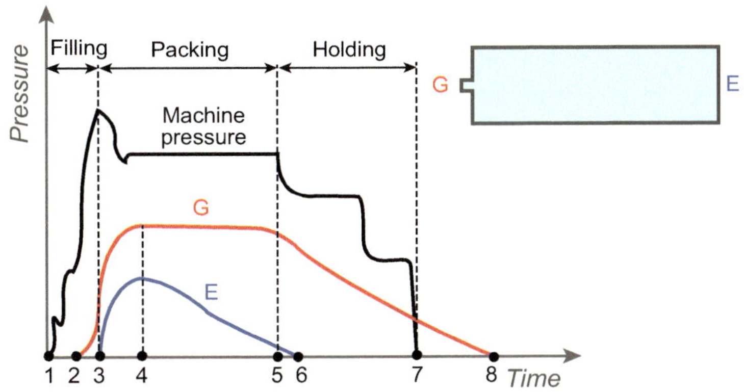

Figure 4.10 Melt pressures at different locations

For acrylonitrile butadiene styrene (ABS) plastic as an example, the processing conditions for melt, mold, and ejection temperatures are 240, 50, and 70 °C, respectively. To briefly show the change of the melt state in each stage of the molding process, only that of a single area is explored. By recording a PVT chart of the melt in the plasticizing stage, we can roughly obtain the state change path as shown in Figure 4.11, where (1) denotes the state of plastics at the hopper under room temperature and atmospheric pressure. After being heated by screw heaters and shear friction, and pressurized, the plastic raw material becomes melt (2) and occupies the front of the screw, to be metered and ready for injection.

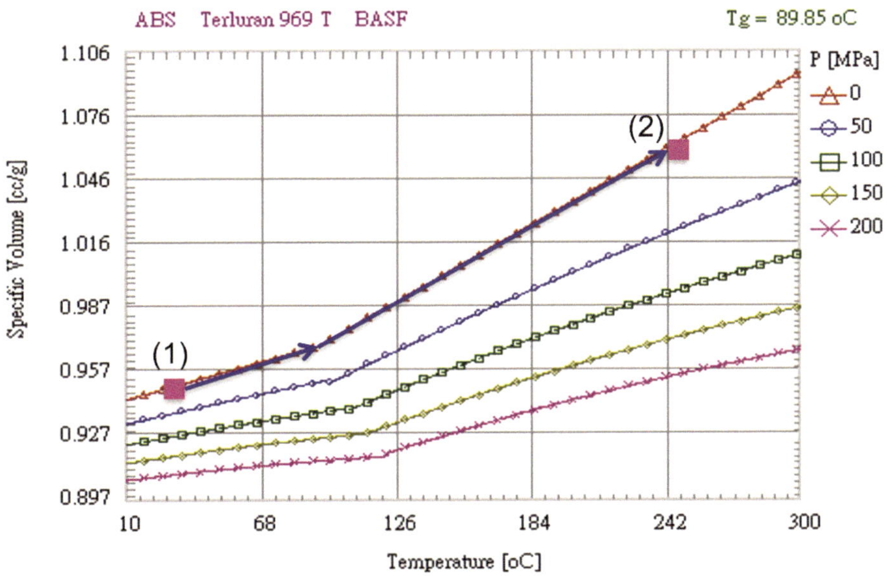

Figure 4.11 ABS PVT states during plasticizing stage

2. Filling Stage

The filling stage of injection is shown in Figure 4.12. The melt is squeezed to a highly compressed condition from the joint effect of a gradually shrinking nozzle and given injection speed.

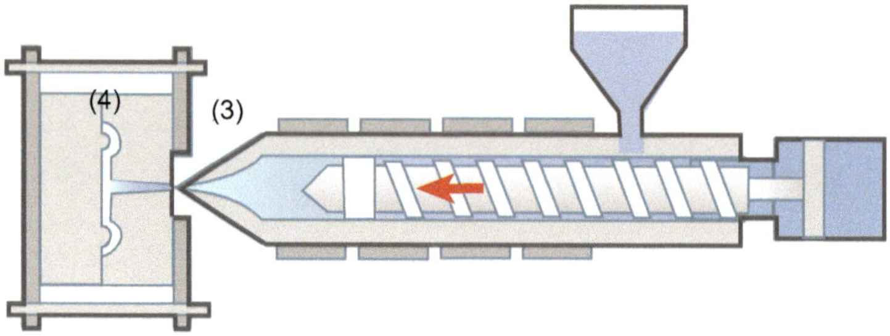

Figure 4.12 Filling stage

Figure 4.13 shows a PVT graph recording the melt state in front of a screw during filling (3) and inside the mold cavity (4). The melt is at a low pressure (2) while in the metering section before being injected. When the melt is squeezed by the injection speed but still stays in front of the screw, it enters a high-pressure state (3). After being injected into the cavity, the melt temperature drops a bit because it contacts the mold. Flow-back of the melt from the cavity to the nozzle should be avoided and the densities of the whole part should be kept as uniform as possible in this stage, which slightly increases the pressure too (4).

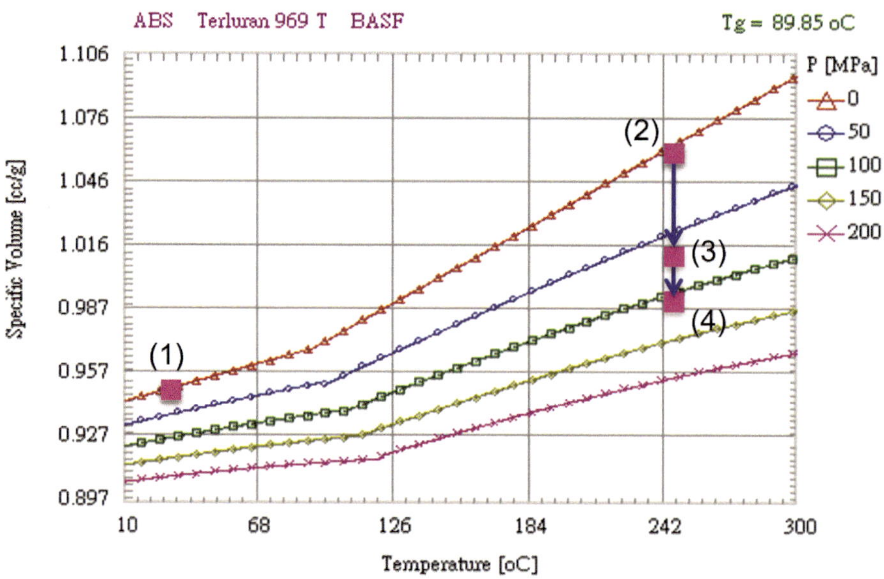

Figure 4.13 ABS PVT states during filling stage

3. Packing Stage

Figure 4.14 shows the packing stage when the mold cavities are almost filled with melt. A packing pressure is applied to keep supplying melt into the cavities to compensate for shrinkage. The pressure increases a bit as the compensated melt enters the cavity. Also, due to the cooling effect from the heat dissipation of the mold, the melt temperature inside the cavity decreases and drops to state (5) in Figure 4.15.

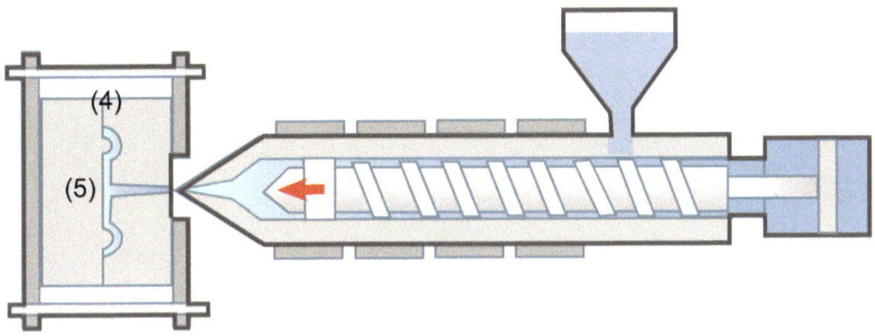

Figure 4.14 Packing stage

As shown in the PVT chart in Figure 4.15 including the packing stage, a certain pressure is maintained, or the pressure increases initially but, as the cooling time increases, the temperature drop becomes larger. Because there is no more melt bringing heat into the mold cavity, the heat entering the mold is much less than the heat taken away by cooling channels, and thus the temperature drops rapidly.

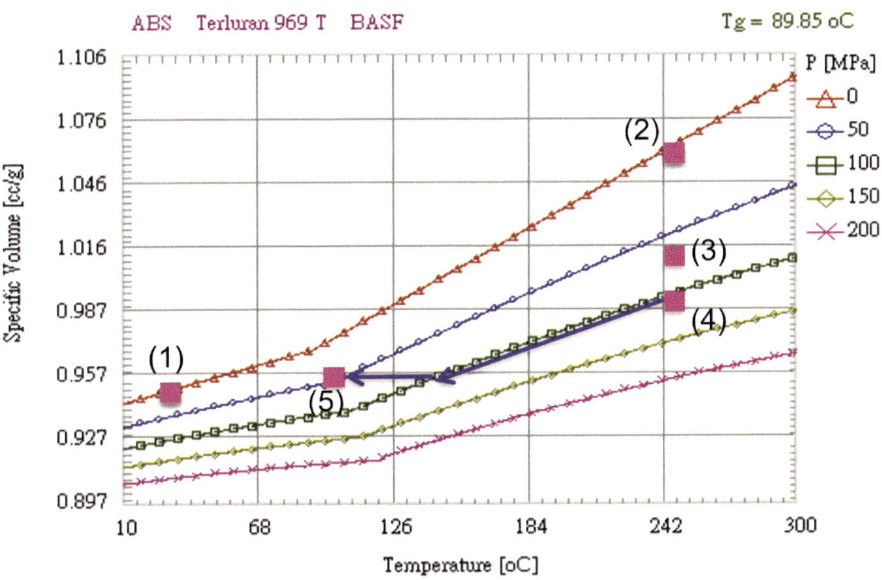

Figure 4.15 ABS PVT states during packing stage

4. Cooling Stage

As shown in Figure 4.16, during the cooling stage, the screw moves back to prepare for the plasticizing of the next shot, while the melt inside the mold keeps cooling until the temperature is low enough for ejection.

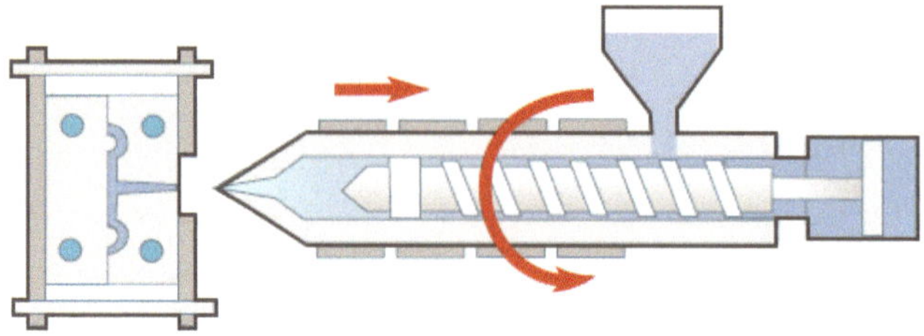

Figure 4.16 Cooling stage

Plastics gradually solidify during cooling and become solid when they return to atmospheric pressure, as shown in state (6) in Figure 4.17. The cooling stage comes to an end when the temperature is low enough for ejection.

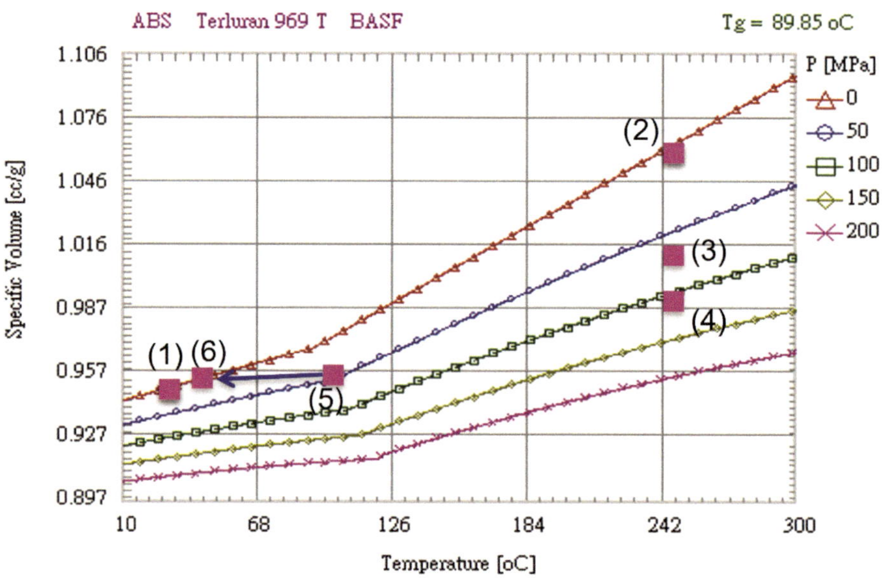

Figure 4.17 ABS PVT states during cooling stage

5. After Demolding

Once the part is cooled down enough to form a thick frozen layer, theoretically there will be no ejector damage. However, as the temperature of the part is still higher than room temperature, the plastic part still cools and shrinks until it reaches ambient temperature, as shown in Figure 4.18. Because there is no constraint for shrinkage after demolding, deformation could also be related to part geometry, gravity, and way of placing.

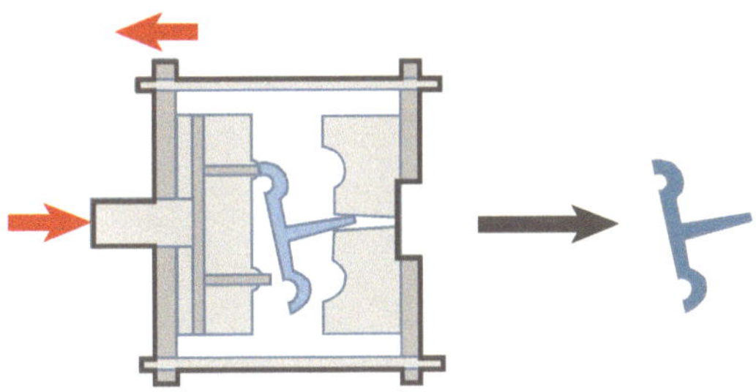

Figure 4.18 The part is ejected

The path of the PVT states of the product after demolding is shown in Figure 4.19. The temperature continues to drop from the ejection temperature (6), as well as the specific volume decreasing (shrinkage), until room temperature (1) is reached.

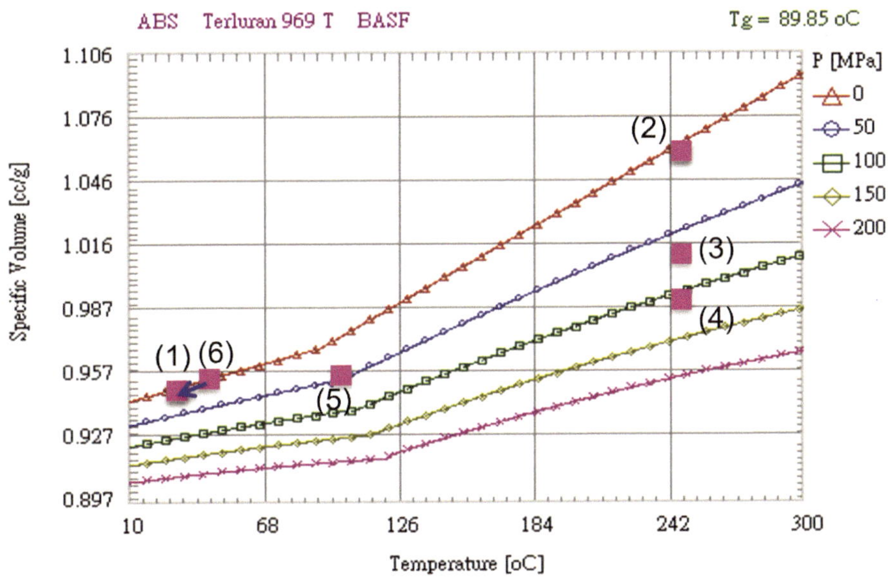

Figure 4.19 ABS PVT states after ejection

From the above description, plastic materials are exposed to high temperature and high pressure and are then brought back to room temperature and low-pressure conditions during the injection process. To control the part weight precisely, we should push into the cavity as much melt as possible before the end of packing.

That way, the specific volume difference between state (5) and state (1) in Figure 4.20 can be reduced to minimize shrinkage.

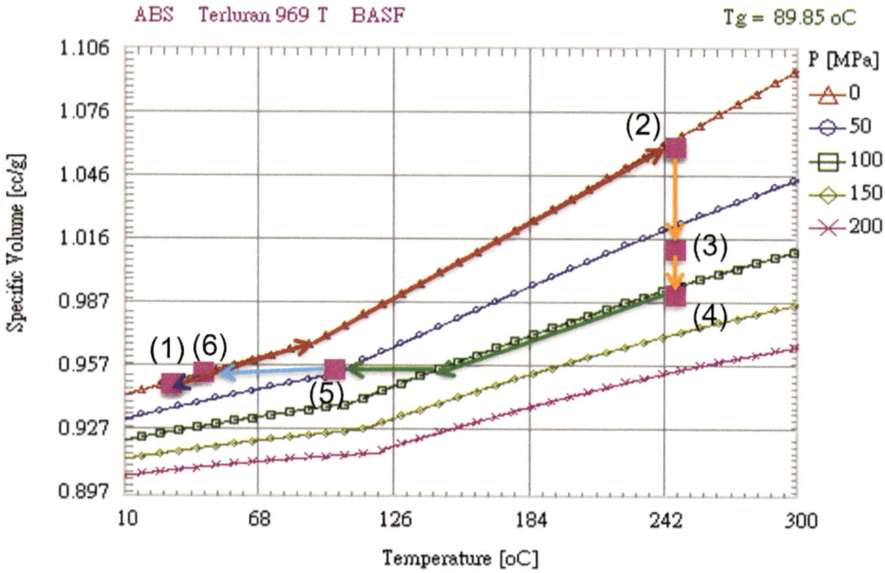

Figure 4.20 ABS PVT state variations during injection molding cycle

Except for the overall average difference in specific volume, the density distribution inside the part should also be considered in the molding process. Different areas of the part can have different PVT state paths as shown in Figure 4.21 for the two PVT variation paths of the gate (G) and filling end (E). The area close to the gate has a better packing and thus shows lower shrinkage, and because the pressure cannot be delivered to the filling end easily, the compensation effect is poorer. These differences can all be causes for warpage.

In order to reduce the impact exerted on the product size by product shrinkage and warpage at different positions, the results for the change in specific volume of each position in the product should be as consistent as possible. Take, for example, a two-cavity-mold car interior product whose dimensions are 1305 mm × 720 mm × 193 mm. Figure 4.22 shows the positions of the sensor nodes, located at the product ends, for observing PVT changes, which are used to reduce the difference in product size between the product ends.

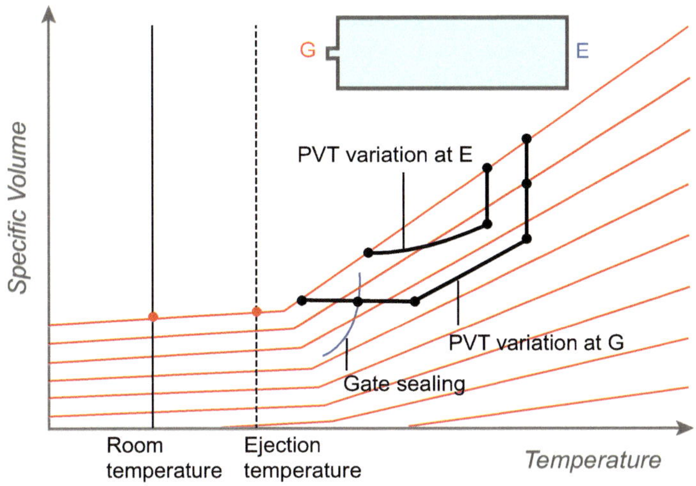

Figure 4.21 PVT states differ with position inside the part

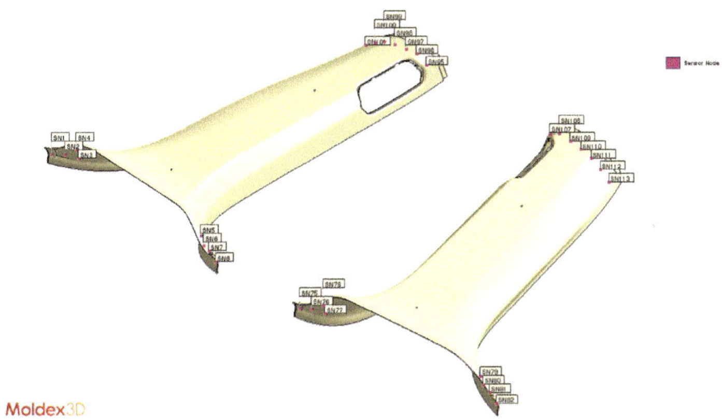

Figure 4.22 Some positions on the part are chosen for the purpose of observing PVT changes

The PVT history curves in Figure 4.23 show that although the PVT states at different positions of the part are not consistent in the beginning, all of them eventually reached a similar specific volume state at the end of injection molding by applying the CAE analysis methods disclosed in patents [1–5].

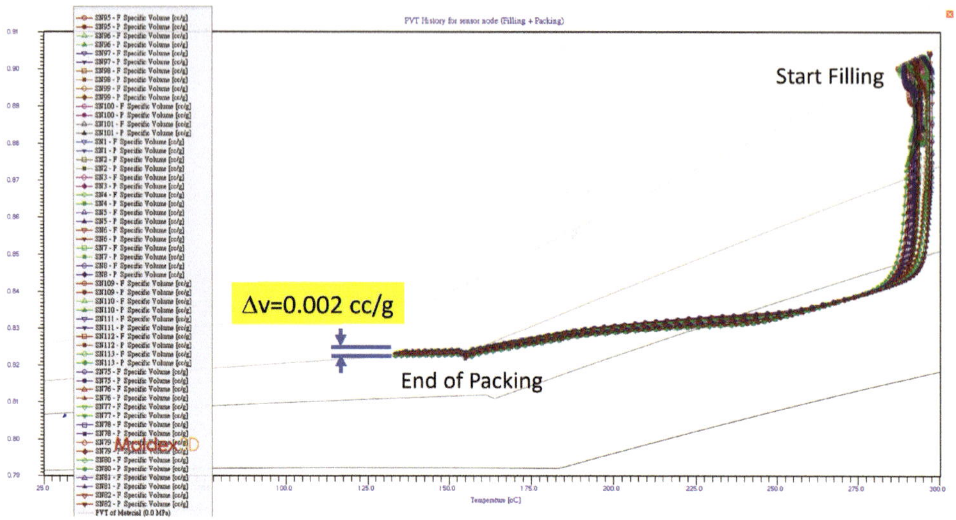

Figure 4.23 PVT history curves for different positions on the part

Therefore, the purpose of this chapter is to understand the controllable molding conditions and the parameters adjustable in each stage, as shown in Figure 4.24. These parameters include screw revolutions per minute (rpm), heater temperature control, back pressure, metering, injection speed and screw stroke, packing pressure, packing time, cooling temperature, and cooling time, etc. What are the meanings of all these parameters and how do they link to simulation analysis? One should understand the effects of the molding parameters first and then adjust these parameters properly.

Among the three major factors of process conditions, part and mold design, and plastic material properties, the process conditions are the most complicated. As mentioned previously, molding can be divided into several stages with different molding conditions to be configured respectively, as shown in Figure 4.25. Because the molding conditions are also interconnected, the effects on a part or plastic of changing these conditions have to be known first, and then the reasons for these changes as well as the consequences therefrom have to be understood.

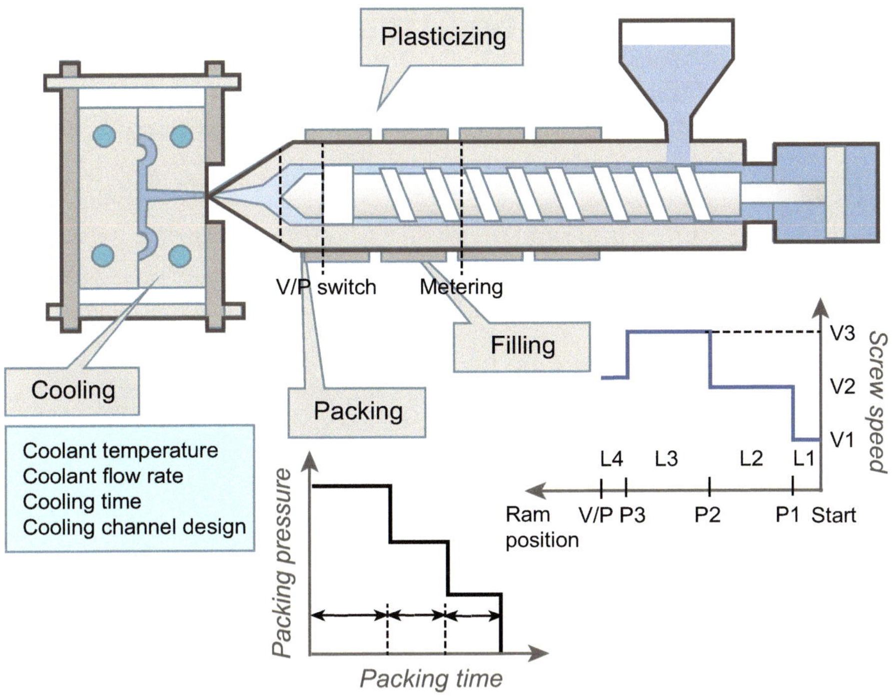

Figure 4.24 Principles of process conditions

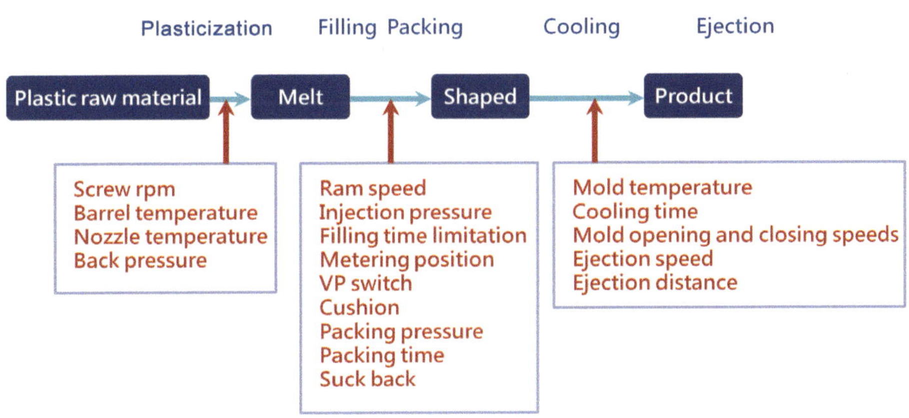

Figure 4.25 Control parameters

■ 4.2 Plasticizing Conditions

4.2.1 Nozzle Temperature and Cylinder Temperatures

Sufficient heat is required to turn solid plastic pellets into a melt state, which makes the temperature settings of the nozzle and the barrel the key points of this step. For example, the temperature can be controlled in three sections for the barrel and in one section for the nozzle with the Nissei injection machine (Figure 4.26).

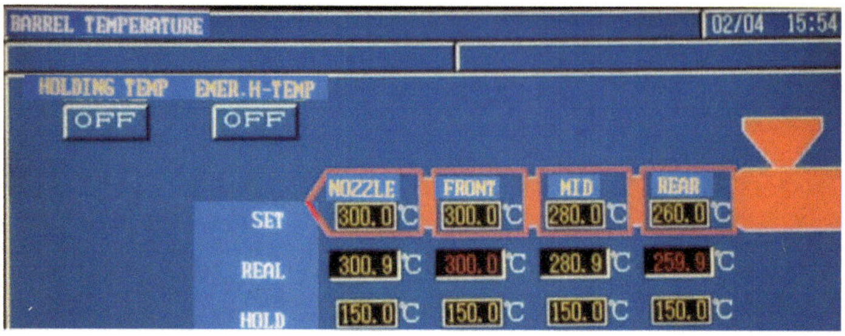

Figure 4.26 Setting interface of nozzle temperature and cylinder temperatures

In general, the processing temperature settings depend on the melt viscosity. As shown in Figure 4.27, because the flowability of high-viscosity materials is poorer, the temperature change should be smoother in this case to avoid nonuniformity of the plastic material caused by large state differences between the screw pitches. In contrast, a larger change in temperature is acceptable for low-viscosity materials. As for the temperature of the nozzle, it should be determined according to site conditions if the melt is dripping.

Figure 4.28 shows a few temperature settings for plastics of different characteristics. Normally, for special materials or plastics that are more thermally sensitive, special attention should be paid to the change of plastic characteristics caused by temperature.

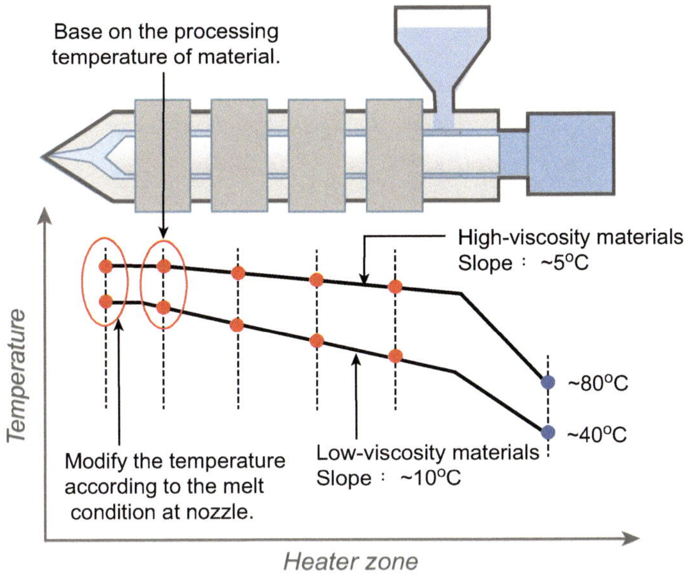

Figure 4.27 Recommended temperature settings in plasticizing system

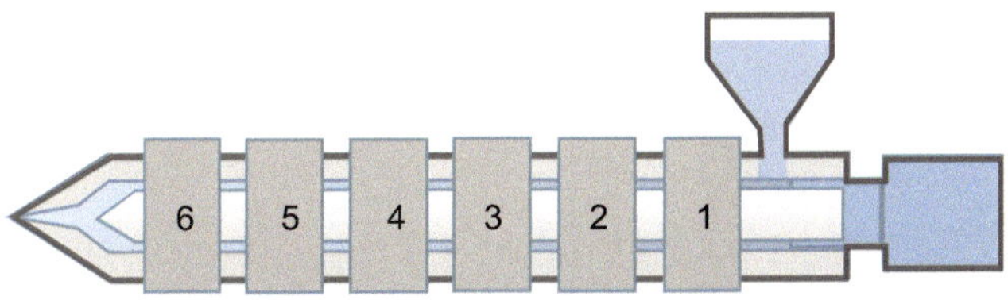

Heater Zone	6	5	4	3	2	1	
Temperature guideline(˚C)	280	280	280	280	280	280	Horizontal
	280	275	270	260	250	240	Ascending
	280	280	285	290	295	300	Descending
	270	280	270	260	250	240	Descending-Ascending

Figure 4.28 Temperature settings for different plastics

4.2.2 Back Pressure, Screw rpm, Suck Back, and Metering Stroke

In addition to the barrel temperature controlled by heaters, plasticization quality is also related to the four parameters back pressure, screw rpm, suck back, and metering. As shown in Figure 4.29, the screw rpm setting (VS) of the Nissei injection machine is set at 60% of the specification of 350 rpm. These four parameters are described as follows:

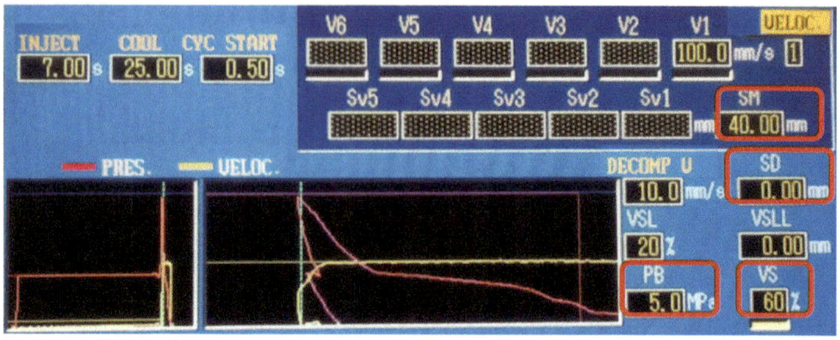

Figure 4.29 Setting interface of back pressure (PB), screw rpm (VS), suck back (SD), and metering (SM)

1. Back Pressure

If the back pressure is too low, metering would be inaccurate and more air will be brought in; however, if it goes too high, shear heating will become significant causing more loading on the screw rotation. As shown in Figure 4.30, a larger back pressure leads to a smaller movable distance of the plastic in the screw per time unit, which can be interpreted as an increase in contact time for the plastics with the screw, so that the backward speed becomes slower and the metering time increases.

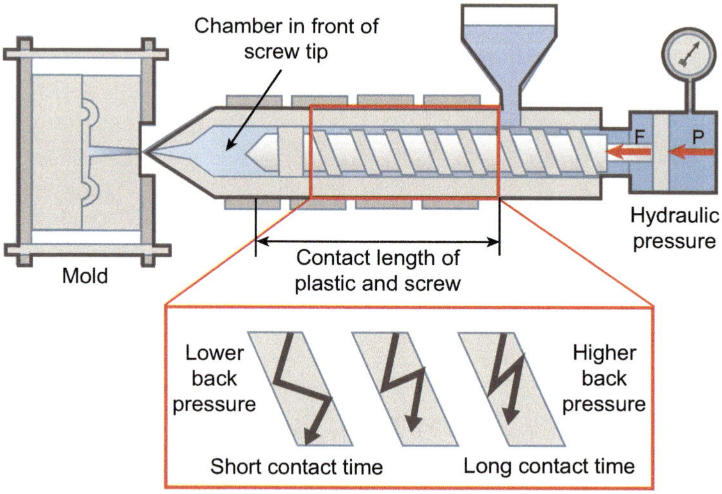

Figure 4.30 Back pressure effects on the contact time between plastic and screw

2. Screw rpm

Normally, the maximum surface speed (tangent speed of screw outer diameter), which is the product of screw revolutions per minute (rpm) and screw perimeter, is set at 5 to 15 m/min. For a single plastic, the rpm value should be adjusted along with the screw size, and the metering stage is a bit shorter than the cooling time, which can be used for rpm adjustment. The screw rpm should be set in the order of high, middle, and low speed along with the feed stroke, as shown in Figure 4.31. A slow speed leads to a better compounding effect and a lower shear rate, and can shorten the residence time of the plastic material in the barrel. However, the metering time becomes longer as the rpm is reduced, which has a negative effect on cycle shortening.

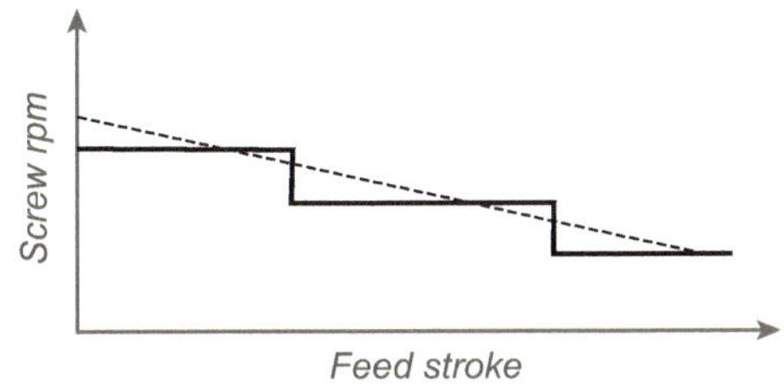

Figure 4.31 Settings of screw rpm

3. Suck Back

Suck back can be divided into forward suck back and backward suck back, which is defined by whether it is applied before or after the stroke of the screw for plasticization. The forward suck back is used for hot runner molds and the backward suck back mainly for redistribution of plasticized material in front of the screw to achieve a uniform density and improve a poor mixing situation, as shown in Figure 4.32. Usually the backward suck back is set to 3 to 7 mm at a speed of 10 mm/s, and is adjusted if dripping at the nozzle occurs or air traces (air pulled in by too much suck back) are seen in the finished part.

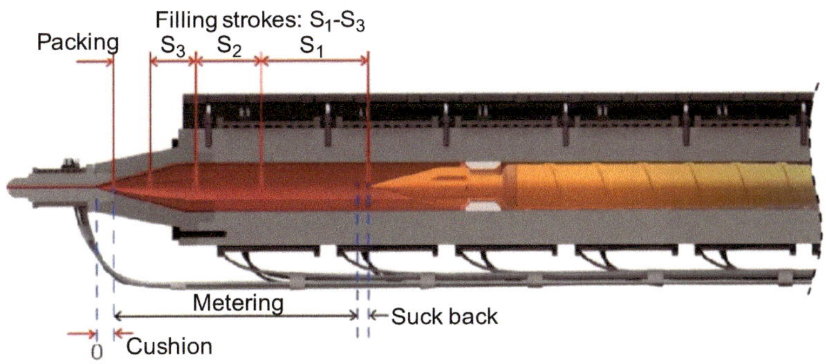

Figure 4.32 Suck back setting to reduce the melt pressure in front of screw

4. Metering Stroke

Normally the metering stroke is set to 2 to 3 times the screw diameter. If the metering stroke is too short (or the screw diameter is too large), the molding precision is harder to control. If the stroke is too long, venting becomes difficult as more air is inside the barrel.

A summary of setting guidelines for plasticizing conditions is shown in Figure 4.33. These are related to the plastic's characteristics. Especially for plastics with high viscosity but poor thermal stability, using a slower screw rpm and a bit lower back pressure is better. Back pressure should be adjusted to the conditions of the finished part. The goal of back pressure is to maintain a uniform feed, whereas the highest priority of suck back is to avoid air entry.

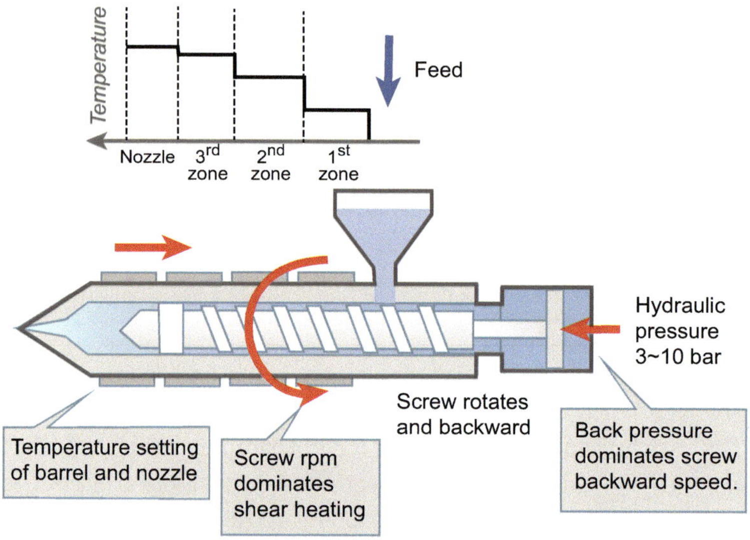

Figure 4.33 Principles of plasticizing conditions

As shown in Figure 4.34, melted and solid layers exist as the plastic is being melted in the screw. In the process of feed, compression, compounding, and metering, plastics are being melted constantly. The simulation tool ScrewPlus can provide the melt states and temperatures of plastics in each stage in the screw.

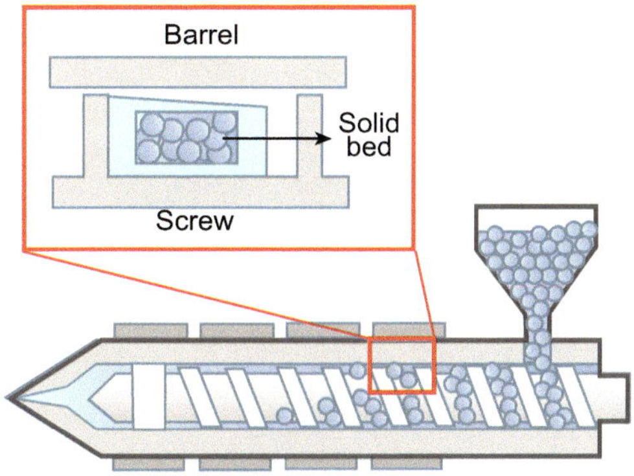

Figure 4.34 Melt states inside the screw at plasticizing stage

Taking an acrylonitrile butadiene styrene copolymer (ABS) as example, Figure 4.35(a) shows that by entering the screw melt temperature and other relevant settings, a nozzle temperature distribution analysis can be obtained, as well as the plastics melt index in the screw and at the exit. As shown in Figure 4.35(b), the plastic material is in more of a solid state as the solid bed ratio gets higher (red), whereas a lower value (blue) means it is melted more completely. Figure 4.35(c) shows the temperature distribution of the plastic material inside the screw when the screw location is at 13.47D, where the y-axis denotes the temperature and the x-axis the depth of the screw (0 means close to screw center and 1 means at the outmost layer). The highest melt temperature can be seen around the middle of depth. Furthermore, ScrewPlus also provides the result of shear stress and pressure distribution, which can be imported into Moldex3D as initial conditions of plastics for molding analysis, and thus a more realistic molding result can be obtained.

(a)

(b)

(c)

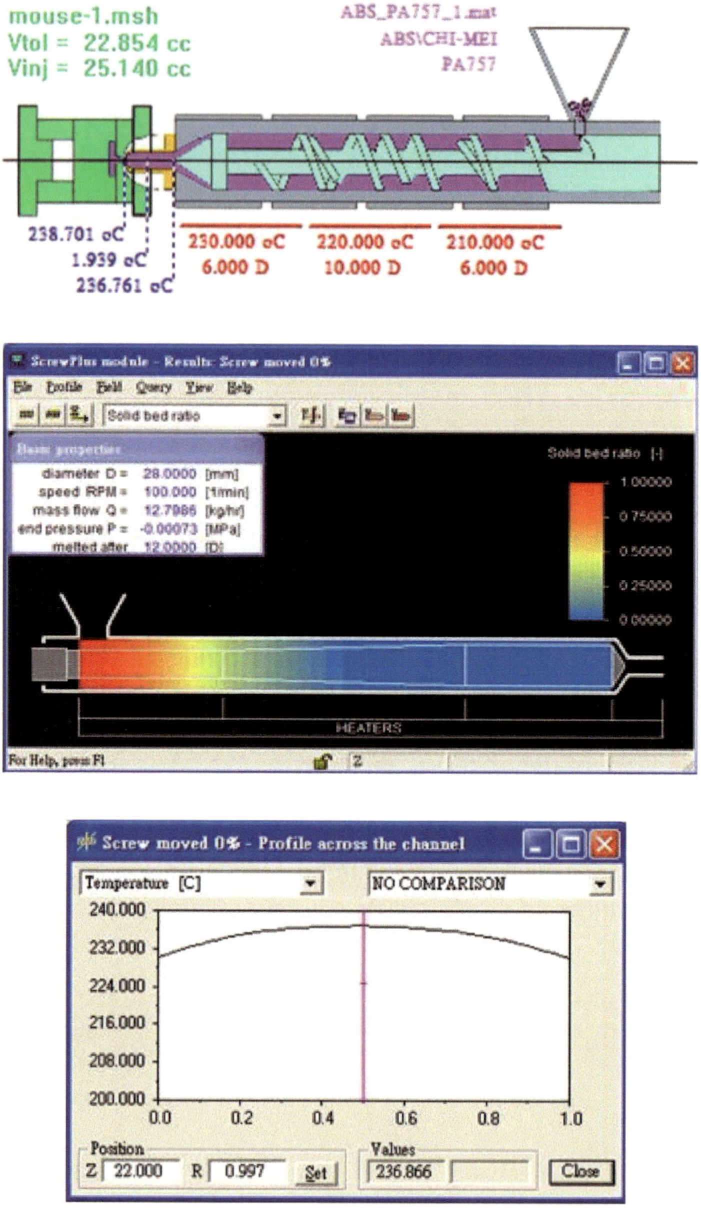

Figure 4.35 ScrewPlus simulation results of ABS: (a) parameter settings and melt temperature distribution at nozzle, (b) solid bed ration at the instant of melt exiting the nozzle, and (c) melt temperature distribution at the radial direction of screw

■ 4.3 Filling Conditions

4.3.1 Filling Time versus Injection Velocity

The curves in Figure 4.36 show the input values of molding conditions and the actual process results of the machine. The actual injection speed is not identical to the setting values, because the setting is an ideal step function and the actual injection speed involves a time delay effect, which results in a delay in both the acceleration from zero to V1 and V1 to V2.

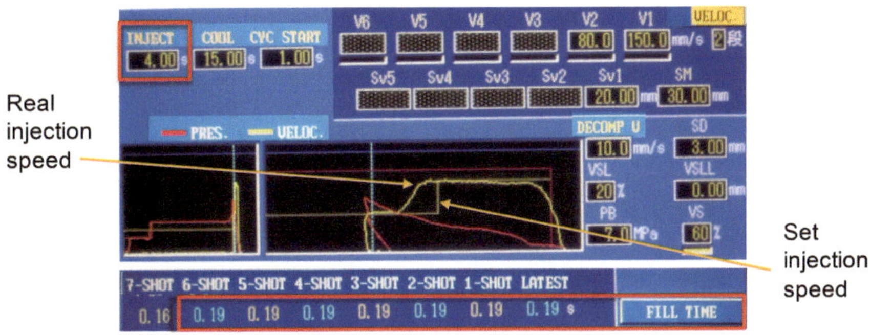

Figure 4.36 Setting and practical values of injection velocity

To make simulation results comply with actual site conditions as much as possible, the response time of the machine should be considered in molding simulation as well. As the blue curve in Figure 4.37 shows, the response time of the machine in the filling stage affects the time required for the injection speed to reach the set value (red curve).

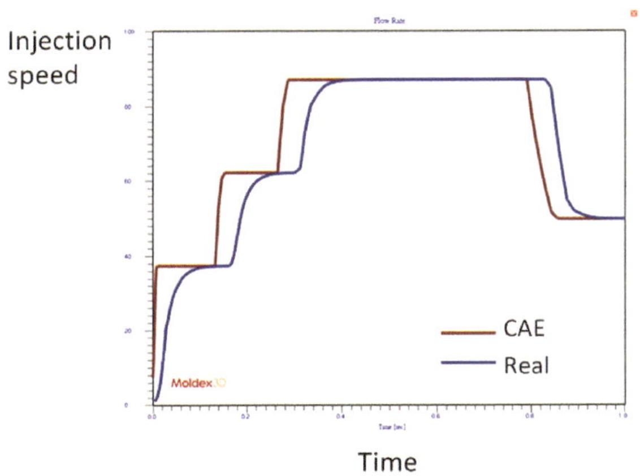

Figure 4.37 Effect of velocity response time

It is generally believed that an electric injection machine has a faster response speed than a hydraulic machine. Ideally, the machine will reach the set values for speed and packing pressure, as shown in Figure 4.38.

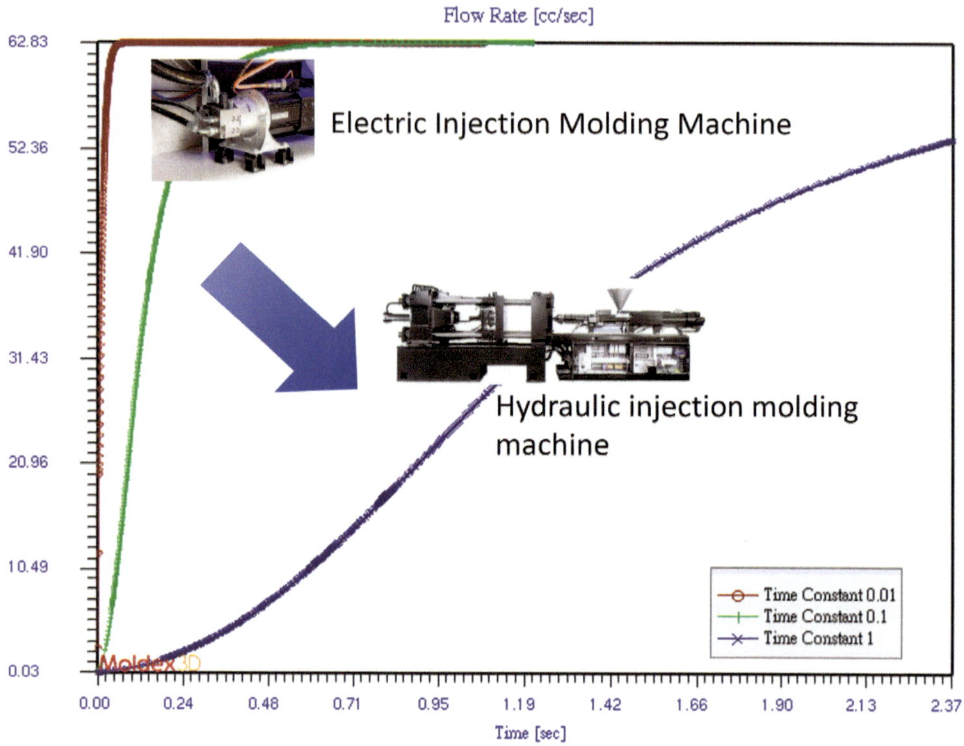

Figure 4.38 Difference in the response speeds of electric and hydraulic injection machines

Two electric injection machines are shown in Figure 4.39, in which the red line is the set speed and the yellow line is the real speed. In Machine #1, a short delay in the speed curve occurs at the beginning, then the speed starts to increase, and decelerates when it almost reaches the set value. In Machine #2, a short delay occurs at the beginning and then the speed starts to increase; however, the speed does not immediately decrease when the set value is reached, but rather slowly falls back after exceeding the set value. Judging from the observed speed responses, the two injection machines have different speed-change patterns. Comparison of the time taken to reach the set value shows that it takes around 0.2 s for Machine #1 and around 0.7 s for Machine #2. The two types of electric injection machines have different response modes.

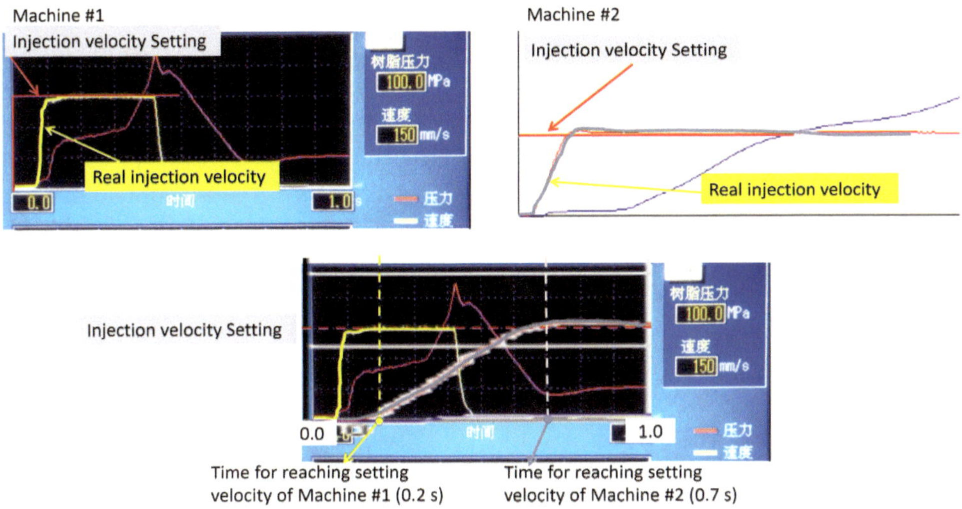

Figure 4.39 Speed response comparison of different brands of electric injection machines

Two hydraulic injection machines are shown in Figure 4.40, in which the set speed is the red line and the real speed is the yellow line. In Machine #1, it can be seen that the real speed fails to reach the set value and is lower. In Machine #2, the real speed is higher than the set value. Although the real speed does not match the set speed, the consistent curves of each shot represent a stable process for the machine and it is able to make good products.

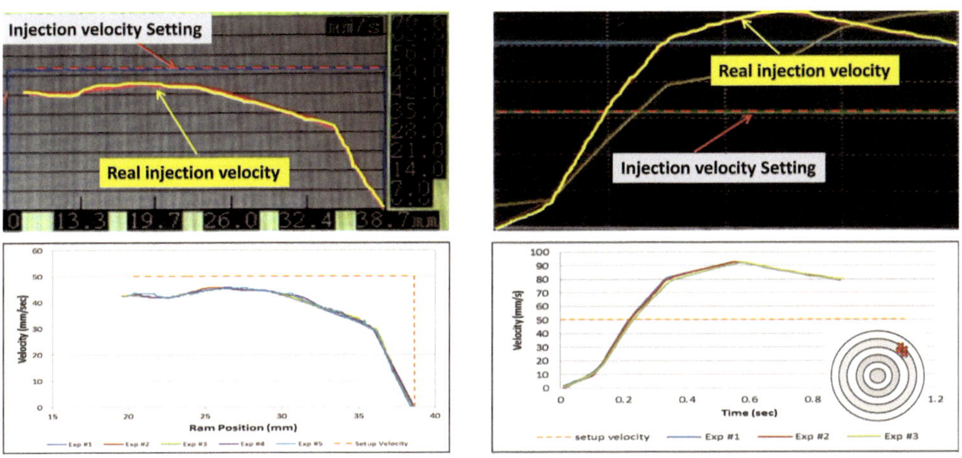

Figure 4.40 Speed response comparison of different brands of hydraulic injection machines

Another screw movement relating to the difference in filling time is suck back, which is often seen before the screw starts to fill, and thus the start position of the screw is no longer the start point for screw metering. Figure 4.41 shows the injection flow rate vs. ram position and time, respectively, where the red curve indicates an ideal condition without suck back setting in CAE, whereas the blue curve shows the process with suck back setting. It can be seen that the curve without suck back setting is shorter, and so is the time of the filling stage, which shows one of the differences between the practical situation and simulation.

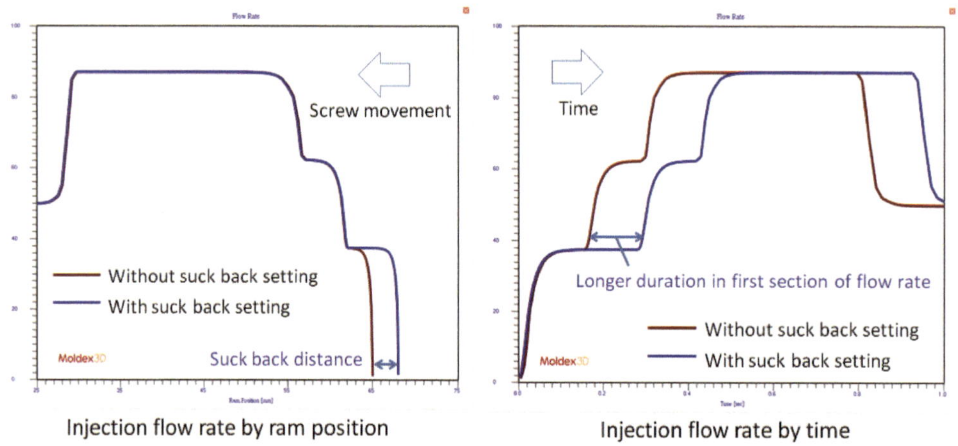

Figure 4.41 CAE results comparison of setting with and without suck back

Figure 4.42 illustrates an injection stroke with multi-sectional speed settings. A slow speed (V1) is used from the start to the gate, whereas most of the volume of the part is filled up at high speeds (V2 and V3) after passing the gate. Lowering the speed (from V3 to V2) before velocity/pressure (VP) switching avoids excessive injection pressure and mold-clamping force in the final filling stage due to the forwarding inertia of the screw.

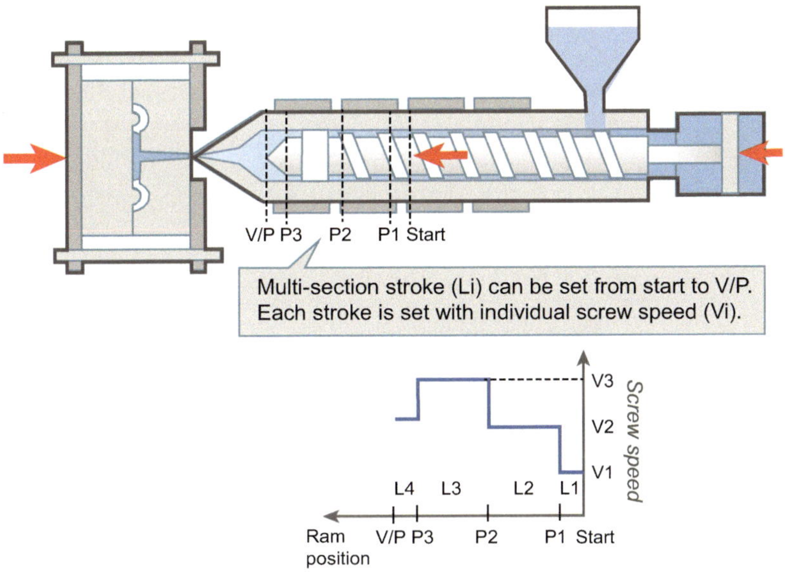

Figure 4.42 Principles of injection velocity profile

4.3.2 Injection Pressure

The injection pressure denotes the pressure applied to the melt from the injection screw, as shown by P_m in Figure 4.43. For a hydraulic pressure injection machine, usually the gauge pressure reading P_h is recorded. The injection pressure of the melt is proportional to the square ratio of the hydraulic cylinder diameter to the screw diameter, and such a ratio is defined as the intensification ratio, which is used to calculate the injection pressure in front of the screw. The ratio of common small- to medium-sized machines is about 10, while the maximum output pressure of a hydraulic cylinder of general specifications is 140 kg/cm^2 (2000 psi).

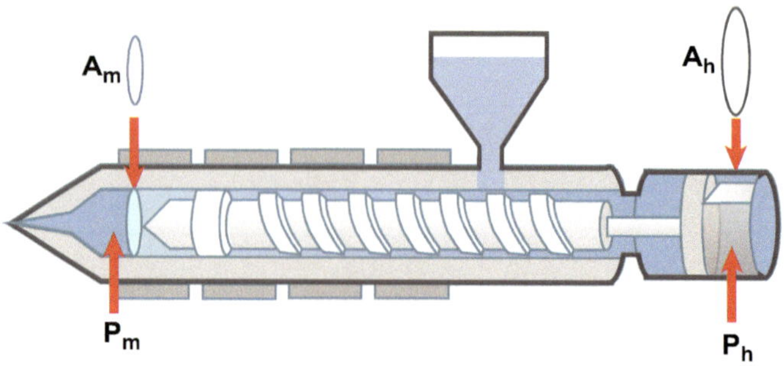

$$P_m \bullet A_m = P_h \bullet A_h$$

$$P_m = P_h \quad A_h / A_m = \text{intensification ratio}$$

Melt Pressure (psi)	Intensification Ratio	Hydraulic Pressure(psi)
5,000	(7:1)	714
6,000	(8:1)	750
7,000	(9:1)	778
8,000	(10:1)	800
9,000	(11:1)	818
10,000	(11:1)	909
10,000	(7:1)	1,429

Figure 4.43 Intensification ratio for hydraulic driving system

With different screw diameters come different intensification ratios, that is, the larger the screw diameter the lower the pressure becomes. As shown in Figure 4.44, a common machine has three screw specifications: 32 mm, 36 mm, and 42 mm, whereas the intensification ratios are 12 : 1, 9.5 : 1, and 7 : 1, respectively. From the chart of melt pressure (y-axis) to hydraulic pressure (x-axis), it can be seen that by application of the same hydraulic pressure, the injection speed of the screw with 42 mm in diameter is less than that of a 32 mm screw system.

Since injection pressure affects melt conditions, viscosity, and other properties, correct machine process data should be obtained for the comparison with simulation results, that is, one should be able to understand the data and charts generated by the machine. Taking Figure 4.45 as an example, the yellow line is the injection speed and the red line is the injection pressure. As the injection speed is subjected to the effect of machine response time, it cannot reach the speed setting value rapidly, which means that the yellow line has a segment of ramping from the start and generates extra filling time. A difference between simulations and real situations thus exists. Furthermore, the peak value of actual filling pressure (red line) is still far away from the limit of injection pressure setting, so that by properly adjusting the injection speed a better filling effect can be achieved.

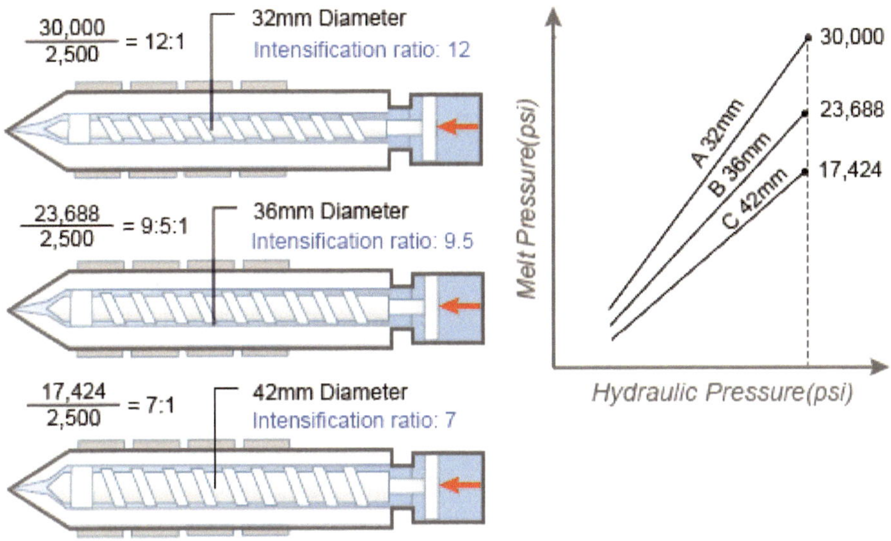

Figure 4.44 Maximum injection pressure for different screw diameters

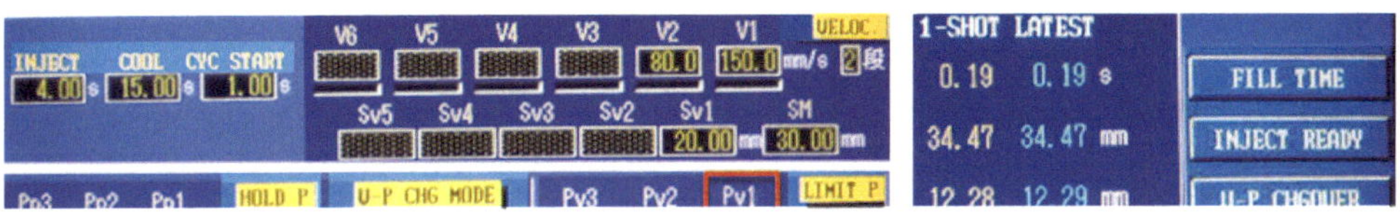

Figure 4.45 Setting and practical values of injection pressure

The filling time has a critical influence on the molding quality. The relationship between filling time and injection pressure is shown in Figure 4.46. The advantages of a short filling time are a short cycle time and small temperature differences among the parts of the finished product (considering the cooling time difference between each area, shorter time leads to smaller difference). The disadvantage is that the injection pressure surges up rapidly due to the high injection speed, and also a high temperature would be generated by the friction in the high-shearing zone that causes greater temperature differences between each area. A longer filling time has the advantage that molecular chains and fibers are less stretched and sheared with the lower shear velocity, and the internal stress induced is smaller. The disadvantage is the longer molding cycle. A too-low injection speed also leads to a higher pressure or even short shot as the viscosity increases due to material cooling, which causes lower flowability.

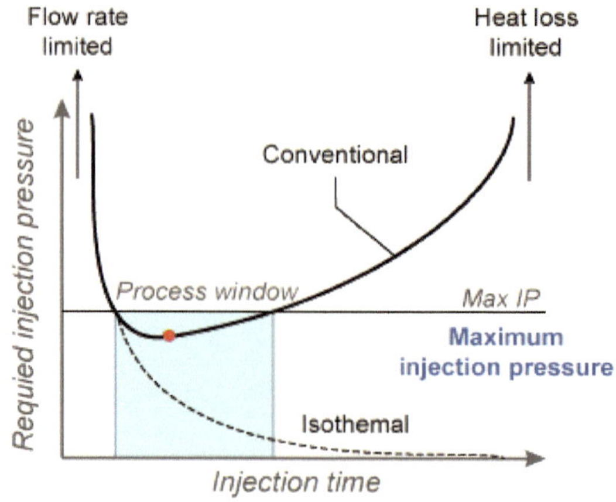

Figure 4.46 U-curve theory

4.3.3 VP Switch

For the velocity/pressure (VP) switch setting in simulation, the program can switch to packing once the filling is finished; however, in practical situations the screw movement is affected by inertia and cannot follow exactly the setting for switching. As shown in Figure 4.47, from the chart and machine records it can be seen that after the screw passes the VP switch value set at 12.30 mm, as indicated by (3), the injection pressure (red line) still keeps increasing until the screw reaches 11.53 mm, as indicated by (4). Due to the inertia of the screw moving forward in the filling stage, filling stops at (4) instead of at the setting value of (3). Therefore, the actual stop position of filling would be different from the setting of the VP switch. Then the packing stage controlled by pressure kicks in until the pressure drops to 80 MPa as indicated by (5).

However, (4) and (3) appear to overlap with each other in simulation, and the time between (4) and (5) is much shorter, which constitutes a difference between simulation and practical situations. Practically speaking, the precise choice of the VP switch is very important and will be introduced in detail in Section 6.1.4.

The relationship between VP switch time and mold pressure is shown in Figure 4.48. When the VP switch is too early, the injection pressure drops suddenly and then increases as the filling percentage goes further up. In case of a proper VP switch time point, the injection pressure curve transforms smoothly. If the VP switch is too late and the screw still moves forward with high speed while the cavity is completely filled, a sudden squeeze cause a surge of internal pressure in the mold cavity.

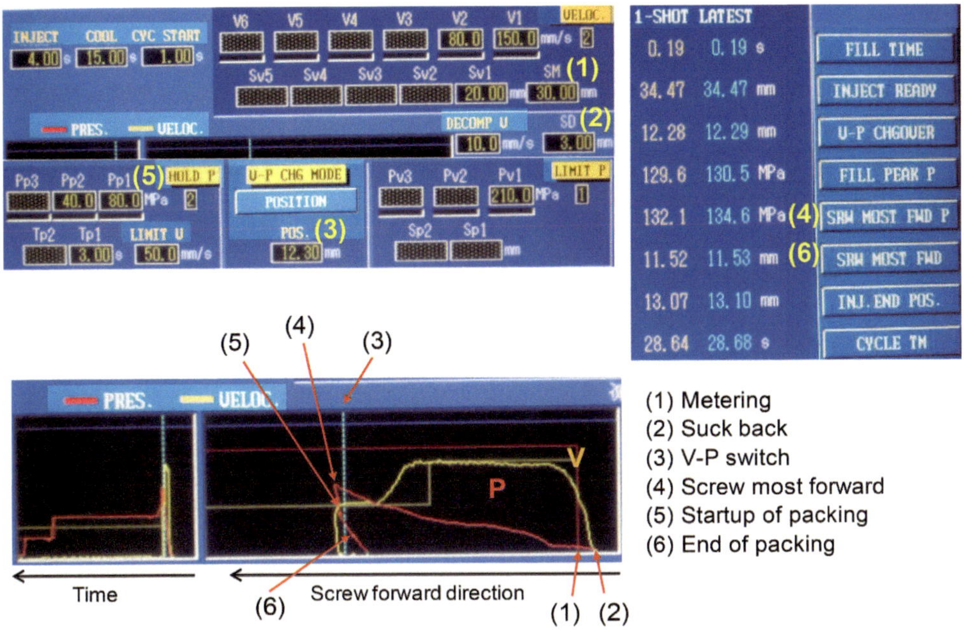

Figure 4.47 Stroke of filling and packing stages

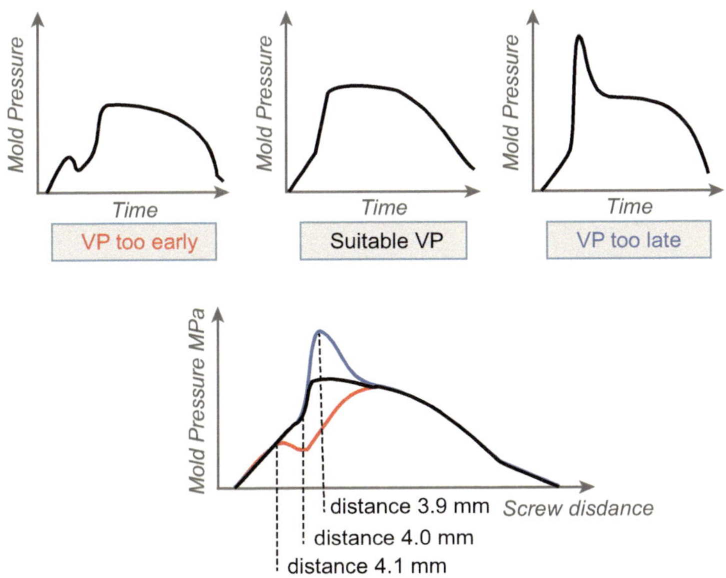

Figure 4.48 VP switch and pressure curve

The ideal VP switch timing should be within a stable speed control stage and be adjusted according to the filling percentage as well. The definition of filling percentage is the ratio of screw stroke in the filling stage to that in the combination of filling and packing stages, as shown in Equation 4.1. Parts with different thickness have different filling percentages. For thick-walled parts, a lower filling percentage (%) is induced by the compressibility of the plastics.

$$\text{Filling}\ \% = \text{Filling stroke} / \text{Total}(\text{filling} + \text{packing})\ \text{injection stroke} \times 100\% \tag{4.1}$$

4.4 Packing Conditions

Packing Time and Packing Pressure

The purpose of packing is to compensate the melt that could not fill up the mold cavity completely in the filling stage. The first packing stage is set to meet the specifications regarding the surface quality of the part. A general rule is not to see any kind of surface defects or sink marks. The second packing stage of lower pressure is to prevent the melt inside the cavity from flowing back, and also to release stress. The packing pressure profile is set as in Figure 4.49.

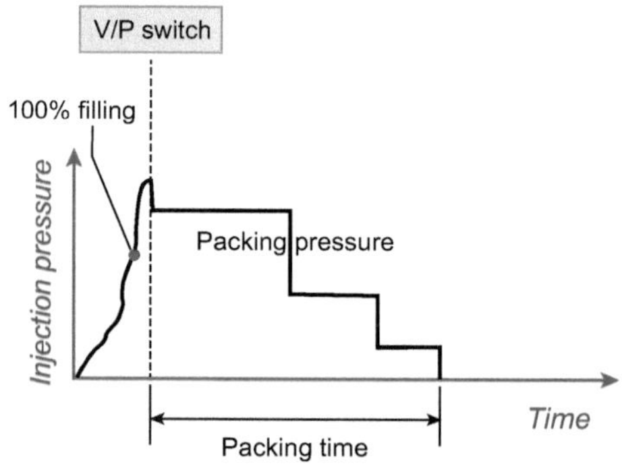

Figure 4.49 Injection pressure at filling and packing stages

Similar to injection speed, the pressure switch at each stage in packing is also affected by the machine response time. As shown in Figure 4.50, a time delay is usually seen (blue line) when the pressure changes; the default condition in simulation is an ideal direct switch (red line).

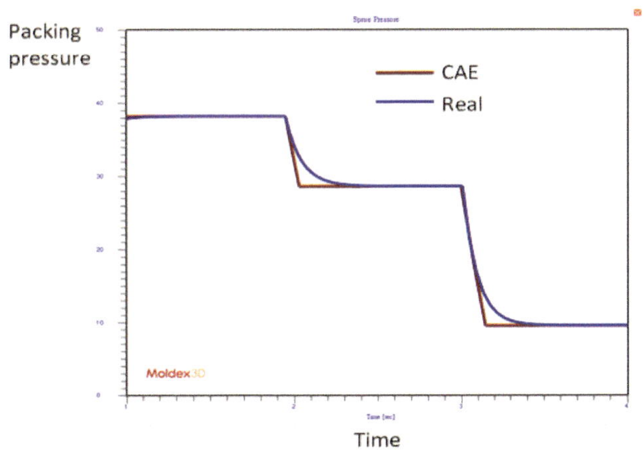

Figure 4.50 Effect of pressure response time

The setting of the packing time is related to the solidification time of the gate. Practically, the weighting method is used to derive the solidification time of the gate under certain specific conditions. As shown in Figure 4.51, an increase of the packing time does not correspondingly increase the part weight.

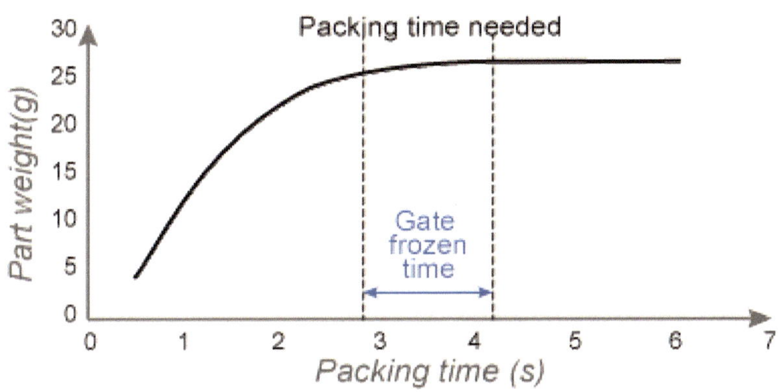

Figure 4.51 The relation between part weight and packing time

4.5 Cooling Conditions

4.5.1 Cooling Time

In an injection molding machine, only the cooling time can be configured for the cooling stage, as shown in Figure 4.52. Control of the mold temperature is done indirectly by an external mold temperature controller with a certain coolant temperature (see Figure 4.53).

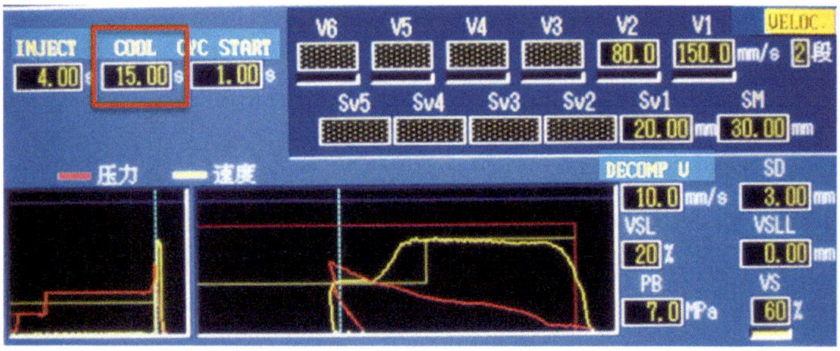

Figure 4.52 Setting interface of cooling time

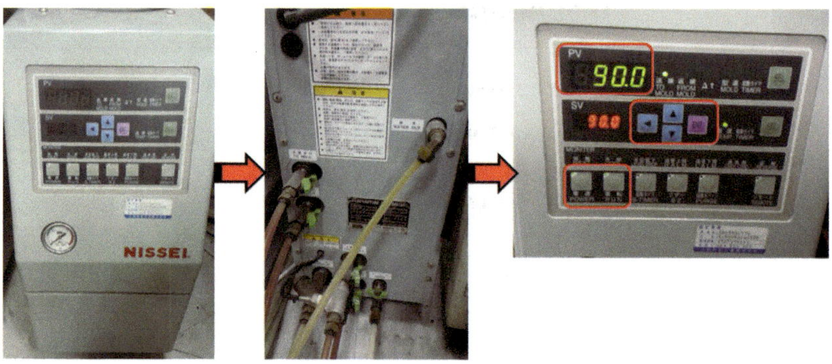

Figure 4.53 Mold temperature controller

A practical methodology for finding the best cooling time is as follows:

1. Injection molding of at least three molded samples under different cooling times.
2. Measurement and data collection of the critical dimensions for each molded sample.
3. Plot of the dimensions vs. cooling time.

4. Data analysis to find out which dimensional value is affected more by cooling time.
5. Decision of a proper cooling time parameter according to the result of target value dimensional compliance.
6. Injection molding of 30 samples with the decided cooling time and summary of stability and satisfaction of product quality, from which the injection manufacturability with this cooling time can be evaluated.

4.5.2 Coolant Flow Rate

To take away the heat brought in by the plastic within the cooling time for each molding time, it should be estimated whether the coolant flow rate is sufficient to guarantee a certain amount of heat exchange within a fixed time.

The type of coolant also affects the cooling effect. Theoretically, a coolant with larger specific heat (e.g., water) takes away more heat. The cooling effect can also be strengthened by a higher coolant flow rate. If the temperature difference between the inlet and the outlet of the coolant is larger than 3 °C, the temperature distribution inside the mold is not uniform, leading to warpage. This sometimes requires other conformal cooling channels to improve the cooling uniformity.

4.5.3 Mold Temperature

The third important control condition of the cooling stage is the mold temperature. However, the mold temperature is not controlled directly, but rather indirectly by a mold temperature controller with coolant flowing through the mold. As shown in Figure 4.53, the actual operation steps are as follows:

1. Check the installation of all cooling channels before operating the mold temperature controller.
2. Turn on the switch for the coolant. The controller shown has two switches: the upper one is for the incoming and the lower one for the outgoing coolant. The cooling channels of the cooling system have to be kept unblocked and without leakage. All control valves have to function normally.
3. Turn on the power to start operation, and set the required mold temperature.

Control of the mold temperature can only be achieved on site through the temperature setting on the mold temperature controller. Other auxiliary equipment such as thermographs can be used to check if the mold temperature reaches the setting value. Figure 4.54(a) indicates that the highest temperature area of the mold is at the part location, whereas Figure 4.54(b) shows that the highest temperature of the part concentrates at the sprue and fan gate. The basic concept of mold temperature control is to manage uniform heating and cooling of the system, which is related to cooling channel design and coolant flow rate.

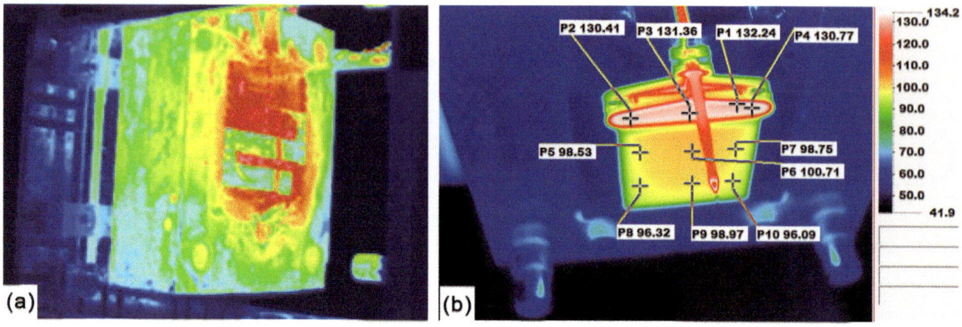

Figure 4.54 Thermal imagings of (a) the mold, and (b) sprue and part

After the melt enters the mold cavity, runners, gates, and the geometric design of the part confine the flow space of plastics, and through a mold cooling system the part is cooled and taken out from the mold. Operators can enter multiple molding conditions through the machine panel to control the machine actions and to finish the process that turns plastic materials into parts in a short time. However, as for how to adjust these parameters, learning in theory and in practical operation are equally important. We hope that through the summary of this book, readers will find a convenient way of understanding important factors controlling the molding process.

■ 4.6 Connecting Smart Design to Smart Manufacturing

In the era of Industry 4.0, industry hopes to extend data beyond the design side to the manufacturing side. Traditional CAE analysis considers product design, mold design, and material characteristics, but also assumes ideal operations on the part of the machine characteristics. CAE fails to take account of actual machine situations and sometimes analytical errors occur. This section will explain step by step how to take the machine information at the manufacturing end into account in CAE analysis, how to connect the manufacturing and the setting ends, and illustrate the complete process in a case study.

4.6.1 Machine Characterization

To understand the influence of the machine's dynamic response, an experimental method of the kind shown in Figure 4.55 is used to collect the response data. The various steps are: select the injection machine, mold, and material used in the factory, collect the experimental results at different speed and pressure settings, and then use control theory to identify the machine's dynamic response parameters.

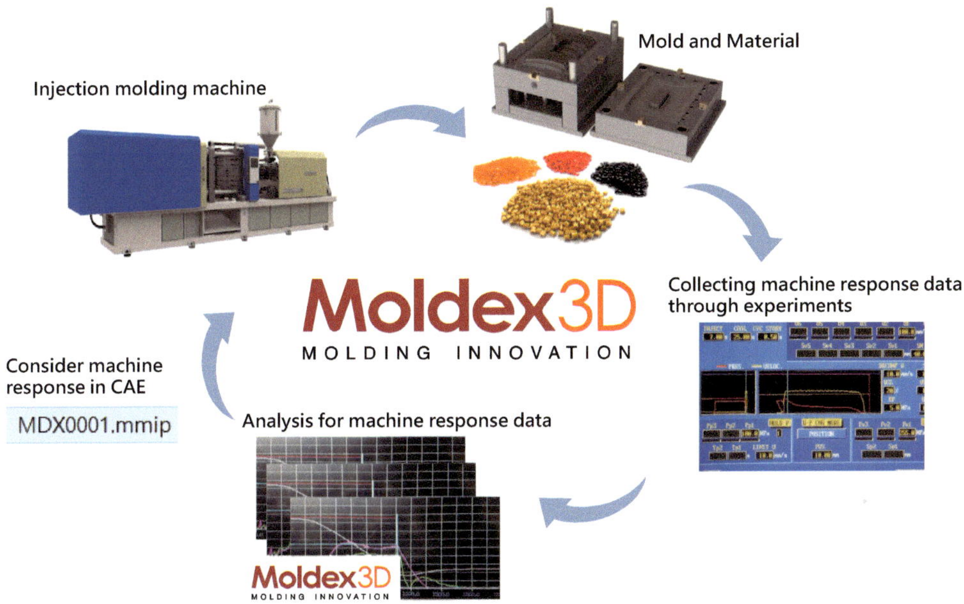

Figure 4.55 The process of machine characterization

Once the file of identified machine characteristics file has been imported into the analysis software, it can be automatically applied to the polymer processing simulation. As the setting conditions are based on the actual machine response, the analysis results are closer to the real situation, as shown in Figure 4.56.

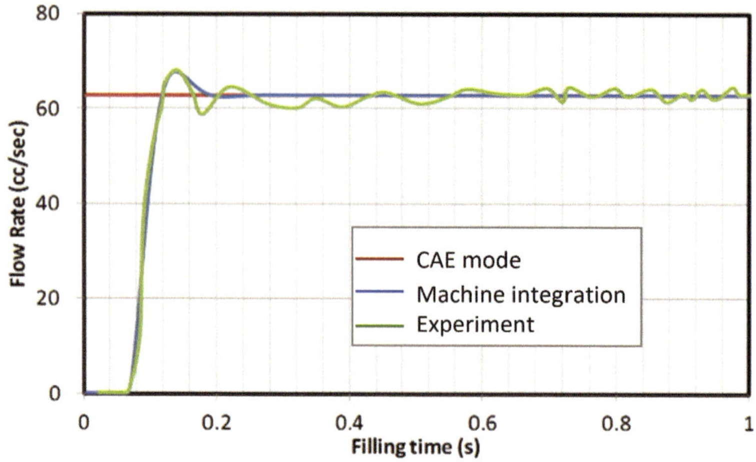

Figure 4.56 Flow rate – as the machine characteristic under consideration

Another actual situation for a hydraulic injection machine is shown in Figure 4.57, in which the real VP switch time is 2.1 seconds and the predicted VP switch time, after consideration is given to the machine characteristics, is 2 seconds. The two results are very close, and the consistency between simulation and practice is also improved. This helps to more closely integrate virtual and real injection molding.

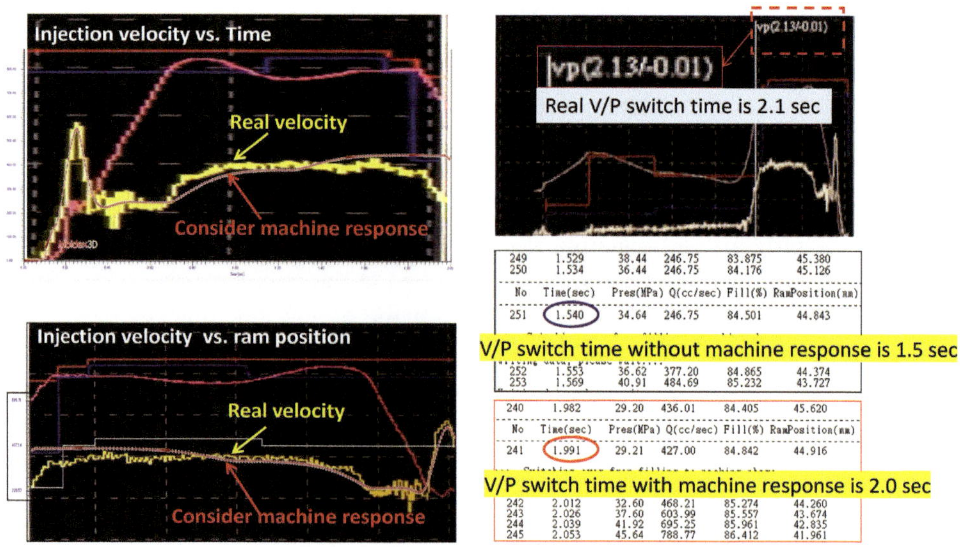

Figure 4.57 Comparison of VP switch time, with and without consideration of machine characteristics

An example of a fan shell model is shown in Figure 4.58, the characteristics of the ABS material are shown in Figure 4.59, and the machine specifications and processing conditions are shown in Figure 4.60. In this case, the influences of ideal speed and the response of the two machines on filling time and injection pressure will be discussed.

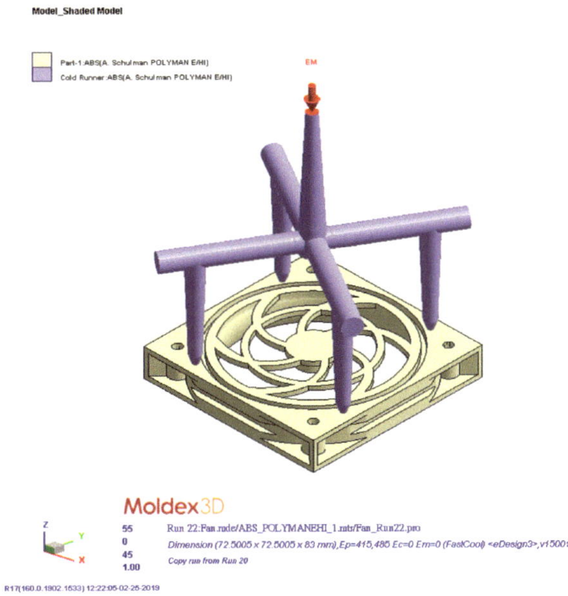

Figure 4.58 Fan shell model

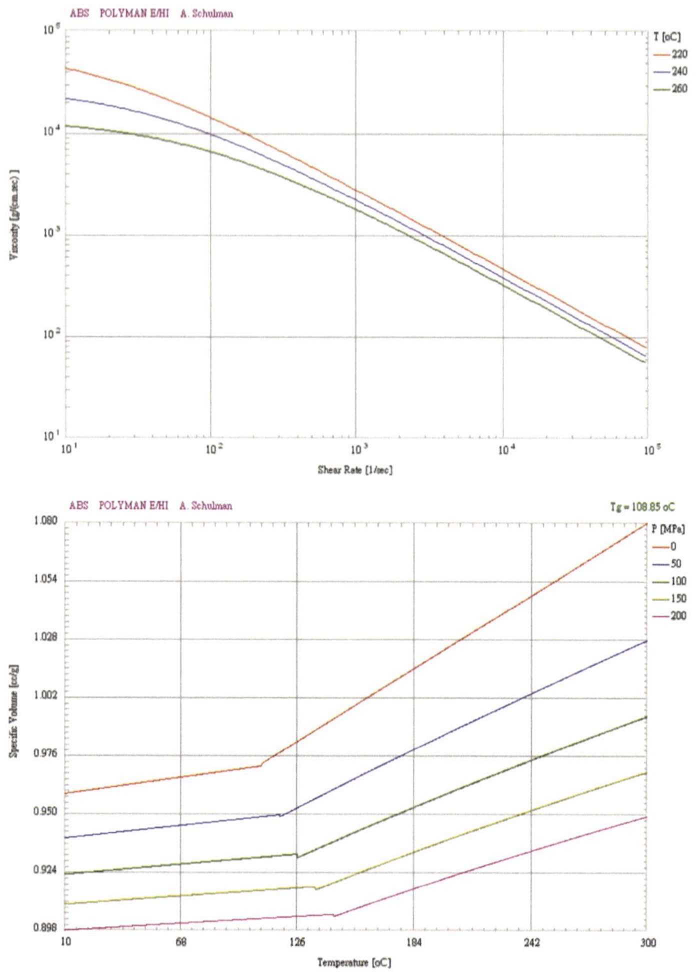

Figure 4.59 Viscosity and PVT of ABS

The ideal speed response curve is indicated by the green line in Figure 4.61. The actual machine responses for two speeds are considered: the actual speed of the blue line in Figure 4.61 is higher than the set speed while that of the red line in Figure 4.61 is lower than the set speed. The results show the delays that occur in the actual machine, i.e. the blue and red lines remain at 0 when the speed is at 0 to 0.1 s. Moreover, the VP switch time of the ideal speed response setting is 0.44 s, that of the actual speed higher than the set speed is 0.54 s, and that of the actual speed lower than the set speed is 0.81 s, due to the delay effect.

Figure 4.60 Process conditions for filling

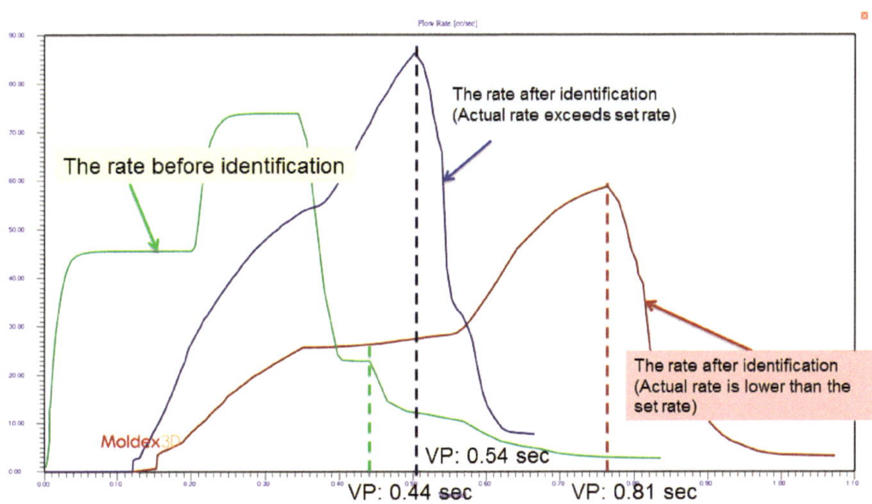

Figure 4.61 Changes in response speed of different machines for the same machine specifications and speed settings

Figure 4.62 shows the changes in filling pressure of different machines with the same settings, where the VP switching pressure is 43.3 MPa under the ideal speed response setting. The VP switching pressure is 55.6 MPa for the actual speed higher than the set speed, while that for the speed lower than the set speed is 45.3 MPa. It can be seen that the result of the entire filling process will be different for different machine responses even under the same settings, which affects the product quality. It is therefore very important in CAE analysis to give consideration to the actual machine response.

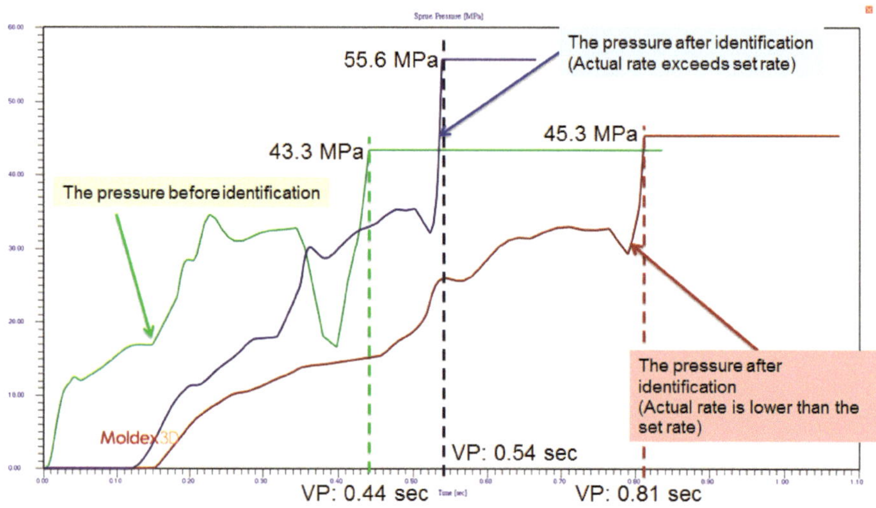

Figure 4.62 Changes in injection pressure with filling time

4.6.2 The CAE Setting Mode in Combination with Injection Machine on Site

Traditionally, the CAE mode for setting the processing conditions is independent of manufacturing, with CAE users and molding engineers separately setting the parameters on their own interfaces. In order to connect the two in series, the traditional approach is to use a trial form to communicate between the two parties. However, misunderstandings often arise when CAE users transfer the CAE conditions onto a trial form for a site, because different machines have different setting units and setting methods. In order to convert CAE correctly, the users must clearly understand the characteristics of each machine. Moldex3D therefore features an integrated injection molding machine control panel, i.e. the Moldex3D Process Wizard has a real injection molding machine setting interface that allows CAE users to set the processing conditions on the CAE side (which they are familiar with). By following simple steps after the analysis, users can easily transfer the CAE interface to the real machine setting interface; this not only eliminates communication gaps, but also automatically transfers the modes from the CAE to the real machines, as shown in Figure 4.63 and Figure 4.64.

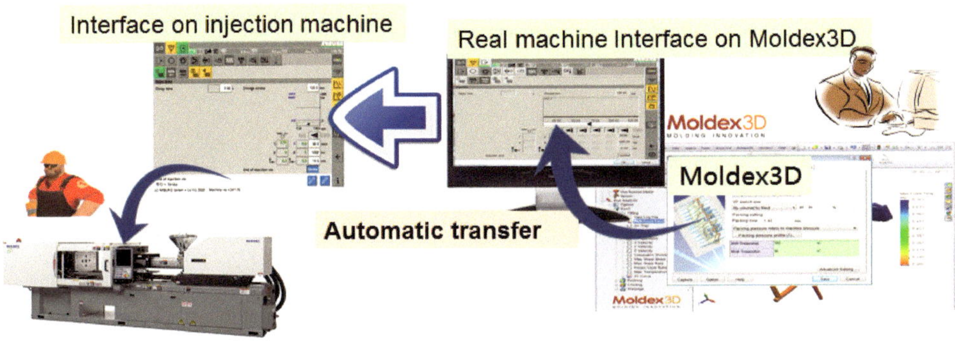

Figure 4.63 Automatic transfer of machine operation interface

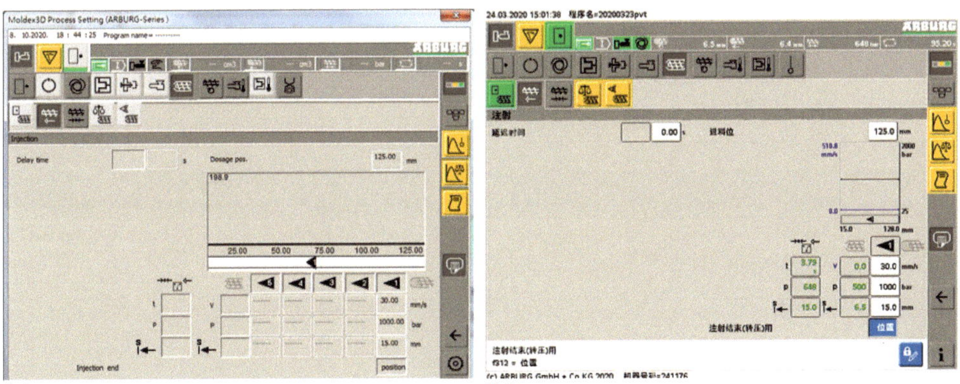

(a) (b)

Figure 4.64 Setting interfaces in (a) CAE and (b) on real machine

By switching between the interfaces to check the difference between the simulation and real machine at any time, users can easily obtain information and improve production efficiency. Moldex3D presently supports more than 20 machine brands and more than 30 controller panels, as shown in Figure 4.65.

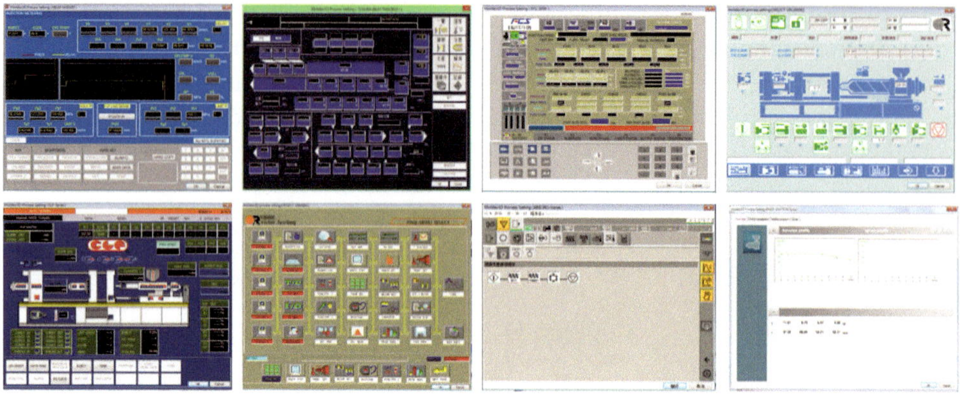

Figure 4.65 Moldex3D CAE setting interfaces emulate real machines

4.6.3 Case Study

The machine in the Moldex3D Molding R&D Center is used for sample verification. The machine is an ARBURG hydraulic-electric hybrid injection molder with a screw diameter of 40 mm and a clamping force of 250 tons. The other machine specifications are shown in Table 4.2.

Table 4.2 ARBURG 630H Specifications

Injection unit	400
Screw diameter, mm	40
Effective screw length, L/D	20
Screw stroke max., mm	160
Calculated injection volume max., cm^3	201
Shot weight max., g PS	184
Material throughput max., kg/h PS	29
Material throughput max., kg/h PA 6.6	15
Injection pressure max., bar	2000
Holding pressure max., bar	2000
Injection velocity, mm/s	512
Injection flow max., cm^3/s	642
Back pressure positive/negative max., bar	350/160
Circumferential screw speed max., m/min	54
Screw torque max., Nm	550
Nozzle contact force max., kN	60
Nozzle retraction stroke max., mm	300

Injection unit	400
Installed cylinder heating power / heating zones, kW	8.8/4
Installed nozzle heating power, kW	0.6
Machine dimensions and weights of the basic machine	
Net weight approx., kg	6050
Electrical connection (pre-fused) motor + heating A	100

Figure 4.66 shows the test mold for a flat part used for verification and measuring 300 mm × 100 mm × 3.2 mm.

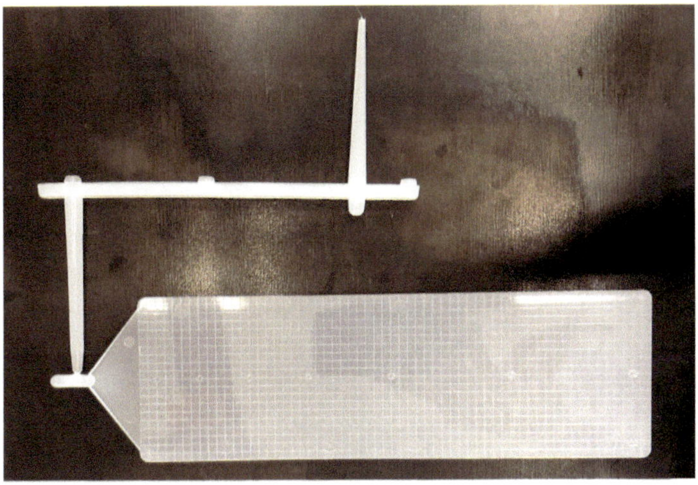

Figure 4.66 Test mold for a flat part

LCY Chemical Globalene 6331(PP) is used for the CAE analysis and the properties are shown in Figure 4.67.

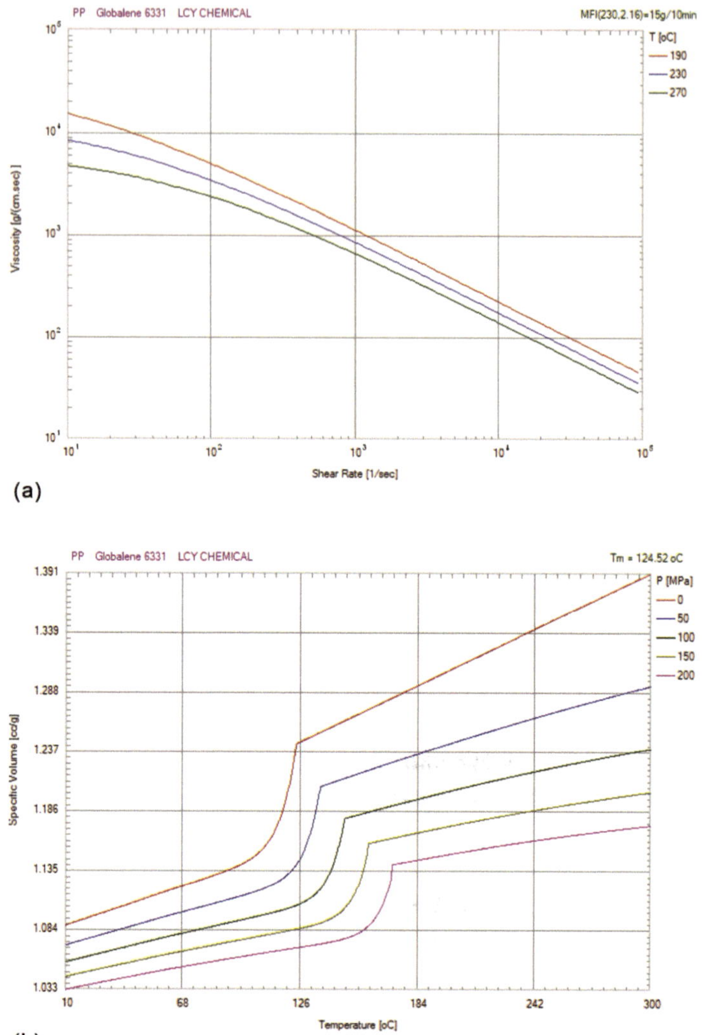

Figure 4.67 (a) Viscosity and (b) PVT of LCY Chemical Globalene 6331(PP)

Based on the mold and material characteristics, the melt temperature is set to 200 °C, the filling speed is 30 mm/s, the VP switching position is 15 mm, and the three-stage packing pressure setting is 500 bar, 250 bar and 125 bar in sequence. The relevant processing conditions set in CAE are shown in Figure 4.68, and the analysis is then performed, with consideration given to the machine response.

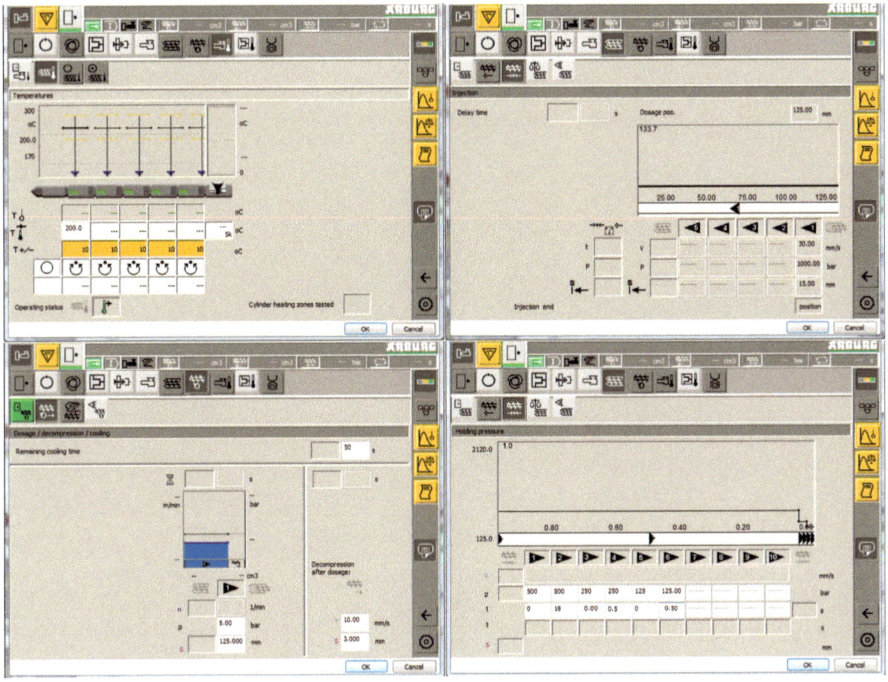

Figure 4.68 Setting interfaces for processing conditions in CAE

Since the CAE provides the same setting interfaces as those used on site, the designer can directly import the same conditions as those set on site. The differences in filling speed, nozzle pressure and machine injection pressure are shown in Figure 4.69, Figure 4.70, and Figure 4.71 respectively. The orange line in Figure 4.69 is the speed change in the CAE and the blue line is the actual speed change of the machine. Since the actual machine response is considered in the CAE analysis, the same trend is found in the two curves.

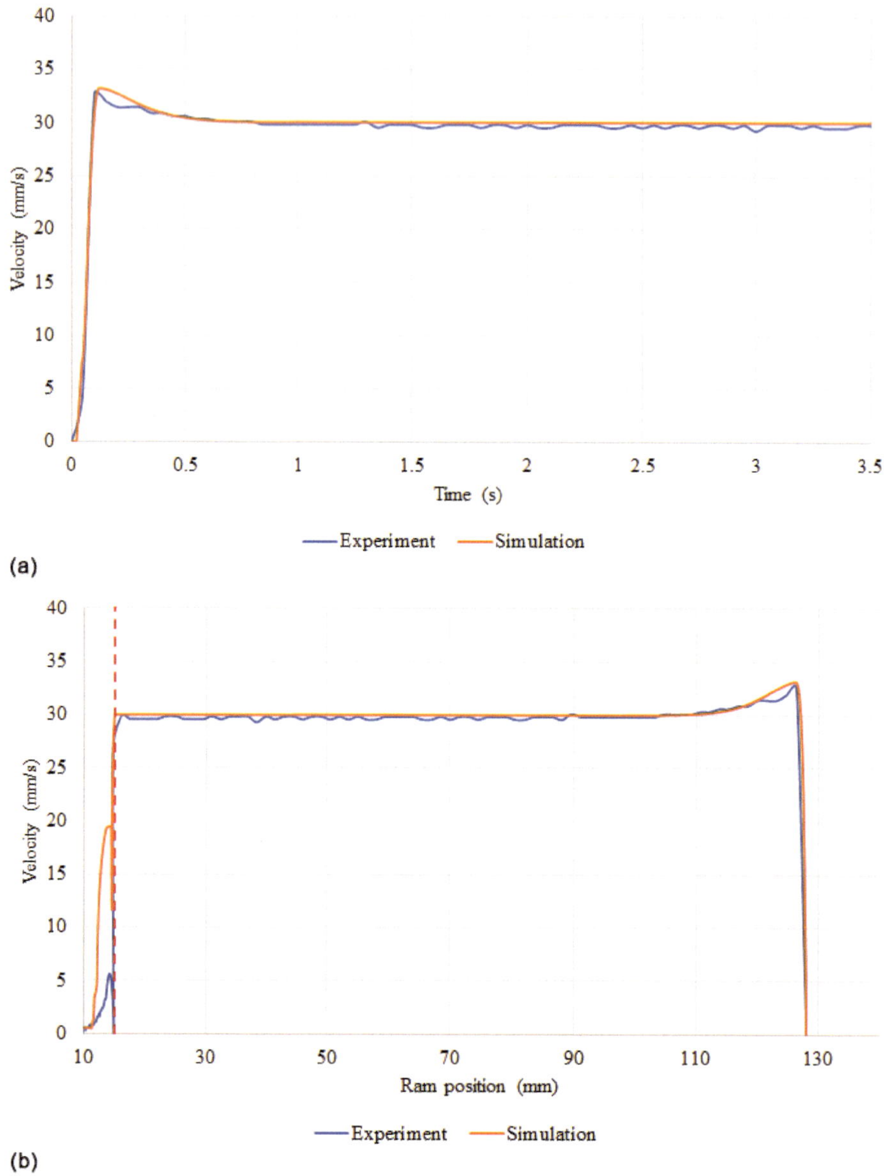

Figure 4.69 Changes in filling speeds as a function of (a) filling time and (b) ram position

Figure 4.70 shows the change in nozzle pressure as a function of ram position, where the orange line is the CAE result and the blue line is the actual value obtained with the machine. At the VP switch position, the nozzle pressure, as analyzed by CAE, is 573.1 bar, while that of the real machine is 585.3 bar; the difference is only 12 bar.

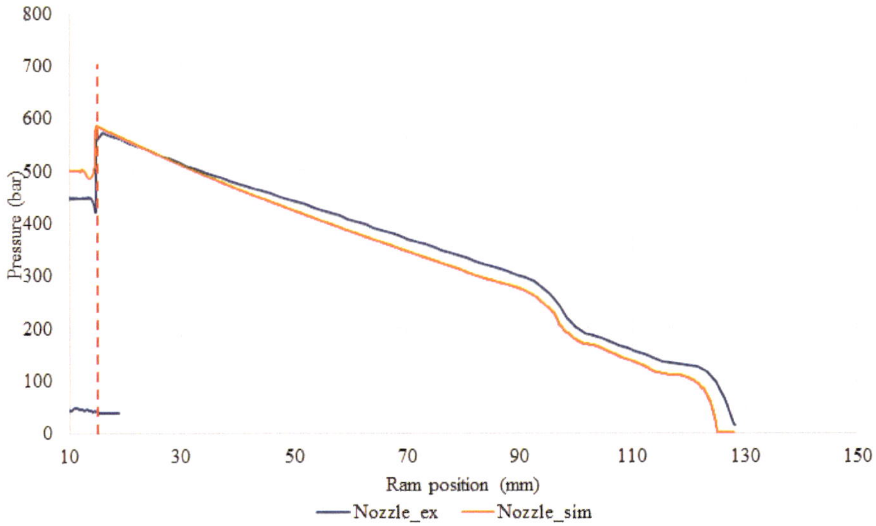

Figure 4.70 Changes in nozzle pressure as function of ram position

Figure 4.71 shows the results of the changes in machine injection pressure, with the mode being changed to packing after the VP switch. Since consideration has been given to the machine response, the profile of the CAE packing pressure is also consistent with the on-site situation.

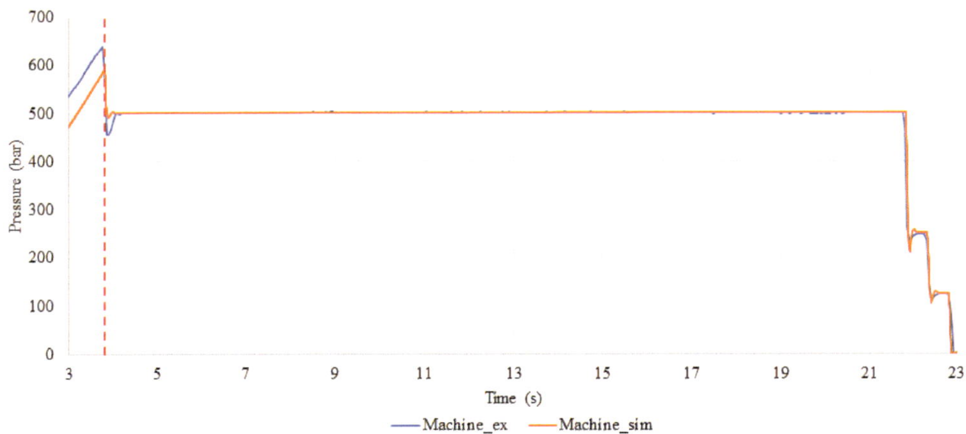

Figure 4.71 Changes in machine injection pressure as a function of time

The very close match between CAE results and actual short shots are shown in Figure 4.72, where the VP switch positions are set at 20 mm and 40 mm. As the CAE analysis results are consistent with the experimental results, all variations in production can be predicted and discussed in advance by CAE to avoid problems during production. The design end and the manufacturing end can communicate via the real machine interface, eliminating the communication barriers between each other and achieving the perfect integration of smart design and smart manufacturing.

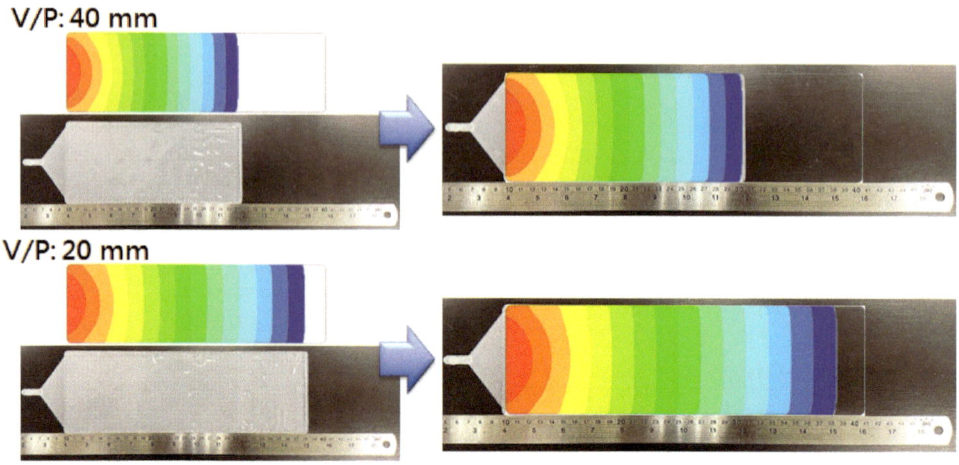

Figure 4.72 Comparison of short shots predicted by CAE and the experimental results

References

[1] Chang, R.-Y.; Hsu, C.-H.; Chang, C.-W.; Chien, C.-C.; Chiu, H.-S.: "Method for Operating a Molding Machine with a Predicted In-Mold PVT Waveform of a Molding Resin". Patent US 9,555,571 B1, 2017

[2] Chang, C.-W.; Chang, R.-Y.; Hsu, C.-H.; Chien, C.-C.; Chiu, H.-S.: "Molding System and Method for Operating The Same". Patent US 9,684,295 B2, 2017

[3] Chang, R.-Y.; Hsu, C.-H.; Chang, C.-W.: "Display Screen with an Animated Graphical User Interface". Patent US D784, 392 S, 2017

[4] Chang, R.-Y.; Hsu, C.-H.; Chien, C.-C.; Chang, C.-W.: "Display Screen with Animated Graphical User Interface For a Molding Process". Patent US D790, 582 S, 2017

[5] Chang, R.-Y-; Hsu, C.-H.; Chang, C.-W.: "Display Screen with an Animated Graphical User Interface". Patent US D817, 970 S, 2018

5 Molding Simulation Methodology

Hsien-Sen (Ethan) Chiu

As shown in Chapter 1 through Chapter 2 to Chapter 4, the quality of a molded plastic product depends heavily on plastic material properties, part mold design, and process conditions. In injection molding, the control of a good molding process results in good product quality. However, the process itself is a black box so that you will never know until the moment of demolding what can possibly go wrong, let along the cause of these problems.

The difficult part is that the adjustment of a single molding parameter can influence the result in different ways due to the nonlinear coupling of the process. For example, short shot can be solved by increasing the injection speed, material temperature, or mold temperature. However, the increase of the injection speed can lead to excessive shear rate, higher molding stress, and more residual stress; the increase of material or mold temperature would result in a longer cooling time for the plastic material and thus prolong the molding cycle and induce more production costs. In addition, there is not much room left for mold modification after the mold has been made, and problems would occur after repeatedly modifying the dimensions in the same mold. These are some of the problems that come with the traditional way of mold modification.

5.1 The Goal of Molding Simulation

For molded products, the purpose of simulations is to hopefully realize the manufacturability of the designed products. Thousands of ideas are developed every day, but only some of them can be manufactured successfully. It is the purpose of molding analysis to ensure that the design of the products or molds is suitable at the design stage, and therefore a good production quality can be achieved.

Molding simulation can be done by using CAE tools, that is, computer-aided engineering software and technical analysis, to assist in the diagnosis and development of complicated injection molding processes. CAE facilitates quick integration of complex rheological, thermal, and mechanical material properties, allowing designers and developers to perform qualitative and quantitative analysis and diagnosis for part and mold design. Thereby, they can thoroughly analyze and diagnose existing molds and operating conditions.

5.1.1 Design Verification and Optimization

5.1.1.1 Overview of Design for Manufacture (DFM)

The difficulty of ensuring design quality and the control of molding stability are the two major challenges in molding industry practice. To understand why such problems occur, we have to go back to the process of mold development, which generally includes the seven steps shown in Figure 5.1:

Figure 5.1 Mold development procedure

1. Kickoff: Establishes the project to start the design of 3D/2D drawings and make decisions on the techniques, machine, materials, and processing methods to be used.
2. Design: Involves finalization of the overall configuration of the mold, and molding simulation analysis, steel, CNC (computerized numerical control) machining, and design drawing approval.
3. Manufacture: Raw steel is machined into the mold according to the design drawings.
4. T1 (first trial): The mold is made, assembled, and loaded on a molding machine for the first trial to check that the mold is made according to the design drawings and meets the requirements. The mold is then refined within the allowable range to improve for a smoother mass production. Follow-up steps should not be commenced until the quality of the trial is approved.
5. Approval
6. Shipping
7. Service

The first common challenge for molding is how to ensure the design quality.

The traditional development process for plastic molding products is shown in Figure 1.24 in Chapter 1. Because no CAE is used for design verification, many molding issues, such as burning, weld lines, warpage, and flashes, cannot be seen before the molding trial or even mass production, but the introduction of CAE analysis tools in the design stage can solve these problems effectively. Figure 5.2 shows an example of an injection molded plastic product. The available CAE tools that can be used according to process sequence are simulation of screw design and mold injection, and structure analysis starting from mold filling. The quality of the products obtained from the initial conditions of the design stage is evaluated. The simulation results are interpreted and taken as the guidance for design modification. The simulation-driven product development cycle can confirm the design parameters and avoid potential molding problems before the mold is cut. A virtual molding trial can be done cost-effectively in the computer to iterate for suitable or even optimal design and process parameters. The mass production can proceed once the quality is approved. If not, design changes and re-examination of the design conditions will be done repeatedly until the quality is satisfactory.

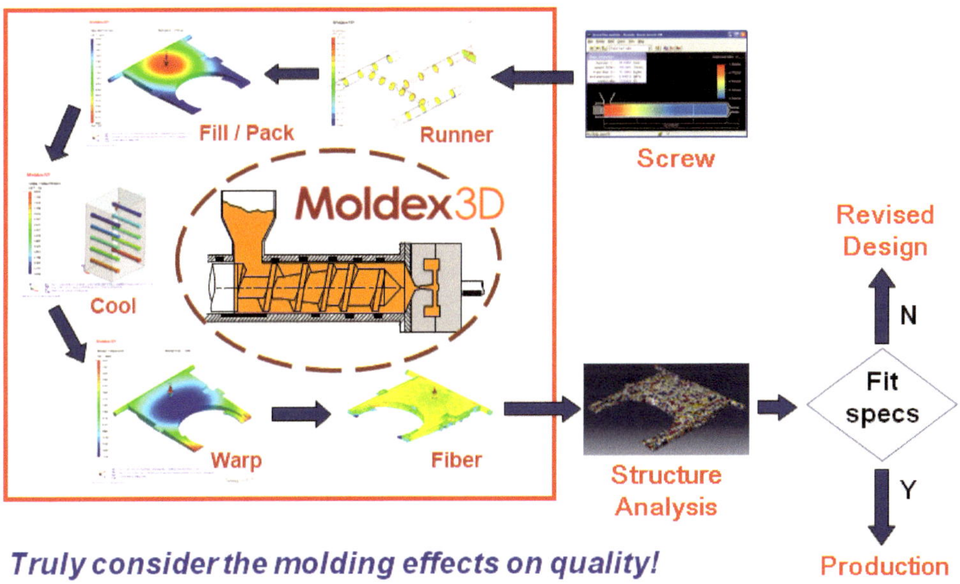

Figure 5.2 CAE applications at the design stage

In particular, when anisotropic materials, such as in a fiber-reinforced-plastic part, are added, the prediction of the orientation and the distribution of the fibers becomes very important, because the fiber orientation and concentration are closely related to the mechanical response and strength of the molded parts. Such molding effects can only be predicted using tools such as Moldex3D. In addition, the information of microscale defects, such as residual stress due to molding, fiber orientation, or pore distribution in foam molding, is transmitted to structural analysis software for more in-depth stress simulation. This enables a more precise simulation of these material properties and therefore predictions that are more realistic can be induced.

5.1.1.2 CAE and DFM: A Practical Case Study

How does flow analysis help improve the design quality? Take the connector shown in Figure 5.3 as an example. The features of a connector are slender geometry and multiple slots. The wall thicknesses of these slots are small and therefore the possibility of short shot should be a concern. Also, as it is necessary to consider the flatness of the product, warpage must meet the required criteria after molding.

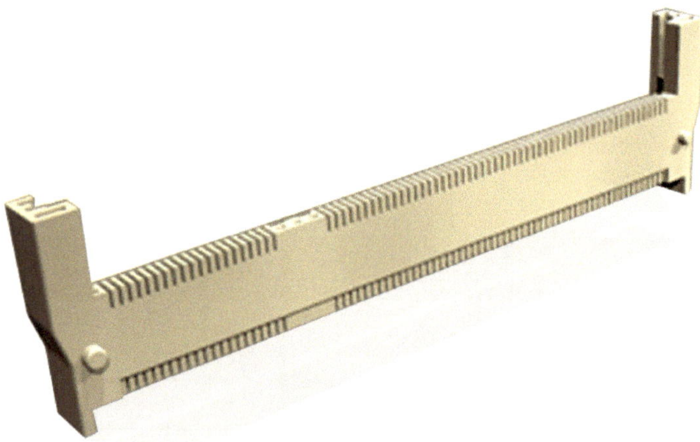

Figure 5.3 Connector case model

In this case the dimensions are 55 × 5 × 15 mm with a thickness of 0.7 to 1.0 mm. A liquid crystal polymer (LCP) is a common material for connectors as a result of its high strength and good thermal resistance and flowability. Unique in this case is the very short filling time of only 0.05 s, which is also a feature in the process conditions of the connector. Because the slot consists of many structures and is designed with a high flow length/wall thickness (L/t) ratio, a high filling speed is required to prevent loss of heat and viscosity increase that leads to short shot. A LCP requires a higher processing temperature; in this case the melt temperature is 340 °C and the mold temperature is 120 °C.

Figure 5.4 illustrates a runner design that introduces single gating with a sprue of 2 to 3 mm in diameter, main runner of 2.5 mm in diameter, and a gate size of 1 × 0.8 mm. With a gate position for single-sided gating like this, it is expected to see large shrinkage and changes in dimension along the flow direction, that is, the X-direction. Because of possible short injection and warpage in parts, it is necessary to examine the simulation results, such as melt front, volumetric shrinkage, displacement, and flatness, before identifying the cause of these problems and possible solutions.

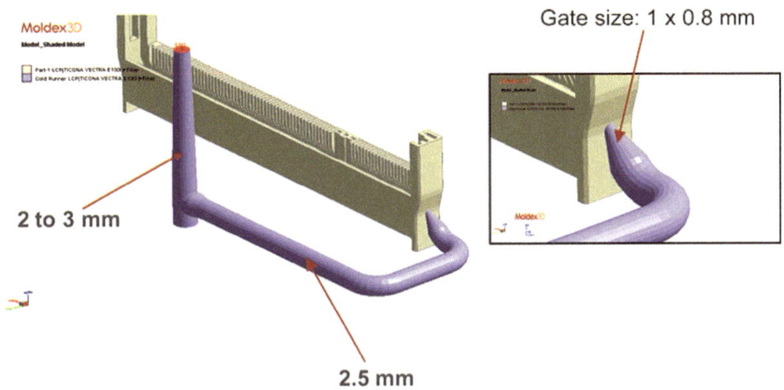

Figure 5.4 Runner and gate design

To find out whether the filling process of the original mold is good or not, the melt front at a few specific points in time is observed as shown in Figure 5.5. The melt passes through the gate at approximately 35% of the filling time, while a race-tracking effect occurs at 50% and keeps growing until 75% of the filling time. At 85%, the faster flowing outside melt reaches the end of the part and starts to flow backward from the end. At 95%, the weld line starts to form, and at 99%, the weld line falls to the side of the part (as indicated). The fact that the injection end is not at the edge of the part may result in difficulty of venting, burning, and air traps.

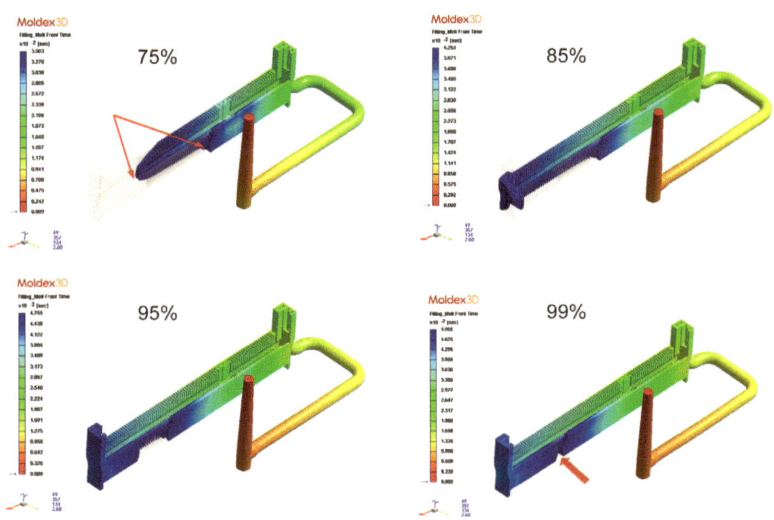

Figure 5.5 Melt front time of the original design

A look at the melt front indicates an issue of uneven flow. The area with such issue is magnified as shown in Figure 5.6, and it is found that the melt flows faster on the −Y side in the beginning of filling, while the flow is slower at the +Y side due to many thin-walled areas. It can be seen clearly on the cross section that the thickness at the −Y side of the part is much larger than that at the +Y side, which is obviously the cause of the race-tracking phenomenon.

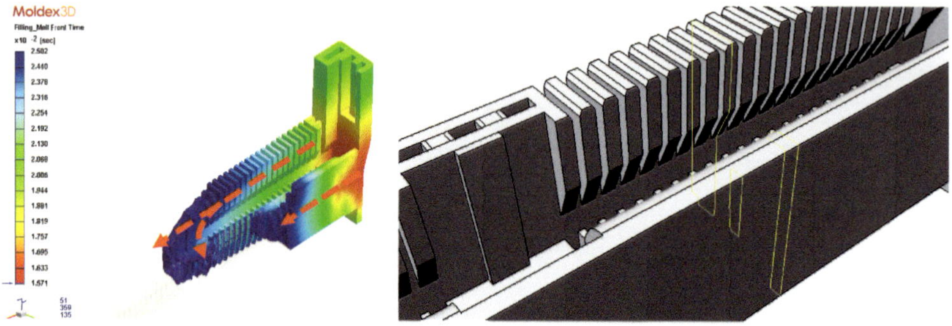

Figure 5.6 Uneven distribution of part thickness

Figure 5.7 shows that, at 85% of the filling time, the part is filled up at the −Y side due to the race-tracking effect, but on the other side, there is still half of the part length unfilled, and a weld line is expected to appear here. Usually, the location of a parting surface has to be taken into consideration in the venting design of a mold. The position of weld line in the original design is unfavorable to venting and associated with potential risk of air trap.

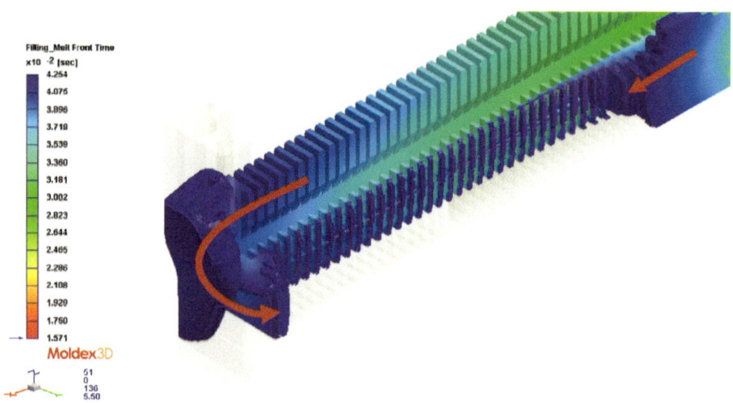

Figure 5.7 85% melt front time of the original design

Apart from flowing, molding analysis is also used to examine other problems caused by uneven wall thickness. For example, Figure 5.8 shows the temperature distribution of a part at the end of packing. An isothermal diagram is used to highlight the areas hotter than 230 °C. It can be seen that a large area of high temperature still exists inside the part, because heat dissipation from the thick-walled areas is difficult. Such hot areas would continue to shrink during cooling or even after demolding, which increases the risk of warpage.

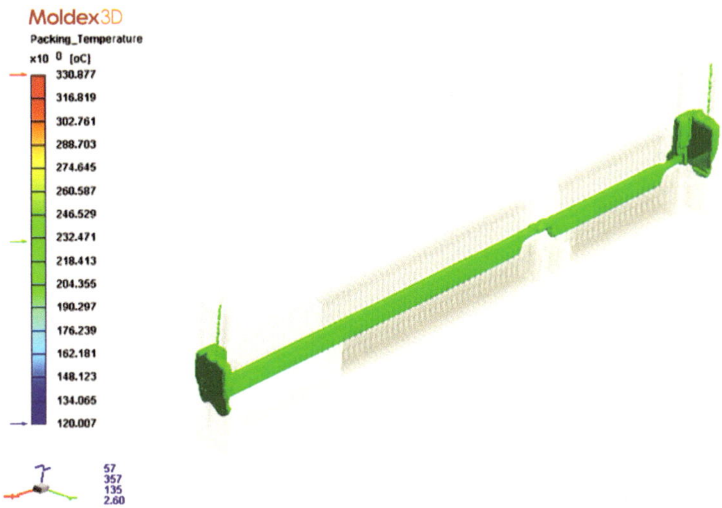

Figure 5.8 Temperature higher than 230 °C

Figure 5.9 shows the difference in warpage due to uneven cooling of the plastic because the cooling channels on top of the product are not dense enough to create the desired cooling rate. The high temperature at the top leads to greater shrinkage and thus convex deformation. This leads to the following question: with this connector as an example, what will the deformation look like?

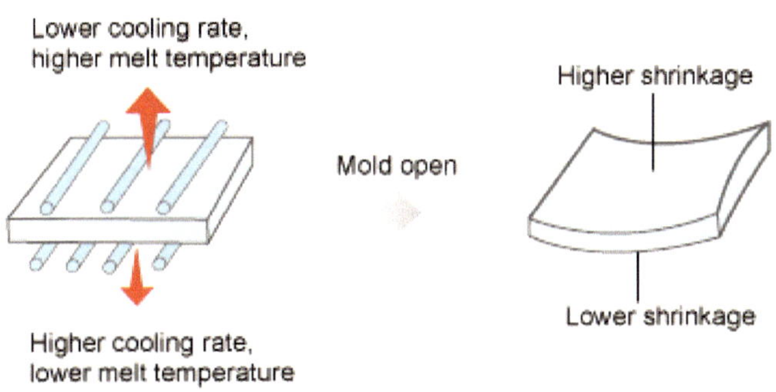

Figure 5.9 Differential cooling rates leads different shrinkage behavior

Figure 5.10 the side view of the displacement in the Y-direction. As the wall thickness is greater in the Y-direction of the part, the shrinkage is greater on the −Y side. Also, it can be seen from the color gradation that the red color at both ends of the part indicates the displacement in the +Y-direction and the blue color in the middle indicates the displacement in the −Y-direction, and thus the distortion shown. The overall displacement is in the region ±0.08 mm.

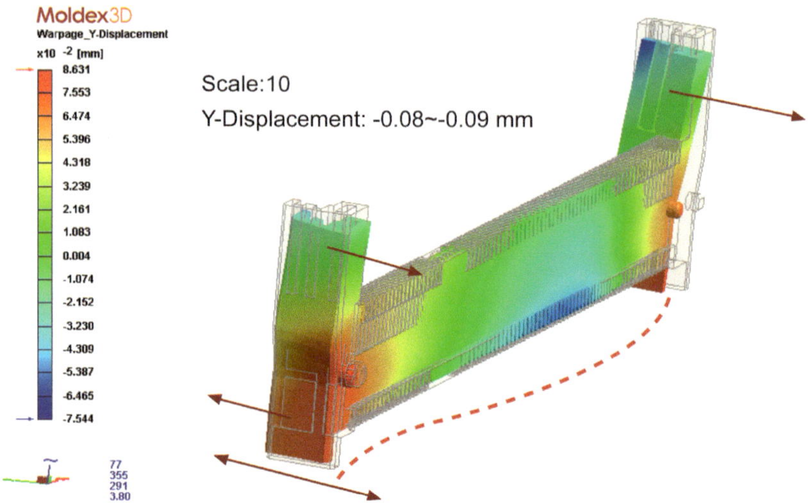

Figure 5.10 Warp prediction in Y-direction displacement: side view/top view

Molding simulation CAE helps predict possible molding defects and causes in the original design. According to the found causes, the design can be changed and simulation is used again to determine whether the problems are solved. As the root cause in this case is the uneven thickness, a design change is carried out by thinning the thick-walled regions, as shown in Figure 5.11.

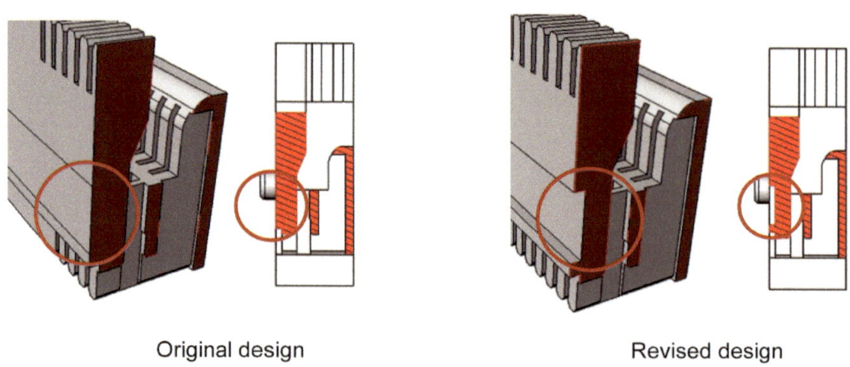

Figure 5.11 Revised design with the reduced thickness

To verify the effectiveness of a design modification, analysis results before and after modification are compared. For the filling of the melt shown in Figure 5.12, it is found that the difference of the melt front is greatly improved at 80% filling time, and the filling end and weld line position are very close to the edge of the product. Therefore, the reduction of the thickness in the changed design improves the imbalanced flowing of the plastic material.

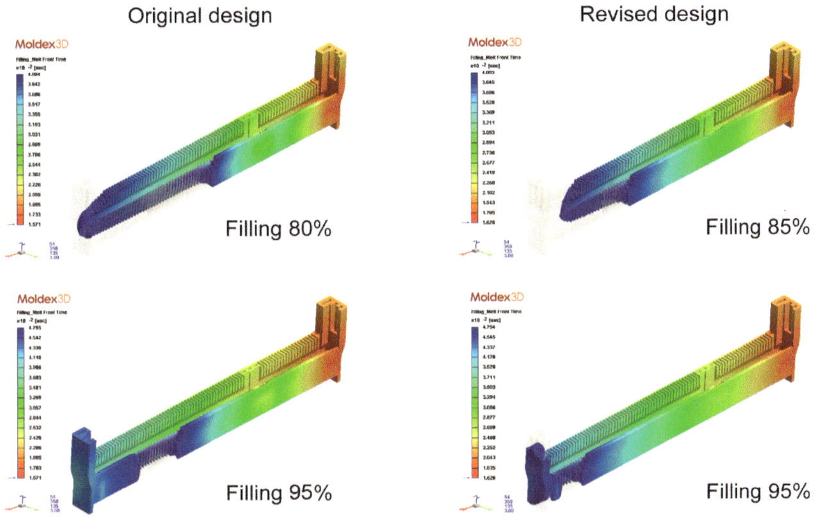

Figure 5.12 Comparison of melt front time

From the comparison of warpage in Figure 5.13, warping in the Y-direction improves significantly as well. The improvement in the melt flow balance as well as the thickness reduction after the modification both contribute to a decrease of consecutive high-temperature areas in the part, which reduces the difference of shrinkage between the +Y and −Y sides, and thus the S-shaped distortion in the initial design is flattened.

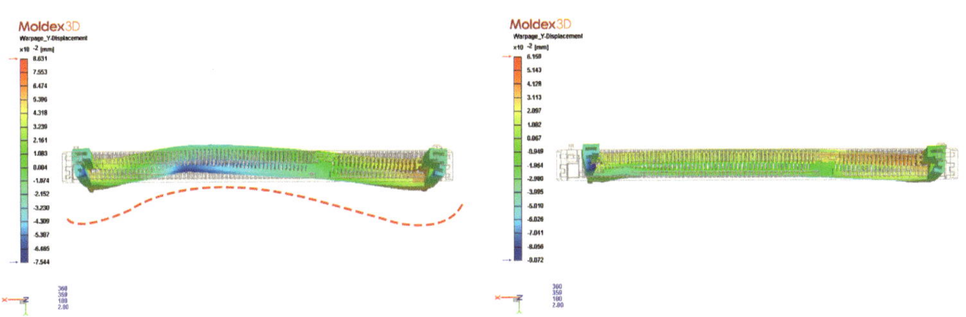

Figure 5.13 Comparison of warpage: Y-direction displacement

In addition to the overall trend of warpage, changes in flatness can be observed as well. As shown in Figure 5.14, with one side of the product defined as the origin, the color gradation shows the middle concave area in blue and the warping up at both ends in red. The flatness of the original design is approximately ±0.13 mm.

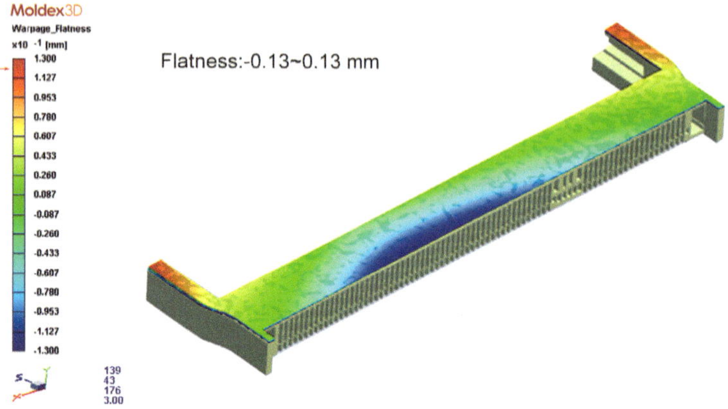

Figure 5.14 Warp prediction in flatness of the original design

The flatness distribution after the design change is −0.06 and 0.03 mm, as shown in Figure 5.15. The color gradation displays a smooth and even distribution, which shows a significant improvement in the large deformation areas shown in blue and red color in the original design.

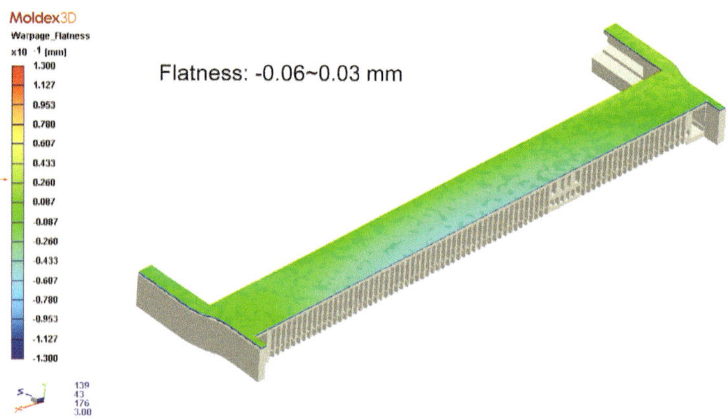

Figure 5.15 Warp prediction in flatness of the revised design

Warping and flatness results are summarized in Table 5.1. The revised design produces improvement in both Y- and Z-directions with the greatest improvement in flatness.

Table 5.1 Comparison of Warp Results between the Original and Revised Design

[mm]	Original Design		Revised Design		Improvement
	Min.	Max.	Min.	Max.	
Flatness	−0.13	0.13	0.06	0.03	65%

This section illustrates how molding simulation tools can be used to validate and improve the design parameters of the part and the mold.

5.1.2 Process Conditions Optimization

5.1.2.1 Molding Stability

In addition to the verification of design manufacturability as discussed above, molding simulation analysis provides effective solutions to molding problems encountered in mass production of existing molds. Here, a practical example should be discussed on the use of molding analysis to solve the instability issues that often arise during production.

First of all, what is molding instability? In a nutshell, it denotes the inconsistent quality between each shot. It could be unstable metering, mold temperature fluctuation, or a narrow processing window so that the plastic injection process varies from shot to shot, varied part weight, or difficult control of the accuracy of the part dimensions. The short shot method is a good way to determine the consistency of the melt front in every shot. The consistency of temperature and pressure can be examined via the data obtained from embedded sensors.

Figure 5.16(a) presents widely spread injection temperature and pressure curves measured at each cavity, whereas Figure 5.16(b) shows the diagrams converging roughly into a single curve, indicating that parts with consistent quality would be obtained because the molding process for each individual cavity is similar.

The quality of a molded plastic product depends on the three major factors of molding: the material used, the design of mold, and the molding conditions. In the case of a poor mold design, for example if the runner is too slender, the gate is poorly positioned, or the difference in thickness is significant, the allowed molding window for this system would be narrowed down (see Figure 4.8 in Chapter 4); that is to say, the range of operable temperature, pressure, and speed conditions becomes smaller. Therefore, it is sometimes difficult to manufacture a good part even with the adjustment of molding conditions, so it can only be practically achieved by revision of the part or mold design.

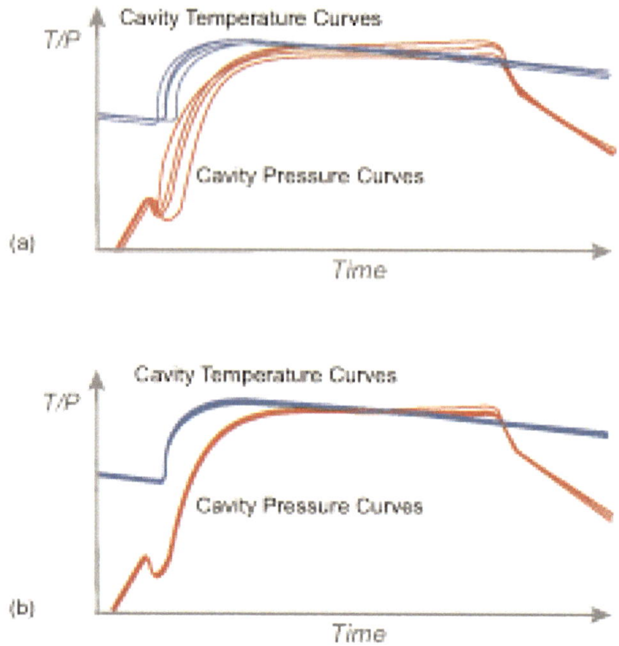

Figure 5.16 (a) Unstable and (b) stable molding conditions

For the issue of molding instability, only a more advanced CAE application can identify the cause of instability by combining molding analysis and practical experience, because the entire molding system consists of the injection machine, the mold, and plastic material in the screw, and the mold cavity has to be taken into consideration. In practice, the molding engineer inputs the process condition parameters into the machine expecting that the properties and the behavior of the plastic material are under control for consistent injection quality every time. That is why the control over material properties and machine performance is critical. In general, electric systems perform relatively more stable compared to hydraulic systems.

Therefore, the solution for molding instability can be broken down into the steps as shown in Figure 5.17:

1. Identify the molding problem.
2. Start with the four major control factors and identify the crucial one with the highest influence on the problem. In case of a poor design, the molding analysis can be used for validating of the revised design and molding conditions to determine the solution with the greatest benefit. The design of experiments (DOE) approach can be used in combining with simulation to optimize the design parameters and process conditions.

3. If the problem cannot be solved by modification of the part or mold design of the traditional injection molding process, the next step will be considering the introduction of other molding processes or technologies, such as the hot runner system or conformal cooling. Variotherm or MuCell are capable of incorporating these applications into analysis, and solutions have been developed with joint development with the suppliers of these physical technologies (e.g., Moldex3D partner solution).

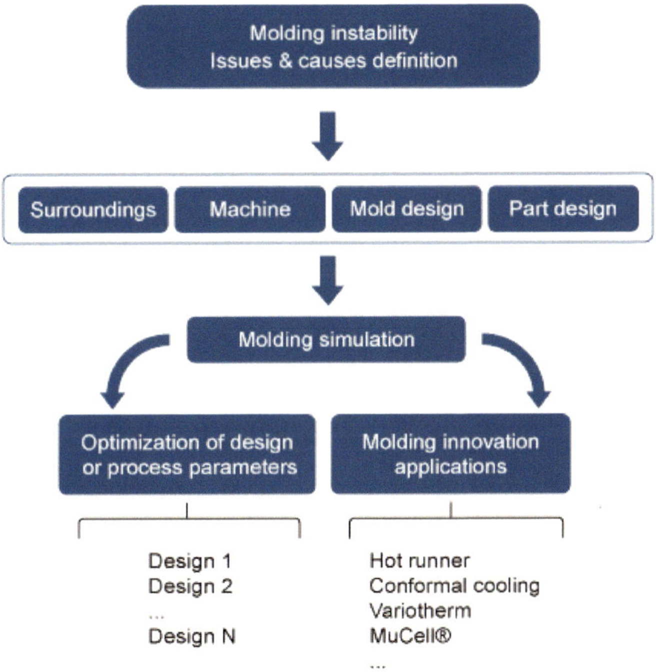

Figure 5.17 Flowchart of solution to molding instability

5.1.2.2 Real Case

Next, a real example is used to show how molding analysis is applied in conjunction with practical experience to deal with the molding instability issue. The case is a light guide component as shown in Figure 5.18. Clearly, the design of the component features three curved lenses of different sizes. The molding problem encountered is the burning close to both ends of the lens with the largest curvature. If that happens, the machine has to be shut down for half a day just to clean the carbon deposit (in the red circles) from the mold, which leads to poor production efficiency and unwanted waste of labor and materials.

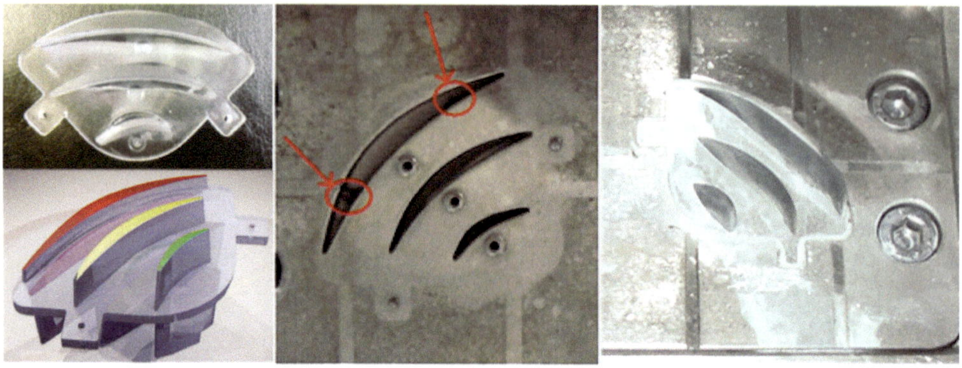

Figure 5.18 Light guide component

In this case, the part dimension is 43 × 42 × 25 mm with a primary thickness of 2 mm and volume of 10.3 cm^3. Poly(methyl methacrylate) (PMMA), a material commonly used for optical products, is used for the injection. The filling time is 0.5 s, the melt temperature 250 °C, and the mold temperature 60°C. Compared to the short filling time of 0.05 s for the LCP connector described in Section 5.1.1, 0.5 s is a common setting. For the injection system, as shown in Figure 5.19, the sprue is 3.5 to 6 mm in diameter with three branch runners of 4 mm in thickness and a secondary runner of 6.6 to 1.5 mm in diameter, and the gate is 1.5 mm in diameter. There are three gates in the design. The arrow points at where the burning occurs, which is the end of the longest lens.

To verify the flow problem mentioned above, let's look at the analysis results shown in Figure 5.19. Just as expected, the melt front indicates that the filling end occurs at both ends of the longest lens, and air can be trapped here due to the tip design, which is also the cause of burning. There are several solutions, for example, reduction of the corner length of the ribs as mentioned in Section 3.1.4. In this case, however, the geometry for the optical part is essential to achieve the desired optical properties and application performance, which makes part design modification infeasible here.

To change the filling behavior of the melt, altering the runner system is the only way because the part design is already fixed. This is a multi-gate injection design, and if we want the filling end to be away from the initial location where air may be trapped, we can change the way of the injection. In Figure 5.20(a), the original design has three gates. What if we incorporate only one or two gates? Figure 5.20(b) shows that the revised design 1 has only gate 1 available and the design 2 shown in Figure 5.20(c) has gates 1 and 3 available.

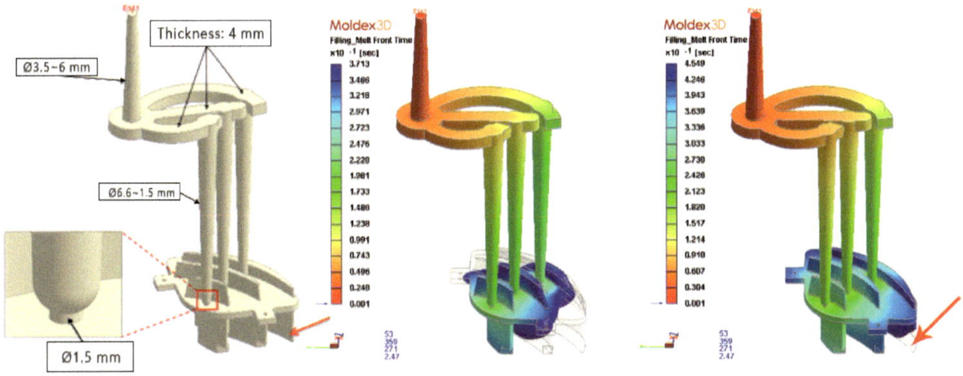

Figure 5.19 Light guide component model and its simulation result

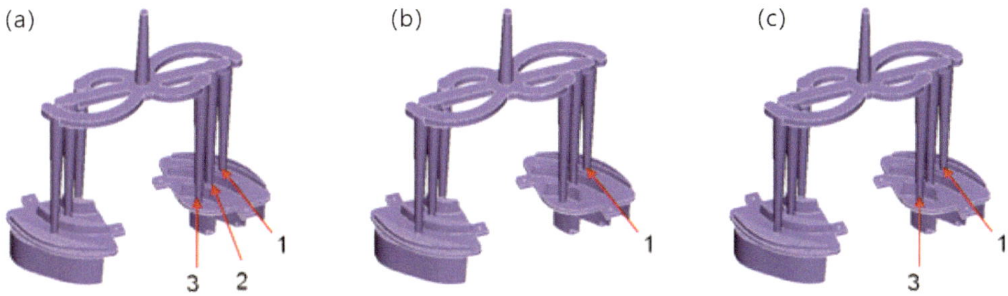

Figure 5.20 (a) Original and revised designs 1 (b) and 2 (c)

Figure 5.21 shows the flow analysis result of the revised design 1. It indicates that the end of flow in the single gate design moved from the original long lens to another area. Although this behavior reduces the risk of air traps, it leads to new problems because all the filling has to be done through a single gate. The excessive flow length limits packing compensation at the filling end, and thus creates larger volume shrinkage.

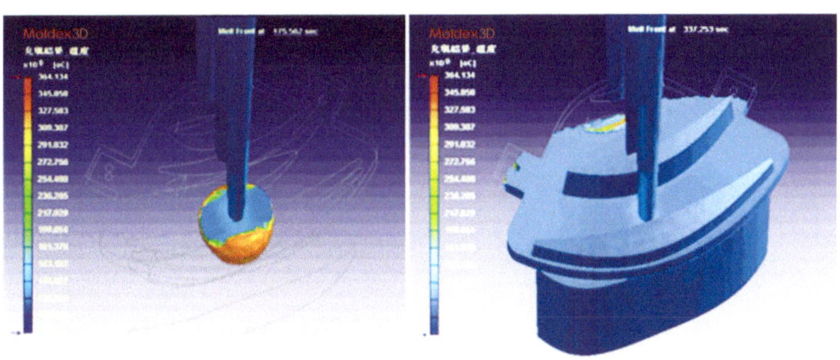

Figure 5.21 Melt front time of the revised design 1

Figure 5.22 shows the filling result in the case of two gates. Not only is the problem of air trap resolved, but the second gate can also maintain the packing on the other end of part. However, it is also observed that the filling of the short lens finishes too quickly, so that there is still a risk of air trapping in the long lens. Further adjustment may be needed to achieve a more uniform filling pattern.

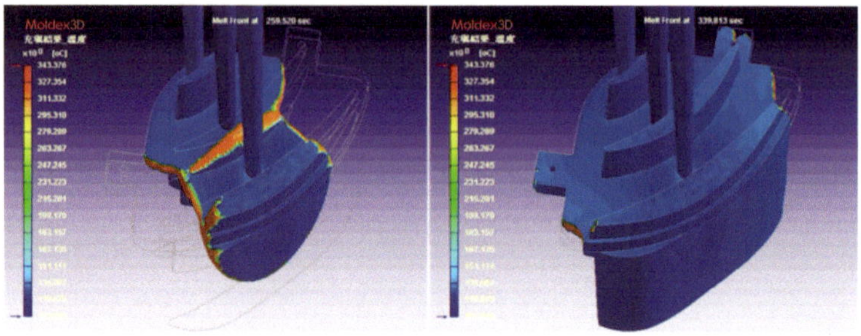

Figure 5.22 Melt front time of the revised design 2

For a non-uniform flow pattern in revised design 2, an attempt to reduce the runner of the gate at the short side can be made. The red circles in Figure 5.23 indicate that the thickness of the revised design 3 is reduced to 1 mm and that of revised design 4 to 2 mm. It is found as shown in Figure 5.23(a) that the flow in the 1-mm runner is completely stagnant, which requires the melt to flow back after the other side is filled. In Figure 5.23(b), the 2-mm runner provides a more uniform flow.

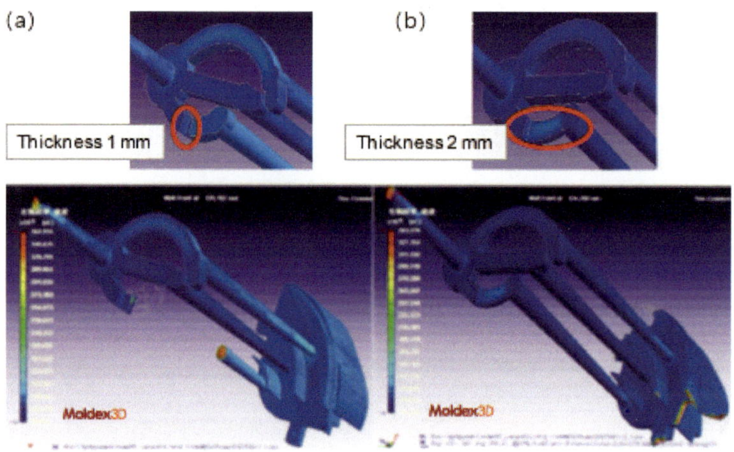

Figure 5.23 (a) Revised design 3 and (b) revised design 4

Ultimately, the revised design 4 shown in Figure 5.23(b) is selected, and molding simulation is then used to examine the benefits of design revision and reduce the cost of mold modification during the process. The original design is revised twice: the number of gates is changed from three to two and the dimension of the runner that corresponds to gate 3 is reduced. With practical mold modification, the venting design at the edge of the product, as shown in the red lines in Figure 5.24, is improved to eventually solve the problem that originally required stopping the machine every 12 hours for mold cleaning. In addition, the production yield is increased by 25%.

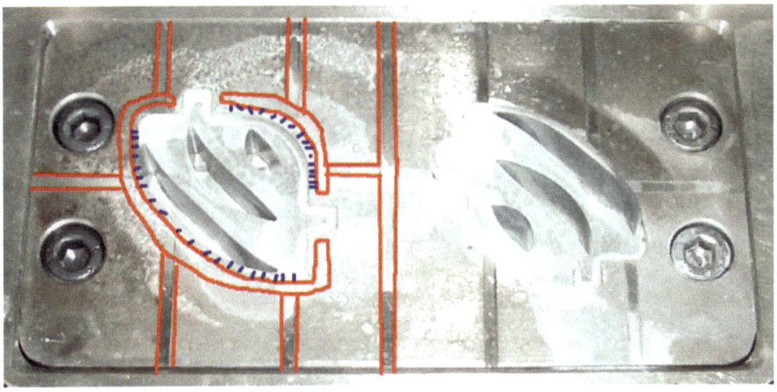

Figure 5.24 Venting design enhancement

This is a good example of molding simulation combining with practical experience to solve molding problems. It shows us that CAE not only plays an important role in the examination of design quality during product development, but also solves molding problems encountered on site. However, to achieve this, molding simulation and practical applications have to be integrated so that the link between simulation results and practical situations can be established correctly.

■ 5.2 Basics of Simulation Equations

The introduction to the purposes and methods of CAE presented above lays a sufficient foundation for general users in the use of CAE in molding simulation. This section will provide additional molding analysis methods for interested advanced users. Section 5.2.1 describes the governing equations for the physical system and Section 5.2.2 introduces the numerical methods to solve problems with computers. From a technical point of view, CAE is a numerical application for the solution of the theoretical model of a physical system. As a result, the accuracy and reliability of the analysis are closely related to the level of emulation of the theoretical model, the accuracy and convergence of the numeric method used, and the mesh model used for analysis.

5.2.1 Governing Equations

The fundamental governing equations of a molding system are the conservation laws, that is, the conservation of mass, momentum, and energy, and can be derived using the conservation of the control volume, which is the smallest unit in a system. As shown in Figure 5.25, by considering the sum of a physical quantity that comes in, goes out, and is generated in an infinitesimal element, the rate of change of the physical measurement over time is derived. By adding the values of all these tiny spaces together, the overall measurement of a macroscopic system is obtained.

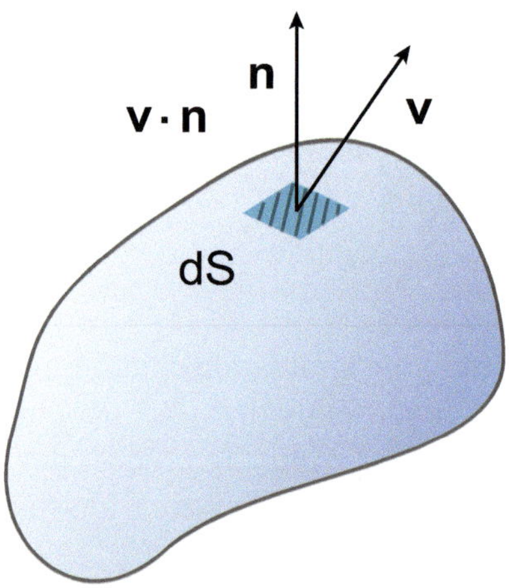

Figure 5.25 Control volume

A universal control volume is represented by a shaded region within a fluid and velocity vector field **v**. dS is the differential surface area element, and is shown along with the corresponding unit outward surface normal vector **n**. For the derivation details of conservation laws, one can refer to reference [1].

Continuity equation:

$$\frac{\partial \rho}{\partial t} + \nabla \cdot (\rho \mathbf{v}) = 0 \qquad (5.1)$$

where ρ is the density, t is the time, and $\mathbf{v}$ is the velocity vector.

Equation 5.1 is a standard continuity equation.

Momentum equation:

$$\frac{\partial}{\partial t}(\rho \mathbf{v}) + \nabla \cdot (\rho \mathbf{v}\mathbf{v}) = -\nabla p + \rho \mathbf{g} - \nabla \cdot \boldsymbol{\tau} \tag{5.2}$$

where $\mathbf{g}$ is the gravity and $\boldsymbol{\tau}$ is the shear stress.

Energy equation:

$$\frac{\partial}{\partial t}\left(\rho C_p T\right) + \nabla \cdot \left(\rho C_p \mathbf{v} T - k \nabla T\right) = \eta \dot{\gamma}^2 + \Delta H \tag{5.3}$$

where T is the temperature field, C_p is the heat capacity, k is the thermal conductivity coefficient, η is the viscosity of the fluid, $\dot{\gamma}$ is the shear rate, and ΔH is the heat of generation.

In general, these three equations are considered for common injection molding processes. For a special process or material, additional equations will be needed to provide an accurate profile of individual physical systems. For example, Chapter 2 "Material Properties of Plastics", Chapter 16 "Resin Transfer Molding", and Chapter 17 "IC Packaging" describe the curing reaction of thermosetting plastic and its influence on the molding processes; Chapter 10 "Fiber Orientation Control" describes fiber orientation behavior using the iARD-RPR model when fibers are introduced into the plastic; Chapter 14 "Foam Injection Molding" describes the theory of cell nucleation and growth; Chapter 15 "Powder Injection Molding" describes the shear-induced diffusion model in particle movement; and Chapter 16 "Resin Transfer Molding" describes Darcy's law for the penetration behavior of resins.

5.2.2 Numerical Approximation

The accuracy of simulation depends on the discretization approach, which determines whether a numerical solution can be obtained successfully, that is, whether the calculation diverges and how efficient the calculation is (the rate of convergence). A good discrete method approaches correct answer more closely when the mesh becomes denser, and presents a result of convergence. Three commonly used numerical methods are [2]: the Finite Difference Method (FDM), the Finite Volume Method (FVM), and the Finite Element Method (FEM).

5.2.2.1 Finite Difference Method (FDM)

1. One-Dimensional Case

This is the earliest numerical method developed where the Taylor expansion is used to expand a known number to determine neighboring unknown ones. Figure 5.26 shows the idea of using the known value of node 2 to determine the unknown values of nodes 1 and 3.

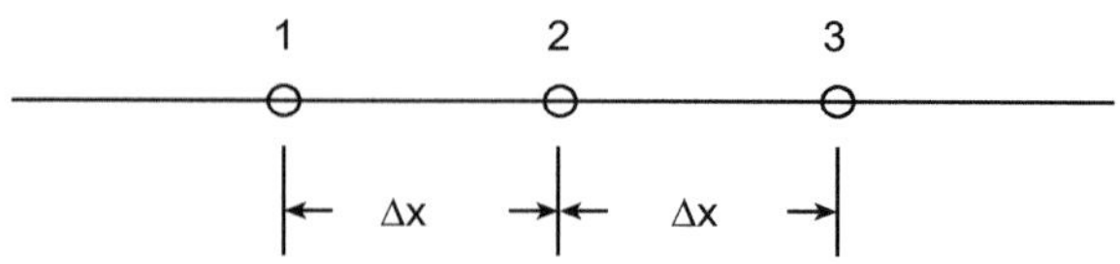

Figure 5.26 One-dimensional mesh

The Taylor expansions for nodes 1 and 3 are:

$$\phi_1 = \phi_2 - \Delta x \left(\frac{d\phi}{dx}\right)_2 + \frac{(\Delta x)^2}{2}\left(\frac{d^2\phi}{dx^2}\right)_2 + O\left((\Delta x)^3\right) \tag{5.4}$$

$$\phi_3 = \phi_2 + \Delta x \left(\frac{d\phi}{dx}\right)_2 + \frac{(\Delta x)^2}{2}\left(\frac{d^2\phi}{dx^2}\right)_2 + O\left((\Delta x)^3\right) \tag{5.5}$$

where $O((\Delta x)^3)$ is the error term derived by approximation and is related to the 3rd order of the grid space Δx.

By subtracting and adding both equations, the first- and second-order differential equations are obtained, respectively:

$$\left(\frac{d\phi}{dx}\right)_2 = \frac{\phi_3 - \phi_1}{2\Delta x} + O\left((\Delta x)^3\right) \tag{5.6}$$

$$\left(\frac{d^2\phi}{dx^2}\right)_2 = \frac{\phi_1 + \phi_3 - 2\phi_2}{(\Delta x)^2} + O\left((\Delta x)^3\right) \tag{5.7}$$

The differential terms can now be expressed as difference terms, and the higher-order terms can be downgraded into algebraic equations when determining solutions for differential equations. Taking a 1D steady state diffusion equation as example, for a temperature field, there are no instability and convection terms in the system and the diffusion coefficient is constant:

$$\Gamma \frac{d^2\phi}{dx^2} + S = 0 \tag{5.8}$$

where Γ is the diffusion coefficient and S is the source term. As node 2 is known, then:

$$S_2 \equiv S(\phi_2) \tag{5.9}$$

Assuming that $O((\Delta x)^3)$ is negligibly small and can be neglected, then:

$$\Gamma\left(\frac{d^2\phi}{dx^2}\right)_2 \approx \Gamma \frac{\phi_1 + \phi_3 - 2\phi_2}{(\Delta x)^2} \tag{5.10}$$

$$\frac{2\Gamma}{(\Delta x)^2}\phi_2 = \frac{\Gamma}{(\Delta x)^2}\phi_1 + \frac{\Gamma}{(\Delta x)^2}\phi_3 + S \tag{5.11}$$

Now we convert the original second-order differential equation into an algebraic equation that can be solved by a matrix calculator. For a multi-node system, the method used above produces multiple linear equations, and such simultaneous equations can be solved by appropriate boundary conditions. FDM is easy to understand and use but does come with several limitations: linearity is assumed between nodes; higher-order terms are neglected and thus the conservation of equations is compromised; and it works only for orthogonal grids but not for non-orthogonal and non-structural grids.

2. Two-Dimensional Case

Figure 5.27 provides an example of a 2D system:

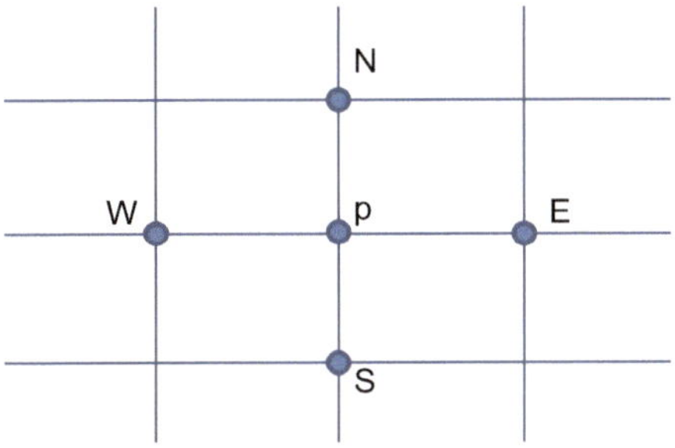

Figure 5.27 Two-dimensional mesh

Consider a 2D form of diffusion equation:

$$\Gamma\left(\frac{\partial^2\phi}{\partial x^2}\right)+\Gamma\left(\frac{\partial^2\phi}{\partial y^2}\right)+S=0 \tag{5.12}$$

Substitute the known boundary conditions in the equation to yield:

$$\Gamma\left(\frac{\partial^2\phi}{\partial x^2}\right)=\Gamma\frac{\phi_E+\phi_W-2\phi_P}{\Delta x^2},\Gamma\left(\frac{\partial^2\phi}{\partial y^2}\right)=\Gamma\frac{\phi_N+\phi_S-2\phi_P}{\Delta y^2} \tag{5.13}$$

Add and transpose, then

$$\left(\frac{2\Gamma}{\Delta x^2}+\frac{2\Gamma}{\Delta y^2}\right)\phi_p=\Gamma\frac{\phi_E}{\Delta x^2}+\Gamma\frac{\phi_W}{\Delta x^2}+\Gamma\frac{\phi_N}{\Delta y^2}+\Gamma\frac{\phi_S}{\Delta y^2}+S \tag{5.14}$$

Similarly, every node in the system can be expressed in the same way, and by combining all the equations of the system together into a simultaneous set, a linear system is established to be solved.

5.2.2.2 Finite Volume Method (FVM)

1. One-Dimensional Case

Finite volume is an extension method of the finite difference method for non-orthogonal grid systems. In this numerical approach, the spatial domain is subdivided into non-overlapping finite volumes or cells. For example, Figure 5.28 shows the cell of primary consideration, cell P, and the neighboring cells W and E, where w is the interface between P and W and e denotes the interface between P and E.

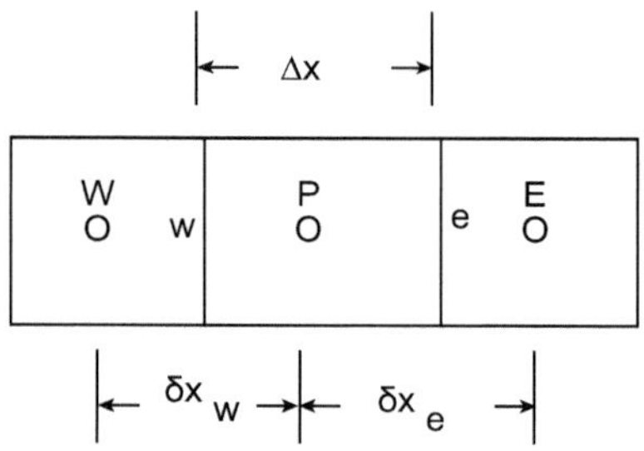

Figure 5.28 Control volumes of a 1D system

Consider a 1D thermal conduction question. Assuming that every cell is a control volume, we have:

$$\frac{d}{dx}\left(\Gamma\frac{d\phi}{dx}\right)+S=0 \tag{5.15}$$

We can define the volume average of the equation by integration:

$$\int_{w}^{e}\frac{d}{dx}\left(\Gamma\frac{d\phi}{dx}\right)dx+\int_{w}^{e}Sdx=0 \tag{5.16}$$

Substitution of the upper and lower limits of integration, that is, the two interfacing conditions of cell P, into the equation above, we get:

$$\left(\Gamma\frac{d\phi}{dx}\right)_{e}-\left(\Gamma\frac{d\phi}{dx}\right)_{W}+\int_{w}^{e}Sdx=0 \tag{5.17}$$

Now the original second-order differential equation is downgraded into a first-order one. The downgrading makes the numerical method easier to manipulate and also provides better control in error calculation.

Also, it is clear in the equation above that FVM features a thorough consideration of changes within a cell and the conservation of flux between cells. Assuming that φ varies linearly between cell centroids, Equation 5.14 can be expanded as follows:

$$\frac{\Gamma_e(\phi_E-\phi_P)}{\delta x_e}-\frac{\Gamma_w(\phi_P-\phi_W)}{\delta x_w}+\bar{S}\Delta x=0 \tag{5.18}$$

where $\bar{S}$ is the average source S of a cell.

Therefore, the algebraic relation of every cell is determined:

$$a_P\phi_P=a_E\phi_E+a_W\phi_W+b \tag{5.19}$$

where

$$a_E=\frac{\Gamma_e}{\delta x_e},a_W=\frac{\Gamma_w}{\delta x_w},a_p=a_E+a_W,\ b=\bar{S}\Delta x \tag{5.20}$$

Therefore, for a system with N cells, a linear simultaneous equation set of N is established and the solutions can be determined by substitution of the boundary conditions into these equations.

2. Two-Dimensional Case

Considering a 2D Laplace problem, FVM can be used for discrete deduction as well. An example of a 2D system is provided in Figure 5.29.

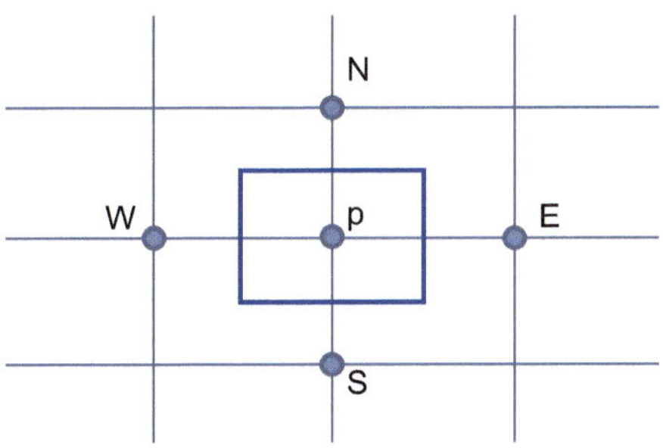

Figure 5.29 Control volumes of 2D system

The governing equation of the 2D system is:

$$\frac{\partial}{\partial x}\left(\Gamma\frac{\partial\phi}{\partial x}\right)+\frac{\partial}{\partial y}\left(\Gamma\frac{\partial\phi}{\partial y}\right)+S=0 \tag{5.21}$$

After integration, the space within the control volume becomes:

$$\iint\frac{\partial}{\partial x}\left(\Gamma\frac{\partial\phi}{\partial x}\right)dxdy+\iint\frac{\partial}{\partial y}\left(\Gamma\frac{\partial\phi}{\partial y}\right)dxdy+\iint Sdxdy=0 \tag{5.22}$$

By expanding the equation:

$$\left(\Gamma\frac{\partial\phi}{\partial x}\right)_E dy-\left(\Gamma\frac{\partial\phi}{\partial x}\right)_W dy+\left(\Gamma\frac{\partial\phi}{\partial y}\right)_N dx-\left(\Gamma\frac{\partial\phi}{\partial y}\right)_S dx+\bar{S}dxdy=0 \tag{5.23}$$

Modifying the above equation via numerical differential calculus gives:

$$\Gamma_e\frac{(\phi_E-\phi_P)}{(\delta x_e)}dy-\Gamma_w\frac{(\phi_P-\phi_W)}{(\delta x_w)}dy+\Gamma_n\frac{(\phi_N-\phi_P)}{(\delta y_n)}dx-\Gamma_s\frac{(\phi_P-\phi_S)}{(\delta y_s)}dx \\ +\bar{S}dxdy=0 \tag{5.24}$$

The following linear system is obtained:

$$a_P\phi_P=a_E\phi_E+a_W\phi_W+a_N\phi_N+a_S\phi_S+b \tag{5.25}$$

where

$$a_E=\frac{\Gamma_e}{\delta x_e}dy,a_W=\frac{\Gamma_w}{\delta x_w}dy,a_N=\frac{\Gamma_n}{\delta y_n}dx,a_S=\frac{\Gamma_s}{\delta y_s}dx \\ a_P=a_E+a_W+a_N+a_S,\ b=\bar{S}dxdy \tag{5.26}$$

The solution can be obtained by solving a set of simultaneous equation of a linear system consisting of each equation of the system that expresses every node in the 2D system in the form above. FVM uses the control volume concept for numerical discretization. The flux concept can be introduced to guarantee mass conservation; therefore, it is generally suitable for computational fluid dynamics (CFD) calculations.

5.2.2.3 Finite Element Method (FEM)

This is another commonly numerical method which determines an approximate solution by starting with an equivalent integral form instead of using the differential equations directly. The weighted residual method is normally used for equivalent integration, among which the Galerkin method is one approximation method [3]. Certain problems have their physical implications, and their variations may be shared with equivalent integration, such as the total potential energy function in the principle of minimum potential energy of solid mechanics.

Generally in a problem of physics or engineering, the unknown function of a field usually requires the satisfaction of both the set of differential equation and the boundary conditions:

$$\mathbf{A}(\mathbf{u})=\begin{Bmatrix}A_1(\mathbf{u})\\A_2(\mathbf{u})\\\vdots\end{Bmatrix}=0 \text{ in } \Omega,\mathbf{B}(\mathbf{u})=\begin{Bmatrix}B_1(\mathbf{u})\\B_2(\mathbf{u})\\\vdots\end{Bmatrix}=0 \text{ on } \Gamma \tag{5.27}$$

where **u** is the unknown field function, **A** is the equation set to be satisfied in the volume W, and **B** is the boundary condition to be satisfied on the boundary of W, G. With W and G satisfying both **A** and **B**, the following equivalent integral equation is developed:

$$\int_{\Omega} \mathbf{v}^{\mathrm{T}} \mathbf{A}(\mathbf{u}) d\Omega + \int_{\Gamma} \overline{\mathbf{v}}^{\mathrm{T}} \mathbf{B}(\mathbf{u}) d\Gamma = 0 \tag{5.28}$$

where

$$\mathbf{v} = \begin{Bmatrix} v_1 \\ v_2 \\ \vdots \end{Bmatrix}, \overline{\mathbf{v}} = \begin{Bmatrix} \overline{v}_1 \\ \overline{v}_2 \\ \vdots \end{Bmatrix} \tag{5.29}$$

Assuming that **u** can be expressed as an approximating function (the shape function) such as:

$$\mathbf{u} \approx \tilde{\mathbf{u}} = \sum_{j=1}^{n} \mathbf{N}_j a_j = \mathbf{Na} \tag{5.30}$$

Then, it is defined as:

$$\mathbf{v} = \sum_{j=1}^{n} \mathbf{W}_j \delta a_j = \mathbf{W}\delta\mathbf{a}, \overline{\mathbf{v}} = \sum_{j=1}^{n} \overline{\mathbf{W}}_j \delta a_j = \overline{\mathbf{W}}\delta\mathbf{a} \tag{5.31}$$

By introducing the integral equation, the following is obtained:

$$\delta\mathbf{a}^{\mathrm{T}} \int_{\Omega} \mathbf{W}^{\mathrm{T}} \mathbf{A}(\mathbf{Na}) d\Omega + \delta\mathbf{a}^{\mathrm{T}} \int_{\Gamma} \overline{\mathbf{W}}^{\mathrm{T}} \mathbf{B}(\mathbf{Na}) d\Gamma = 0 \tag{5.32}$$

Because δa_j is an arbitrary parameter, the integral equation is converted into a linear equation to determine $\boldsymbol{a}$:

$$\int_{\Omega} \mathbf{W}^{\mathrm{T}} \mathbf{A}(\mathbf{Na}) d\Omega + \int_{\Gamma} \overline{\mathbf{W}}^{\mathrm{T}} \mathbf{B}(\mathbf{Na}) d\Gamma = 0 \tag{5.33}$$

This is called the weighted residual method. The weighting function **W** is determined by the method used, such as the Galerkin method, and by setting **W** = **N** the corresponding matrix is obtained.

Assume an equation of the fluid mechanics P, which has an extreme value when the variation $\delta\Pi$ of Π is 0:

$$\Pi = \int_{\Omega} F\left(u, \frac{\partial u}{\partial x}, \ldots\right) d\Omega - \int_{\Gamma} G\left(u, \frac{\partial u}{\partial x}, \ldots\right) d\Gamma \tag{5.34}$$

If $\delta\Pi$ is deduced into the following form, the solution can be determined using the Galerkin method.

$$\delta\Pi = \int_{\Omega} \delta\mathbf{u}^{\mathrm{T}} \mathbf{A}(\mathbf{u}) d\Omega + \int_{\Gamma} \delta\mathbf{u}^{\mathrm{T}} \mathbf{B}(\mathbf{u}) d\Gamma = 0 \tag{5.35}$$

By comparing FEM and FVM, it is clear that FEM delivers an element-based shape function that determines the accuracy of the solution, and there are certain principles of mathematical manipulation to follow. Therefore, the method is suitable for solving problems of structural mechanics. However, for a fluid system, there are not many physical meanings similar to the principle of minimum potential energy mentioned above. A fluid system can only be characterized appropriately by considering the conservation of flux within a cell and on the interface between cells. That is why FVM is more suitable for solving problems of fluid mechanics.

Presented above is a simple introduction to the theoretical formulas and numerical methods for molding analysis. Refer to reference [4] for the more theoretical discussions.

■ 5.3 What Is Molding Simulation?

Section 5.1 provides details of the purposes of simulation analysis along with several examples of molding analysis, including inspection of design quality and solution for molding instability. To help readers understand what data and indices are needed to correctly obtain and interpret simulation results, this section goes back to the source of simulation and presents the knowledge and fundamental theories for the analytical work.

5.3.1 Brief History of Molding Simulation

Molding simulation analysis has its origin with Dr. Kamal of McGill University, Canada, in the 1970s. In 1978, Colin Austin developed the first commercial program that performed flow path analysis by laying a 1D thin flat wall using the layflat method, as shown in Figure 5.30.

The evolution of mesh techniques, as shown in Figure 5.31, started from the layflat method in the 1970s. Then, the 2.5D, called midplane, and shell methods were developed in the 1980s, followed by the quasi-3D and dual domain models in the 1990s, until the true 3D mesh models arrived in the 2000s. Because of the enormous growth of computational performance of computers, the number of mesh grids that can be processed has increased significantly. In theory, the 3D approach produces more accurate results.

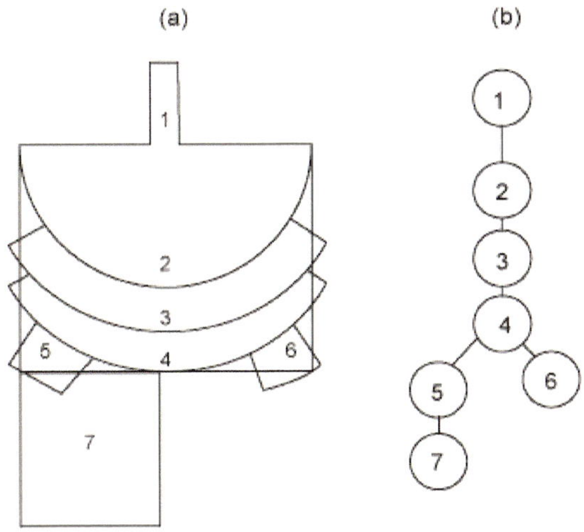

Figure 5.30 Schematic representation of runner-gate-cavity system in terms of (a) primitives and (b) various flow paths [5]

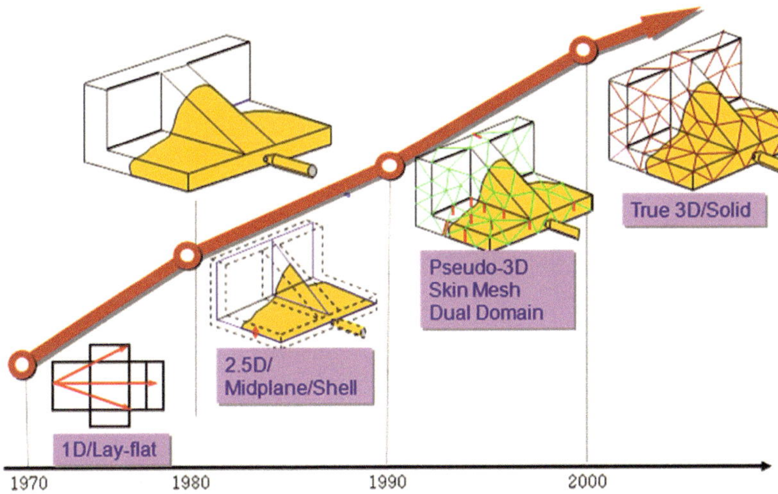

Figure 5.31 Brief history of mesh models for simulation

After the layflat method, Cornell developed the 2.5D thin-shell mesh technique in the Cornell Injection Molding Program (CIMP), as shown in Figure 5.32(a). The basic assumption is to simplify a 3D model into 2.5D one and to take the geometric middle plane as the calculation elements. The theoretical basis of this 2.5D mid-plane method comes from the Hele-Shaw equation [6] (or generalized Hele-Shaw). The theory utilizes the feature that the cavity of the injection mold is thinner in the direction of thickness and ignores the pressure gradient in the direction of thickness, that is, a simplified 2D pressure Poisson equation is used for the conservation of mass and momentum. For the conservation of energy, although the thermal convection in the direction of thickness and the thermal conduction in the flow plane are neglected as shown in Figure 5.32(b), the 3D changes of the temperature in the system still have to be considered. As a result, the entire system is simplified into a problem of 2D pressure field coupled with a 3D temperature field, and thus it is called “2.5D.”

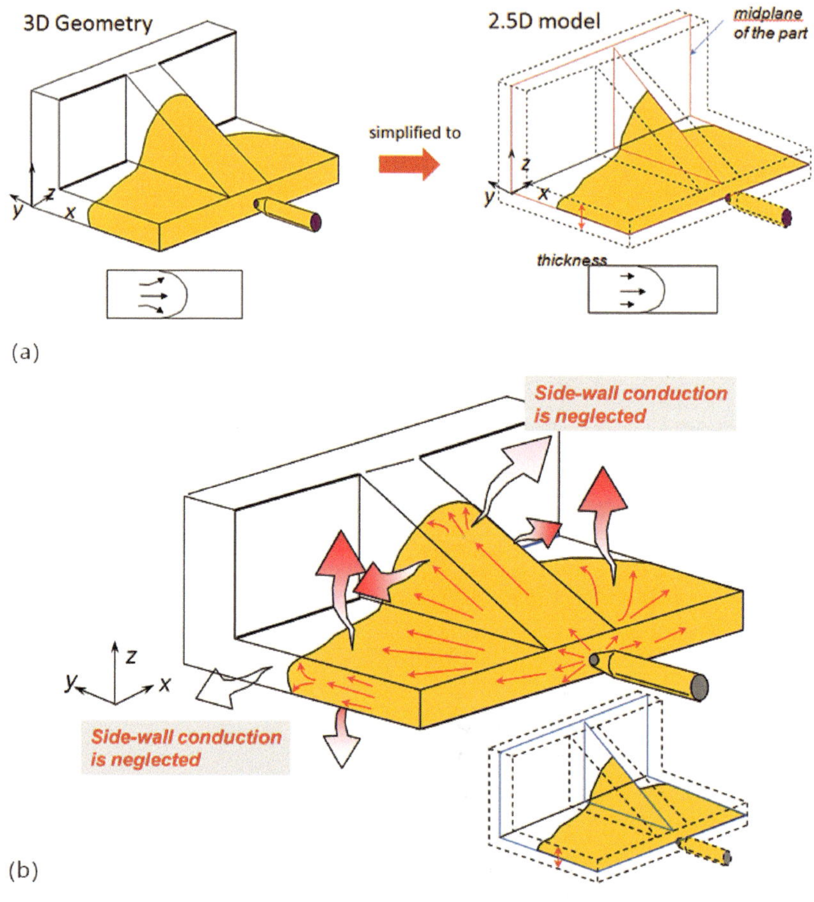

Figure 5.32 (a) 2.5D/midplane/shell and (b) limitations of 2.5D approach

The numerical methods appropriate for solving a 2.5D problem are FEM, which is used to solve the pressure equations, and FDM, which is used to deal with the temperature field in the direction of thickness. However, in certain flow areas as shown in Figure 5.33(a), such as the injection flow area at the melt front, T-shaped areas like ribs, or where there is a sudden change in thickness, the assumption that ignores the pressure in the direction of thickness in Hele-Shaw equation becomes false and would generate errors to a certain degree. These errors can be very influential to predictions of fiber-reinforced plastics, because movements of these fibers, such as twists and turns in the 3D space, have to be considered. As shown in Figure 5.33(b), the 2.5D model cannot predict the out-of-plane direction of fiber orientation precisely in the zones with thickness change. In addition, the thermal convection in the direction of thickness and the thermal conduction in the flow plane are ignored due to the assumption of a Hele-Shaw flow. The 2.5D model does not provide a complete picture for parts with inserts or pores where side-wall heat conduction can be significant. Another problem of the 2.5D approach is the creation of the midplane model from the 3D part, which is usually very time-consuming.

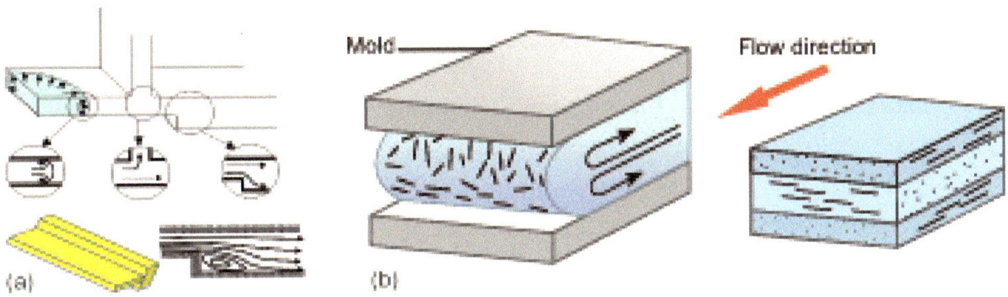

Figure 5.33 Limitations of 2.5D approach: (a) melt front and thickness variation zones, and (b) flow behavior of fiber induced by melt

The dual-domain method, as shown in Figure 5.34(a), was born to improve the limits of the 2.5D technique by using the surface meshes of the 3D model directly. In this approach, the 2.5D solver is applied to the surface (skin) mesh of the model. The flow pattern on both the core and cavity sides of a mold are calculated and introduction of additional linking coefficients corrects the double-sided flow behavior of the model. By doing so, the model produces results that look like 3D and there is no need to take the midplanes out.

Simple as it may be, the dual domain method still has a risk when it comes to parts that are difficult to define the thickness. For a geometry rectangular on the inside and circular on the outside such as the one shown in Figure 5.34(b), errors in thickness judgment at the rectangular corners can easily result. Errors such as discontinuity of melt may also occur with the dual domain method due to the connector approach [7]. In comparison, only the true 3D mesh can produce correct simulation results.

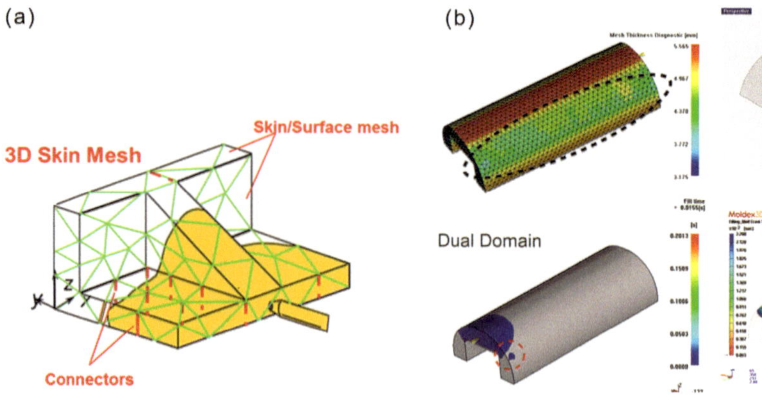

Figure 5.34 (a) Dual domain and (b) thickness determining issues of dual domain model

Of the evolved mesh techniques up to now, the 3D mesh model is still the most accurate method because it requires no geometrical simplification, as shown in Figure 5.35(a), and the governing equations (see Section 5.2.1) describe exactly the 3D space of the system. By building products and runners using the 3D mesh model as shown in Figure 5.35(b)–(c), the simulation results can be closer to the real situation. The key to the use of this meshing method is the solver's stability and performance.

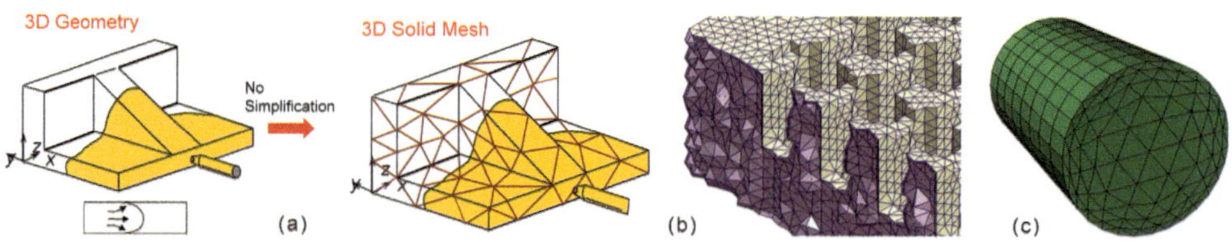

Figure 5.35 True 3D/solid model: (a) no simplification is needed, (b) 3D mesh model of part, and (c) 3D mesh model of runner

For example, in Moldex3D the following three-dimensional types of mesh are basically used: hexahedron, prism, pyramid, and tetrahedron, as shown in Figure 5.36. The triangular mesh is frequently used for discretizing irregular shapes.

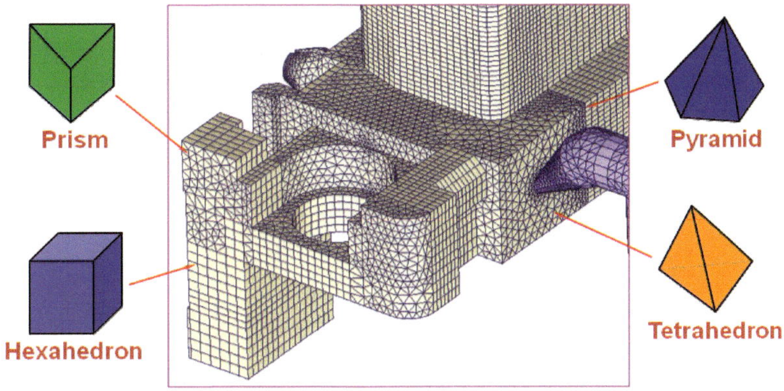

Figure 5.36 Hybrid meshing technology

Among all of these meshes, tetrahedron, prism, and hexahedron are commonly used for molding analysis. Table 5.2 lists the commonly used mesh types and the time for building the model with a certain mesh type. Tetrahedrons are suitable for thick products, whereas hybrids perform well for both thin and thick parts. There are two types of hybrid: one is based on hexahedrons with the advantage of better calculator performance, but the downside is that the model has to be built manually, and thus requires experience and time. The other hybrid type, for a more convenient way of building grid models, is the boundary layer mesh (BLM). It can be used for products of any shape. The main feature is that the prisms are used on the surface of the product, and the inside is created by tetrahedrons. The advantages of having a better resolution at the walls of the model and a medium number of elements help produce better predictions in a relatively short calculation time. For example, the temperature distribution in the direction of thickness as presented lastly shows accurate simulation of the temperature rise in the shear layer.

Having a high-quality mesh that best represents the geometric characteristics of a complex model is very desirable for attaining accurate mold-filling analyses. Thus, deciding the appropriate type of meshing for mold-filling analyses has become essential if the accuracy is of the highest concern. BLM is a powerful pre-processing tool for complex geometries. It includes multiple layers of prismatic meshes out of the triangular surface mesh of the object, followed by filling up the internal space with a quality-controlled tetrahedral mesh. CAE solvers including mold-filling software can capture the variable distributions more accurately with a much lower number of elements of a BLM than an isotropic tetrahedral mesh.

Table 5.2 Comparison of 3D Mesh Types in Moldex3D

	Tetrahedron	Hybrid (Hexahedron + Prism)	Boundary layer (Tetrahedron + Prism)
Suitable for	Thin part only	Both thick and thin	Both thick and thin
Elapsed meshing time	3-6 hours	1-3 days	4-8 hours
Element layer count	Cannot control	Manually (3 or 4) (uniform distribution)	Automatically (at least 3) (biasing distribution)
Solver accuracy	Good	Excellent	Excellent
Solver stability	Excellent	Excellent	Good
Solver efficiency	Good	Excellent	Good
Temperature distribution	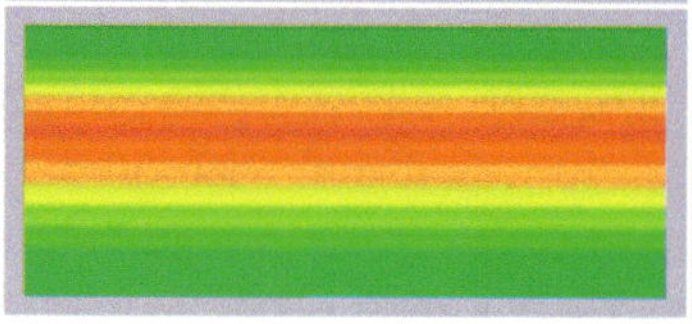	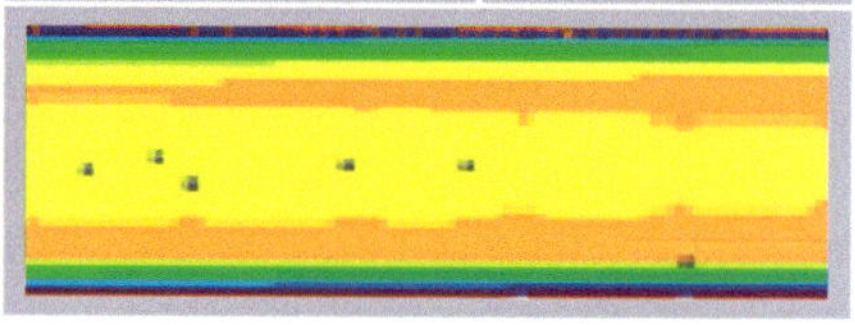	

For example, constructing a satisfactory mesh for a complex conformal cooling channel is not an easy task. Also, due to the unique feature of a conformal cooling design, as shown in Figure 5.37(a), it poses a great difficulty to fully capture the curvatures in meshing. Figure 5.37(b) shows that the BLM approach can be applied to solve this problem. A BLM can precisely capture the drastic changes of temperature and velocity near the cavity during the filling process, so that users can accurately detect viscous heating and warpage problems in advance. That is to say, a BLM is a powerful pre-processing tool for complex geometries.

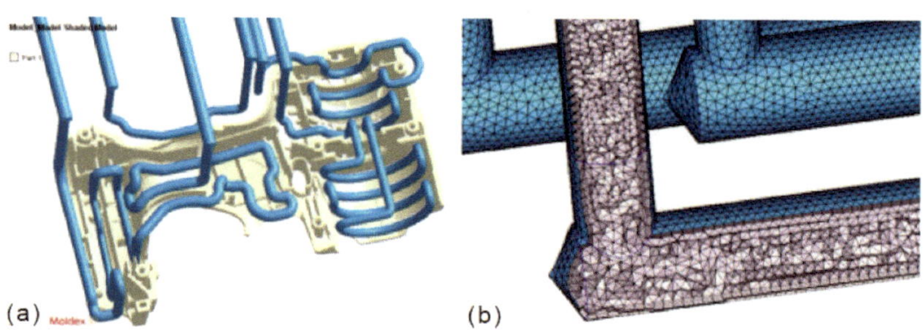

Figure 5.37 (a) A complex conformal cooling design and (b) its 3D solid BLM for accurate mold-filling analyses

5.3.2 Simulation Workflow

The development of molding analysis software has been going on for decades. The analysis steps can be summarized in a flow chart, as shown in Figure 5.38. First of all, we need the three major items in a molding system: the overall geometric model of the system, the selected material, and the processing conditions, which are transmitted to the numerical engine through the user interface. The governing equations are used as the algorithm logic, and the appropriate material theory model is selected for numeric simulation.

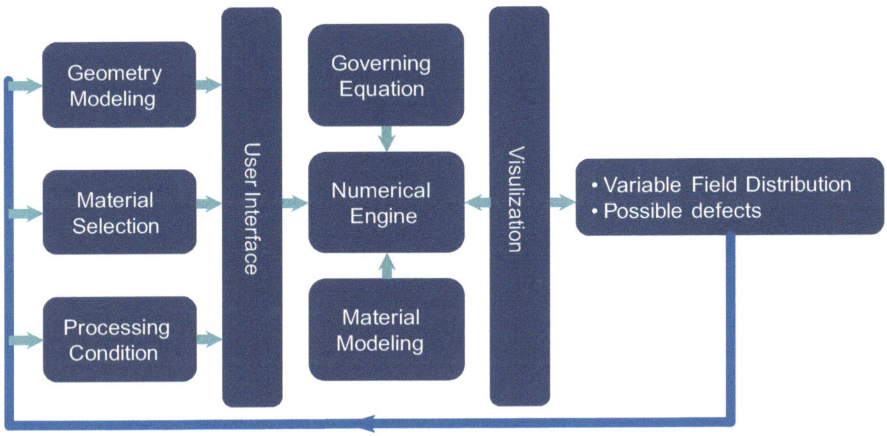

Figure 5.38 General workflow of CAE simulation

So what do we look at in a flow analysis? Figure 5.39 lists a few basic indices for interpretation, including (a) will the part fill? (b) is the flow balanced? (c) where will weld lines be and how is the quality? (d) will there be any air trap? (e) how large is the maximum stress level? (f) is the clamping force too large? (g) are there any hesitation or race-tracking issues? (h) how significant is the viscous heating effect? and (i) is there an over-pack issue? If any of these criteria fails, the design has to be modified and the analysis redone until the problem is solved. Things like these will point the direction for interpretation of the results for beginners.

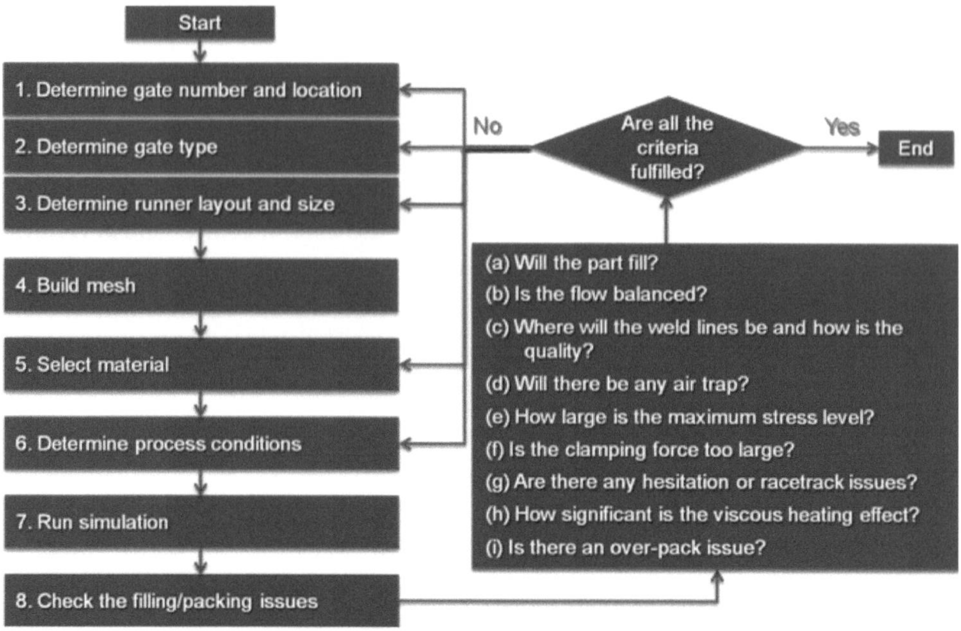

Figure 5.39 Workflow of filling and packing analysis

Figure 5.40 shows the cooling stage after the flow analysis, and this requires the definition of cooling channels, mesh model, cooling material properties, and molding conditions. The analysis results will give us answers to questions such as: (a) is the cooling time to long? (b) where is the hot spot? (c) is the part surface temperature uniform? (d) is the mold temperature difference larger than 10 °C? (e) is the temperature difference of coolant inlet and outlet larger than 3 °C? and (f) what are the cooling channel efficiencies? Similarly, if any of these criteria fails, the design has to be modified and the analysis redone until the problem is solved.

Apart from the simulation of the molding process, molding analysis tools also provide deformation predictions after demolding. As shown in Figure 5.41, the software captures the results of flow analysis and cooling analysis. The warping analysis reveals the level and trend of warping deformation caused by molding effects, the cause of warping, and the value of residual stress.

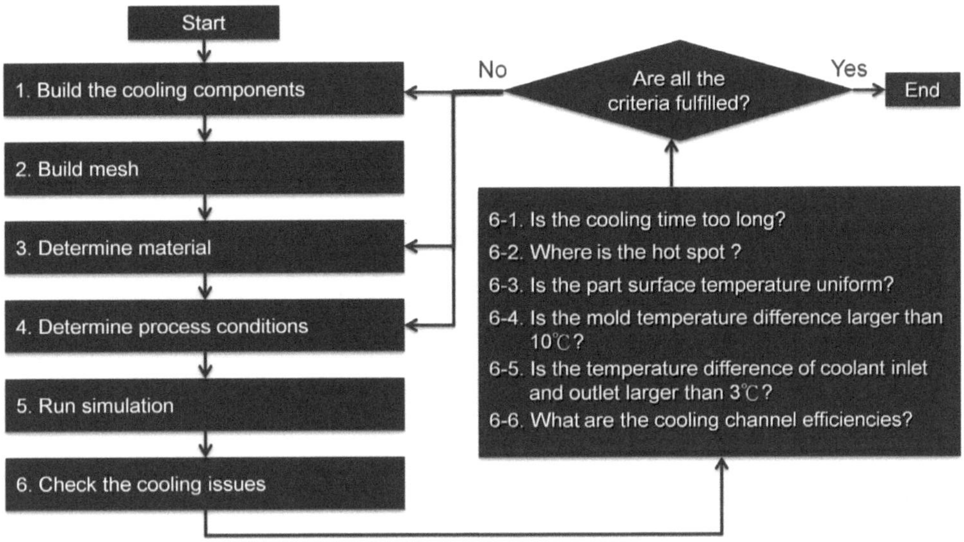

Figure 5.40 Workflow of cooling analysis

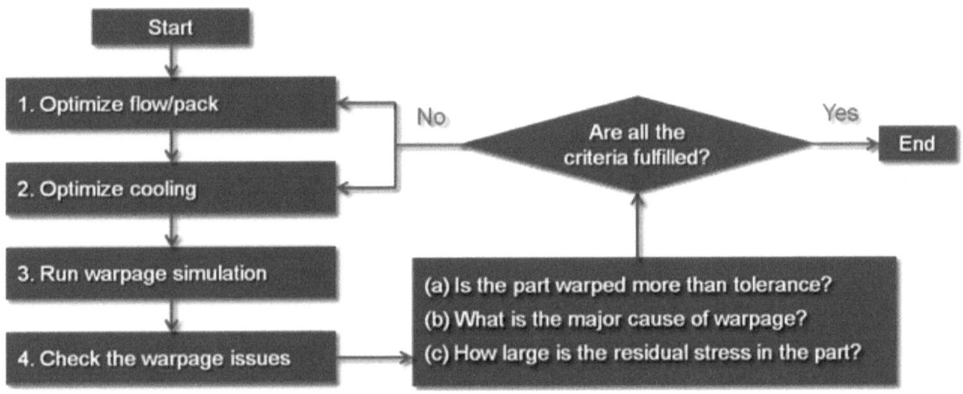

Figure 5.41 Workflow of warp analysis

References

[1] Bird, R. B.; Stewart, W. E.; Lightfoot, E. N.: *Transport Phenomena*, Revised 2nd ed., New York, USA: John Wiley & Sons, 2007

[2] Murthy, J.; Mathur, S.: Draft Notes of ME 608 Numerical Methods in Heat, Mass, and Momentum Transfer, School of Mechanical Engineering, Purdue University, 2002

[3] Zienkiewicz, O. C.; Taylor, R. L.: *Finite Element Method*, 5th ed., vol. 1, Woburn, USA: Butterworth-Heinemann, 2000

[4] Chang, R.-Y.; Yang, W.-H.: Numerical Simulation of Mold Filling in Injection Molding Using a Three-dimensional Finite Volume Approach, *International Journal for Numerical Methods in Fluids*, vol. 37, 2001, pp. 125–148

[5] Isayev, A. L. (Ed.): *Injection and Compression Molding Fundamentals*, New York, USA: CRC Press, 1987, p. 450

[6] Kennedy, P. K.: *Flow Analysis of Injection Molds*, New York, USA: Hanser, 1995

[7] Yu, H. G.; Thomas, R.: "Method for Modelling Three Dimension Objects and Simulation of Fluid Flow", US Patent US6096088, 2000

6 Flow Consideration versus Part Features

Wen-Hsin (Debbie) Weng

6.1 Basics

The filling stage of an injection molding process denotes the course of pushing the melt into the mold cavity with external pressure. The pressure induced in the process drives the melt front forward until the mold cavity is filled. The source of the pressure inside the cavity comes from the gate, which is the location under the most pressure in the whole part.

As the distance from the gate increases, the pressure drops gradually. While the fluid is filling the cavity, friction also causes the polymer to undergo shear heating and melt even more. The melt front is under the least pressure. The pressure difference inside the melt is the main force driving the thermoplastic melt to move forward.

In an injection molding manufacturing cycle, the filling stage determines the whole framework of the molding performance. Improper processes or designs lead to poor product quality, whereas the flow behavior of the material during the filling process directly affects all kinds of defects in the part.

6.1.1 Flow Behavior of Plastic Melt in the Cavity

During melt filling, the melt flow front generates the fountain effect while heat conduction gives rise to the skin effect. As shown in Figure 6.1, these two effects are described as follows:

1. Fountain effect: Because a melt is a fluid of high viscosity, it forms a laminar flow when moving forward in the mold cavity. As the flow rate is higher in the center of the melt and lower on the sides, the central part of the melt flows sideways when it moves to the frontmost end and thus forms a fountain-shaped melt front.

2. Skin effect: The mold is a region with relatively low temperature in comparison to the high temperature of the melt. The surface layer of the melt material is cured upon contact with the mold of lower temperature, and a solid, non-flowable region is thereby formed, which reduces the actual flowing space of the melt.

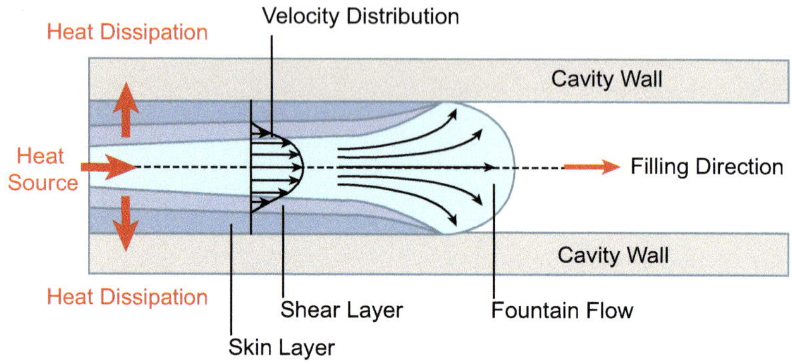

Figure 6.1 Polymer melt filling progress

The flow behavior of a plastic cannot be controlled when it leaves the nozzle tip. The plastic flows to wherever there is lower resistance, which can only be changed by adjusting the various flow rate stages. It is not easy to control the flow of plastic because the plastic at the front is always colder than the hot plastic at the back. As the plastic flow is laminar and not turbulent, there are layers in the distribution of velocity and temperature, and thus issues such as stress are generated.

Because the plastic has high viscosity, it forms a steady laminar flow field upon flowing, as shown in Figure 6.2. The immobilized region close to the wall is a frozen skin layer. The shear layer lies more inside.

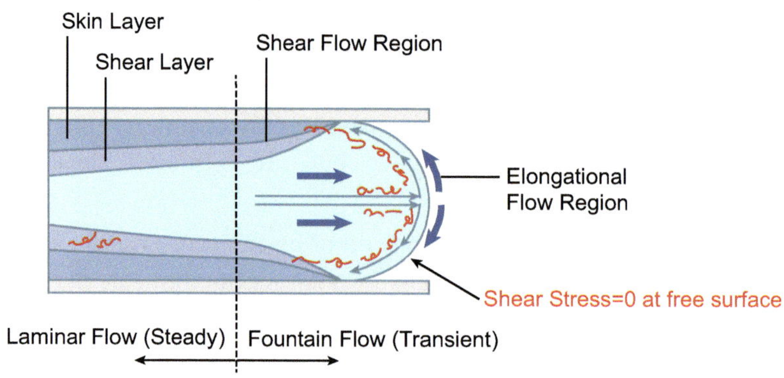

Figure 6.2 Fountain flow field of melt front

Because the velocity is fastest in the center of the fountain, the melt is pushed forward and creates a fountain-shaped flow field of the melt front, which causes an elongation effect on the frontmost melt.

As shown in Figure 6.3, the melt velocity of the shear layer has a higher speed difference to the skin layer (zero velocity); the melt adopts the flow behavior of a high shear rate and the molecular chains become highly oriented, causing the material viscosity to be lower in the shear layer. However, product quality is seriously affected by the high shear rate in the shear layer. Therefore, more care needs to be taken with the injection speed, which is the major injection factor influencing the shear rate.

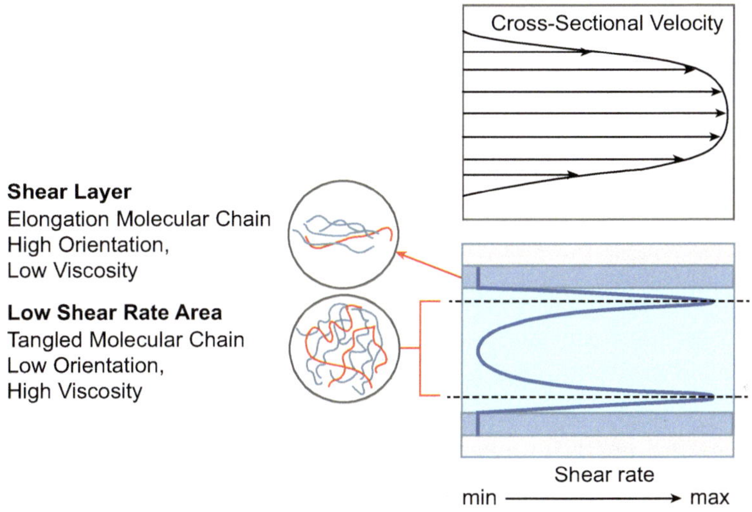

Figure 6.3 Steady laminar flow field of polymer melt

In addition to the injection speed, the thickness of the product and the temperatures of the mold and the melt all affect the thickness of the shear layer during molding. When the thickness of a product decreases, the velocity and shear stress increase and induce shear heating, leading to a thinner frozen skin layer. Also, an increase in the mold and melt temperature directly weakens the dissipation of melt heating, and thus the skin layer becomes thinner as well.

Figure 6.4 shows the viscosity chart of a plastic material. The x-axis denotes the shear rate (1/s), and the y-axis is the viscosity (g/(cm s)). The curves in different colors represent the viscosity at different temperatures.

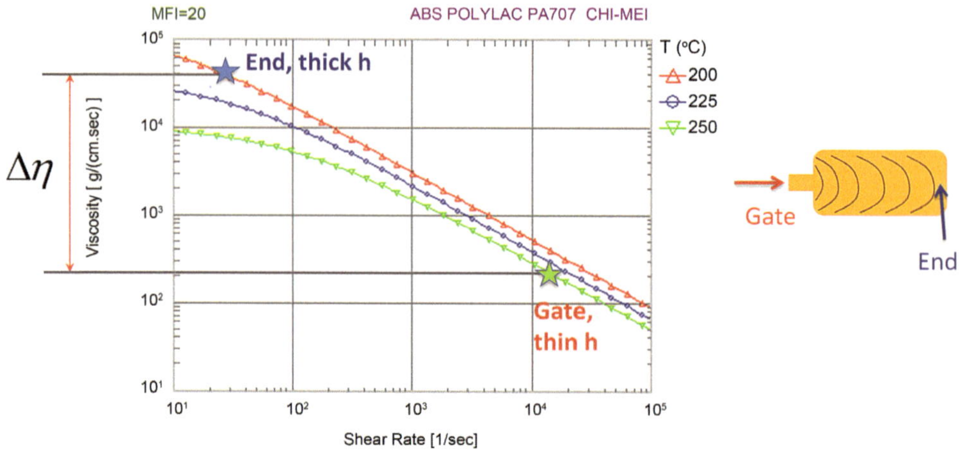

Figure 6.4 Injection process versus material properties of viscosity

At a constant temperature (i. e., a curve of one color), the viscosity decreases as the shear rate increases; this shows that the polymer material has a shear-thinning property; shifting the focus to a constant shear rate (for the same vertical line), the viscosity decreases as the temperature increases. The material shear rate during the process ranges from 1×10^1 to 1×10^4 1/s.

As the cross section at the gate is the smallest, it also has the highest shear rate in the whole mold cavity (around 1×10^4 1/s) and thus the temperature increases significantly. Usually, when the melt reaches the filling end, the screw speed is reduced to a very low level and thus the shear rate is relatively lower (around 1×10^2 1/s), and so is the temperature.

Figure 6.5 shows a C part of a notebook keyboard frame, which is an injection molding that uses hot runners to connect cold runners and 10 gates injecting at the same time. The bottom right corner of the part appears to be thinner, because the cross section is smaller for the melt to pass and thus the velocity difference in this extremely small height forms a high shear rate of melt, which reaches 4×10^3 1/s. Also, as the high shear rate effect leads to the decrease of viscosity, the melt viscosity in this region is smaller than 1×10^3 poise.

According to the description in Figure 6.6, a product specification includes size, impact strength, tensile strength, chemical properties, optical properties, etc., which have a direct relationship to temperature, pressure, viscosity, thermal properties, and density during plastic molding. However, injection molding machines on site have no means to directly control these product specifications, and neither can they control the plastic properties of processing. Therefore, product quality is controlled indirectly through the control of screw temperature, mold temperature, injection speed, injection pressure, VP switch, pack pressure, pack time, cool time, and other conditions on the machine, by which the material properties are affected and hence the part specifications are affected afterward. In addition, the conditions are highly nonlinear relative to each other.

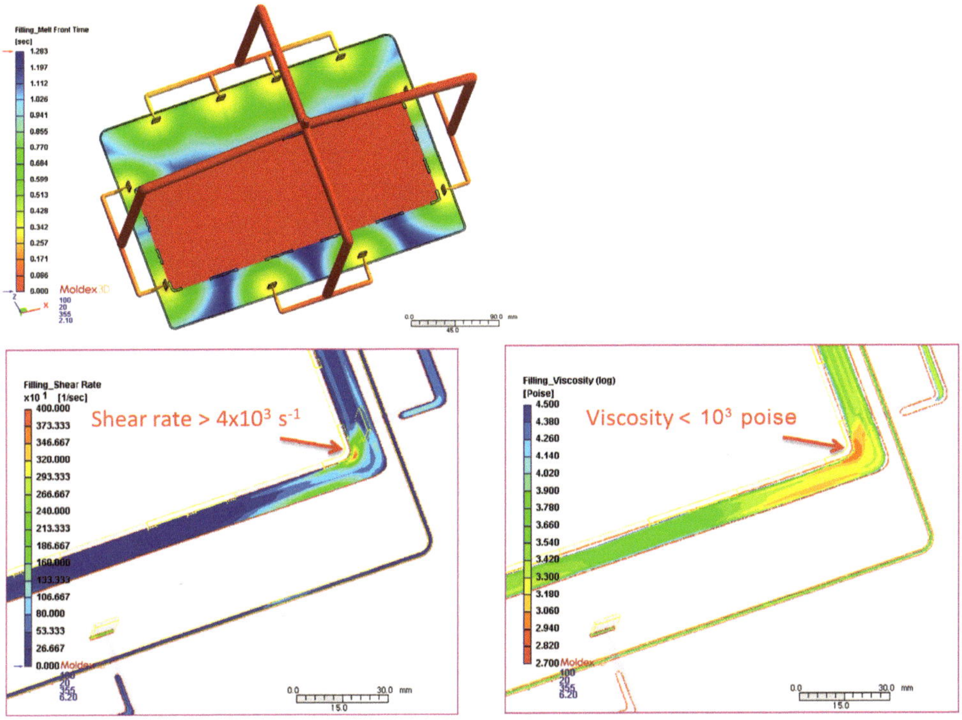

Figure 6.5 Shear rate and viscosity at thin thickness location

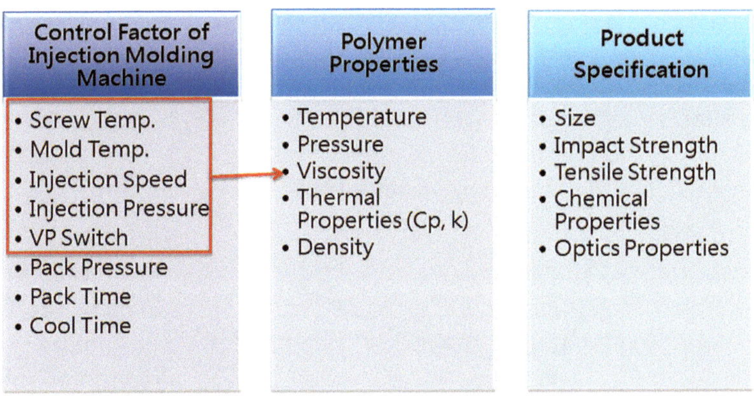

Figure 6.6 Indirect product quality control of injection molding

Because injection molding is a batch production process, an injection molding machine should basically be capable of plasticizing at a stable temperature, performing stable metering, filling, and packing, and providing stable injection speed, pressure, and time.

6.1.2 Effects of Filling Time

The filling time has a critical influence on molding quality. The relationship between filling time and injection pressure is shown in Figure 6.7.

1. Short filling time:

   Advantages: A short cycle time and short filling time result in a smaller reduction in melt temperature, with only small differences between each region (temperature loss in a short filling time is rare).

   Disadvantages: The injection pressure required for extremely high injection speed soars rapidly and may create high temperature due to friction in high shear rate regions, which instead causes high temperature differences between each region.

2. Long filling time

   Advantages: Molecular chains and fibers are less prone to be stretched and sheared as the shear rate is lower, which generates smaller internal stress.

   Disadvantages: There is a longer molding cycle. An excessively low injection speed would cause a significant drop in material temperature, which increases the viscosity and filling hesitation. Thus, the molding may need an even higher injection pressure or a short shot may occur.

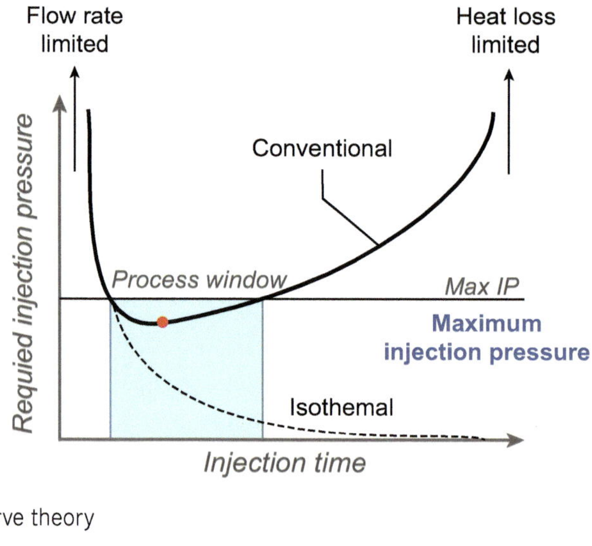

Figure 6.7 U-curve theory

U-curve theory expresses the relationship between injection pressure and injection time. When the injection time is very short, higher injection pressure is required to overcome melt flow resistance. As the injection time increases, the injection pressure decreases. When the injection time is too long, the decrease in melt temperature causes the viscosity to rise rapidly and the injection pressure increases. The curve of the pressure versus time becomes U-shaped.

Other important machine parameters include amount of injection, injection pressure, clamping force, back pressure, suck back, pressure loss, barrel residence time, etc.

A simple estimation of filling time can be made based on the injection volume, and the fact that the higher the cavity volume, the longer is the injection time needed for filling. And the materials are divided into three major categories - low, medium, and high viscosity - the one of high viscosity does not flow easily and thus the recommended filling time should be longer.

6.1.3 Flow Rate versus Injection Pressure

A smooth flow front of constant velocity in the mold can lead to injection-molded parts of better quality. However, most part thicknesses are not even, and filling cannot be done under constant injection speed; instead, a multi-step injection speed control is required to generate a smoother flow of the plastic, as shown in Figure 6.8.

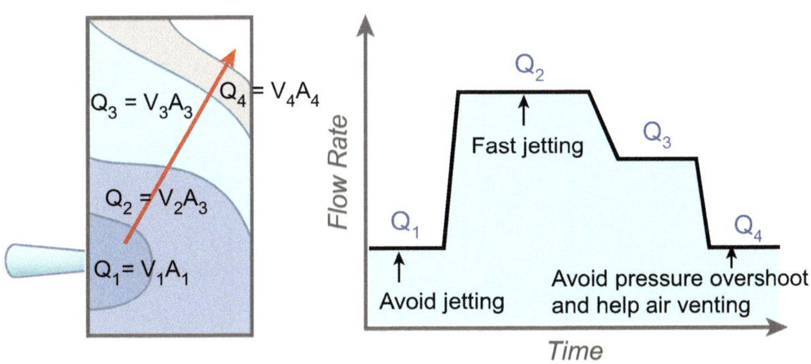

Figure 6.8 Principle of multi-step injection rate

6.1.3.1 Flow Rate Curve Setting

The injection rate denotes the maximum volume of plastic melt able to be injected by the nozzle in a unit time. The velocity of plastic melt passing through the nozzle can be represented by (cross sectional area of injection screw × linear screw speed) or (injection volume/injection time); the unit is cm^3/s.

Normally, a higher injection rate is better, so that a higher volume of material can be injected in a short time, which avoids insufficient filling resulting from the loss of flowability of the melt inside the mold due to overcooling under an excessively low injection speed in the thinner part. However, with a high injection rate, a plastic of poor heat stability may thermally decompose due to excessive shear heating, which often leads to yellowing, brown-black stripes, and other defects in appearance.

Generally, injection machines under 100 tons are often chosen to manufacture small and medium-sized parts, and the injection time is better controlled in less than one second. An ideal condition is one in which the melt moves forward at a constant speed throughout the whole injection process.

As shown in Figure 6.9 for a disk part, because the gate locates in the center, the melt front area (MFA) expands along with the forward moving of the melt front, leading to a gradually increase of the flow rate from Q_1 to Q_3. However, the last filling stage of the flow rate should be decreased before the part is completely filled up to allow the air enough time to vent out the mold.

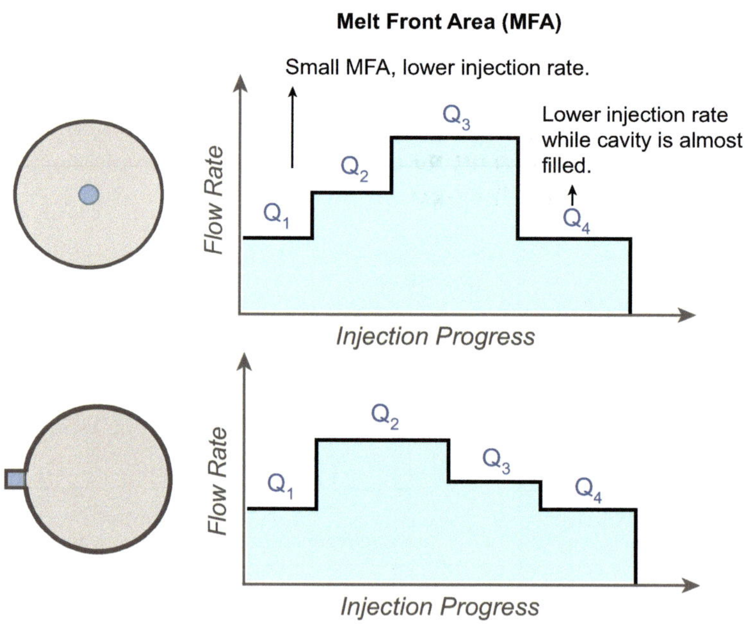

Figure 6.9 Example: multi-step injection rate

In case the gate is located on the side of the disk part, the initial cross section of the melt front is smaller with the flow rate of Q_1. As the cross section increases gradually the flow rate moves to the second stage, a faster Q_2. When it comes to the second half of the injection process, the cross section starts to decrease and thus the flow rate gets lower to Q_3, which continues to drop to Q_4 before the mold cavity is finally filled up.

Concerning the filling stability of a plastic material, as the thickness of a part varies, the melt front velocity changes between each region, which could further affect the quality of the plastic part. Hence, upon encountering resistance from different thicknesses, proper adjustment of the filling flow rate improves the stability of the melt front velocity.

With designs differing in various runner and gate types, the concept for the multi-step setting of filling rate also differs. The setting of a multi-step filling rate is generally divided into the following few parts if no special purpose is required:

1. Runner region.
2. Near the gate (before and after entering the gate, the melt front area is about twice its original size).
3. Main regions in mold cavity (60–90%).
4. End of mold cavity: normally a lower flow rate is set to avoid excessive high pressure.

Take the part of a mold with four cavities as an example for a multi-step screw velocity setting in five stages V1–V5, as shown in Figure 6.10:

1. 1st stage V1: from sprue to runner.
2. 2nd stage V2: for passing the gate, screw velocity is decreased.
3. 3rd stage V3: as the structure of the part is simple and the cross section of the melt front is fixed, the injection speed is not reduced until the part is nearly filled up.
4. 4th stage V4: the cross section of the part starts to shrink, and thus the injection speed starts to decrease.
5. 5th stage V5: the melt front almost reaches the product end, and the injection speed is further reduced to facilitate venting.

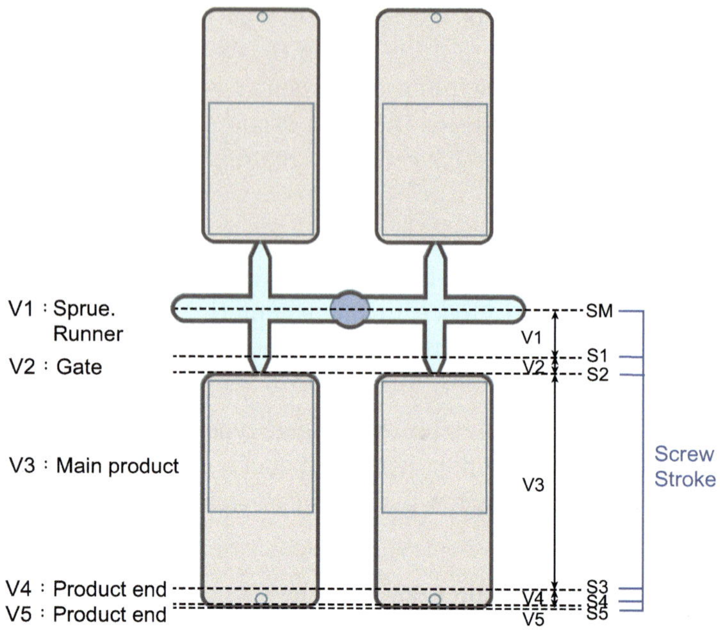

Figure 6.10 One mold with four cavities

Usually, the setting of the filling flow rate can serve different purposes. As shown in Figure 6.11, the part quality can be highly affected by the injection speed:

A – Under a very low injection speed, flow marks, or short shots might occur.

B – Under the optimum injection speed the part quality is assured.

C – When the injection speed is higher, gate marks (cloudy) appear around the gate.

D – As the injection speed becomes even higher, melt fracture is generated.

E – Under an extremely high injection speed, jetting and burning with silver streaks and flash occur in the mold cavity.

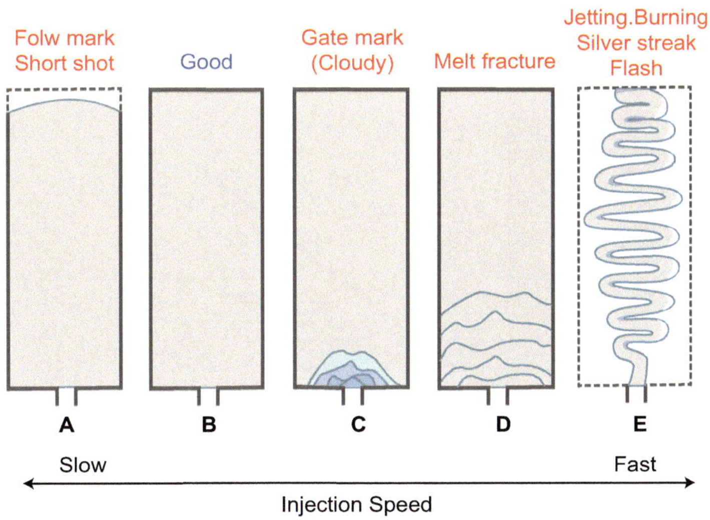

Figure 6.11 Flow rate of molten polymer through gate

To reduce flow marks in the part and other exterior defects, the filling speed is reduced as the plastic material flows through the gate.

6.1.3.2 Relationship of Injection Rate and Injection Pressure

The machine response cannot completely meet the injection speed setting, and thus a delay in response is seen. As the injection speed gets higher, the deviation between the real injection speed and the setting value of the machine increases. For example, as shown in Figure 6.12, the real injection speed and the setting value of the machine are very close when the setting is at a low speed of 200 mm/s; as the injection speed setting reaches 400 mm/s, the real injection speed of the machine cannot keep up and appears to be 350 mm/s only; as the setting value continues to increase, the real injection speed of the machine can only reach 700 mm/s when the setting value is as high as 1000 mm/s.

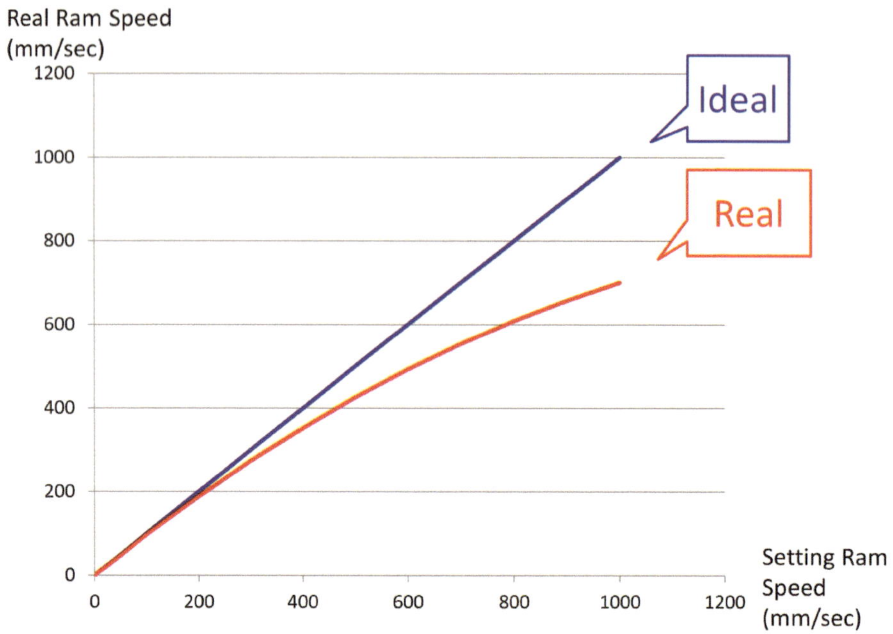

Figure 6.12 Response of injection machine: setting speed versus real speed [1]

Figure 6.13 shows a screen of the actual setting and the response curve of an injection machine, in which two stages of injection speed – 150 and 80 mm/s – are configured. The upper pressure limit is set at 210 MPa. The actual filling speed represented in the yellow curve shows a delay from the setting value, and results in a final injection pressure of 130 MPa.

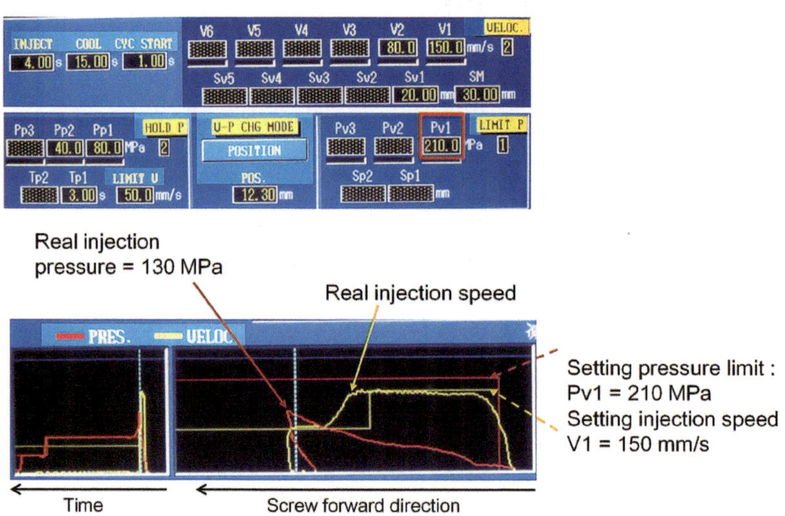

Figure 6.13 Relation between injection speed and injection pressure

During an injection molding process, the flow resistance increases as the mold is being filled gradually. Therefore, the sprue pressure should be increased to provide sufficient pressure to push the melt forward. A typical sprue pressure variation is shown in Figure 6.14.

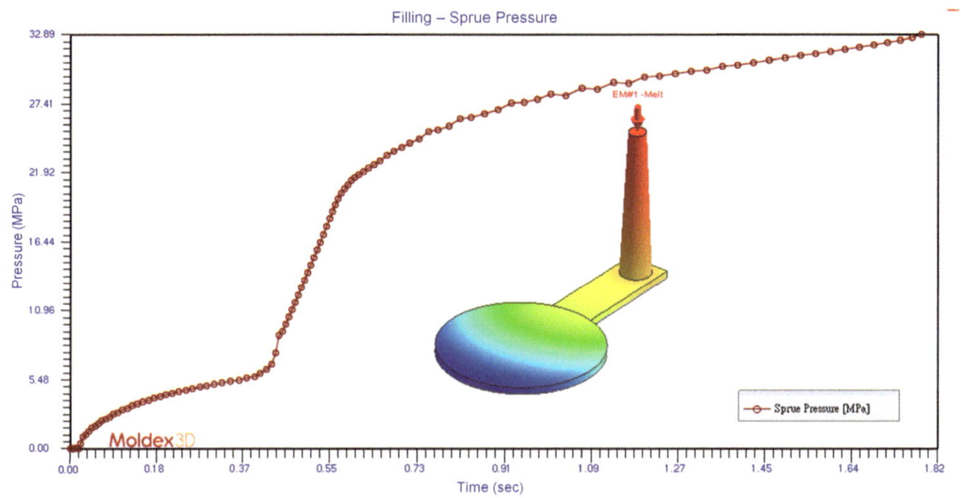

Figure 6.14 Typical sprue pressure variation

A check of the maximum peak pressure via the sprue pressure during the process reveals the flow imbalance condition. As shown in Figure 6.15, there is a significant pressure surge at 16 to 22% filling.

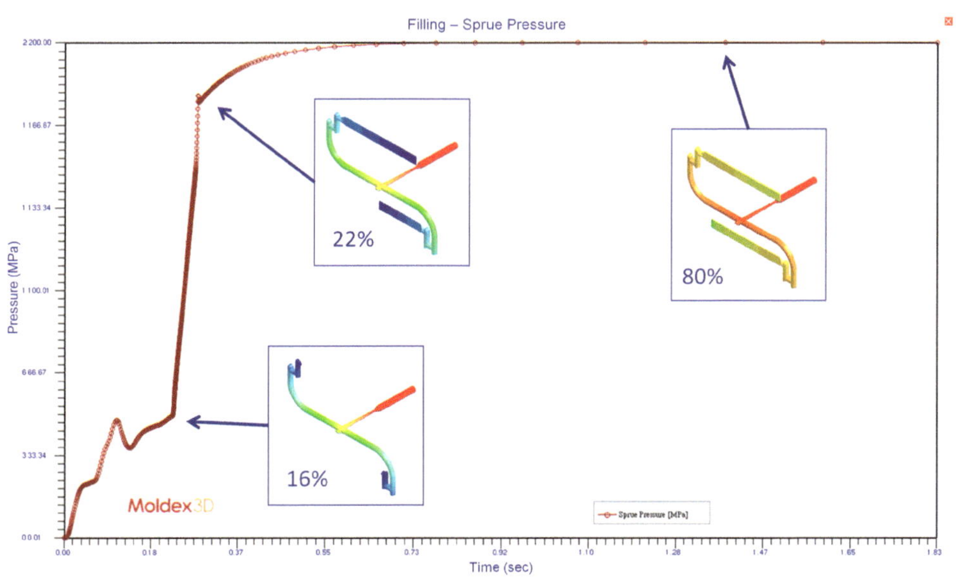

Figure 6.15 Sprue pressure at 85–95% filling

In an injection process, there is a big difference between the actual response values and the setting values of the pressure and the speed of the machine. Taking Figure 6.16 as an example, the two horizontal dotted lines show constant limit values of both pressure (red) and speed (blue). The actual injection pressure curve (Real P) starts from zero at the beginning of filling, then gradually increases as the filling process proceeds, and keeps at a constant value when it reaches the pressure setting value. The blue solid line (Real V) represents the actual speed curve, which also starts from zero at the beginning of filling and gradually increases to the speed setting value. However, due to the inertia of the screw, the actual speed cannot precisely stop at the setting value; instead, it moves up a little and then comes back to the setting value.

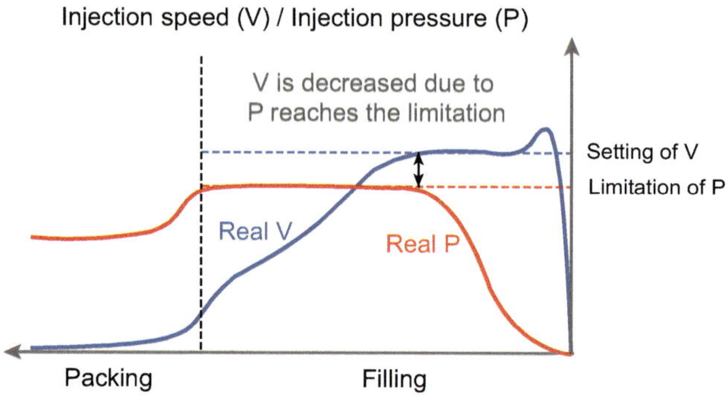

Figure 6.16 Insufficient injection pressure setting value leads to decreasing injection speed

If the mold is filled at a constant speed, the actual injection pressure goes up accordingly. However, the pressure reaches the upper limit pressure, the filling speed no longer fills at the set speed and starts to drop. At the packing stage, the pressure setting value is further reduced. As for the speed curve (blue solid line), because only a very small flow rate is used to compress the melt inside the cavity. Once the cavity is filled up, the value drops to nearly zero.

What can be done to avoid reaching the upper pressure limit too early, which leads to a drop in injection speed? As shown in Figure 6.17, by increasing the pressure limit the setting value is increased (red dotted line), the red solid line showing the actual pressure curve does not reach the pressure limit before the second half of the filling process. As for the blue solid line showing the injection speed curve, the setting value can be maintained for a longer time without dropping until the pressure limit is hit in the second half of the filling process.

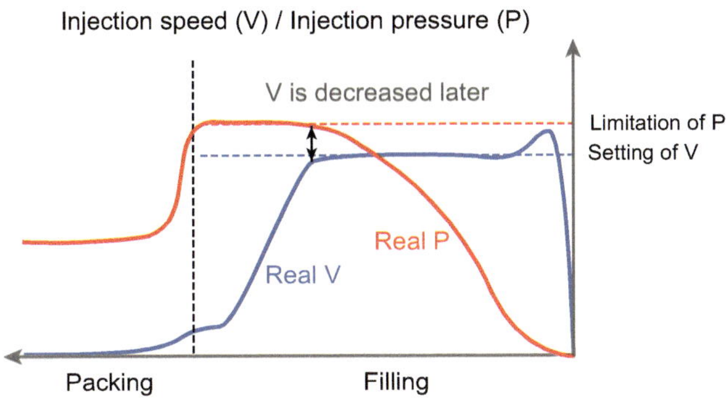

Figure 6.17 Increasing the injection pressure setting value increases the injection speed

The pressure limit can be set further up as shown in Figure 6.18 so that the pressure does not exceed the setting value throughout the whole filling stage. It can be seen that during filling, the blue solid line showing the injection speed curve rises a little bit over the setting value, then returns back to target, and remains constant. Meanwhile, the injection pressure (red solid line) rises steadily before it switches to the packing pressure when the VP switch point is reached, and concurrently the injection speed curve (blue solid line) also drops rapidly to enter the packing stage. The injection pressure does not reach its limit and hence the injection speed can follow the set value well.

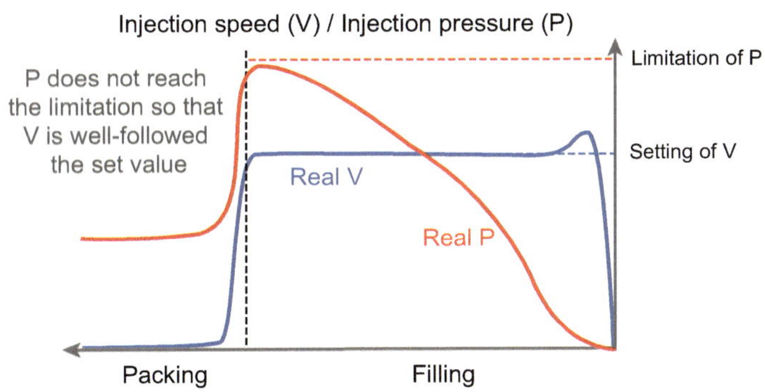

Figure 6.18 Sufficient injection pressure setting value: injection speed increases until VP switch

We can change the injection speed setting value if the pressure of the injection stage is to be controlled. For instance, the injection speed, as shown in Figure 6.19, is set in three phases of the filling stage: in the first to high, in the second to medium, and in the third to low injection speeds. During the first phase, the pressure curve (red solid line) rises up constantly; when the second phase of medium injection speed kicks in, the pressure required for pushing the melt decreases, and thus the pressure curve (red solid line) decreases once but gradually rises back up as the filling volume increases. As the third phase of low injection speed is reached, the pressure curve (red solid line) decreases further. Hence, a gradual decreasing setting of the injection speed keeps the injection pressure under its limit.

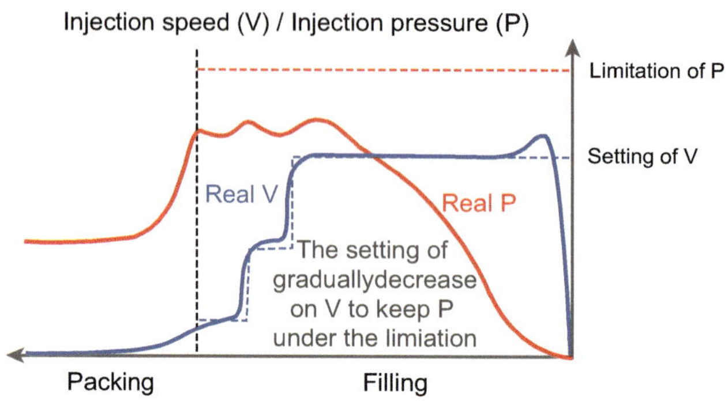

Figure 6.19 Decrease injection speed to decrease injection pressure

Figure 6.20 compares two different speed and pressure control settings.

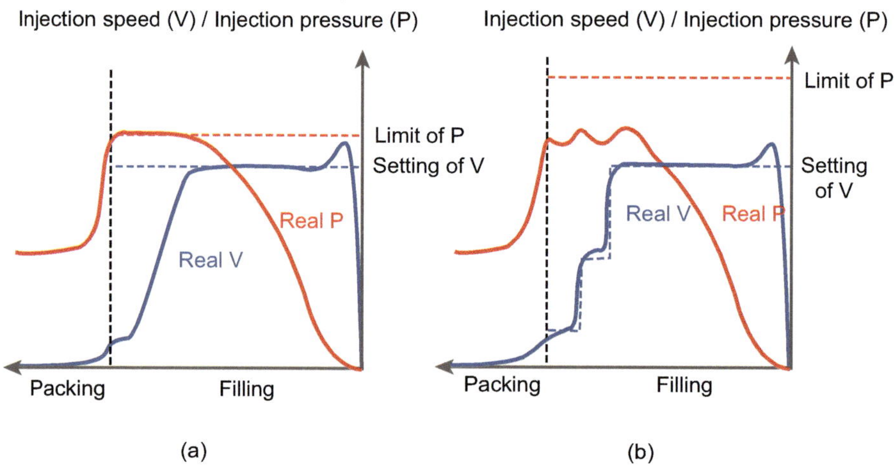

Figure 6.20 Comparison of two different speed and pressure control settings

As shown in Figure 6.20(a), in an injection process where the injection pressure (red solid line) reaches the set upper limit, the actual injection speed starts to decrease. However, this decrease is a feedback from the injection pressure limit, which cannot be controlled by the machine any more.

Before the injection pressure (red solid line) reaches the set upper limit, as shown in Figure 6.20(b), the injection speed is switched and the filling is proceeded by the setting value according to the injection speed curve (blue solid line), while the pressure is maintained under the limit. Everything is under control of the machine.

Figure 6.21 compares the influence on the viscosity of the melt under two different control settings:

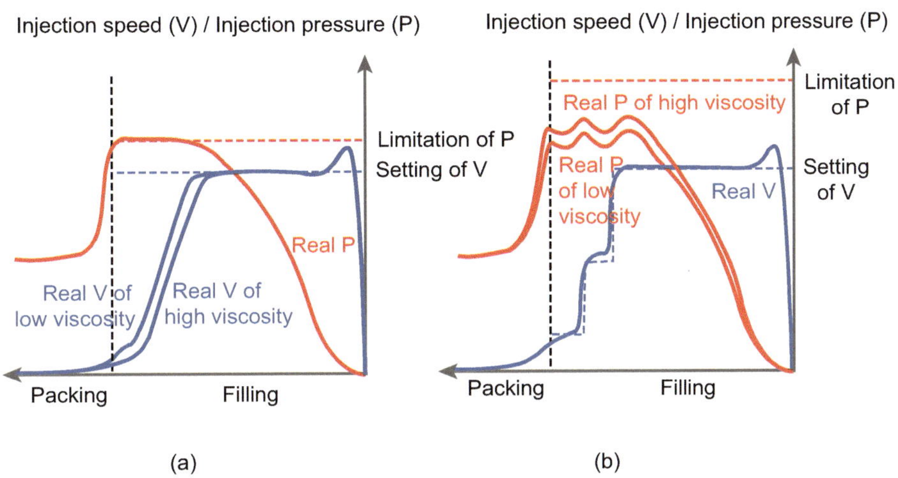

Figure 6.21 Melt viscosity versus injection speed and pressure

Under the filling condition where the pressure limit is reached, as shown in Figure 6.21(a), the injection speed (blue solid line) decreases and cannot be controlled, and the speed of a high-viscosity material decreases faster than that of low viscosity due to its larger flow resistance.

Under the filling condition of speed control, as shown in Figure 6.21(b), the injection pressure (red solid line) does not exceed the pressure limit, and filling can proceed independently of the viscosity of the plastic material within the control range according to the set injection speed curve.

6.1.4 VP Switch and Cavity Pressure

The injection pressure results from driving of the screw which moves forward according to the configured injection speed of the machine. The filling stage is controlled by the injection speed, during which the pressure limit setting is a protection mechanism. As shown in Figure 6.22 on the right side of the graph, the injection speed is configured as several time steps for the injection process, and the filling pressure is generated as a response value according to the injection speed. After the VP switch position, the packing stage, shown on the left side of the graph, is at a slow speed and high-pressure condition. Because the mold cavity has been filled up, the screw speed reduces to a very low value that continues to compressively push the melt into the mold cavity with a high pressure, so that the material density can be increased and compensate for volumetric shrinkage during the cooling process.

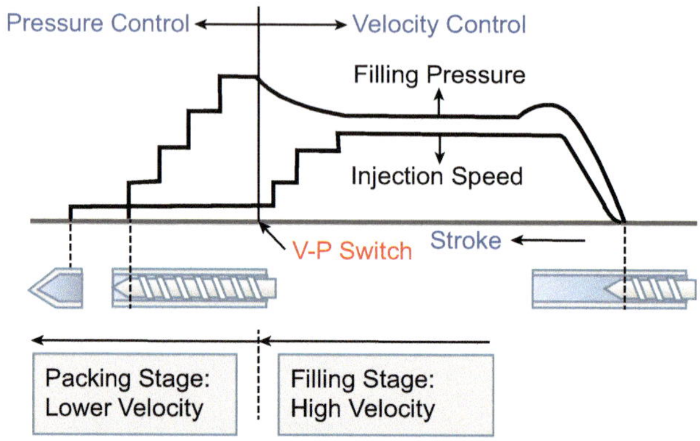

Figure 6.22 Velocity and pressure control during injection molding process

The filling process denotes the transfer of the plastic material into the mold cavity through runners and gates. An ideal condition for this process would be a consistent forward speed of the melt front at each region to achieve flow balance. However, as the flowing of fluids is driven by pressure difference, there should be a higher pressure close to the gate, while a lower pressure appears at the flowing end. As shown in Figure 6.23(a), there is more plastic material squeezed into the same volume close to the gate inside the part, and thus the density is higher, inducing lower shrinkage, whereas the plastic density becomes lower in the low-pressure region away from the gate, resulting in higher shrinkage.

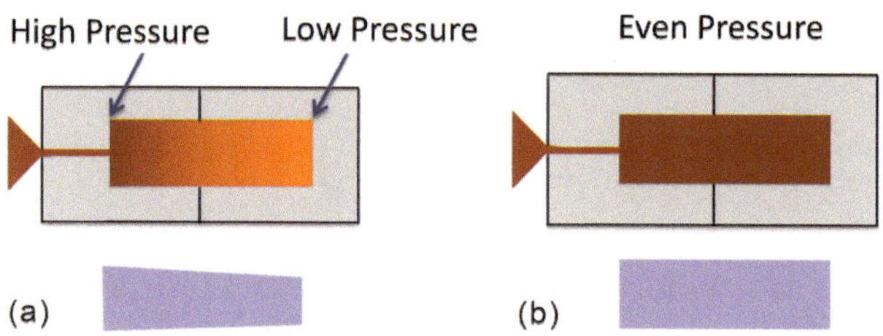

Figure 6.23 Pressure in filling versus packing stage

As shown in Figure 6.23(b), the packing process provides the pressure after the melt has filled up the mold cavity to withstand the shrinkage behavior of the plastic induced by the cooling process. Because the plastic now shows a slow flowing behavior, the pressure difference inside the mold cavity is lower than that in the filling process. As the density difference of the plastic in the mold is smaller, the shrinkage is more uniform. The key point in the packing process is to constantly provide pressure before the melt in the gate location has cured, because the pressure cannot be transferred once curing is done.

As in the filling stage of the injection molding process, the plastic stays in a high-temperature condition and hence it is bound to shrink when being cooled to room temperature. The purpose of packing is to squeeze in more plastic material during the molding process to compensate for possible sink marks or dimensional contraction due to contraction of the plastic. The packing process can improve the shrinkage problems, part flash, and control the size of the part. Also, high stress and thermal deformation or part breakage can be avoided.

The timing of VP switch is shown in Figure 6.24. If the VP switch comes too early, the injection pressure would drop suddenly and then rise back up as the filling percentage grows; with a proper VP switch timing, the injection pressure curve runs smoothly; however, if the VP switch comes too late, and the screw still moves forward with high speed when the mold cavity is completely filled up, the instant squeezing then leads to a pressure surge in the mold cavity.

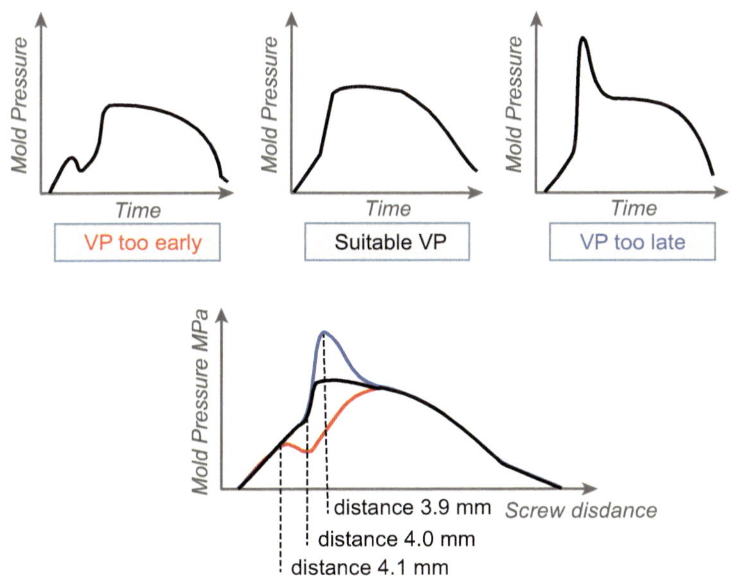

Figure 6.24 VP switch and pressure curve

Normally, the pressure limit setting would affect the filling speed and the pressure curve in the VP switching process. Figure 6.25 shows a typical example of a VP switching process. The blue curve shows the injection speed and the red one the injection pressure.

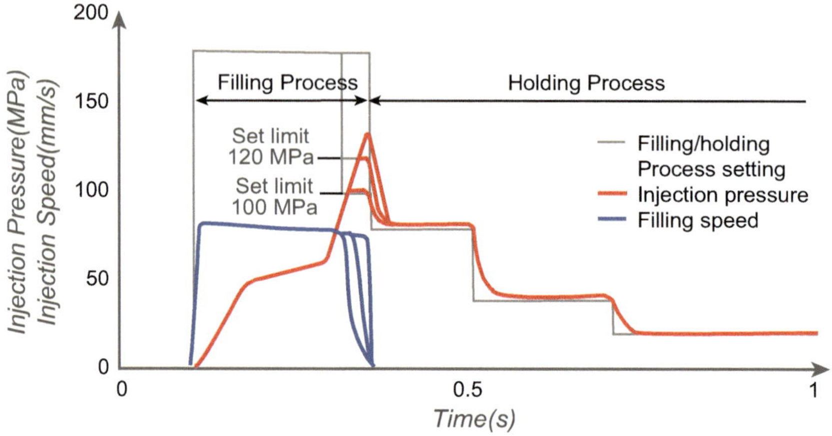

Figure 6.25 VP switch and pressure setting limit

When the pressure limit is set to 100 MPa, the injection pressure reaches the limit earlier, and then decreases after the VP switch; when the pressure limit is set to 120 MPa, the injection pressure reaches the limit later, and then decreases after the VP switch; when the pressure limit is set so high that the whole pressure does not exceed the limit throughout the whole injection stage, the pressure keeps increasing until it reaches the VP switch and then decreases thereafter. When the pressure reaches the upper limit, the filling rate decreases. With the upper limit of pressure increases, the filling rate can be maintained for a longer period of time.

If the molding does not work well even if the injection pressure has reached the pressure limit of the machine, the problem can be solved by increasing the melt temperature or raising the initial injection speed to reduce the pressure, or by pulling ahead the VP switch point. For instance, by moving the VP switch from 5.0 to 7.0 as shown in Figure 6.26, the shaking behavior due to uncontrollable injection speed by hitting the pressure limit can be avoided, and the pressure can smoothly decrease to the packing stage. So the injection speed before VP switch point can be controlled to a fixed value. After VP switch point, it changes to pressure control phase.

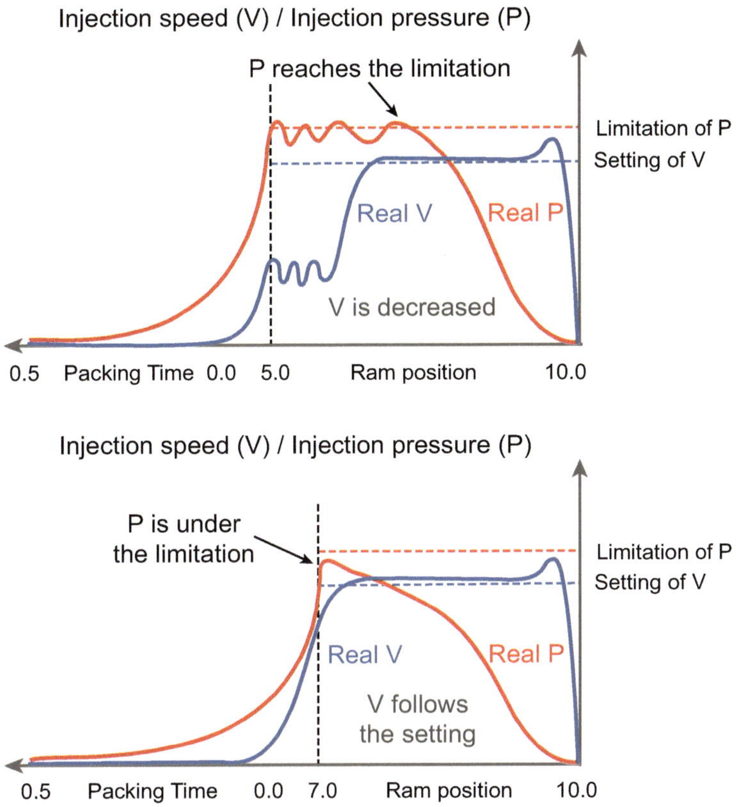

Figure 6.26 Example: modification of process based on machine response

Another important factor is the pressure transmission and distribution status of the melt inside the part, which affects the amount of shrinkage of the plastic material and the weight and size of the part. Therefore, the pressure setting and transmission mechanism of an injection system should be understood, as shown in Figure 6.27. The injection pressure is provided by the hydraulic system of an injection machine. In the case of a full electric injection machine, the power for the screw to move forward is provided by the servo motor on the back side. As the cross section of a hydraulic cylinder is different from that of a screw, which is usually 10 times larger than the latter, and thus the melt pressure at the screw front is 10 times larger than the pressure of the hydraulic oil in the cylinder. The pressure here is under control of the machine, which is irrelevant to the mold and part design.

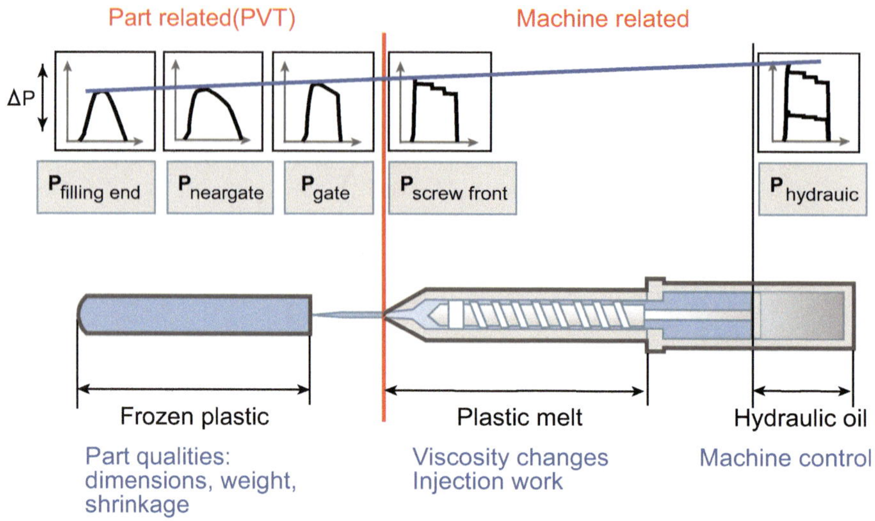

Figure 6.27 Pressure setting and delivery in an injection molding system

Once the melt flows through the runner system and reaches the gate, the pressure has been consumed a bit and becomes lower than that at the screw front. As it passes the gate and reaches the inside of the part, the pressure further drops with the increase of flowing distance. It is expected that the pressure difference inside the part can be reduced by packing for shrinkage compensation.

The pressure inside a mold, and not the injection pressure of the machine, is the key that determines the quality of a finished part. Figure 6.28 shows how the pressure inside a mold responds to the variation of injection time, which is more important in the packing stage. As for the filling stage, the difference of pressure variation between the injection pressure of the machine and the pressure inside the mold can be compared. During filling, the pressure inside the mold increases gradually, but surges once it passes the VP switch, which makes it necessary to monitor the pressure distribution in the mold during the packing and cooling stage. In contrast to the pressure detected at the nozzle and screw front, it is the pressure inside the mold of a part that mostly determines the quality of a finished part.

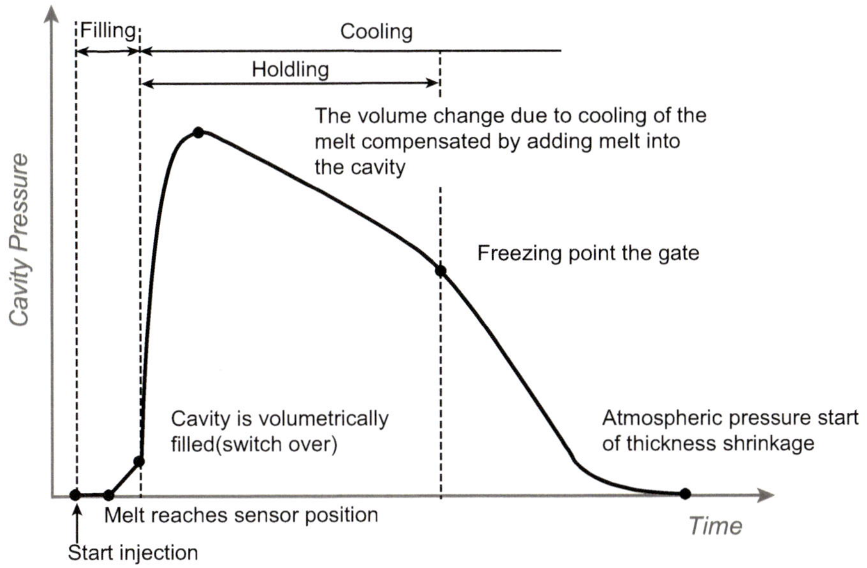

Figure 6.28 Pressure curve inside the cavity

Using a built-in sensor to record the molding pressure of the melt is a good way, however, the sensing location should be paid attention to (frozen layer or inside). For small/thin parts, the pressure curves of the machine may be different, although there is no pressure difference inside the part, which makes the pressure of the mold cavity a real factor that affects the molding of the part.

The VP switch timing has a direct relation to the part quality. By observing the pressure curve inside the cavity versus time, the appropriateness of VP switch timing can be confirmed. Figure 6.29 shows an illustration of the relationship between the VP switch timing and the pressure inside the mold, from which it can be seen that the VP switch timing indeed affects the pressure inside the mold, and consequently affects the part quality.

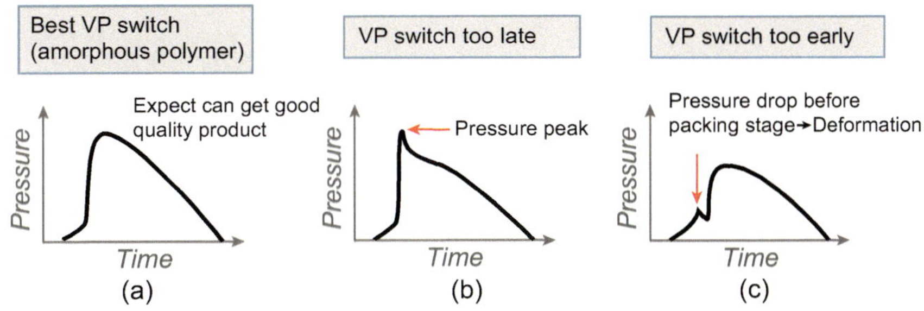

Figure 6.29 VP switch versus cavity pressure

Figure 6.29(a) shows an optimum setting of VP switch, which has a very smooth pressure versus time, and a good part quality can be expected. Figure 6.29(b) shows a too-late VP switch timing, from which the packing pressure cannot smoothly catch the filling pressure and therefore a peak is generated. Figure 6.29(c) shows a too-early VP switch timing before a sufficient volume is filled, and therefore the pressure decreases, which then increases along with the increase of the filled volume thereafter.

The pressure versus time chart shown in Figure 6.30(a) can be divided into three stages: filling stage, packing stage, and holding stage, where the red line denotes the pressure curve close to the gate and the blue one the pressure curve at the filling end.

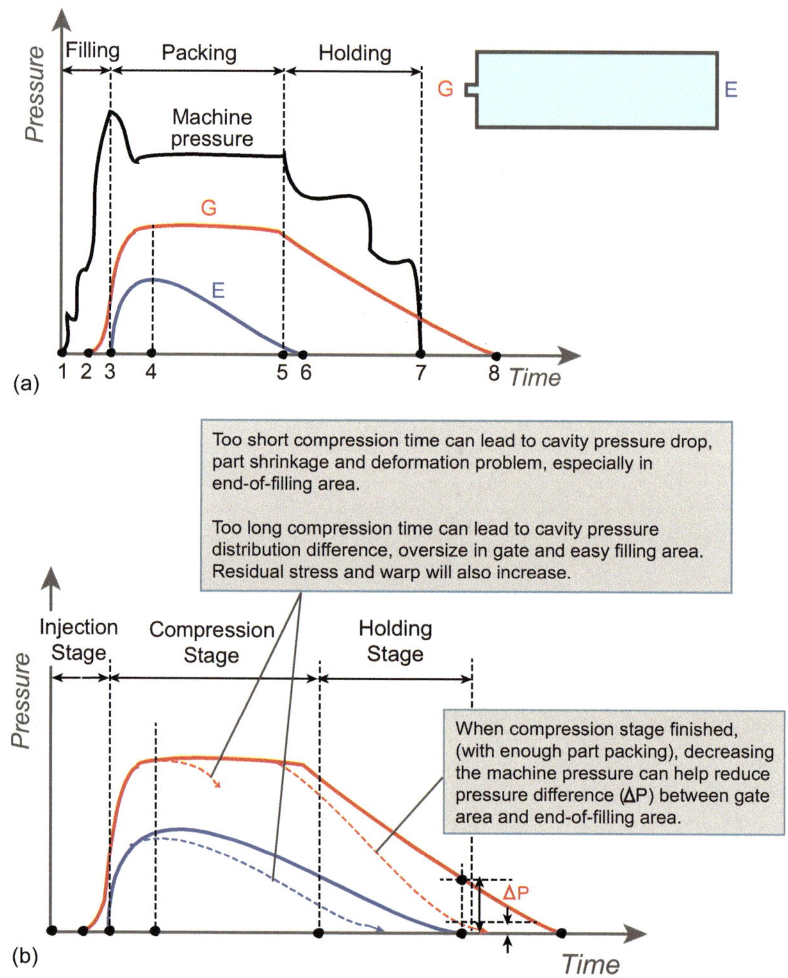

Figure 6.30 (a) Machine pressure versus cavity pressure and (b) packing time versus packing effect

In Figure 6.30(b), if the compression (packing) stage is too short, as the left dashed red line and dashed blue line indicated, the pressure would drop rapidly and cause too much product shrinkage, resulting in an insufficient size of the part. However, if the compression stage is too long, as the solid red line and solid blue line indicated, the pressure difference in the mold would become too much and the dimensional and residual stress at the gate would increase as well, and similarly, the pressure difference ΔP between the near and far end regions inside the mold cavity would be increased and thus warpage would be induced. With a proper setting of the holding stage time, as the left dashed red line indicated, the pressure inside the mold can be decreased reasonably, and so the difference ΔP between the near and far end regions inside the mold can be decreased as well as warpage.

Figure 6.31 shows a CAE analysis on the difference of the packing pressure at each region of the part. Because the structure of the part and the runner design cannot be made perfectly in symmetry, the pressure inside the mold during the packing stage would appear as a distribution of high and low pressure. Comparing the four corners most distant from the center regarding the pressure process chart, the top right corner SN6 has the smallest pressure, and there is a pressure difference of 12.18 MPa to the bottom left corner SN2, which has the largest pressure. Overall, the packing pressure of the top half is lower than that of the bottom half, and therefore the top half is expected to have a larger volumetric shrinkage and dimensional change.

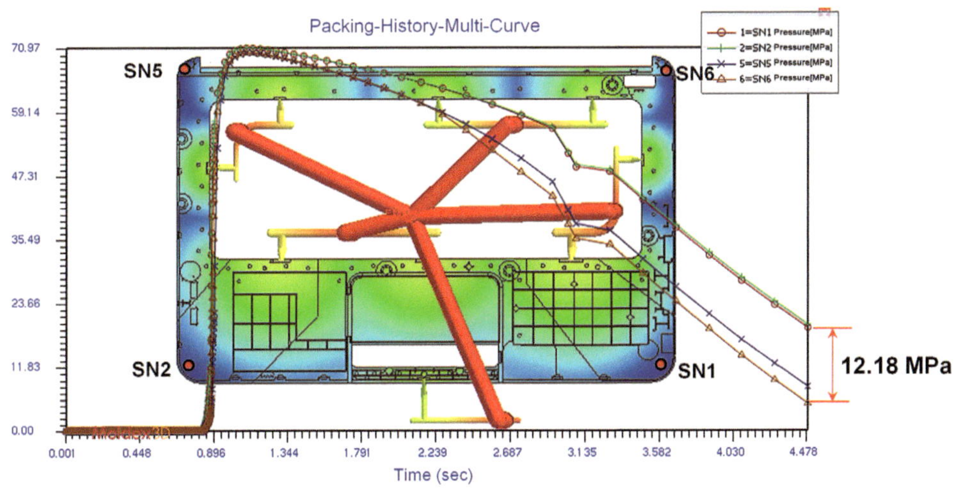

Figure 6.31 C part of laptop: pressure difference in cavity

In a CAE analysis for size prediction, as shown in Figure 6.32, if the shrinkage of the part is non-uniform, causing a shrinkage difference between the top side (top of the part, A side) and the bottom side (bottom of the part, B side), deformation occurs. The shrinkage ratio of the A side is 0.49% with a longitudinal shrinkage of 1.69 mm, and the shrinkage ratio of the B side is 0.44% with a longitudinal shrinkage of 1.50 mm; therefore, a difference of 0.19 mm exists in between.

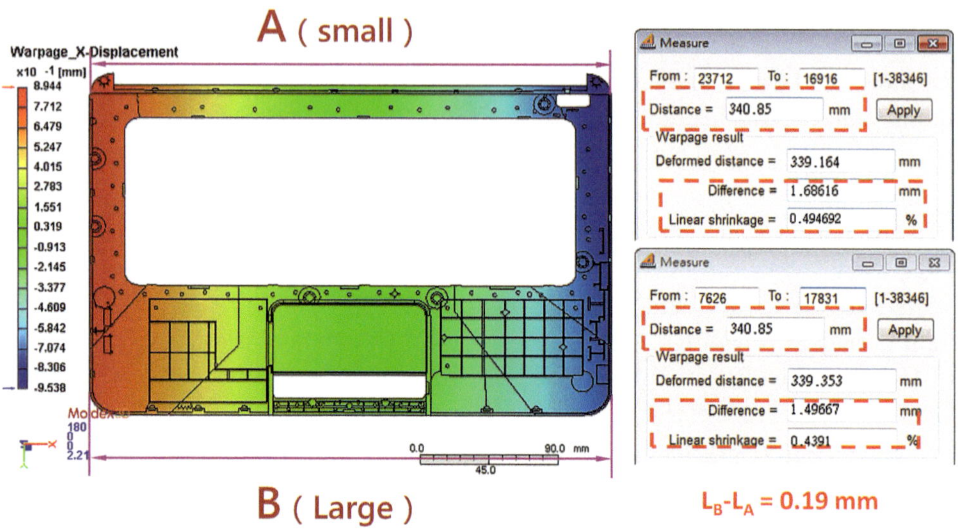

Figure 6.32 C part of laptop: shrinkage difference in top and bottom sides

6.1.5 Effects of Part Thickness

From the description above we know that as the thickness decreases, the cooling is faster, the viscosity becomes higher, the resistance grows bigger, the pressure transmission becomes tougher, the required injection pressure is higher, and short shots are more likely to happen.

As shown in Figure 6.33, the reason for short shot is an insufficient driving power generated by the injection machine to push the melt to fill up the mold cavity system.

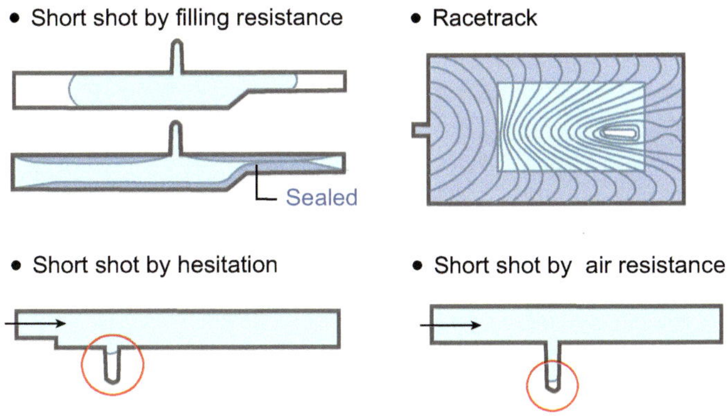

Figure 6.33 Causes of short shot

Possible causes are described as follows:

1. Short shot caused by high flow resistance: the part probably has a high flow length/wall thickness ratio (L/t ratio) that leads to a higher filling resistance, or the running path is blocked by a thick frozen layer at a thin region.
2. A race-track phenomenon due to uneven thickness of the main plane, which leads to air traps and an insufficient filling.
3. Ribs that have a relatively slow flow velocity compared with the main plane would result in material curing over a long time and consequently an insufficient filling.
4. A poor venting design that compresses air and obstructs venting, which induces resistance on the melt and limits the melt front from moving forward.

In the case of an air trap issue caused by a thickness difference, it can be solved by changing the thickness of the part.

Examples are shown in Figure 6.34. The upper part with a thinner bottom design than the four sides tends to generate flow hesitations and forms air traps; however, by modifying the thickness to a uniform one, such a problem can be solved. For the middle part in Figure 6.34, the melt front reaches the four corners later because they are further away from the center where the gate is located; however, this issue can be solved by accelerating the melt front via a flow leader that increases the thickness of the part to reach all four corners earlier. For the bottom circular part in Figure 6.34, an air trap tends to occur by a race tracking race effect due to a lower thickness in the center compared to that on the circumference, and by properly notching out the circumference, forming flow restrictors to restrain the melt flow, a better balance can be achieved to avoid the air trap issue.

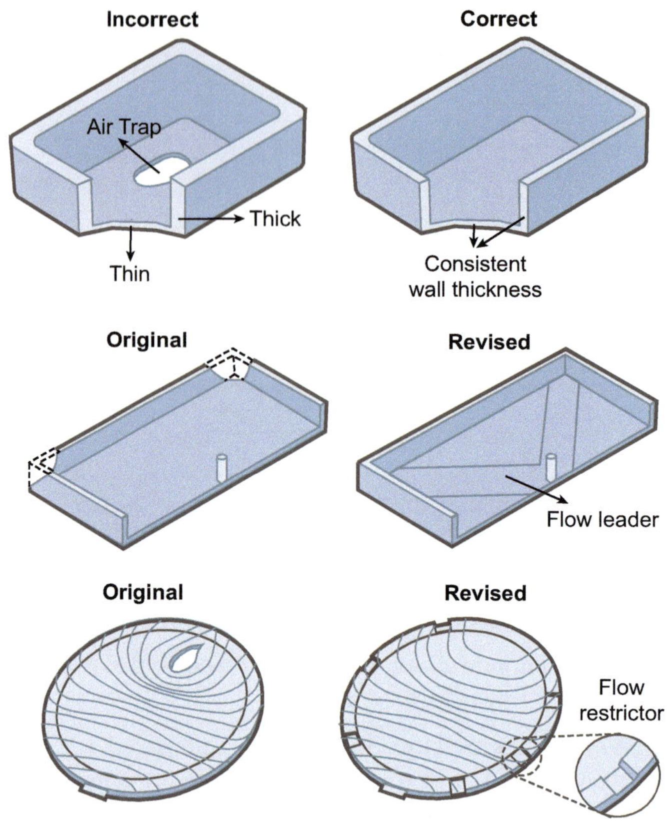

Figure 6.34 Revised design: adjust part thickness and use flow leaders and flow restrictors

Being able to control the flow behavior requires understanding that the flow length is a critical characteristic affecting the moldability of parts, which is related to the following factors:

1. Flow properties of plastics: Materials with low viscosity can flow longer than those with high viscosity.
2. Part geometry: The part thickness has the most significant influence on the flow length. The thicker the part the smaller the resistance becomes and therefore the flow length increases.
3. Molding conditions: High melt and mold temperatures are better to decrease the viscosity of the plastic material and thus elongate the flow length or increase the injection speed. Normally a high injection speed increases the shear rate, reduces the viscosity of the plastic material, and increases the flow length.

A correct determination of the number and location of gates depends on a proper prediction of the flow length. When the part is of a size that exceeds the flow length, more gates should be incorporated for a proper filling. However, the flow length of the melt should be known first so that all regions in the part can be reached by the melt from all gates without exceeding the flow length, by which a proper filling can be assured.

As an example, Figure 6.35 shows the relationship between part thickness and flow length under constant melt and mold temperature and pressure conditions for polycarbonate (PC). When the thickness is 1.5 mm, the corresponding flow length is about 160 mm; at a thickness of 3.7 mm, the flow length reaches 550 mm.

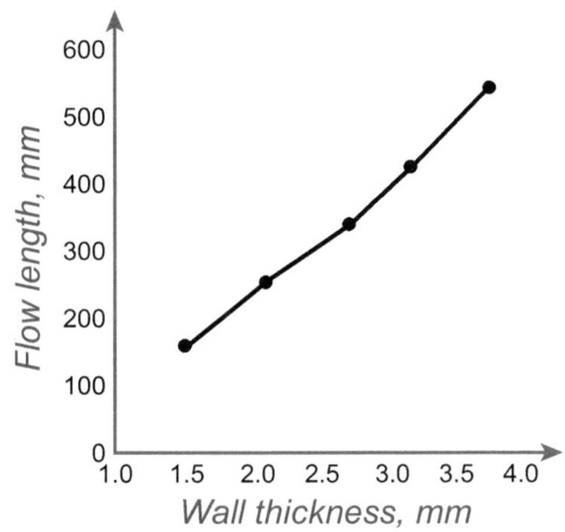

Figure 6.35 Filling capability limitation: flow length

The significant difference of the viscous properties of the different types of plastic materials directly affects the L/t ratio and clamping force. Table 6.1 shows reasonable design values for the L/t ratio. Table 6.2 lists common plastic types and the clamping force required in the unit area of the part. For example, the L/t ratio of acrylonitrile butadiene styrene (ABS) ranges between 50:1 and 200:1, whereas that of PC with a higher viscosity ranges between 30:1 and 100:1. The clamping force per unit area is 2.5–4.0 ton/in^2 for ABS and 3.0–5.0 ton/in^2 for PC.

Table 6.1 Reasonable Design Values for the L/t Ratio

Abbreviation	Material	Flow:Wall Thickness
ABS	Acrylonitrile butadiene styrene	100–200:1
ASA	Acrylate styrene acrylonitrile	180–230:1
EVA	Ethylene vinyl acetate	200–300:1
HIPS	High impact polystyrene	250–340:1
HDPE	High density polyethylene	200–270:1
LDPE	Low density polyethylene	200–300:1
LLDPE	Linear low density polyethylene	180–250:1

Table 6.1 Reasonable Design Values for the L/t Ratio *(continued)*

Abbreviation	Material	Flow:Wall Thickness
PA6	Nylon (polyamide) 6	160–300 : 1
PA66	Nylon (polyamide) 66	180–300 : 1
PA11/PA12	Nylon (polyamide) 11 & 12	180–220 : 1
PBT	Polybutylene terephthalate	140–220 : 1
PC	Polycarbonate	30–110 : 1
PEI	Polyetherimide	70–140 : 1
PES	Polyether sulfone	60–140 : 1
PETP	Polyethylene terephthalate	220–350 : 1
PETG	Polyethylene terephthalate glycol	50–90 : 1
PMMA	Poly(methyl methacrylate)	110–170 : 1
POM-CO	Acetal copolymer	100–250 : 1
POM-HO	Acetal homopolymer	100–250 : 1
PP	Polypropylene	230–340 : 1
PPO-M	Polyphenylene oxide	100–200 : 1
PPS	Polyphenylene sulfide	120–185 : 1
PS/GPPS	Polystyrene (general purpose)	150–200 : 1
PSU	Sulfone polymers	60–120 : 1
PVC (flex) (PPVC)	Plasticized polyvinyl chloride	200–250 : 1
PVC (rigid) (UPVC)	Unplasticized polyvinyl chloride	80–190 : 1
SAN	Styrene acrylonitrile	170–200 : 1

Table 6.2 Polymer Type versus Clamping Force [2]

Resin	[tonnes/in^2]	[tonnes/cm^2]	[MN/m^2]
PS (GPPS)	1.0–2.0	0.155–0.31	15.4–30.9
PS (GPPS) (thin walls)	3.0–4.0	0.465–0.62	46.3–61.8
HIPS	1.0–2.0	0.155–0.31	15.4 - 30.9
HIPS (thin walls)	2.5–3.5	0.388–0.543	38.6–54.0
ABS	2.5–4.0	0.388–0.62	38.6–61.8
AS (SAN)	2.5–3.0	0.388–0.465	38.6–46.3
AS (SAN) (long flows)	3.0–4.0	0.465–0.62	46.3–61.8
LDPE	1.0–2.0	0.155–0.31	15.4–30.9
HDPE	1.5–2.5	0.233–0.388	23.2–38.6
HDPE (long flows)	2.5–3.5	0.388–0.543	38.6–54.0
PP (homo/copolymer)	1.5–2.5	0.233–0.388	23.3–38.6

Resin	[tonnes/in²]	[tonnes/cm²]	[MN/m²]
PP (h/co) (long flows)	2.5-3.5	0.388-0.543	38.6-54.0
PPVC	1.5-2.5	0.233-0.388	23.3-38.6
UPVC	2.0-3.0	0.31-0.465	30.9-46.3
PA6, PA66	4.0-5.0	0.62-0.775	61.8-77.2
PMMA	2.0-4.0	0.31-0.62	30.9-61.8
PC	3.0-5.0	0.465-0.775	46.3-77.2
POM (homo/copolymer)	3.0-5.0	0.465-0.775	46.3-77.2
PET (amorphous)	2.0-2.5	0.31-0.388	30.9-38.6
PET (crystalline)	4.0-6.0	0.62-0.93	61.8-92.6
PBT	3.0-4.0	0.465-0.62	46.3-61.8
CA	1.0-2.0	0.155-0.31	15.4-30.9
PPO-M (unreinforced)	2.0-3.0	0.31-0.465	30.9-46.3
PPO-M (reinforced)	4.0-5.0	0.62-0.775	61.8-77.2
PPS	2.0-3.0	0.31-0.465	30.9-46.3

6.1.6 Material Viscosity an Flow Behaviour

The basic principle for setting the injection speed and pressure in the filling stage is that high speed and high pressure facilitate the filling to be finished in a short time to avoid fast cooling of the melt, which would increase the flow resistance. The reason for using a high pressure is to avoid a drop of injection speed due to insufficient injection pressure, which may lead to deviation from the original parameter setting.

When the plastic material is filling, an internal pressure (counteractive force) is generated by the flow resistance as the plastic is compressed. A faster pressure rise denotes a higher flow resistance, and thus the pressure loss denotes more cavity filling. By increasing the melt temperature to reduce the viscosity of the material, the flow resistance can be reduced, and a lower flow resistance in the filling stage leads to a smaller increase of the internal pressure.

Possible causes for a rising flow resistance, leading to high internal pressure and high pressure loss, are as follows: thinner thickness, long flowing distance, lower mold/melt temperature, slower injection speed, and high viscosity.

Because the flow resistance is proportional to the melt viscosity, it is still possible to change the viscosity and reduce the flow resistance in the same mold and same material through modification of the molding conditions. Polymer materials are non-Newtonian fluids and hence their viscosity not only changes at different temperatures but also depends on the shear rate.

To understand the influence of the shear rate on the viscosity curve of a plastic material is important. The viscosity affects the flow resistance, and then the filling pressure, as well as the specifications and capacity of machine required. The viscosity difference of the plastic material between each region in the mold cavity also affects the pressure transmission, the flow balance, and the follow-up packing pressure transmission, which has a direct relationship to the mass and quality of the part.

Therefore, the quality control of the part is an important factor to understand the relationship between viscosity variation and molding conditions of plastic materials. The viscosity variation of polystyrene (PS) is shown as an example in Figure 6.36.

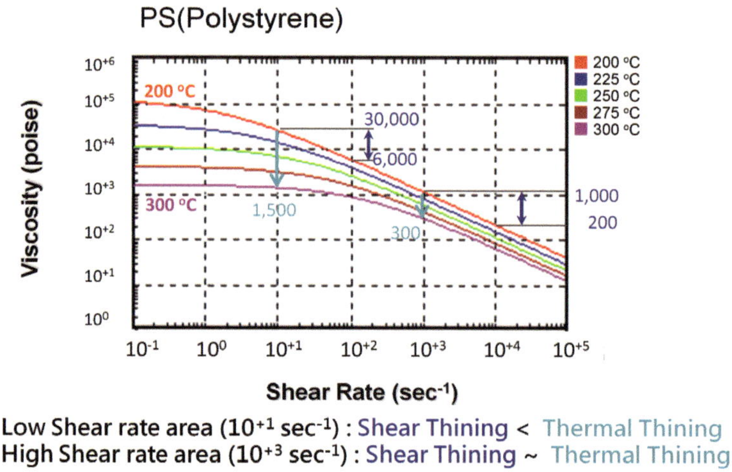

Figure 6.36 Viscosity of polystyrene

In the low shear rate zone where the shear rate is $1 \times 10^1\ s^{-1}$ and the temperature is 200 °C, by increasing the temperature from 200 to 300 °C, the viscosity is reduced from 30,000 poise to 1500 poise. By increasing the shear rate from $1 \times 10^1\ s^{-1}$ to $1 \times 10^2\ s^{-1}$, the viscosity can be reduced from 30,000 poise to 6000 poise. Therefore, the influence of the shear rate on the viscosity is relatively smaller than the influence of the temperature.

In the high shear rate zone where the shear rate is $1 \times 10^3\ s^{-1}$ and the temperature is 200 °C, by increasing the temperature from 200 to 300 °C, the viscosity is reduced from 1000 poise to 300 poise. By increasing the shear rate from $1 \times 10^3\ s^{-1}$ to $1 \times 10^4\ s^{-1}$, the viscosity can be reduced from 1000 poise to 200 poise. Therefore, the influence of the shear rate on the viscosity is close to but slightly larger than the influence of the temperature on the viscosity.

Regarding various effects of viscosity, the viscosity curves of three different groups of plastic materials are compared in Figure 6.37 to explore the sensitivity of viscosity versus temperature and shear rate.

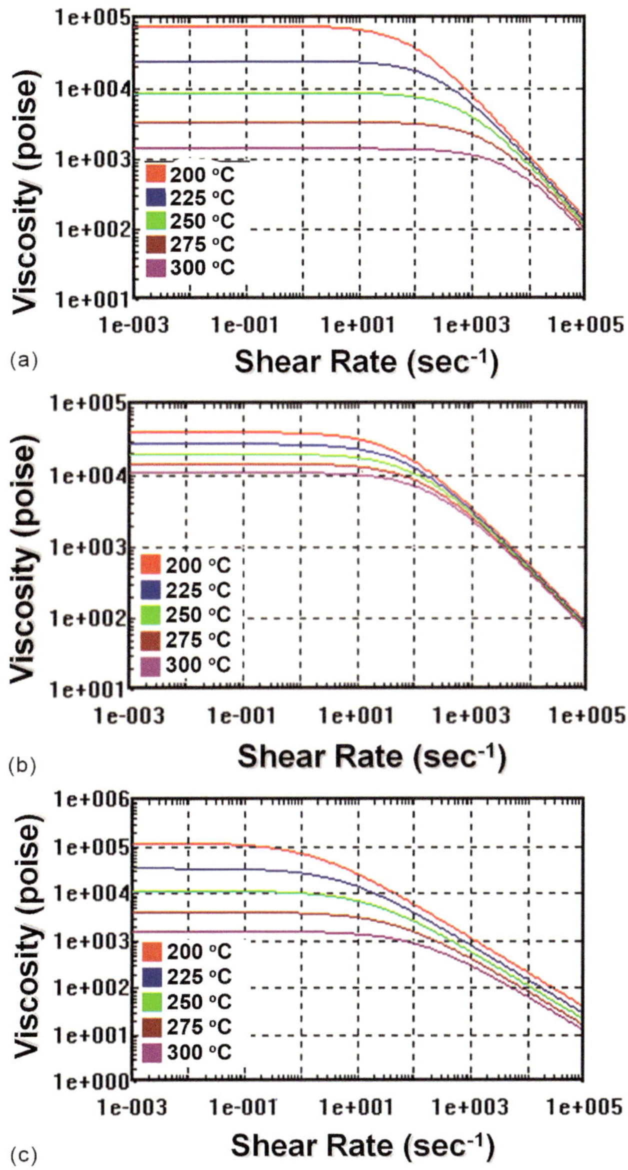

Figure 6.37 Viscosity: (a) temperature sensitive, (b) shear rate sensitive, and (c) temperature and shear rate sensitive

Figure 6.37(a) shows a bigger gap between each temperature curve, indicating a significant effect on decreasing the viscosity by raising the temperature but an insignificant effect on decreasing the viscosity by raising the shear rate, which shows that the viscosity of the material is sensitive to temperature. Figure 6.37(b) shows a smaller gap between each temperature curve, indicating an insignificant effect on decreasing the viscosity by raising the temperature but a significant effect on decreasing the viscosity by raising the shear rate, which shows that the viscosity of the material is sensitive to shear rate. Figure 6.37(c) shows a bigger gap between each temperature curve and also the decrease of viscosity with the increase of shear rate is significant, which shows that the viscosity of the material is sensitive to both temperature and shear rate.

Figure 6.38 shows a comparison of the injection pressure under different melt temperatures (250 and 280 °C) for poly(methyl methacrylate) (PMMA). As the melt temperature rises, the viscosity of the material drops and thus leads to a decrease of the flow resistance as well as the injection pressure. When the melt temperature is 250 °C and the filling time is 1 s, the injection pressure reaches 100 MPa; when the melt temperature is 280 °C and the filling time is 1 s, the injection pressure reaches 70 MPa.

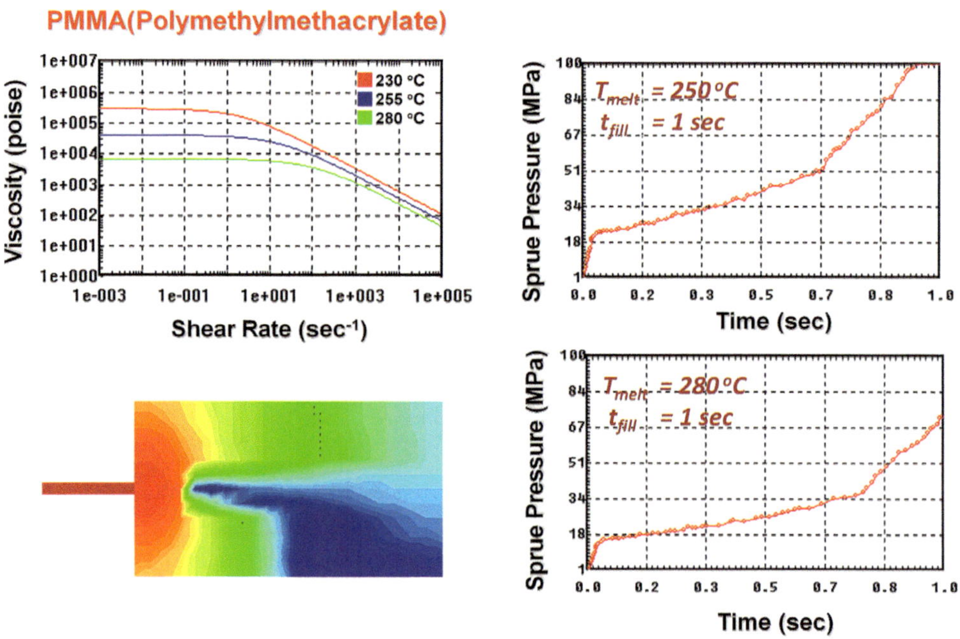

Figure 6.38 Melt temperature versus injection pressure

Different materials have different viscous characteristics. The influence on the injection pressure of three different plastic materials is compared in Figure 6.39. By setting the same filling time at 1.0 s, the injection pressure for polypropylene (PP) is 32 MPa, and 28 MPa for polystyrene (PS), whereas poly(methyl methacrylate) (PMMA) requires 100 MPa to complete the filling process. Furthermore, when the filling time reaches 0.8 s for PMMA, the pressure already hits the upper limit of 100 MPa, which indicates that the pressure control mode already kicks in and the system cannot maintain the original fixed flow rate for filling.

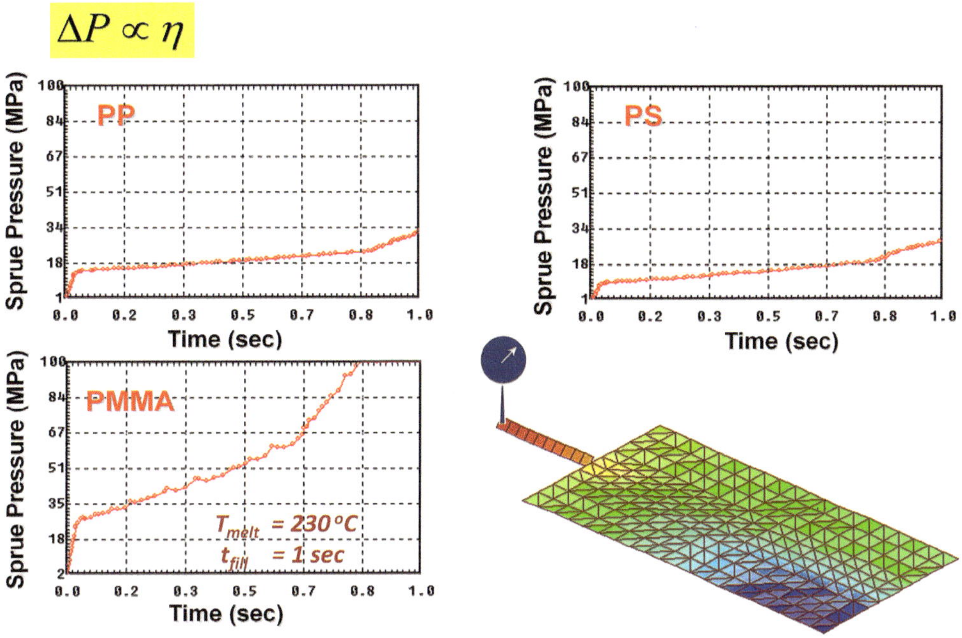

Figure 6.39 Injection pressure for three different plastic materials

6.1.7 Summary

The basic rule for the configuration of parameters is to set the pressure and the injection speed as high as possible. If the injection speed is set too low, the flow speed of a plastic would be limited by the flow resistance and therefore the filling cannot proceed with the original flow rate setting.

The gradient of pressure drop in the packing stage is closely related to mold temperature and the requirements on the appearance of the part. If the pressure goes down too quickly, insufficient packing would occur and thus the size close to the gate becomes smaller or dented. In contrast, if the pressure rises too fast, over-packing near the gate occurs, and the size would grow larger with issues arising such as larger residual stress or warpage.

Sometimes elimination or reduction of exterior defects of finished parts becomes more effective by properly raising the melt or mold temperature, and less residual stress could probably also be achieved at the same time.

The configuration of the filling stage is basically for high pressure and high injection speed. The rule "slow → fast → slow" for the injection speed setting should be followed and the filling time should be kept as short as possible.

Particular attention should be paid to the precision of the VP switch timing.

An improperly set injection speed is likely to affect the appearance of the finished part and the balance of the flow front. If the injection speed is too high, issues such as jetting or flow mark could be generated, whereas a too low injection speed could lead to abnormal surface curing. Therefore, it is necessary to properly define the stage location, the injection speed, and the switch timing according to the individual conditions of the molds and properties of plastic materials.

■ 6.2 Practical Applications

In the filling stage, many problems could be encountered, such as air traps, weld lines, flow imbalance, etc. A few practical examples are introduced in this section to describe how these problems are solved in industry using CAE.

6.2.1 CAE Solution to Stress Mark in a Phone Shell

Figure 6.40 shows an example of the back cover of a cellphone. The thickness of the part ranges from 0.6 to 1.2 mm. The thickness of the main plane is 1.0 mm, but a rectangular region in the center is extremely thin, with only 0.7 mm in thickness.

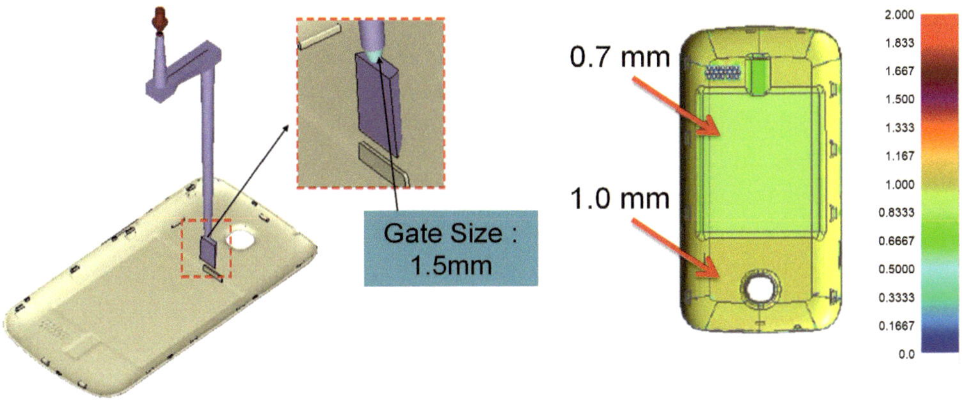

Figure 6.40 Case information

For the back cover PC is used as plastic material, and the filling time, the melt temperature, and the mold temperature are set to 0.4 s, 320 °C, and 110 °C, respectively. The flow front distribution versus time is shown in Figure 6.41. The red area, that is, the gate, is the location where the melt front reaches the part first, whereas the blue area denotes the filling end of the part, located at the bottom right of the figure.

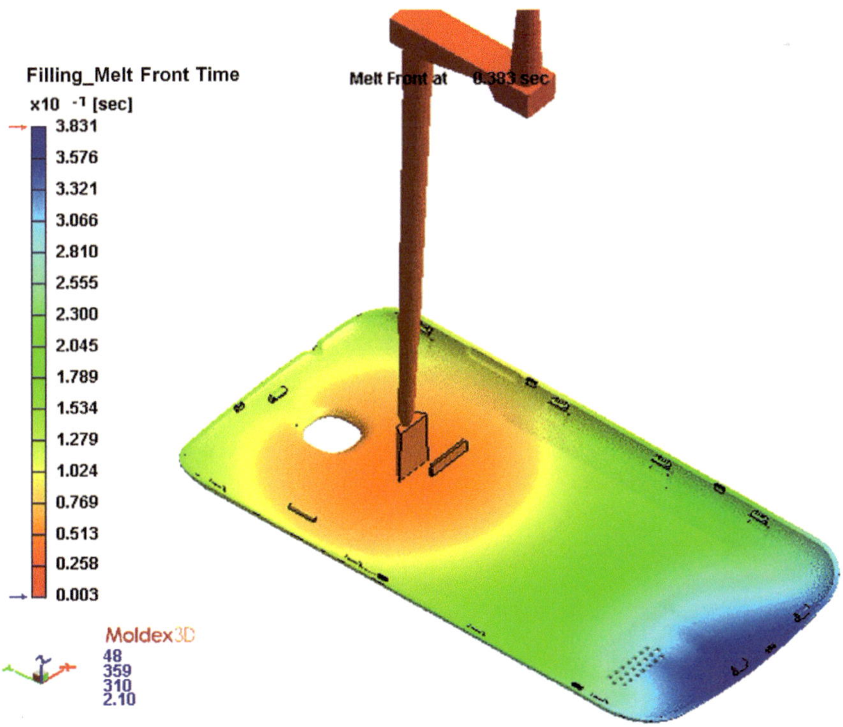

Figure 6.41 Melt front time

Figure 6.42 shows a flow front comparison between the CAE analysis result and the actual part with a short shot. Figure 6.42(a) is an initial condition when the melt just enters the part, whose top side is a hole structure where the lens of the cellphone lies, and hence the melt is blocked by the steel structure of the mold. As more melt flows into the part as shown in Figure 6.42(b), it is observed that the melt front flows faster on both sides than in the center. This is a result caused by a relatively slow flow front in the central area, which has a thickness of only 0.7 mm and induces a higher flow resistance than both sides, which have a thickness of 1.0 mm. When the melt moves further, the tendency of faster flowing on both sides than in the center becomes more significant, as shown in Figure 6.42(c). In Figure 6.42(d), as the melt in the center passes an even thinner area, which generates a bigger flow resistance, the hesitation phenomenon appears more severely.

The comparison between the figures and the pictures indicate a high consistency between the CAE analysis result and the actual part with a short shot.

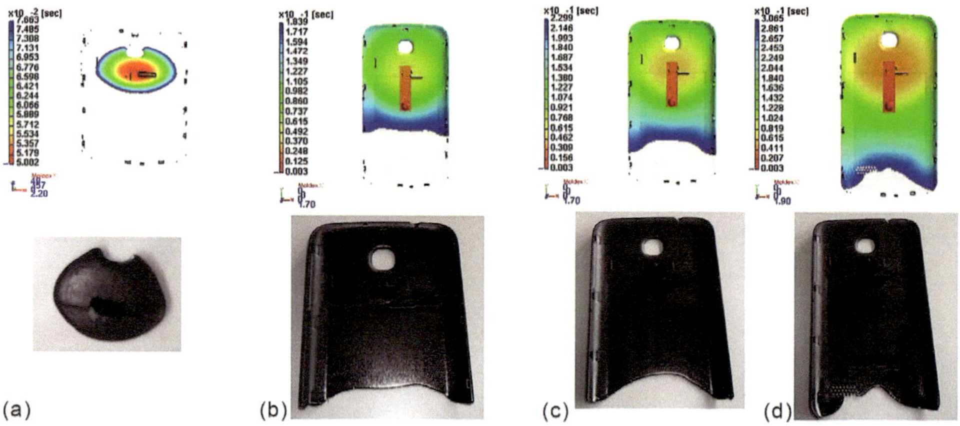

Figure 6.42 Consistency between CAE analysis and experiment in melt front time

The surface quality of an actual part can also be assessed through an analysis result. For example, by observing the surface quality of the injection-molded part as shown in Figure 6.43, a significant mark defect is seen at the boundary between the thickness of 0.7 mm and 1.0 mm in the center, where a considerable shear stress difference also exists.

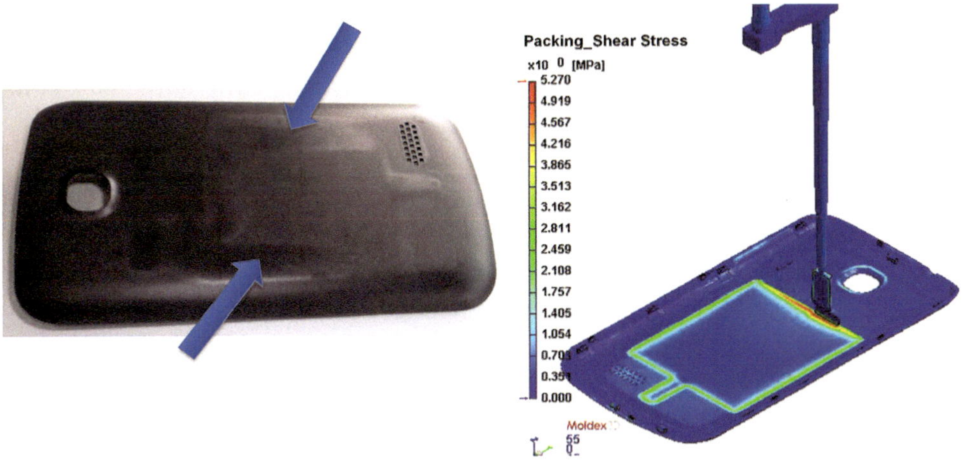

Figure 6.43 Mark defect versus shear stress

As shown in Figure 6.44, the surface quality of the part can be improved by changing the thickness of the design. Figure 6.44(a) is the original design and Figure 6.44(b) is the revised one. The main change is to design a smooth thickness transition area at the boundary where the thickness of 0.7 mm and 1.0 mm meet.

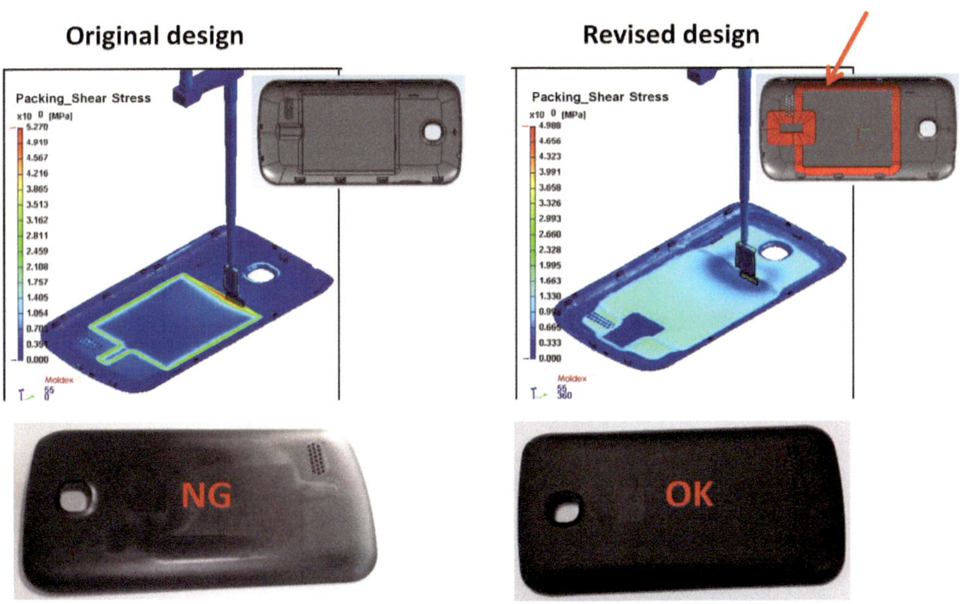

Figure 6.44 Broaden the thickness transition area

From the packing result of shear stress distribution analyzed by CAE as well as the picture showing the part surface, the frame-shaped distribution of shear stress in the central area no longer exists, and the actual surface quality of the part is indeed significantly improved.

6.2.2 Flow Rate Effect on Injection Pressure of Laptop Product

Concerning the injection stroke setting, the runner stroke and the mold cavity stroke are calculated respectively. The runner stroke can be converted from the runner volume, and the mold cavity stroke can be estimated by using 60–90% of the mold cavity volume. Figure 6.45 shows an example using Moldex3D to describe how a multi-step flow rate setting is configured (reference example). The specifications are listed as follows:

- Part model: D part of a laptop PC (bottom plate)
- Part volume: 122.16 cm^3
- Part dimensions: 315 × 225 × 20 mm
- Main thickness: 2 mm
- Material: PC/ABS
- Runner system: hot runner
- Capacity of injection molding machine: 450 tons

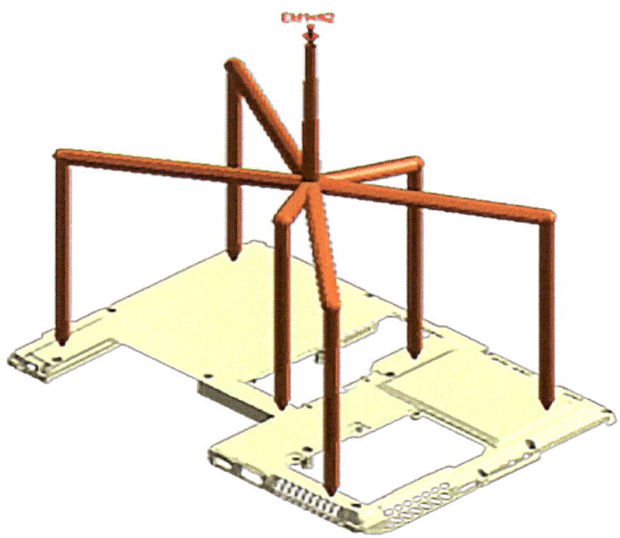

Figure 6.45 Product model

Figure 6.46(a) shows a single flow rate and Figure 6.46(b) illustrates a multi-step flow rate.

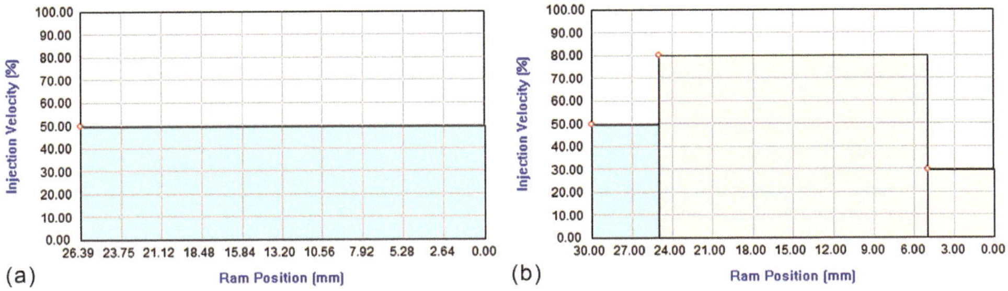

Figure 6.46 (a) Single flow rate and (b) multi-step flow rate

By examining the relationship of flow rate setting to injection pressure and shear rate at the gate as shown in Figure 6.47(a), where the red line denotes a single flow rate, while the blue line denotes a multi-step flow rate, it is found that a multi-step flow rate is able to reduce pressure loss effectively. Also in Figure 6.47(b), where the red line denotes a single flow rate, while the blue line denotes a multi-step flow rate, a multi-step setting is able to reduce the shear rate at the gate as well.

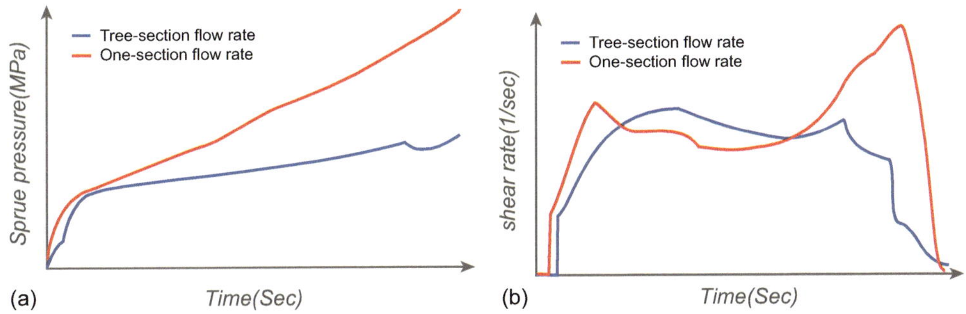

Figure 6.47 (a) Sprue pressure and (b) shear rate

■ 6.3 CAE Case Study

Following the discussions of this chapter and the corresponding CAE verification case, an exercise model, as illustrated in Figure 6.48, is provided in this section for readers to practice CAE application in runner and gate design.

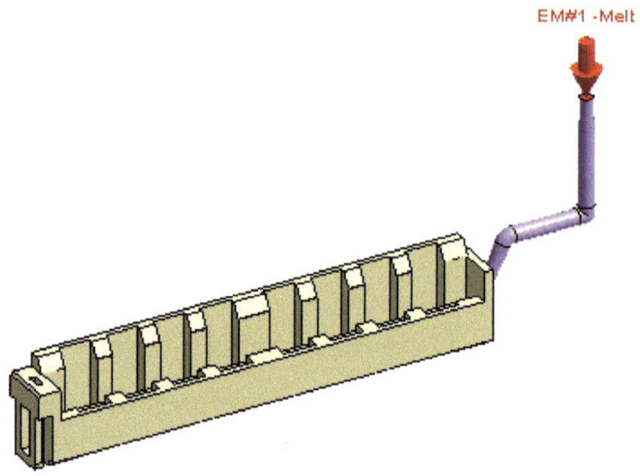

Figure 6.48 Case study sample

Please download the analysis files for CAE Case Study 6.3 from the following website:

https://moldex3d.box.com/s/zr6fvc1vlhbi4ocx11jwd3wmxt4ooif

Then open the .m3j or .mvj file in Moldex3D and answer the questions below according to the analysis results.

Questions:

1. Pre-analysis of the model
   a) What is the maximum L/t of the case? If taking PA6 as the material, does short shot occur? Why?
   b) Where will be the filling ends and weld lines?
2. Comparison of different plastic materials
   a) Run the simulations of flow analysis with “PA Ultramid 8202 BASF” and “PC+ABS CYCOLOY C7410 SABIC(GE)”. Why is the sprue pressure of

PC+ABS much higher than that of PA? Try to explain the reasons with the comparison function of material wizard.

b) What makes the differences in the simulation results of viscosity and shear stress at EOF (end of filling) between PA and PC+ABS? Try to explain the reasons with respect to material properties.

c) Explain how the high-viscosity material affects the molding results, including molding processes and product quality.

3. Comparison of different process conditions: filling time

   a) The default filling time is 0.1 s. Compare the sprue pressure results for different filling times from 0.04 to 0.2 s. Here we set a single flow rate profile. (Hint: check the XY-plots of sprue pressure curves.)

   b) Figure 6.49 shows the internal viscosity results of filling time of 0.04 and 0.2 s. Why does the viscosity distribution differ with the filling time? (Hint: check the temperature distribution.)

4. Comparison of different process conditions: flow rate

   a) The default flow rate is three-segment profile. Compare the XY-plots of sprue pressure curves of different flow rates with single- and three-segment profiles.

   b) Compare the shear rate distributions at the gate of different flow rates with single- and three-segment profiles (by clipping function).

5. Comparison of different process conditions: VP switch

   a) The default VP switch is at 98% filling volume. Compare the XY-plots of sprue pressure and clamping force curves of different VP switch points at 90%, 98%, and 99% filling volume. Try to explain the reasons for the difference.

   b) Compare the locations of weld line and welding temperatures. Try to explain the reasons for the difference.

6. Comparison of different wall thicknesses as Figure 6.50

   a) Compare the XY-plots of sprue pressure and try to explain the reasons for the difference.

   b) Try to catch the melt front time result with the set perspective of 90, 0, 90 on “view control panel” to show the filling difference caused by thickness change.

   c) Compare the shear stress distributions at EOF and try to explain what kind of bad influence to the part might be caused by changing thickness.

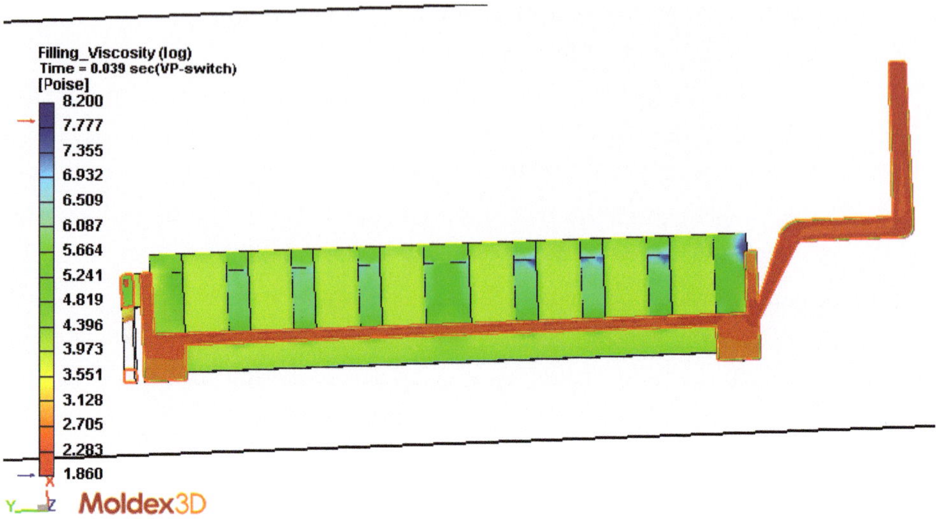

(a)

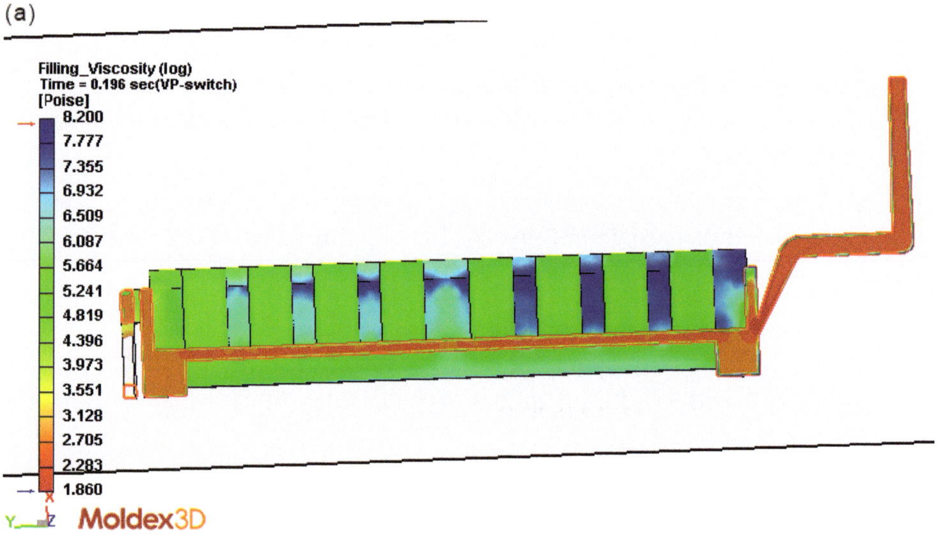

(b)

Figure 6.49 Internal viscosity results (by the clipping function with the plane of z = 1) at VP switch for the filling time of (a) 0.04 s and (b) 0.2 s

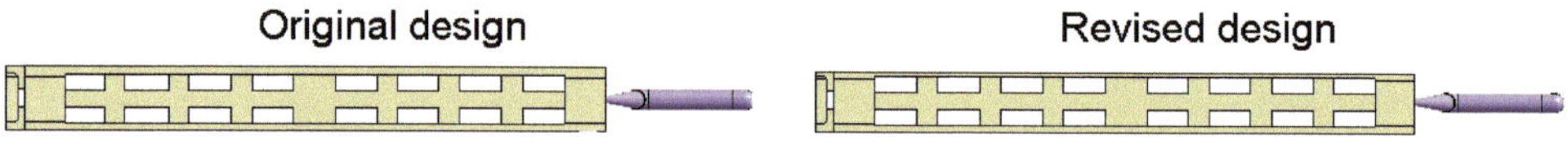

(a) (b)

Figure 6.50 The wall thickness: (a) the original design of 0.633 mm, and (b) the revised design of 0.316 mm

References

[1] "Tools That Help with Processing", Scientific Molding [Online]. *https://www.scientificmolding.com/index.php/processing-tools*

[2] "How to Select Plastic Injection Moulding Machine" [Online]. *https://dlscrib.com/queue/how-to-select-plastic-injection-moulding-machine_58cc47c3dc0d60b71dc346b2_pdf?queue_id=59630e7fdc0d608f182be30f*

7 Runner and Gate Design

Yao-Chen (Cloud) Tsai, Yao-Wei (Willie) Chuang

A runner system is a transferring mechanism for spreading the melt delivered by the screw into the mold cavity. The design of the runner system determines the functionality, manufacturability, and economy of injection molded parts. Such a mechanism consists of sprues, hot runners, cold runners, and gates. Injection molding is a technique to control the flow of the melt inside the mold. The design of the runner system directly affects the appearance, physical properties, dimensional precision, and molding cycle of the finished products.

7.1 Basics

The main function of a runner system is to guide the melt injected from the nozzle quickly and evenly into the mold cavity for molding. Melt enters the runner system by injection through a nozzle, passing sprues, primary runners, branch runners, and gates. An optimized melt injection design should consider gate locations, injection techniques (hot runners/cold runners and procedures), gates/runner types, dimensions, quantity, and the relative dimension of each part of the gating system.

7.1.1 General Design Guide of Runners

The flow balance of molding analysis can identify the optimum runner dimensions for a good runner system using appropriate pressure drop to evenly fill the runners and cavities. A well-designed runner system comes with following advantages:

1. High-quality part production with the maximum-possible number of cavities.
2. Cavities can be filled up completely. Part defects can be minimized.

3. A balanced filling in a multi-cavity system can be achieved to minimize differences between cavities.
4. Filling time, packing time, and molding cycle time can be controlled. Product quality and manufacturing efficiency can be assured.
5. Minimized waste generation.

For the runner system design, determination of the proper number of cavities should be explored first. The determination of cavity amount is based on the production time, the part quantity required for each batch, the injection amount of the injection machine, the plasticization ability in each cycle, the clamping force, the shape and the size of the finished product, and the mold cost, i. e., an all-round consideration is required.

The design of the sprue should consider a convenient and reliable demolding and should not lead to earlier freezing of the plastic material in the sprues than in other parts, by which a stable transmission of packing pressure can be guaranteed. The recommended dimensional specification of the sprue design is shown in Figure 7.1. The root of the sprue should be filleted with a radius of r_2, which facilitates the flow of the plastic material.

$$D_{co} \geq t_{max} + 1.5(mm)$$

$$D_s \geq D_n + 1.0(mm)$$

$$\alpha \geq 1°\sim 2°$$

$$\tan \alpha = (D_{co} - D_s)/2L$$

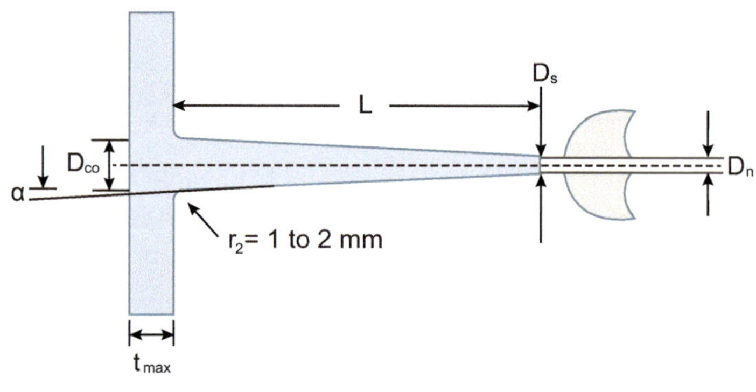

Figure 7.1 Sprue dimension design recommendations

Different cross sections of a runner system have different flow behaviors. The cross section of the runner is determined on the basis of its flow efficiency and the machining conditions used for the mold.

Figure 7.2 shows the recommended values for the design of the geometric dimensions for multi-cavity runners as well as the corresponding area of different runner cross sections.

	Recommended diameters	
	Thin-walled part, large volume, high viscosity	Thick-walled part, small volume, low viscosity
A	t + 2	t + 1
B	t + 3	t + 1.5
C	t + 4	t + 2
D_1	→ ≧ C	
D_2	→ ≧ A	

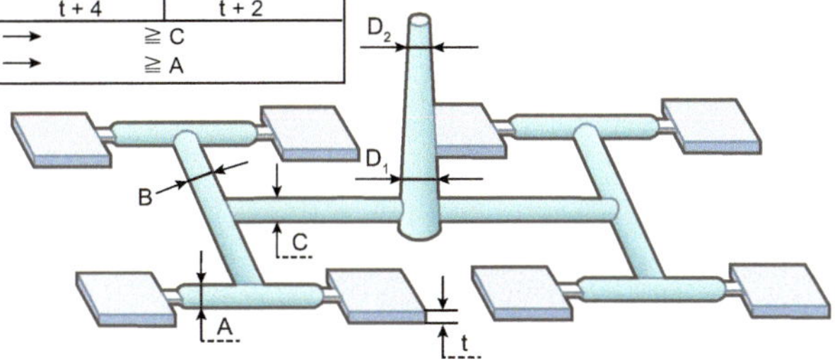

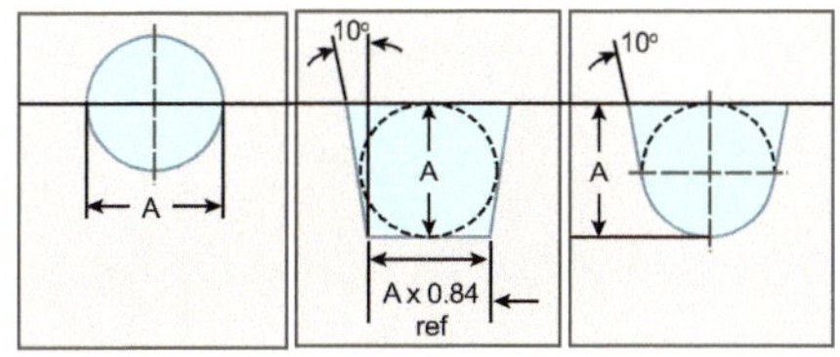

Figure 7.2 Runner dimension design recommendations (A: diameter of tertiary runner, B: diameter of secondary runner, C: diameter of primary runner, D_1, D_2: diameters of sprue, t: part average thickness)

Figure 7.3 shows common types of cross sections of runners, including full round, parabolic, trapezoidal, half round, and rectangular. For the selection and adoption of a runner, the full-round, parabolic, and trapezoidal runners are usually recommended. The full-round runner is the best cross section design because it has the highest volume-to-surface area ratio, which provides better pressure transmission rate and lower flow resistance and heat loss. However, a runner with full round cross section requires machining on both sides of the mold and thus increases cost and difficulty of mold machining in addition to the consideration of alignment between the half round of each side upon mold combination.

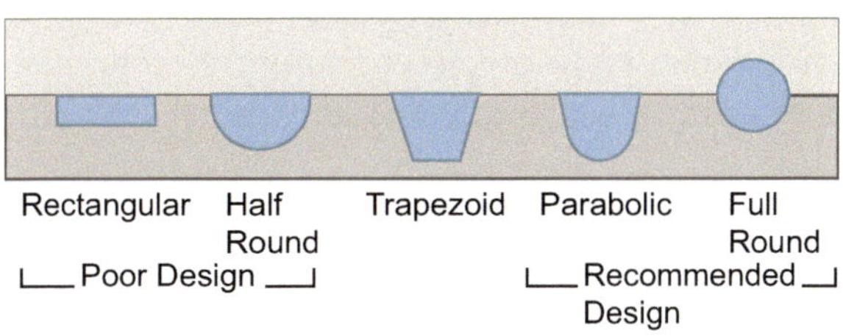

Figure 7.3 Runner cross sectional size and shape

The hydraulic diameter D_h can be used to evaluate the flow resistance index in runners of different cross section geometries, as shown in Figure 7.4, where S is equal to H and R. The larger the hydraulic diameter is, the lower the flow resistance becomes. The hydraulic diameter is defined as follows according to the cross-section area of a runner (A) and the perimeter of the cross section (P):

$$D_h = \frac{4A}{P} \tag{7.1}$$

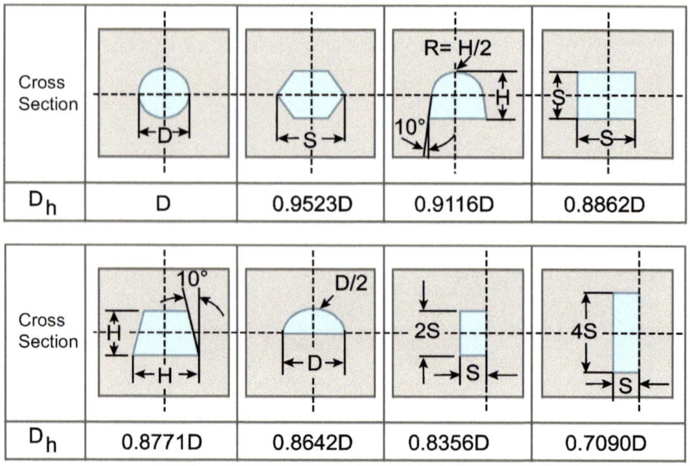

Figure 7.4 Equivalent hydraulic diameters of different runner cross sections

The diameter and length of the runner system affect the flowing of the melt. A runner system with larger flow resistance leads to a higher pressure drop during the filling process. To reduce the flow resistance, the runner diameter can be increased. The empirical equation can be roughly defined as follows:

$$D = \frac{W^{0.5} L^{0.25}}{3.7} \tag{7.2}$$

where D is the runner diameter, W is the part weight, and L is the runner length.

CAE can be used to assist with the design of the runner system. In the Moldex3D runner wizard, the flow-style setting pages for different gate types can be used to quickly create a two-platen or three-platen runner system, and the sizes of the runners and branches can be set simply and quickly, as shown in Figure 7.5.

Moldex3D also allows symmetrical runners configurations to be set, which can simplify the design of multi-cavities to just 1/4 of the runners and cavity objects, as shown in Figure 7.6. The fewer the objects for simulation, the faster the analysis results are obtained.

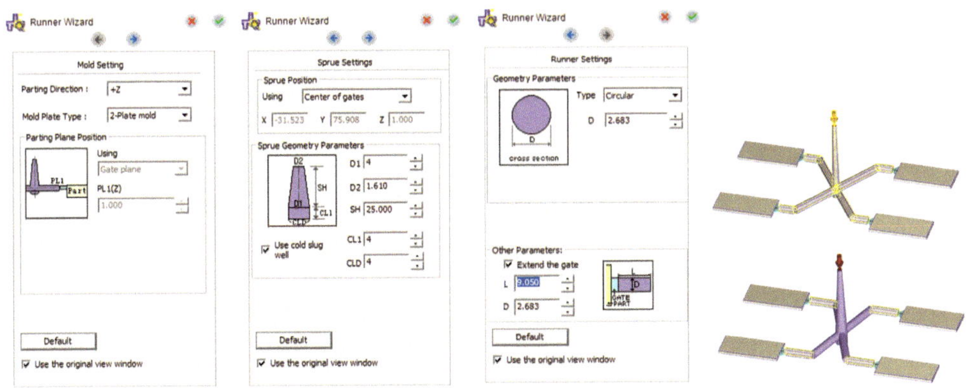

(a)

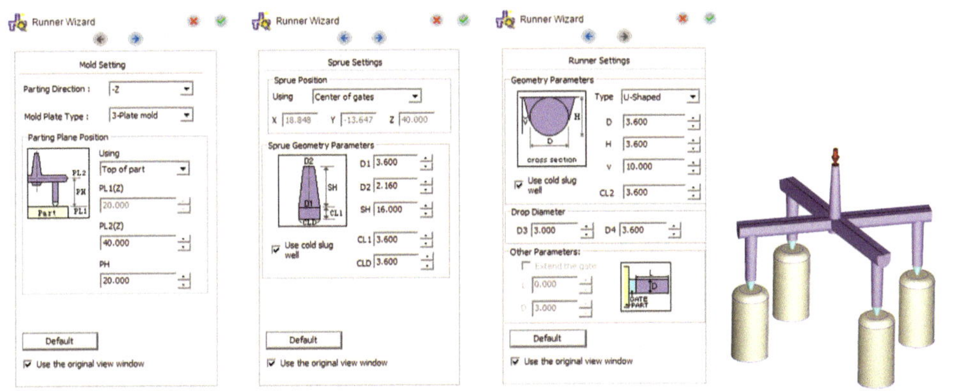

(b)

Figure 7.5 (a) Two-platen mold and (b) three-platen mold

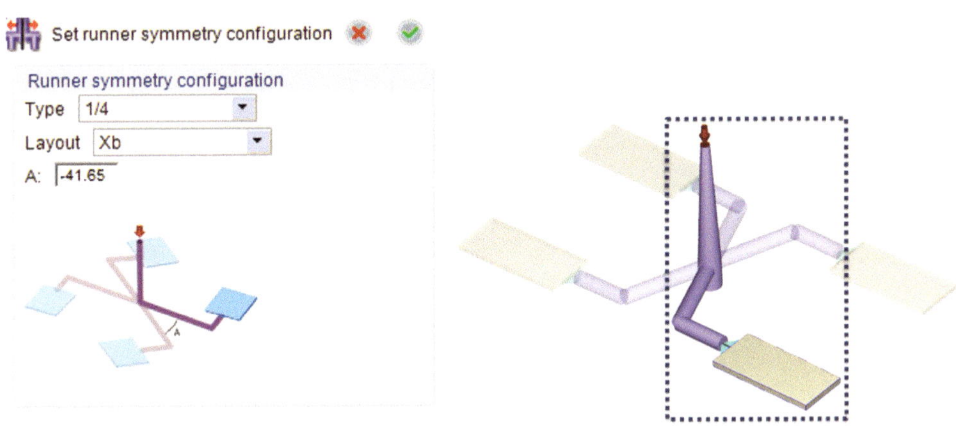

Figure 7.6 Setting the runner symmetry configuration

7.1.2 General Design Guide of Gates

Gates are channels for the melt to enter the mold cavity through runners and generally of the smallest cross-sectional area and the shortest flow length in a runner system. A single-gate system should be used unless a multi-gate system is inevitable for the case that the flow length to wall thickness ratio (L/t) of the melt exceeds the limitation of the material and temperature. Using a single-gate system can not only reduce the issue of weld line, but also assures even distribution of material, temperature, and pressure as well as helps to obtain better molecular chain orientation. It can also reduce the generation of runner waste and thus reduce the cost of production.

The following items should be considered before deciding on a gate location:

1. Injection from the thicker areas of the part to avoid generation of surface shrinkage marks.
2. Gate location is away from critical aesthetic surfaces.
3. Flow path of melt (L/t ratio).
4. Isometric and balanced filling paths.
5. Amount of volume to be filled (filling time/max. shear speed).
6. Location of weld line.
7. Effect of molecular orientation in the flow field.
8. The location of the gate should make post processing (cutting of the gate) easy.
9. Simultaneous feed at each gate of a multi-cavity mold as well as a simultaneous end of filling in each mold cavity.

Figure 7.7 describes how a weld line, a commonly seen cosmetic defect, is formed. Normally the structural strength at this point is weaker. A weld line is a surface flow mark formed by the merging of two plastic melt fronts, also known as fusion line, fusion mark, or joint mark.

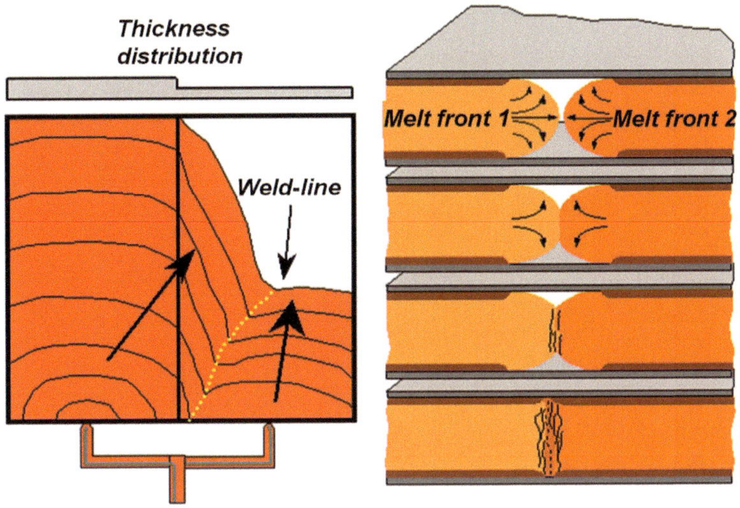

Figure 7.7 Cause of weld line

For the improvement of weld lines, the gating method and the gate number should be evaluated. Flow balance is a good way to solve the problem, which is to make the flow area of the melt injected by each gate of similar size; this makes the number of weld lines as low as possible within a possible range. The number of gates and location of gating should be changed to avoid excessive flow lengths or cooling of the injected plastic from a single gate. The flow length should be as short as possible. Regarding the issue of weld line, the best result is to completely prevent it from happening. However, for a multi-gating plastic part, weld lines are usually inevitable and thus the following considerations are required:

1. Cosmetic: Move weld lines to non-cosmetic sides, inner sides, hidden sides or textured parts where cosmetic properties are not first priority.
2. Final usability: Move weld lines to parts with no force subjected, to a thicker area, or a structurally reinforced area, to prevent breakage due to stress concentration, and compensate possible weakening of welding strength. The way a weld line generates is especially important for plastic parts with mounting holes. Figure 7.8 shows a difference in the generation of weld lines during molding caused by a difference of gate number and gating location of a plastic part with a round hole.

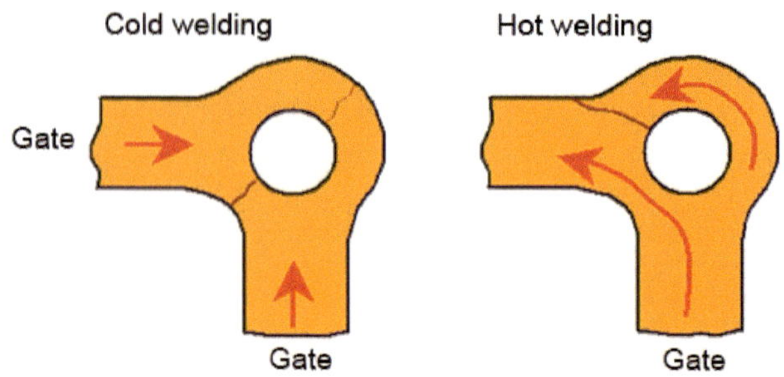

Figure 7.8 Different gating locations induce welding variation near a hole feature

Figure 7.9 shows that to reduce the generation of jetting flow and flow marks, the gate is better designed perpendicular to the runner for injection. By contacting the melt to the mold wall instead of directly injecting it into the mold cavity the phenomenon of jetting flow can be improved. The size of gate should be designed large enough to provide a moderate filling pressure and speed for the melt. Large products are usually designed with multi-gate injection and an overlength of gate spacing should be avoided to prevent excessive pressure loss. Make the length of gates as short as possible.

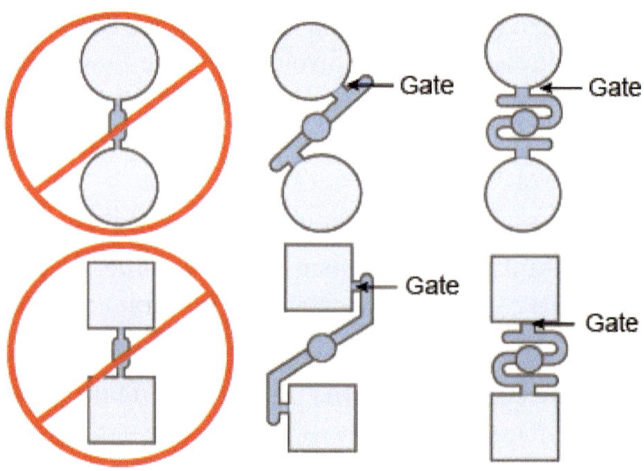

Figure 7.9 Gate design to avoid jetting flow

Compared to plastic parts and runners, the cross section of the gate is usually smaller, around ⅔ of the wall thickness of a plastic part, and thus the gate would be easily removed without leaving a trace of the gate on the plastic part. Figure 7.10 illustrates how the size of the gate affects the solidification speed of a plastic, and further affects packing. If the gate is designed at the thin area of the part, sink marks would occur in the thicker areas and even the defect of internal voids would be generated. A gate with larger cross-sectional area has a smaller pressure drop, which has a positive effect on the appearance around the gate, stress concentration, and dimensional precision. However, if the gate is too big, it would increase the difficulty in trimming off the gate as well as the time for cooling, and thus affect product efficiency consequently.

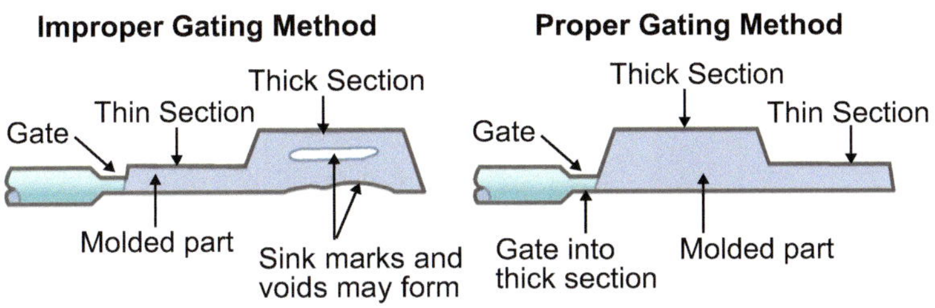

Figure 7.10 Gating location in proper area

The types of manually trimmed gates are as follows: sprue gate, tab gate, edge gate, overlap gate, fan gate, disk gate, ring gate, spoke gate, and film gate.

By summarizing the design key points in a result-oriented manner, the following items should be checked:

1. Stress: The area around the gate is the place where stress concentrates the most. Avoid locating the gate where external force could be applied or it could be damaged easily.
2. Pressure: Avoid excessive pressure loss by designing the gate at the thick-walled area to guarantee a better packing effect.
3. Flow direction: Avoid molecular orientation from occurring in the melt, which would cause severe warpage.
4. Weld line: Equalize the flow length of the melt inside the mold cavity to avoid generation of air traps and weld lines.
5. Filling: A proper location design can increase turbulence and reduce jetting flow and flow marks.

The types of gates can be categorized as follows:

1. Direct/Sprue Gate

A sprue gate enters the part directly. The gate has a taper that goes from big to small as shown in Figure 7.11.

Advantages: Simple structure, less pressure loss, good filling effect, precise dimensions, and good quality.

Disadvantages: Traces would be left after post processing of the gate, which affect the appearance of the part. Molding of one part at a time.

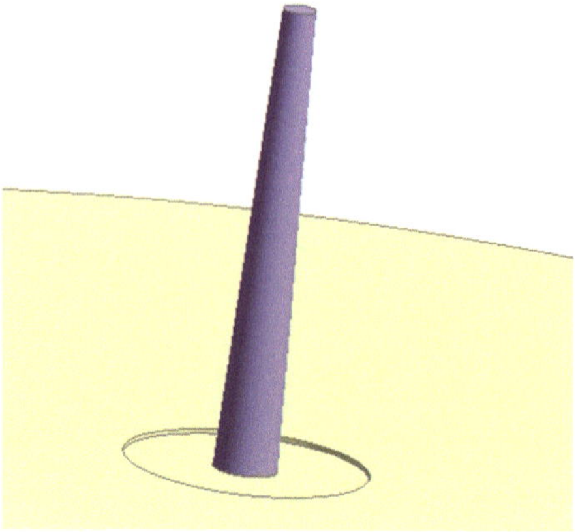

Figure 7.11 Direct/sprue gate

The melt enters the mold cavity from the sprue gate directly. For the injection molding of a single cavity, the injection is done in the center of the mold and thus a mark of the gate is left in the center of the finished part.

The recommended design of a sprue gate is shown in Figure 7.12. The diameter of the entrance is directly related to that of the nozzle of the injection machine; it should be larger than the nozzle by 1 mm in diameter. A standard sprue bushing has a taper of 2.4° (α) opening toward the part.

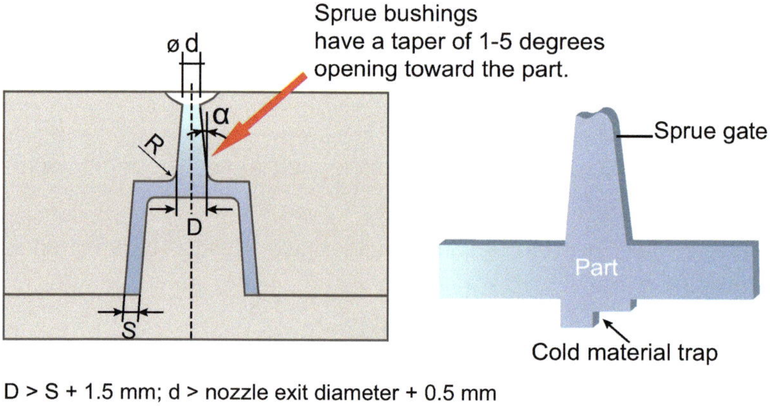

Figure 7.12 Direct/sprue gate design recommendation

2. Side Gate/Edge Gate

A side gate has about 50 to 75% of the thickness of the part, and can be designed as a fixed thickness or as a decreasing thickness gradient, as shown in Figure 7.13. It is usually used in flat plastic parts or thin parts, and can reduce the shear stress inside the mold cavity.

Advantages: Easy for separating gates from finished parts; prevents plastic reverse flowing; shear-heating can be induced at the gate to further increase the melt temperature; facilitates filling; easy for machining of gate dimension.

Disadvantages: Large pressure loss, poor flowability, and flow marks tend to occur easily in finished parts of flat plate shape or with large area.

3. Overlap Gate/Lapped Gate

The overlap/lapped gate is a transformation of the edge gate, as shown in Figure 7.13. The gate is overlapping with the side or surface of the plastic part. Jetting flow can be avoided, and surface defects can be reduced by using this gate type. The recommended dimensions are similar to those for a side gate, i.e. the gate thickness is 60 to 75% and the gate width is 2 to 10 times the part thickness, respectively.

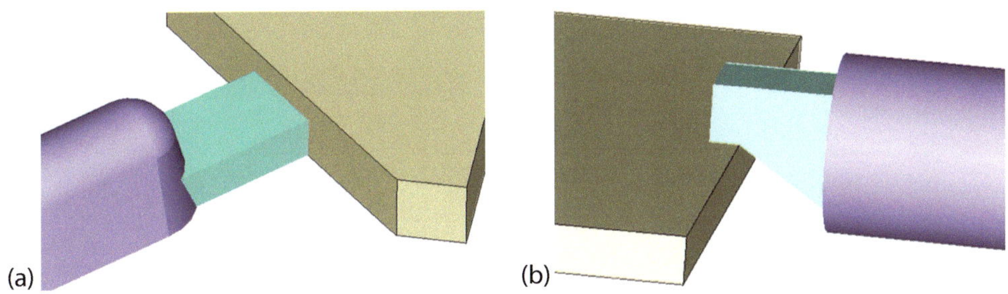

Figure 7.13 (a) Side gate/edge gate and (b) overlap/lapped edge gate

4. Fan Gate

A fan gate has a wide edge that can balance the flow of the melt front, so it enters the part in a stable manner, as shown in Figure 7.14(a).

Advantages: Even filling that can prevent deformation of the finished part; a finished part with good appearance can be obtained; hardly any defects occur.

Disadvantages: Difficult for post processing and trimming off gate; a noticeable gate mark is inevitable.

The wide gate with a thickness gradient has a large area of filling and can fill the melt into large plastic parts rapidly. Warpage and dimensional stability are critical for large plastic parts. A fan gate can fill up the mold cavity of a large plastic part evenly with the melt front. The width and thickness of a fan gate are designed with a taper, so that the area of the melt front can be maintained to ensure a fixed melt velocity that fills the cavity under the same pressure along the width of the gate. Similar to other manual trimming gates, the maximum thickness of a fan gate should not exceed 75% of the wall thickness of the plastic part, and the width is 25% larger than the length of the side of the mold cavity.

5. Film/Flash Gate

The functions and purposes of film gates are similar to that of fan gates. Film gates are applied to acrylic plates with large area and uniform cross section, as shown in Figure 7.14(b).

Advantages: Better balancing of the flow; reduces weld lines and prevents warpage and deformation of the finished part.

Disadvantages: Difficult to remove and affects the appearance of the part.

A film gate is also called a flash gate and is lapped to the geometrically straight side of the part, and the gate width may cover across the edge of the whole mold cavity or partially. A film gate is able to balance the melt front for even filling and keep warpage of the part at a minimum. Although the size and space used by a film gate is lower compared to a fan gate, this type is very sensitive to gate thickness, runner diameter, and filling speed, etc. The recommended thickness of the gate is 50 to 70% of that of the wall thickness, while the length should be short.

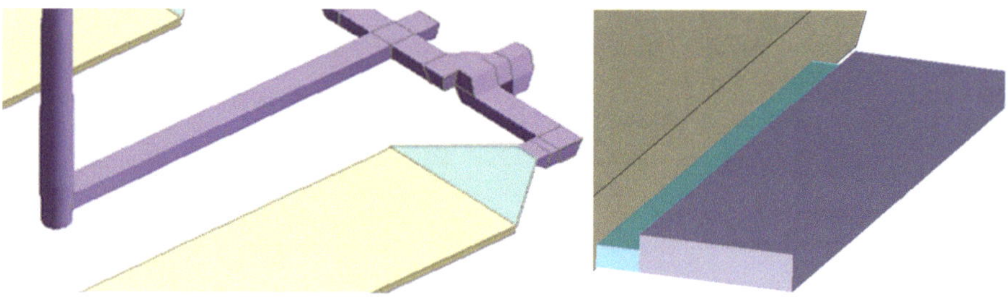

(a) (b)

Figure 7.14 (a) Fan gate and (b) film/flash gate

6. Disk Gate/Ring Gate/Spoke Gate

A disk gate, also called a diaphragm gate, as shown in Figure 7.15, is often used in cylindrical parts with openings on both sides, or circular or other geometrically coaxial plastic parts with openings on the inside. Disk gates are used in three-platen molds. The center thickness is recommended to be 1.25 times the thickness of the part where it contacts the gate and the thickness of gating should be 50 to 70% of the part thickness.

The scope of applications for ring gates, as shown in Figure 7.15, includes cylindrical, circular, or other geometrically coaxial plastic parts, and is similar to that of disk gates. When a ring gate is employed in the design of a two-platen mold with cold runners, the melt flows along the center of the mold and down the wall of the cylinder that has to be filled. The recommended dimensions are the same as those for a film gate.

A spoke gate, also called a four-point gate or a cross gate, as shown in Figure 7.15, is mostly used in multi-cavity, three-platen molds or hot runner systems. Spoke gates are similar to disk and ring gates and are mostly used for tubular plastic parts. However, weld lines occur easily and also the roundness of the part geometry is poorer. The geometry of spoke gates is either mostly rectangular and of constant cross-section or that of a shrinking circular gate. The recommended dimensions depend on the types of branching gates.

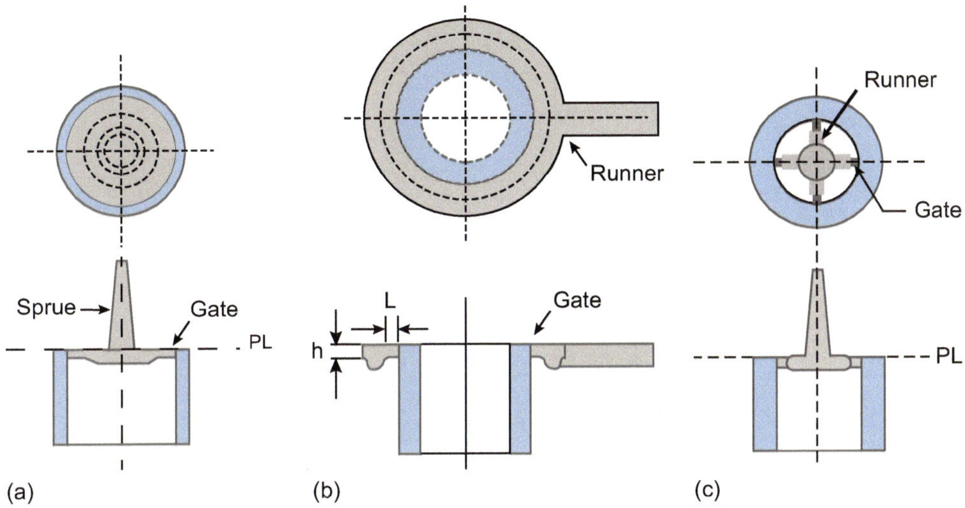

Figure 7.15 (a) Disk gate, (b) ring gate, and (c) spoke gate

Automatically trimmed gates work along with the motion of the mold, and can be removed along with runners from the geometry of the part upon demolding and ejection of the plastic parts. Automatically trimmed gates include the following types: pin gate, submarine gate, cashew gate, hot runner gate, valve gate and in-mold cutting.

7. Pin/Pinpoint Gate

Pin gates, as shown in Figure 7.16, can be separated from parts along with the runners upon demolding and ejection.

Advantages: The finished part and the runner/gate system can be separated automatically upon demolding and ejection without manual handling. Multiple gates can be introduced and a better location for injection can be chosen.

Disadvantages: The pressure drop is higher when the plastic enters the mold cavity; plastic solidification is faster at gates; the packing effect for finished parts is weakened; more waste is generated from runners; and mold cost is higher. The diameter should be 40 to 50% of the part thickness.

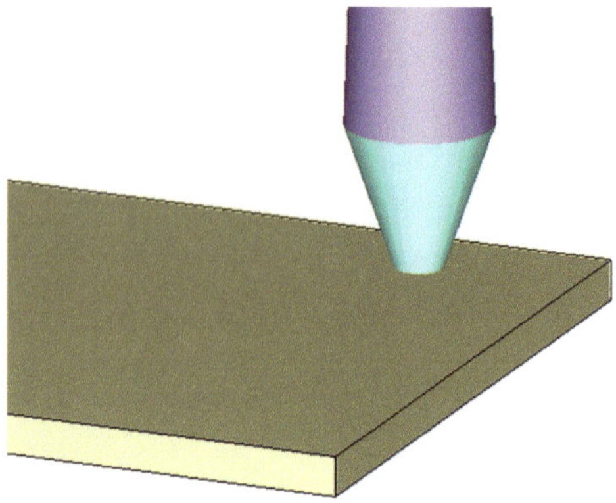

Figure 7.16 Pin/pinpoint gate

8. Submarine/Tunnel/Chisel Gate

A submarine gate, as shown in Figure 7.17, is also called a tunnel or a chisel gate.

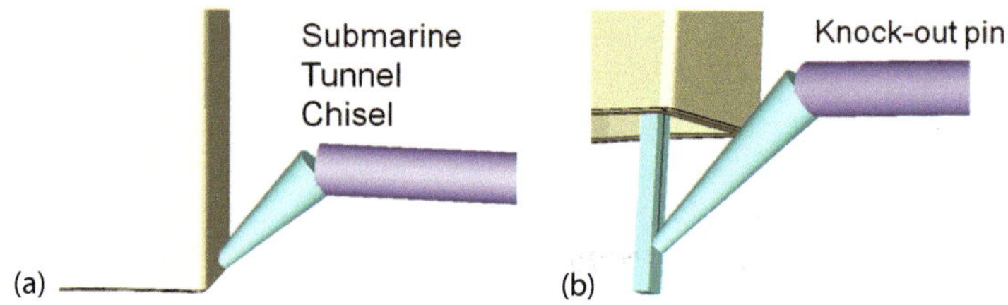

Figure 7.17 (a) Submarine/tunnel/chisel gate and (b) knock-out pin gate

Advantages: The separation of gates and plastic parts can be achieved automatically; the gate mark is small; and the gate location can be allocated at the inner or outer side freely.

Disadvantages: Large pressure drop and can be applied to simple plastic parts only.

Figure 7.17 is an illustration of a submarine/tunnel/chisel gate. This gate type is mostly applied in the design of two-platen cold runner molds of small mold cavities. The runner locates at the parting line of the two-platen mold, and a slant taper gate is made between the end of the runner and the mold cavity to facilitate filling of the melt into the mold cavity. After the melt is fully cooled, the gate is separated from the plastic part automatically via motion of the ejection system. The recommended diameter of a typical submarine gate is 40 to 70% of the wall thickness; however, the size of the gate is still closely related to the plastic material used and application forms. Another gate design with knock-out pin is shown in Figure 7.17.

As the gating usually corresponds to the ejection location, it is especially used the following design change if gate marks are not allowed to be shown on the appearance of a part.

9. Cashew/Banana Gate

The cashew gate, as shown in Figure 7.18 is difficult for machining, and the diameter is 30 to 50% of the part thickness .

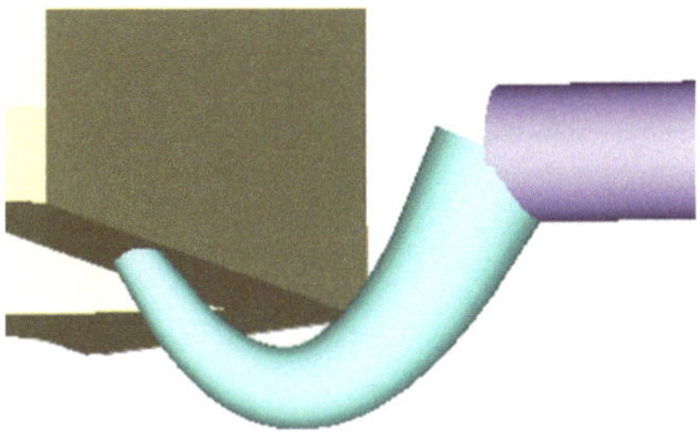

Figure 7.18 Cashew/banana gate

Advantages: Easy to post-process and automatically trim off.

Disadvantages: Large pressure loss, complex mold structures, and high cost.

10. Hot Runner Valve Gate

In a hot runner valve gate system, a movable hidden needle bar can be installed to control the opening and closing of the gate, as shown in Figure 7.19.

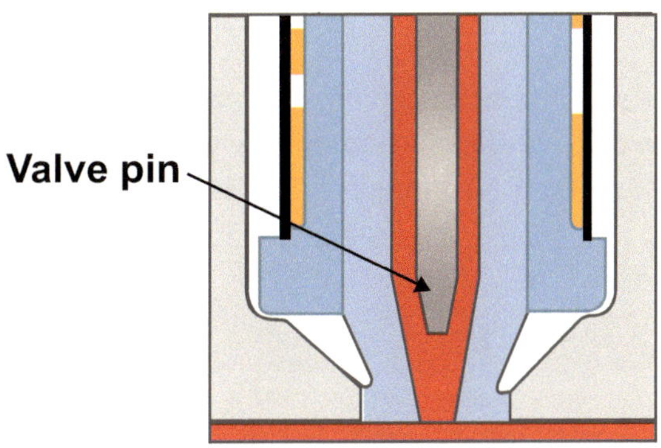

Figure 7.19 Hot runner valve gate

Advantages: Saves material and contributes to a short molding cycle.

Disadvantages: Marks are left on the surface of the part.

The valve pin is opened for filling and packing until the plastic material is fully filled into the mold cavity, while the gate is closed right before sealing; thus, the melt in the runner can be reduced and the step of gate trimming can be spared. The opening and closing of the valve pin can lead to versatile process methods; however, due to the difficulties in design and manufacturing of valve pin gates as well as the higher cost in mold development, they are mainly applied to large-sized parts which emphasize quality.

After the type of gate has been determined, CAE can be used to assist with the gate design. For example, the Moldex3D gate wizard can be used to quickly create a variety of common types of gates, and the gate can be simply and quickly placed in the desired product position of the part via the preview function, as shown in Figure 7.20.

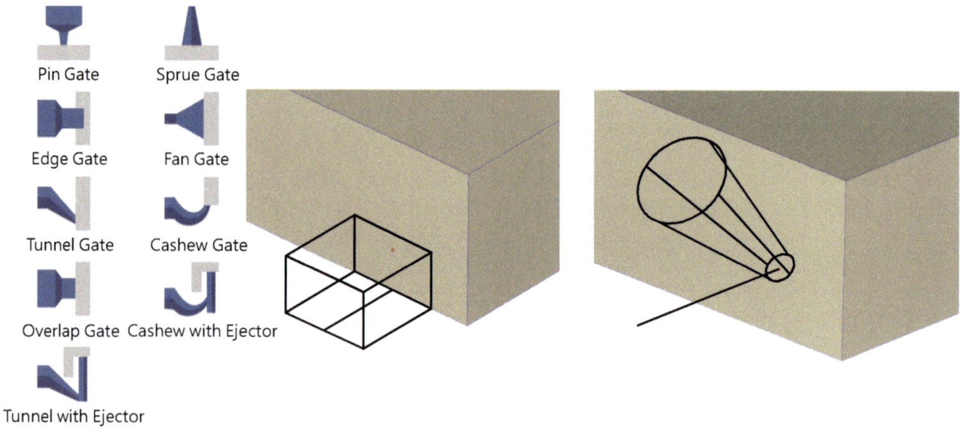

Figure 7.20 The preview function in the gate wizard

Figure 7.21 shows the different parameter adjustment pages for each type of gate. As the cross-section size can be conveniently adjusted, different gate designs can be finished quickly and their advantages and disadvantages analyzed.

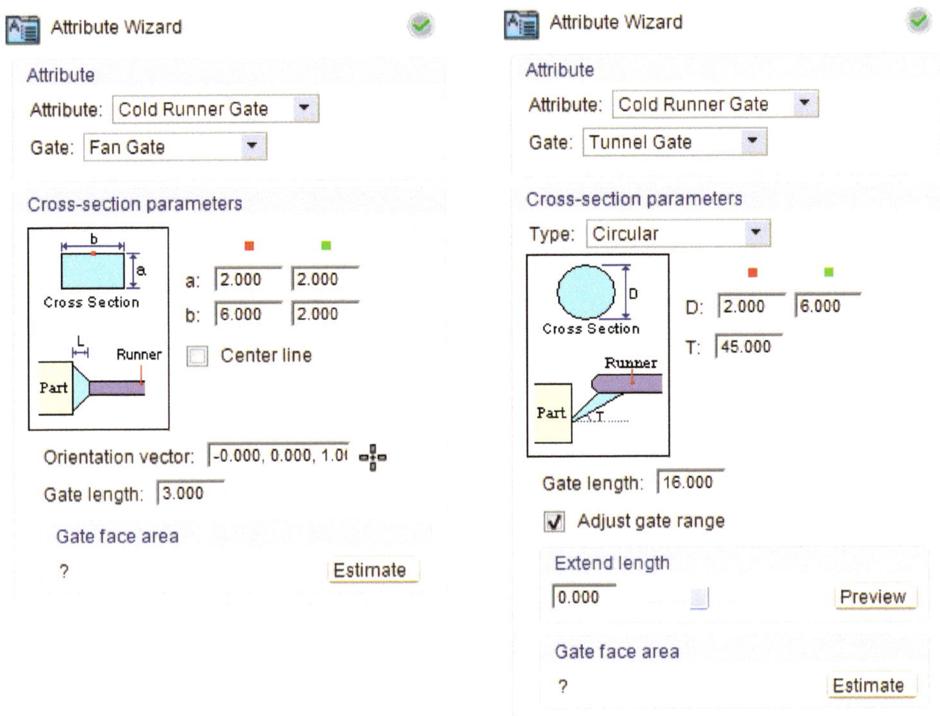

Figure 7.21 Adjusting the gate dimensions via parameters

In addition to the cross-section size, the Moldex3D gate wizard has a function for estimating the actual gate cross-section area of the product. Figure 7.22(a) shows that the calculated cross-section areas of the gates are different because of the different shapes connected to the parts, even if the two gates are set to the same cross-section dimensions. Moreover, the influence of different parameters on the gate cross-section areas is shown in Figure 7.22(b).

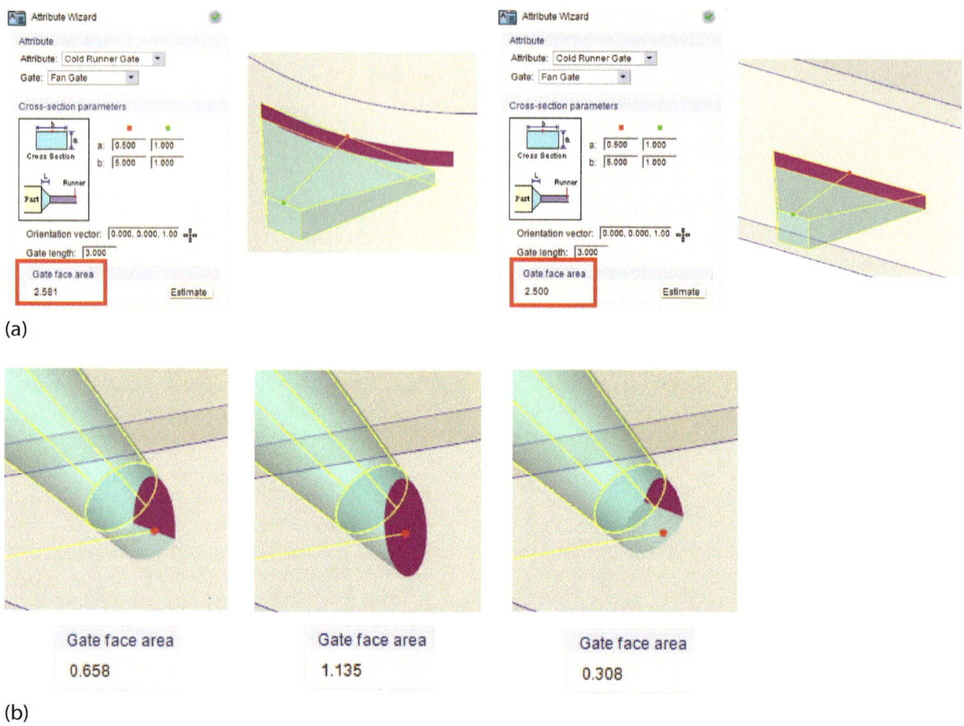

Figure 7.22 The function for estimating the gate cross-section area: (a) The same parameters are used for the cross-section areas, but the cross-section areas are different, and (b) the influence of parameters on the cross-section areas

7.1.3 Gate Sealing

Packing is a parameter that directly affects part quality. The setting of packing conditions depends on the size of the gate, and the pressure inside the mold can be used to facilitate judgment of the appropriateness of the parameters. Figure 7.23 shows the variation of the pressure detected by the pressure sensor inside the mold in an injection molding cycle, whereas Figure 7.23 illustrates the variation of the specific volume of the material in the cycle. In the packing stage, the loss of kinetic energy is trivial, and the pressure distribution in the mold cavity comes from the interaction between pressure, volume, and temperature (PVT). When the melt temperature drops to the extent that the flow stops, which means gate sealing, the cavity pressure decreases with temperature under a constant specific volume.

Remember at all times in packing time setting that the effective packing time is equal to the time for gate freeze. The weight of the part remains unchanged after the gate is frozen, and the packing is limited to the area around the gate and no melt is further injected into the part, which would characterize an ineffective packing process. Generally speaking, the setting of the packing time is based on user experience, but a better setting can be obtained through revision according to analysis results. The flow rate at the gate is a factor for the judgement of the effective packing time. Figure 7.24 shows a reference for the packing time setting by identifying the time the flow rate reduces to nearly zero at the gate, that is, gate sealing.

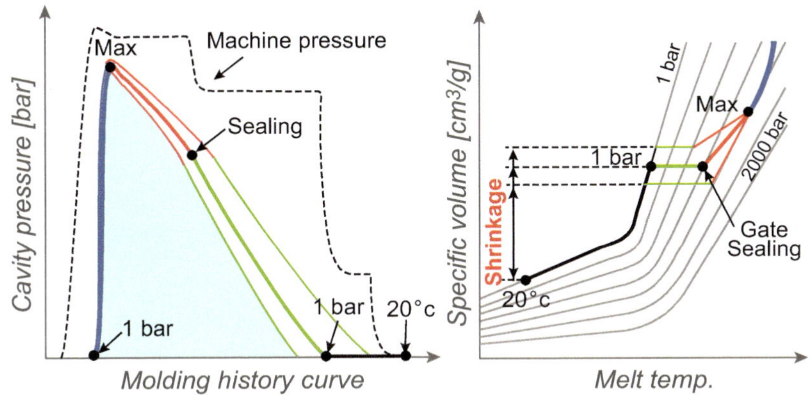

Figure 7.23 Cavity pressure variation during molding process

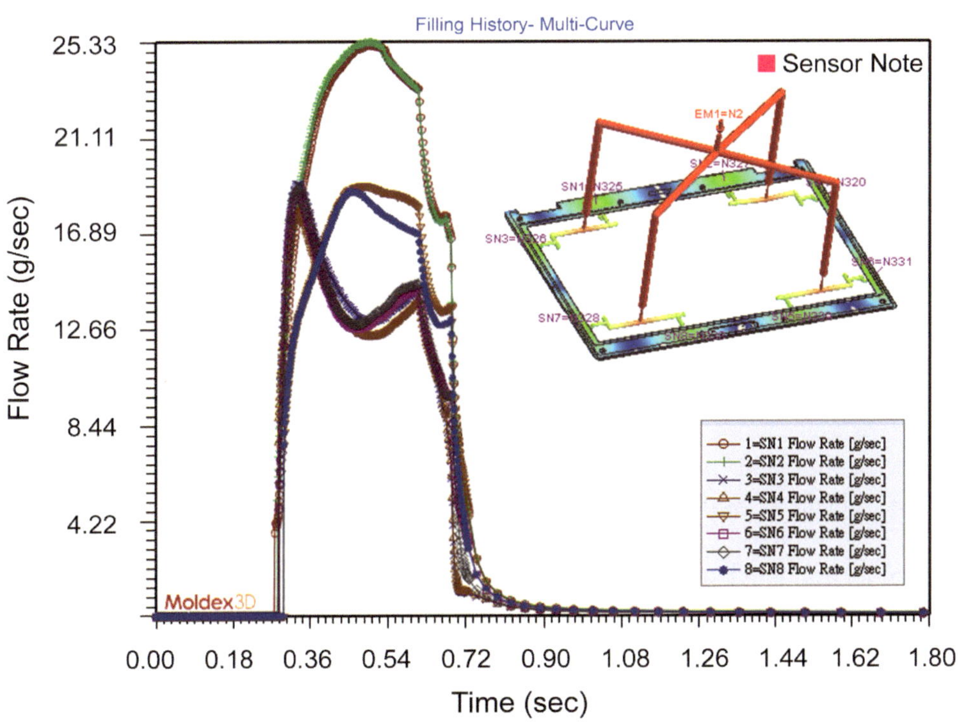

Figure 7.24 History curves of flow rate

Figure 7.25 shows a way to determine whether the gate is sealed or not. The melting core is used to display which areas are not sealed yet and still flowable. In a CAE analysis, a longer packing time can be set first and checked at the end of the analysis to measure the thickness of the frozen layer or to see the melt state at the end of packing.

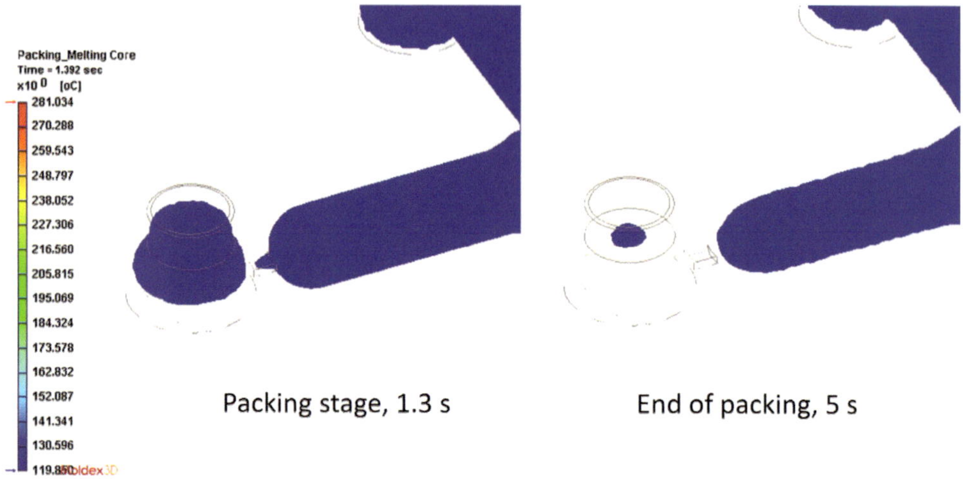

Figure 7.25 Melting core conditions at 1.3 s and 5 s (gate sealing)

7.1.4 Flow Balance

For a multi-cavity mold system, a runner system which balances the filling of multiple cavities is one of the most important issues. The filling quality and efficiency of multi-cavity systems can thus be addressed simultaneously.

As an example, in a system with eight cavities in one mold, samples of short shot are compared with each other regarding the difference in melt front, as shown in Figure 7.26. It can be seen that the difference in weight varies gigantically due to flow imbalance, but it is drastically improved with the application of the MeltFlipper® technique and the weight of each sample of short shot is very uniform (also shown in Figure 7.33(b) and Figure 7.38(b)). Such a result indicates that the MeltFlipper® technique helps to deliver uniform quality and is beneficial to maintaining production efficiency.

The flow of a plastic in runners is very complicated; for example, the viscosity of the melt is unevenly distributed along the whole cross section of the runner under the influence of shear rate, temperature, and other factors. Figure 7.27 shows the distribution of the shear rate in the filling of the runner. The highest shear rate is located at the outer area of the runner, whereas in the center it becomes zero.

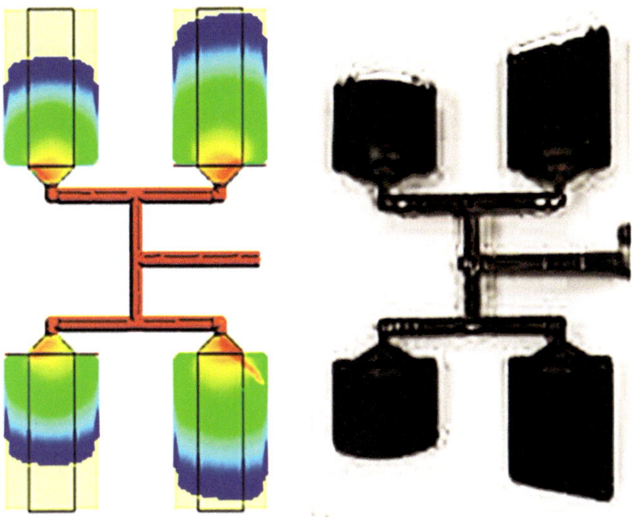

Figure 7.26 Flow imbalance in multi-cavities [1]

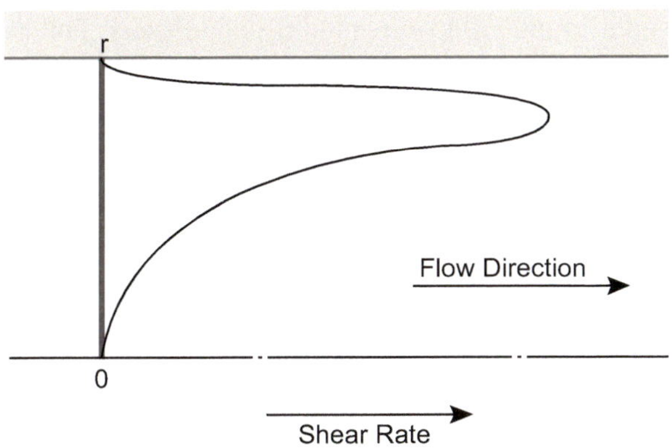

Figure 7.27 Shear rate profile in a runner cross section

The imbalanced melt in the runner system not only causes the imbalanced filling of different mold cavities, but also affects the filling of a single mold cavity, because the melt at one side of the mold cavity comes from a laminar flow with high temperature and shear, whereas that of the other side comes from a laminar flow with lower temperature and shear, as shown in Figure 7.28. The resulting warpage tends to be mistaken as a consequence of uneven cooling of molds, which affects shrinkage, warpage, and mechanical properties of plastic parts.

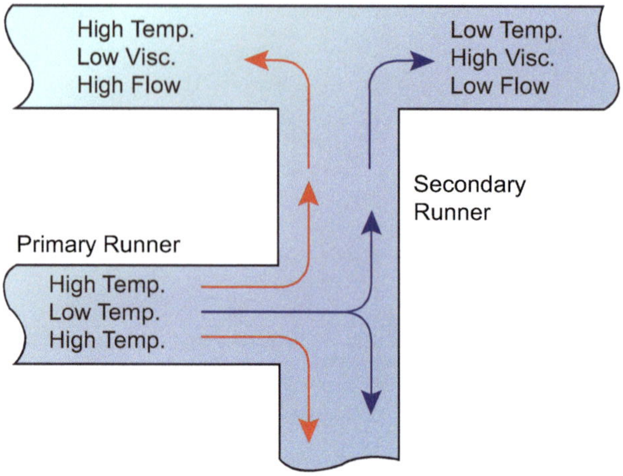

Figure 7.28 Melt flow situation of temperature, viscosity, and flow rate in the runner

Mold analysis software can assist in verifying the sensitivity of the filling process of the runner system under different molding conditions. For example, different filling results would be obtained by using different injection rates in a fishbone runner system. With a slow filling rate process, the mold cavity further away from the sprue would be filled first, whereas the cavity closer to the sprue would be filled first with a high filling rate, as shown in Figure 7.29.

The "fishbone" runner system has the imbalanced-filling problem, namely that some mold cavities are filled first while others are not due to the difference in the melt flow length. Since the discovery of this geometric imbalance problem, mold designers attempt to design geometrically balanced runner systems, where the flow length of each mold cavity in the runner system is identical. Although such "naturally balanced" runners are applied, differences in the filling rate of each mold cavity are still seen. Those inner mold cavities, compared to the outer mold cavities, are always filled first, and thus the plastic parts coming from the inner mold cavities are larger and heavier than those from the outer mold cavities. This phenomenon is commonly mistaken for deformation of mold platens or mold temperature deflection.

Basic runner layouts of multi-cavity systems are standard or fishbone, H-bridge, branching, radial, or star, as shown in Figure 7.30.

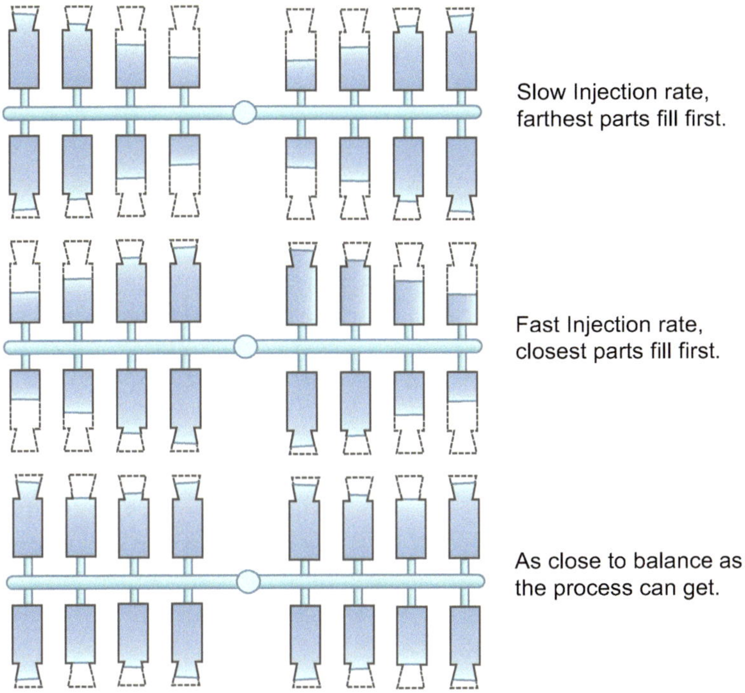

Figure 7.29 Melt flow patterns in an imbalanced runner system resulting from different injection rates

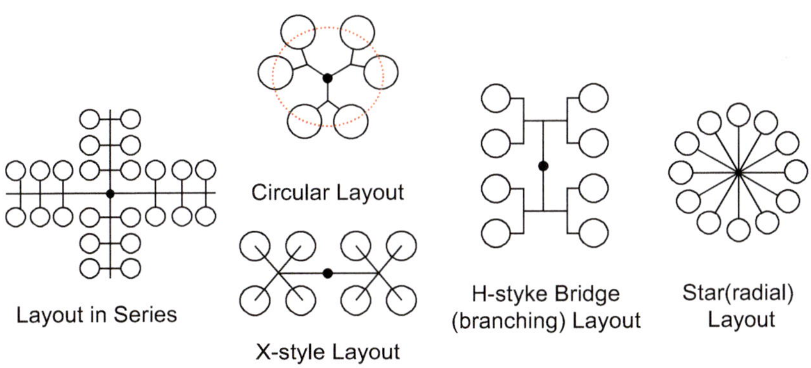

Figure 7.30 Common multi-cavity layouts

For a geometrically balanced runner system, as the flow length for the melt to reach each gate is the same, the flow is more balanced, the pressure drop in each cavity is similar, and the packing effect is more balanced, so the quality of part from each cavity is therefore more even. Figure 7.31 shows that a non-balanced runner system can be improved by adjusting the runner and gate size.

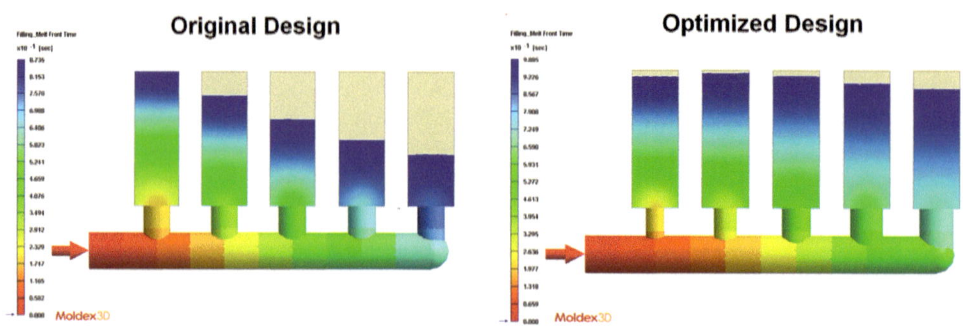

Figure 7.31 Fishbone runner layout optimization with dimensional adjustment

To make the design of an artificially balanced runner system faster as well as to save time and development cost, molding analysis software can precisely simulate the filling result of a runner design in a short time. However, a naturally balanced runner system is still a more stable runner system, which is a better choice in the design of a multi-cavity mold that produces multiple parts of the same type, Figure 7.32 illustrates the difference in the flow balance of melt filling between a manually modified non-balanced runner system and a naturally balanced one.

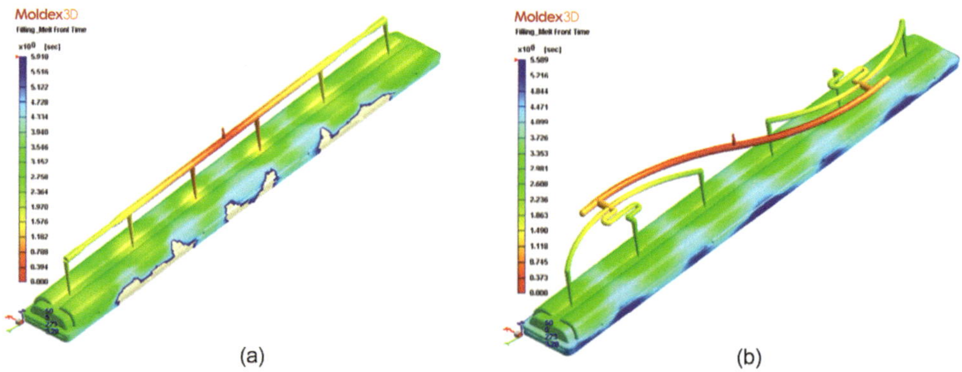

Figure 7.32 Flow balance conditions of (a) manually modified non-balanced runner and (b) naturally balanced runner

To resolve the flow imbalance situation induced by shear in a multi-cavity system, the easiest way is to use the melt flipping technique to manage the varying plastic behavior caused by shear, e.g. the MeltFlipper® technology, patented and sold by Beaumont Technologies. The MeltFlipper® technology can be applied to almost any type of cold and hot runner systems via a non-intrusive method to manage the imbalanced flow of the melt in the runners. This commonly used technology is now even evaluated as a countermeasure against the shear stress variation observed in the runner system during the profile extrusion process and blow molding. Figure 7.33 shows a comparison between a conventional multi-cavity runner system and one applied with MeltFlipper® technology. The flow balance between each mold cavity is improved after modification.

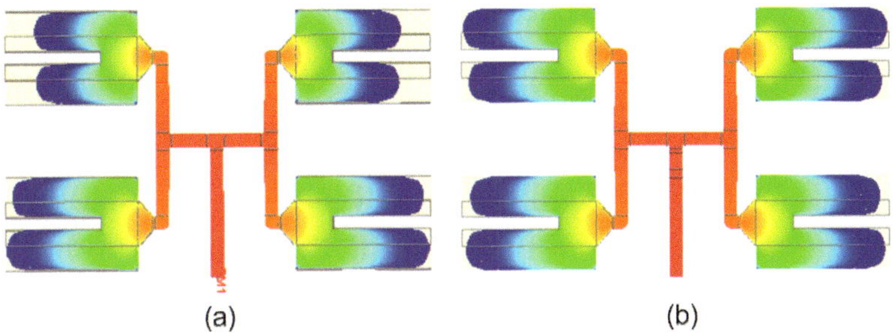

Figure 7.33 Flow pattern of (a) conventional runner and (b) MeltFlipper® runner design

The issue of flow balance occurs not only in multi-cavity runner systems, but generally in the case of large objects, such as front and rear bumpers of vehicle and dashboards, where the parts cannot be completely filled without the use of multiple gating for a single mold cavity. As plastic material has a flow limit, the ratio of flow length to thickness often serves as the criterion for determining whether additional gates should be designed or not.

Figure 7.34 illustrates an example of multiple gating in a single-mold cavity of a laptop housing. From analysis, we can obtain the gate contribution, which shows the volume of the injected plastic melt at each gate at the current timing. The gate contribution is shown in percentage. Normally, the best result cannot be obtained unless the average gate contribution is derived.

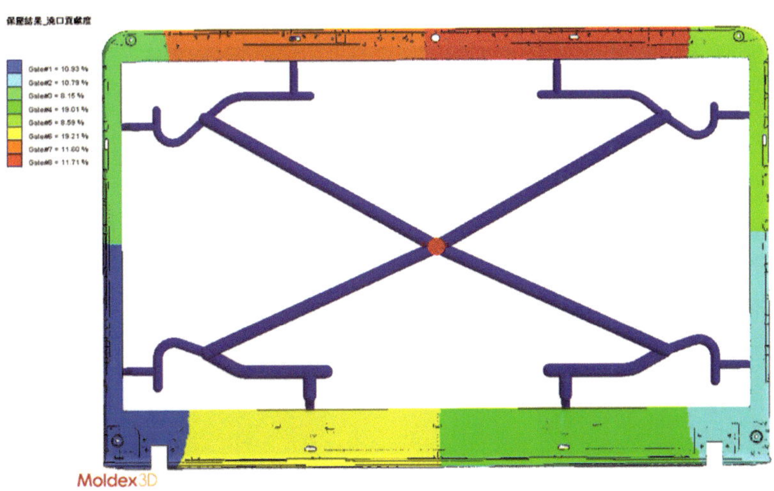

Figure 7.34 Gate contribution

As shown in Figure 7.35, the CAE analysis result for the revised gate dimension shows that the more uniform the pressure distribution is at the time when packing ends the better is the outcome; this indicates that the volumetric shrinkage inside the part is very similar, and by mitigating the imbalanced shrinkage the possibility of warpage can be further reduced.

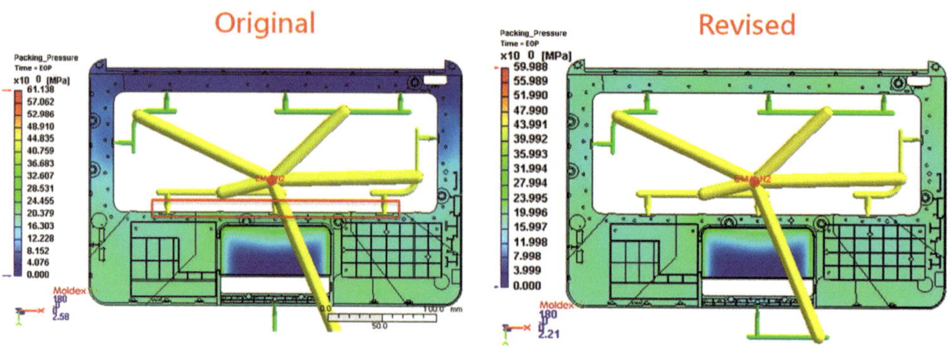

Figure 7.35 Flow balance and packing pressure distribution

Whether or not flow balance is required in the notebook PC industry, which demands precision, depends on the directly related part warpage. Figure 7.36 shows that in the gating design with a higher flow balance, the shrinkage along the long side is more uniform.

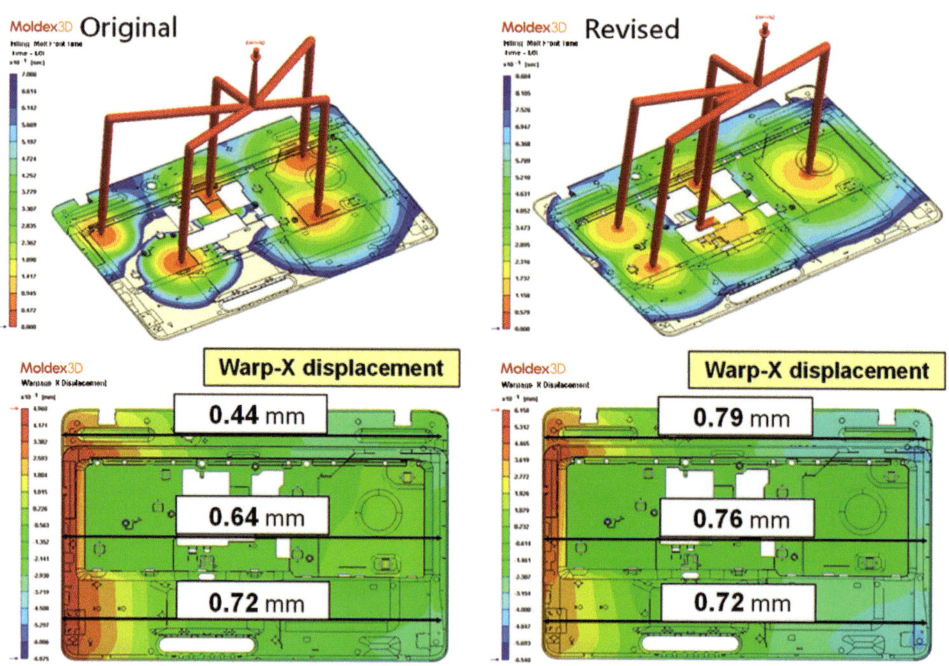

Figure 7.36 Flow balance versus warpage

7.2 Practical Applications

7.2.1 CAE Verification on MeltFlipper® Design

Figure 7.37 shows an example of how to utilize CAE to present the flow imbalance in a part with eight cavities in one mold, and how it is improved by using the runner flipping technology. The relevant design inputs are as follows: product dimensions 140.3 × 69.2 × 2.4 mm^3; material PMMA; filling time 0.3 s; melt temperature 210 °C; and mold temperature 70 °C.

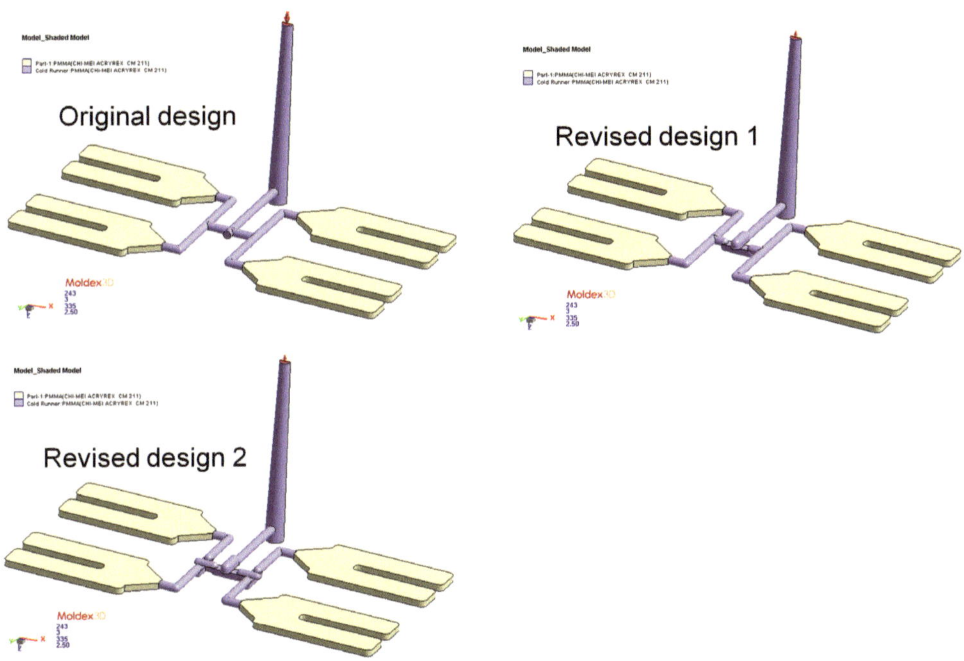

Figure 7.37 Flow imbalanced conditions with original and revised designs

MeltFlipper® provides a way to solve the imbalanced distribution of temperature and shear rate using special devices in the runner. Although such a design can actually solve the flow imbalance issue, as shown in Figure 7.38 the process of melt flow variation still remains a mystery.

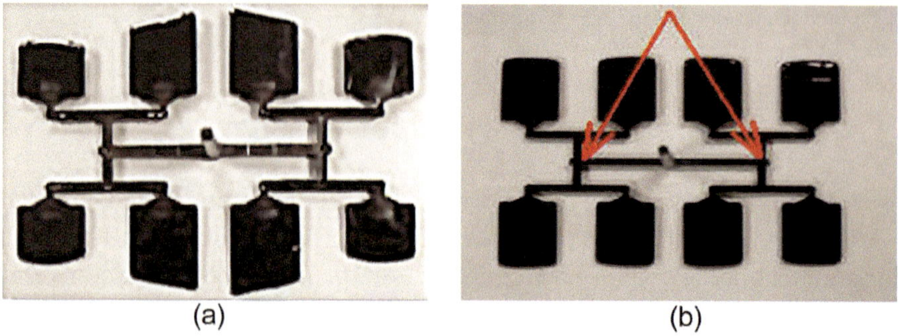

Figure 7.38 Short shot of multi-cavity mold for (a) original design and (b) revised design (MeltFlipper®) [1]

To explore the flow imbalance phenomenon of multiple cavities, a 3D practical model is built according to the actual dimensions and shape of a multi-cavity part and its runners. Via the visualization process of CAE, the causes of flow imbalance can be clearly understood, and thus the way by which the melt is flipped for achieving flow balance can be obtained.

To obtain a reasonable analysis result, the resolution of numerical analysis should be considered concurrently with the preprocess of mesh generation. The innovative preprocess in Moldex3D enables users to automatically generate the hybrid mesh of the runner by merely drawing the runner location with the defined geometric dimensions, as shown in Figure 7.39. With such a technology, not only can the number of layers of the solid mesh be adjusted easily, but also mesh quality and quantity can be considered simultaneously. With a suitably detailed numerical analytic mesh of the runner, the flow imbalance issue caused by shear generated in the runner can be treated precisely.

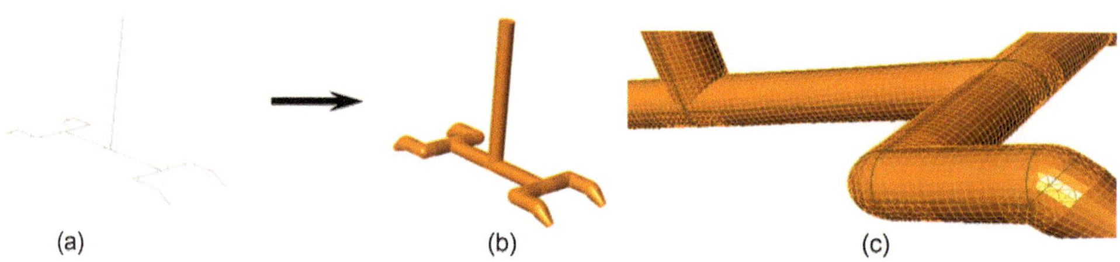

Figure 7.39 Runner generation method: (a) line geometry, (b) solid mesh, and (c) hybrid mesh

As an instructive example, we want to analyze the two models given in Figure 7.40 one is the original design and the other is the revised one. The original design shown in Figure 7.40 has 314,865 solid elements in the hybrid mesh, with 62,640 hexa elements, 93,312 prism elements, 600 pyramid elements, and 158,313 tetrahedral elements, and the number of layers along the thickness even exceeds six. The revised design flips the plastic material in the runner as shown in Figure 7.40 and the hybrid mesh is introduced as well.

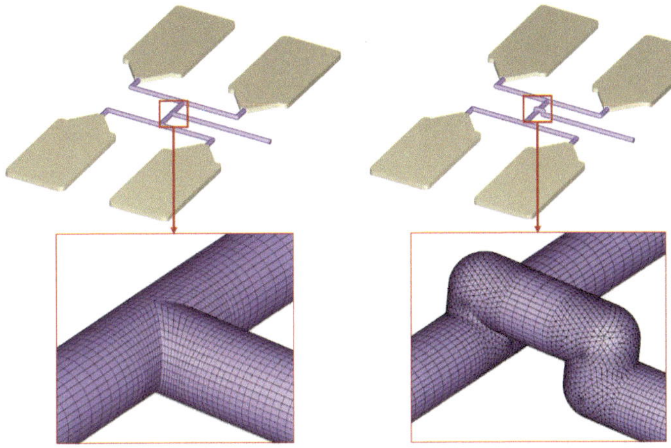

Figure 7.40 (Left) Original and (right) revised design

Regarding the variation of the temperature of the plastic material in the mold during the filling process in the original design, it can be seen in Figure 7.41 that before the melt reaches the corner, a high-temperature ring is formed in the runner close to the surface due to shear heating, and when the melt flows through the first corner, such a high-temperature ring is forced to be separated and thus a left-right asymmetric temperature distribution is generated.

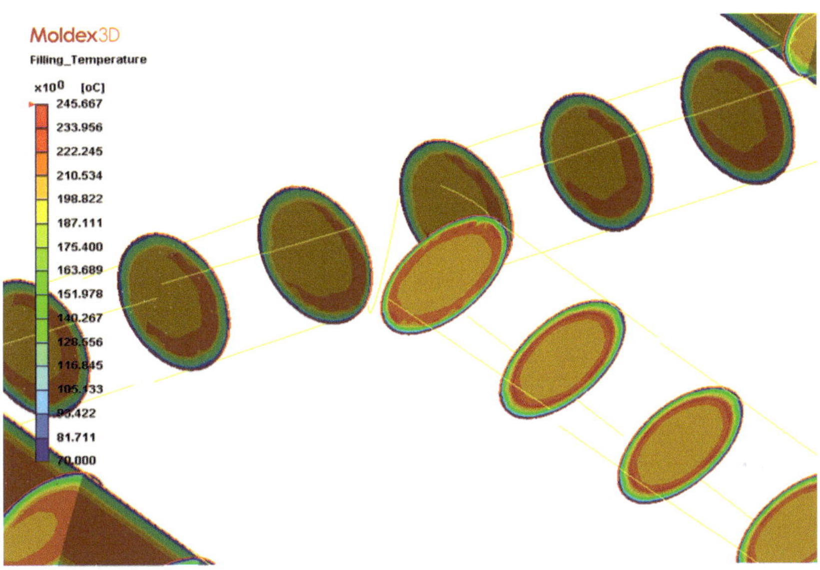

Figure 7.41 Temperature distribution of x- and y-cross sections at the first corner of the original design

Figure 7.42 shows that the hotter plastic material close to the outer edge of the primary runner flows along the right side of runner A, whereas the colder plastic material close to the center of the primary runner flows along the left side of runner A. A similar situation happens on the other side. The temperature asymmetry will occur in melt between the cavities, i.e. the melt of lower temperature has a higher viscosity which leads to a slower filling rate in the cavity, whereas a higher filling rate is obtained for the melt of higher temperature.

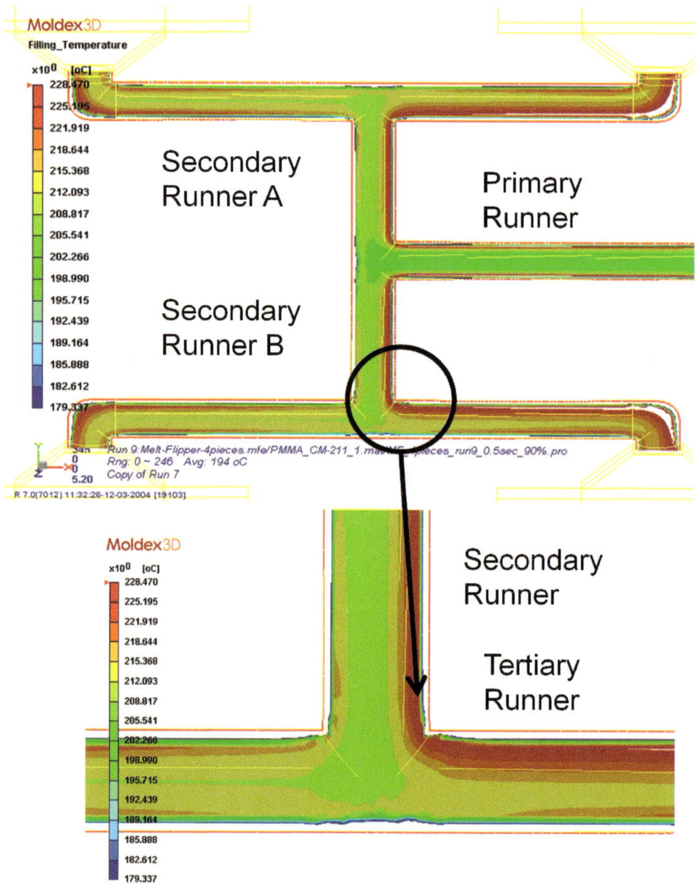

Figure 7.42 Temperature distribution of z-cross section at the second corner of the original design

The results derived from the simulation of the flow imbalance of the original design demonstrate high compliance with the actual experimental result, as shown in Figure 7.43. It can obviously be seen that the plastic material in the mold cavity (right side) closer to the primary runner has a higher temperature compared to the plastic material in the mold cavity on the left side and thus the viscosity for filling is lower as well as the flow resistance, which results in a faster filling rate.

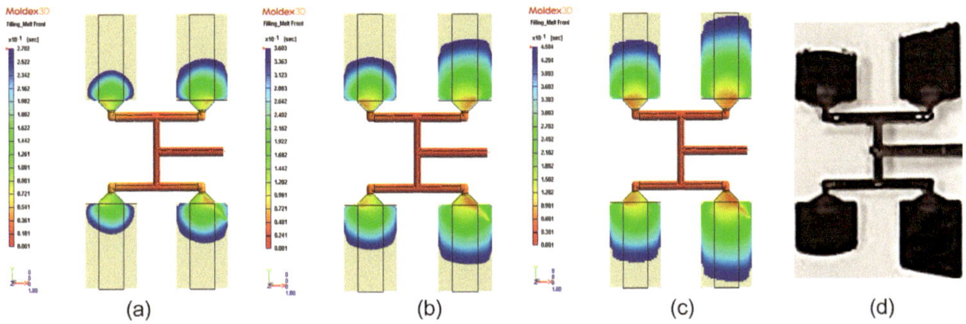

Figure 7.43 Melt-front comparison of simulated and experimental result of the original design: (a) 40%, (b) 60%, (c) 80%, and (d) actual experimental result [1]

Figure 7.44 shows a CAE analysis result of the revised design. After the melt passes the particular flipping structure and enters the first corner, the left-right imbalanced temperature of the melt is transformed into the up-down imbalanced temperature distribution.

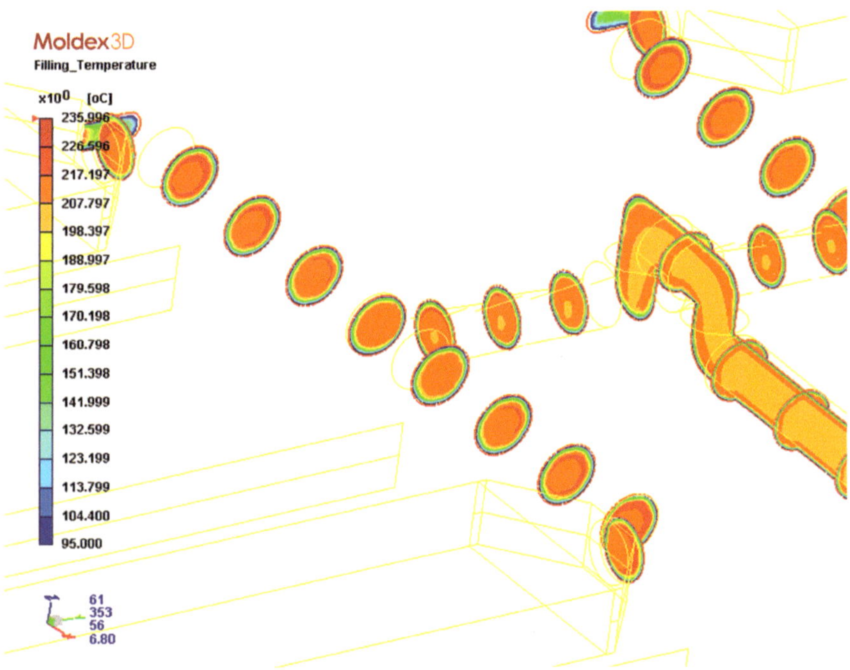

Figure 7.44 Temperature distribution of runner cross sections of the revised design

After the melt flows through the second corner, the up-down imbalanced temperature distribution in the secondary runner would not generate an asymmetric temperature distribution of the melt between cavities, and thus the imbalanced flow between cavities would disappear. Therefore, the flow balance condition can be obtained in both runners and cavities, as shown in Figure 7.45.

Figure 7.45 Temperature distribution of z-cross section at the second corner of the revised design

The analytic technology and hybrid mesh adopted in Figure 7.46 clearly display the flow balance phenomenon of melt in a geometrically symmetric runner system. Through the visualization of CAE, it can be understood how the plastic is flipped in the runner by the MeltFlipper® device to achieve flow balance by flipping the inner-outer asymmetric condition of the melt temperature to an up-down asymmetric condition. The design parameters for such a device would directly affect the temperature distribution and shear rate, which is also a key point to be considered. Therefore, optimum design parameters for such a device can be derived quickly by using CAE tools and thus the time for product development can be shortened.

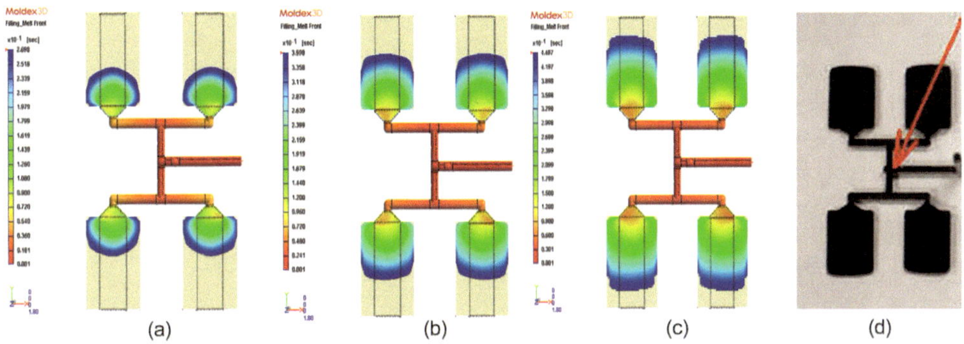

Figure 7.46 Melt-front comparison of simulated and experimental results of the revised design: (a) 40%, (b) 60%, (c) 80%, and (d) molded part (MeltFlipper®) [1]

7.2.2 CAE Verification of Multi-Cavity Systems

The following example is a one-mold-32-cavities LED lead-frame design, and the purpose of analysis is to increase production yield and productivity. The analysis emphasizes the runner design and the issue of filling caused by flow imbalance. In addition, potential issues that are likely to happen during molding of the part are also proposed for exploration to obtain the optimum improvement scenario. The design is shown in Figure 7.47, and the relevant geometric, material and molding conditions are as follows: part dimensions 7.40 × 7.40 × 1.85 mm^3; material liquid crystal polymer (Ticona Vectra E130i); filling time 1.3 s; melt temperature 335 °C; mold temperature 100 °C; and the material of the metal lead frame is copper.

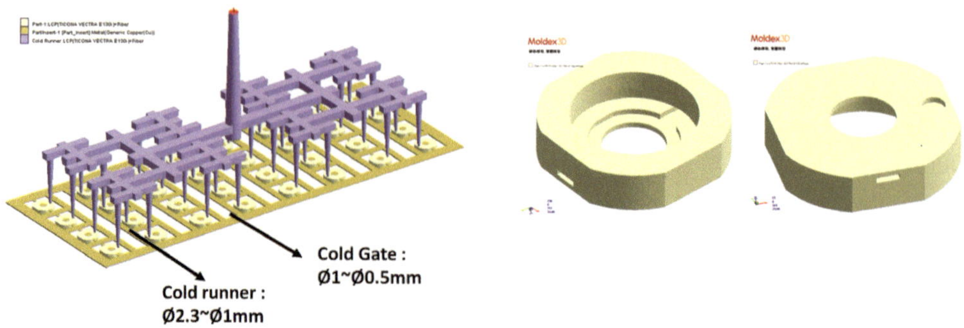

Figure 7.47 One-mold-32-cavities LED lead-frame case study

The plastic materials injection molded on a stamping and electric-plating metal material slit strip into embedded molding parts are nylon polyphthalamide or LCP with heat resistance, and usually there are over 100 parts in one mold, which become LED products after die-bond and packaging.

Key points in molding analysis are flow balance, heat resistant nylon polyphthalamide processing window, and deformation of the embedded part.

Conventional process: lead frame → injection molding → die bond → glue pouring.

From the flow chart or animation of the melt front, the filling conditions of the plastic material in each mold cavity at each point in time can be observed, which enables the prediction of the location of weld lines and air traps and the occurrence of short shot, and provides recommendations for venting channel allocation.

It can be seen in Figure 7.48 that the analysis results clearly display early short shots of the melt flow front, and the difference is generated at 75% filling time.

Figure 7.49 shows a significant result of flow imbalance at 97% filling time (see within the rectangle drawn with a red line).

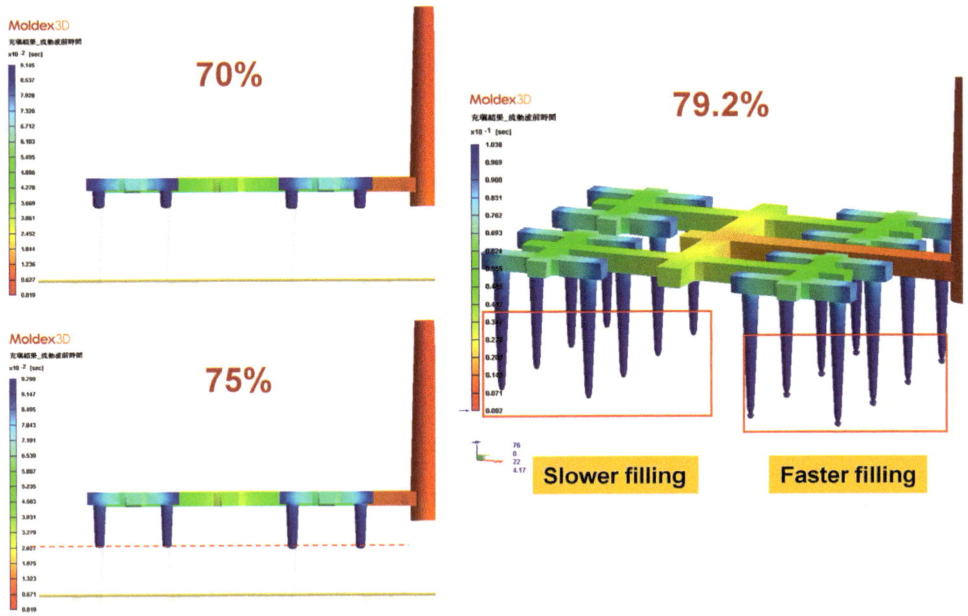

Figure 7.48 Melt front at 70%, 75%, and 79.2% filling time

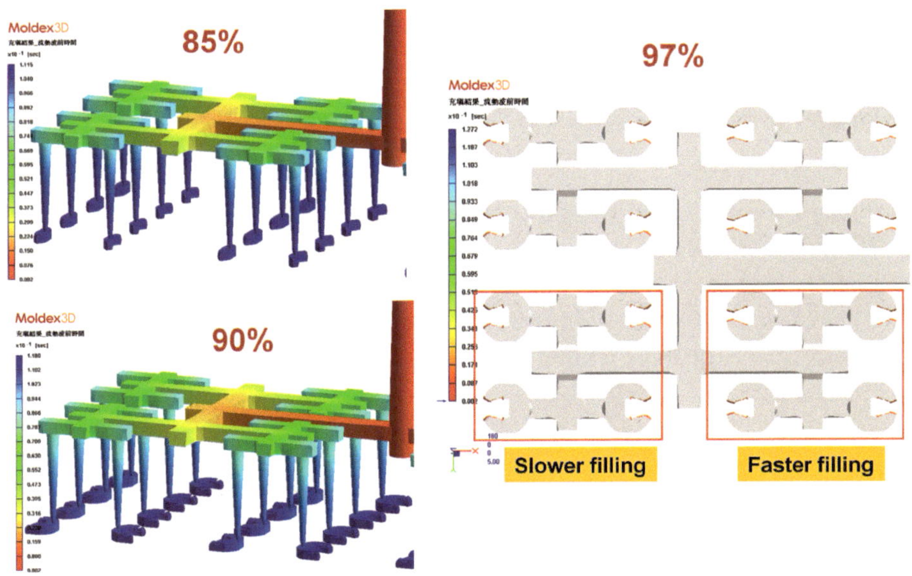

Figure 7.49 Melt front at 85%, 90%, and 97% filling time

Figure 7.50 shows that the left cavities are filled slower, and the difference after merging of the melt fronts in the first group shows consistency with the actual molding time on site.

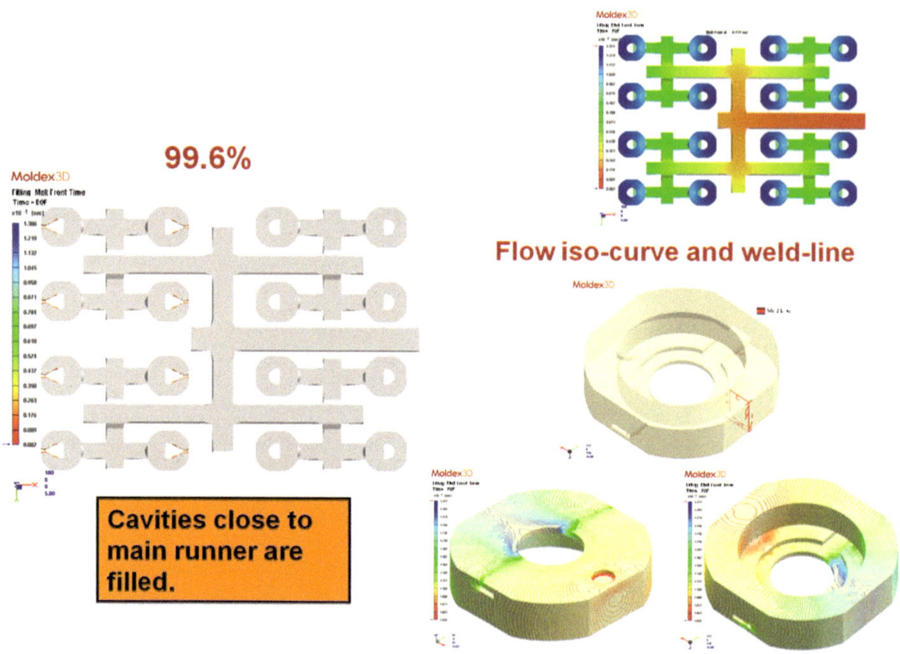

Figure 7.50 Melt front at 99.6% filling time

As illustrated in Figure 7.51, the cross section of the internal pressure distribution of the part shows a lower pressure distribution locally in the mold cavity at the time the filling ends, and the density of the area closer to the primary runner is significantly higher. Figure 7.51 also displays the temperature distribution in different colors at the end of filling (EOF). As the surface of the plastic part has a temperature close the mold temperature, a low temperature condition is often seen.

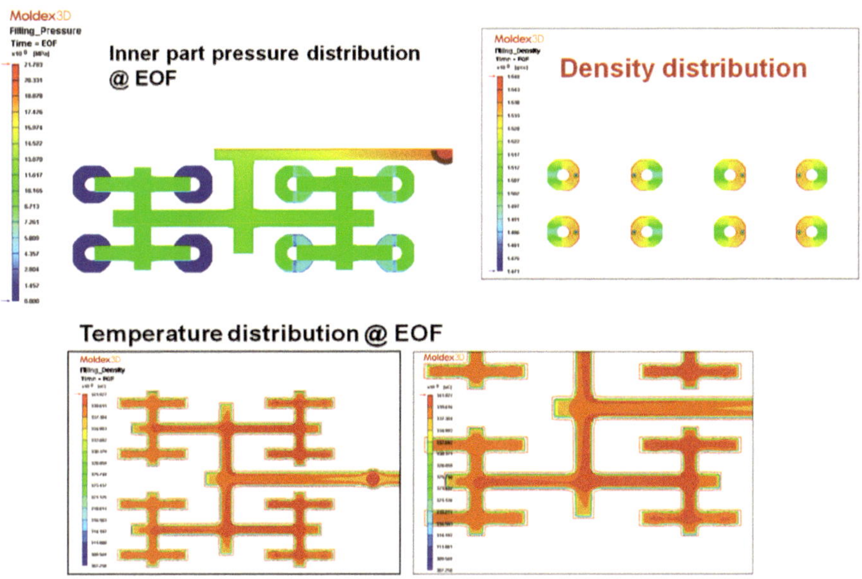

Figure 7.51 Pressure and temperature distributions at the end of filling

■ 7.3 CAE Case Study

Following the discussions of this chapter and the corresponding CAE verification case, an exercise model, as illustrated in Figure 7.52, is provided in this section for readers to practice CAE application in runner and gate design.

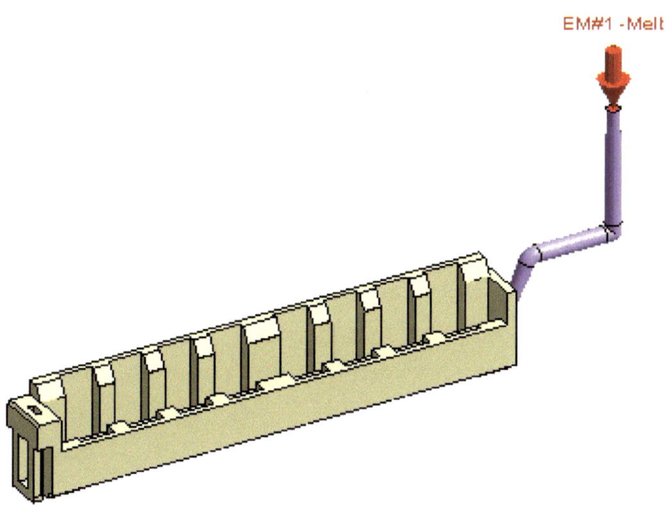

Figure 7.52 Case study sample

Please download the analysis files for CAE Case Study 7.3 from the following website:

https://moldex3d.box.com/s/zr6fvc1vlhbi4ocx11jwd3wmxt4ooif

Then open the .m3j or .mvj file in Moldex3D and answer the questions below according to the analysis results.

Questions:

1. Comparison of different gating positions
   a) Compare the different gating injection pressure profiles (sprue pressure XY plot), and explain the reasons for the differences.
   b) How much is the pressure drop in the runner? Explain why the pressure drop of difference is minor between both?
   c) Explain the reason for the difference of the melt front temperature.
   d) Which has the better structural strength? Why? (Hint: check the weld line temperature and shear stress.)
2. Comparison between the four different-cavity-number systems: one-, two-, four- and eight-cavity systems
   a) Are the default settings of flow rate (flow rate XY profile) in these four designs reasonable? Why?
   b) There are merely minor differences of injection pressure in these four designs. Try to explain this (more reasons are better).
   c) Try to explain the clamping force differences in these four designs.
   d) How can you check the flow balance between the four-cavity and eight-cavity designs?
3. Different gate size: small, original design, and large
   a) Try to measure the size of each gate. (oval length, short axis)
   b) Compare the injection pressure history (sprue pressure XY profile) and record the relationship between maximum pressure and gate area.
   c) Compare the gate shear rate, temperature, and viscosity at 25% fill time, and explain the cause-and-effect relationship.

d) Taking the original design as an example, how can you confirm that the default packing time is appropriate? If not, use the CAE analysis to find the appropriate setting of packing time.

e) Compare the volumetric shrinkage, try to explain the reasons for the differences, and use the CAE analysis to propose the improved program of the revised design.

4. Runner design: eight cavities, H-shaped and fishbone

   a) Compare runner volume, the flow rate, injection pressure, and clamping force of each cavity. Try to explain the reasons for the difference.

   b) Compare the melt front time, temperature, pressure, and volumetric shrinkage. Try to explain the cause-and-effect relationship.

   c) From the above results, do we know what design is better? Why?

References

[1] Beaumont Technologies, Inc. "Beware of the "Naturally Balanced" Runner System" [Online]. *https://www.beaumontinc.com/beware-of-the-naturally-balanced-runner-system/*

8 Cooling Optimization

Hung-Chou (Kent) Wang

In this chapter, the heat transfer principle of an injection molding system, the mold temperature variation effect toward molded part quality, a general design guide of the cooling channel, the relation between cooling efficiency and coolant flow, and the basic theory to estimate cooling time are first introduced. Next, some results from computer aided engineering (CAE) to optimize the cooling design are explained and demonstrated, as well as the use of 3D cooling computational fluid dynamics (CFD) analysis for a novel cooling technology called conformal cooling. Finally, three case studies to illustrate the benefits of using CAE in real application practice for a conformal cooling system are described.

8.1 Basics

In an injection molding process, the heat from the polymer melt is released and removed through convection of the coolants in the channel, and heat conduction of the mold. This process lasts until the whole product is solidified and strong enough for ejection. After that, the temperature of the ejected product continues to decrease to room temperature by dissipation of thermal energy into the environment. In most situations, the cooling time comprises 70 to 80% of the entire molding cycle as shown in Figure 8.1. The cooling effect greatly influences the solidification of the part, which affects potential deformation occurring during the mold ejection.

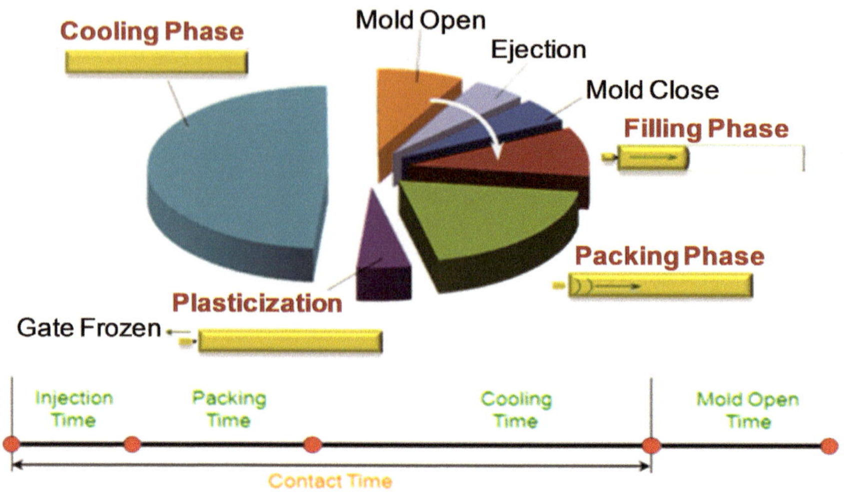

Figure 8.1 Conventional injection molding cycle

8.1.1 Heat Transfer Mechanism

A mold cooling system typically consists of the following items: temperature controlling unit, pump, supply manifold, hoses, cooling channels in the mold, and collection manifold. The mold itself can be considered as a heat exchanger, with heat from the hot polymer melt taken away by the circulating coolant. Figure 8.2 illustrates the components of a typical cooling system. For every different product and mold design, the cooling system should be rationally designed, so it can maximize the cooling efficiency of the different elements in the cooling channel during the specified cooling cycle.

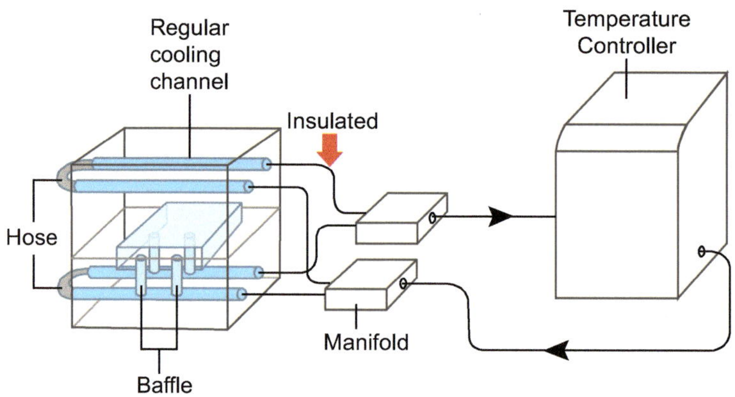

Figure 8.2 A typical cooling system for an injection molding machine

Figure 8.3 illustrates the basic heat transfer in the melt flow. The heat is transferred from the hotter melt to the colder mold wall via the melt front contact with the mold surface due to fountain flow. Therefore, a thermal energy balance situation is established when the melt is heated and cooled, and this means that a stable molding cycle has been achieved. And the mold temperature can be regarded as the average temperature of the mold cycle.

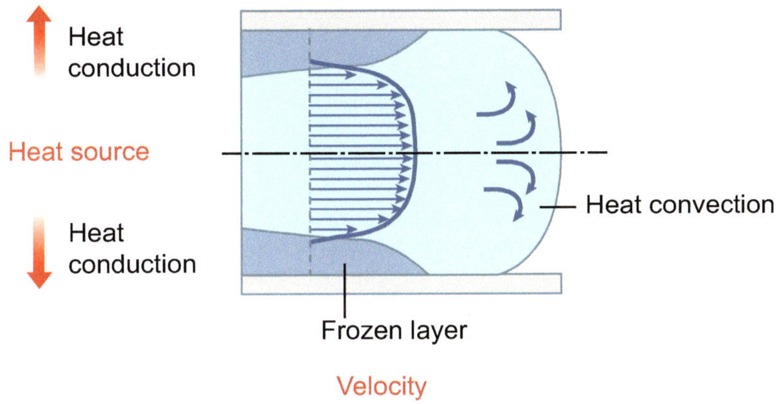

Figure 8.3 Melt heat transfer mechanism

The heat transfer mechanism can be divided into three modes: convection, conduction, and radiation. Convection is the heat transfer due to bulk movement of molecules within fluids such as gases and liquids. Convection takes place through advection, diffusion, or both. Conduction is heat transfer between two solid objects in contact each other. Radiation transfers heat without any media, and heat can be directly spread to a lower temperature object, as shown in Figure 8.4. In the injection molding process, the following heat transfer modes are considered:

1. Heat transfer to the mold base and cooling channel via heat conduction.
2. Removal of heat from the mold base by the coolant via convection.
3. Mold-base surface conduction, convection, and radiation heat transfer.
4. Heat loss during ejection.

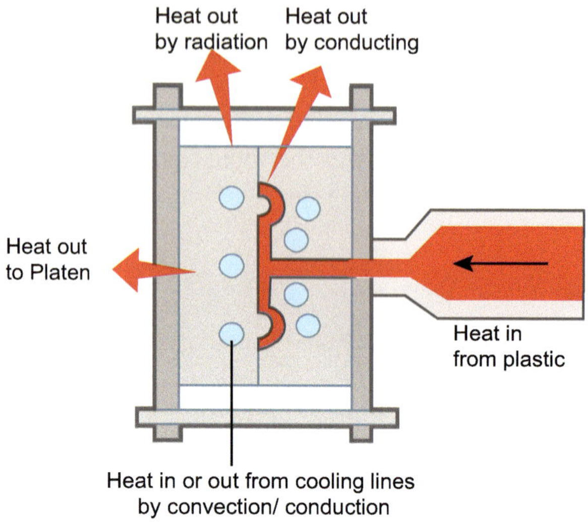

Figure 8.4 Mold heat transfer mechanism

In an injection molding process, the heat dissipation path and temperature distribution between part and cooling channel are determined by the following factors: heat taken when melt contacts with the mold-base wall, heat taken when coolant contacts with the mold-base wall, heat radiated from the mold-base surface to the surroundings, heat taken by air circulation between the mold-base surface and the surroundings, and heat taken by the ejected part after mold opening. The typical temperature distribution is shown in Figure 8.5.

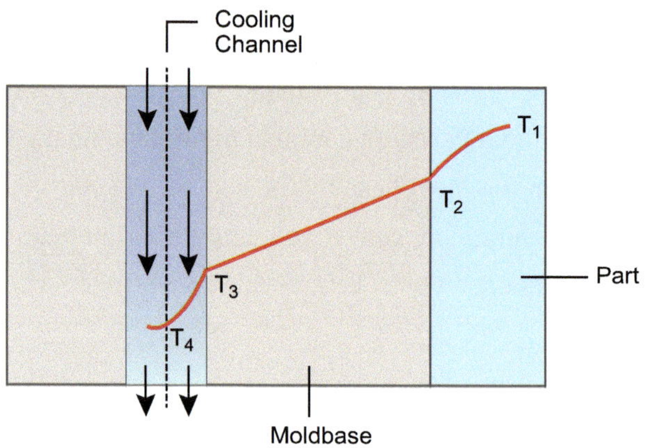

Figure 8.5 Temperature distribution in a standard cooling path

In reality, the mold temperature fluctuates periodically with time during the processing, and it will show as a cycle motion in the steady state. During the melt filling stage, the mold temperature rises because the heat of the melt is transferred to the mold wall via heat conduction. The part temperature is then reduced to the target value. A typical mold temperature variation cycle is illustrated in Figure 8.6. At the start of a molding process, it is unsteady because the mold is starting to warm up. Temperature variation is significant for the first few melt shots, until the molding process reaches the steady state. Then, the mold temperature will not deviate from the average temperature too much; usually, the deviation can be less than 5 °C, sometimes significantly so.

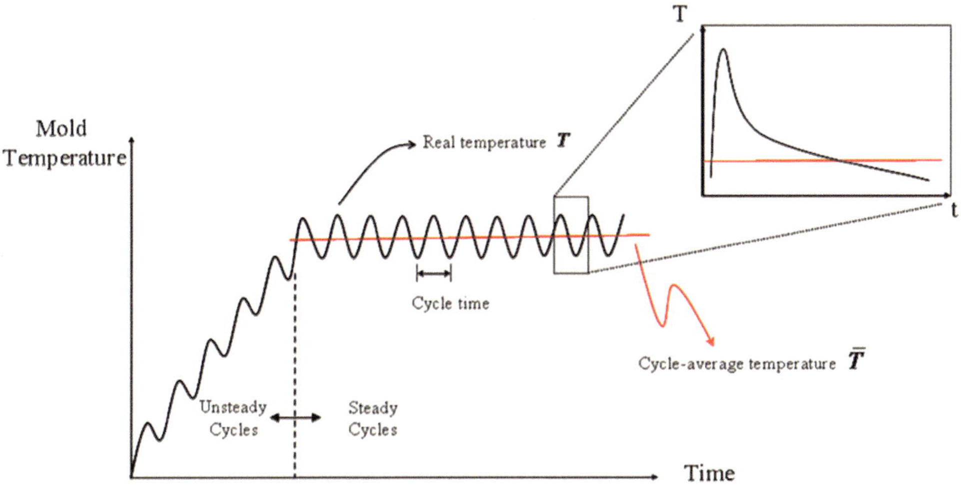

Figure 8.6 Typical mold temperature variation cycle

8.1.2 Design Golden Rule: Uniform Mold Temperature

The different cooling rates of plastic parts due to thickness variation will lead to residual thermal stress. Moreover, residual thermal stress is also generated if the cooling rates on both sides of the cavity are different. From plastic parts with a cross section, the asymmetric tensile stress and asymmetrically distributed compressive stress are caused by the bending moment. The part will bend toward the hotter side so it warps, as shown in Figure 8.7.

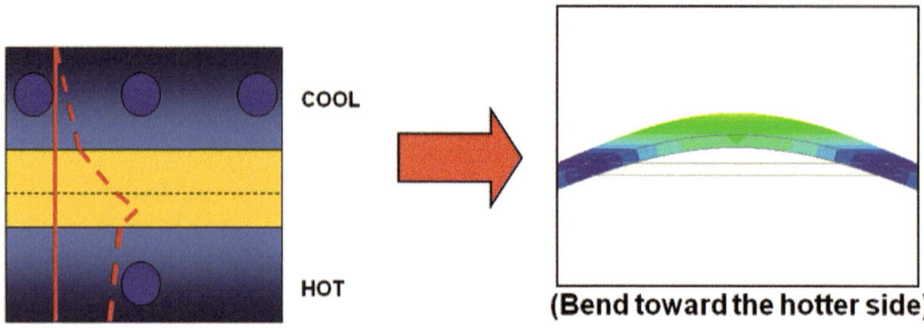

Figure 8.7 Asymmetrical cooling efficiency induces deformation

Key factors governing cooling-induced residual stress include the following:

1. Unbalanced Cooling

Variation in the cooling rate of the part at the mold wall relative to its center can cause thermally induced residual stress. Furthermore, asymmetrical thermally induced residual stress can occur if the cooling rate of the two surfaces (cavity and core sides) is unbalanced. Such an unbalanced cooling will result in an asymmetric tension-compression pattern across the part, causing a bending moment that tends to cause part warpage. This is illustrated in Figure 8.8, which shows that CAE (Moldex3D) cooling analysis can calculate the real temperature distribution and the deformed condition.

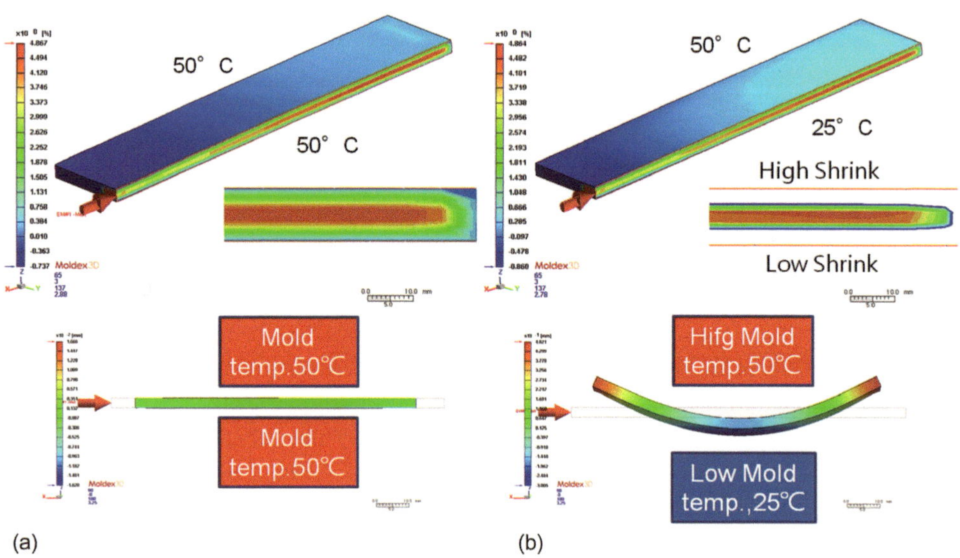

Figure 8.8 Comparison of flat part deformation: (a) symmetrical and (b) asymmetrical mold temperature

2. Cooling Effect Induces Warp

An example of a cooling channel layout design flaw shown in Figure 8.9, where the core and cavity temperatures differ because of an imbalanced cooling channel design. Due to system restrain, four cooling channels are placed on the bottom side of the model, but only three on the top side. The flat plate is of uniform thickness, and therefore it can be predicted that top side is hotter than the bottom side. After mold opening and part ejection, the part is expected to warp into a "smile" shape, similar to Figure 8.8(b), because the mold side with higher temperature tends to have higher shrinkage.

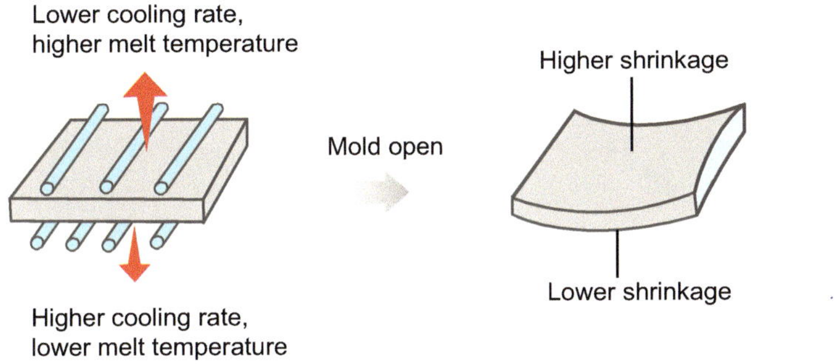

Figure 8.9 Imbalanced cooling channel design

Another imbalanced design of the cooling system is shown in Figure 8.10. Heat accumulation occurs much more easily near the corner of the plastic part due to the difference of heat dissipation rate between core and cavity sides. As the result, the volumetric shrinkage is higher on the core side and lower on the cavity side, which leads to warpage.

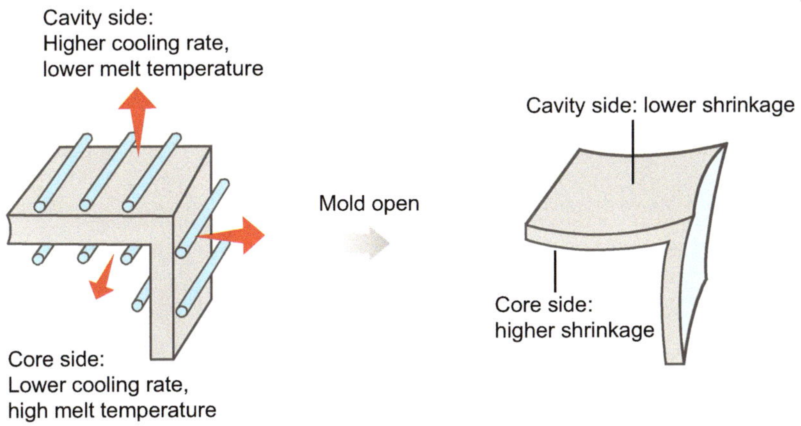

Figure 8.10 Heat accumulation in corner geometry

The sample model in Figure 8.11 demonstrates how different mold temperatures affect deformation and control warpage. The two sides of the mold are adjusted to different coolant temperatures of 70 °C and 80 °C, respectively. The results are shown in the following Figure 8.12 and Figure 8.13.

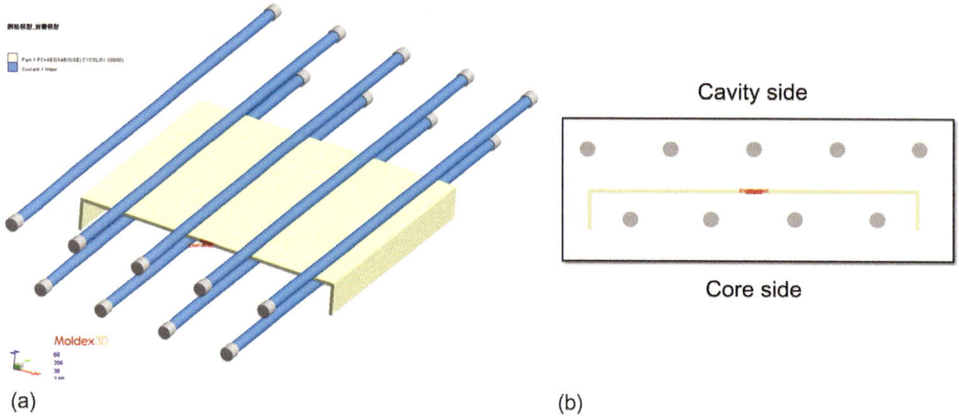

Figure 8.11 Cooling channel layout: (a) trimetric view and (b) front view

First of all, a 10 °C difference in coolant temperature is used to observe whether this is sufficient to maintain part flatness. Molding conditions are set such that the water temperature on the cavity side (80 °C) is higher than that on the core side (70 °C). The mold-base temperature distribution results are shown in Figure 8.12.

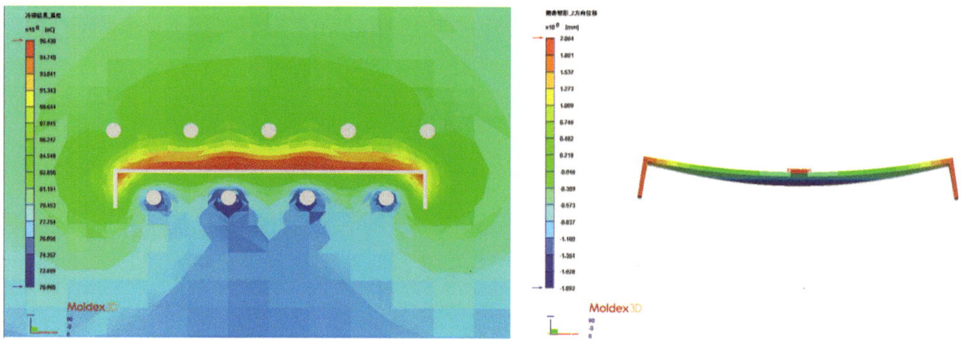

Figure 8.12 Mold temperature distribution and part deformed result (mold temperatures are set as 80 °C and 70 °C for the cavity and core, respectively)

Let's try to reverse the temperature setting: the cavity side coolant temperature (70 °C) is now lower than the core mold coolant temperature (80 °C). With the same test model, the core side temperature is higher than the cavity side, leading to the warpage (bending) toward the core side of the part, where heat accumulation occurs (Figure 8.13).

Figure 8.13 Mold temperature distribution and part deformed result (mold temperatures are set as 70 °C and 80 °C for the cavity and core, respectively)

8.1.3 General Design Guide of Cooling Channel

In principle, the distances between cooling channels should be identical, as shown in Figure 8.14(b), and the spacing between the cooling channels and the mold wall should also be constant, whereas the cooling channel design in Figure 8.14(a) will lead to a non-uniform temperature distribution. However, to maximize cooling efficiency, this should be adjusted considering the uniformity of heat distribution. Designs of straight cooling channels can avoid coolant flow rate slowing resulting in dead zones. It is difficult to achieve heat dissipation around small structures and corners; if necessary, a high conductivity material (such as beryllium copper alloy) can be used, or consider increasing the number of thermal pins or adopting a conformal cooling channel design.

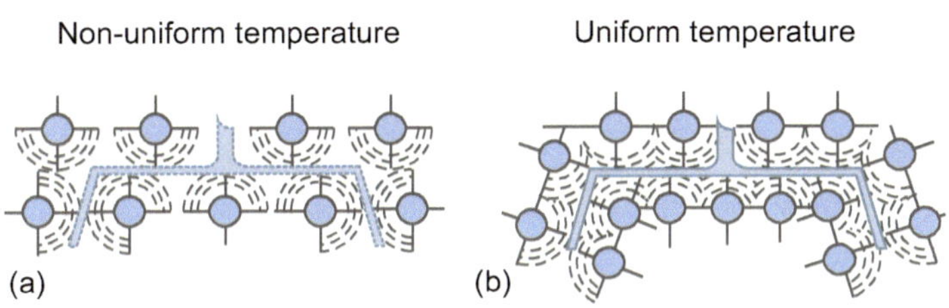

Figure 8.14 (a) Different and (b) same spacing between the cooling channel and the mold wall

The design parameters for the cooling channel are also very important. They include the number of cooling channels, the cooling channel diameter, and the distance between each cooling channel, accounting for mold strength and machining considerations (Figure 8.15). The distance between each cooling channel should be about three times the cooling channel diameter; the distance above and below the part, based on the design guide, is suggested to be two to three times the cooling channel diameter.

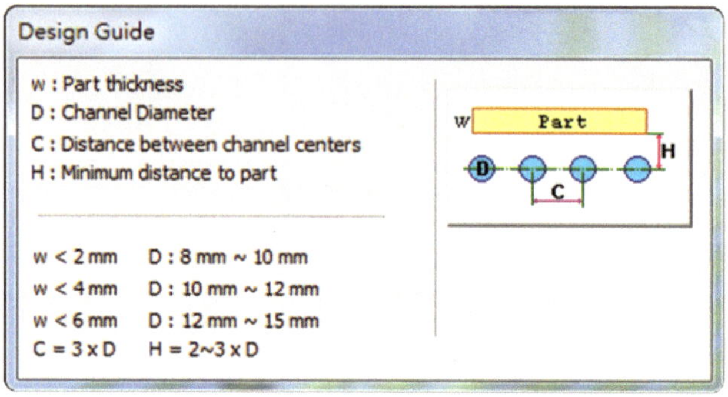

Figure 8.15 Cooling channel design recommendations

The cooling channel layout and connections also have a significant effect on the cooling efficiency. When cooling channels are connected in series, the longer flow distance would easily result in non-uniform cooling. A longer flow distance also tends to cause a higher flow resistance. In addition, a longer channel can build up scale more easily and is harder to clean. When the cooling channels are connected in parallel, the shorter flow distance helps toward uniform cooling and low flow resistance. Also, the connection water scale problem is diminished. Typical cooling channel layouts for connecting in series or parallel are shown in Figure 8.16.

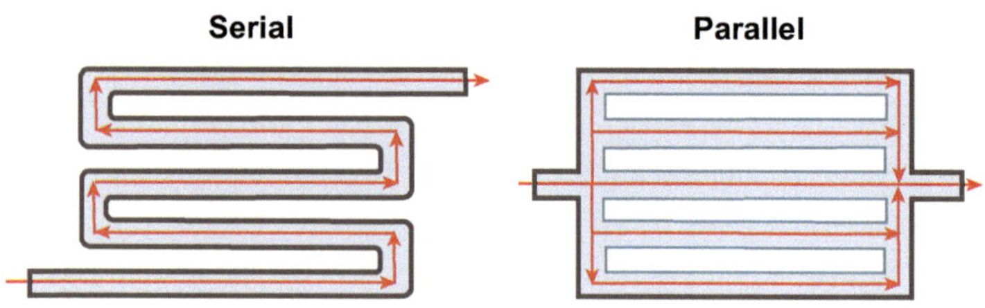

Figure 8.16 Cooling channel layout: serial and parallel types

The cooling system can be constructed in many different ways depending on the molding system and part design. Nevertheless, the main components include the mold temperature controller, regular cooling channel, manifold, hose, and other cooling components such as baffles or bubblers. Specific cooling channel types are described below.

1. Baffle Cooling Channel

The additional components for the cooling system are baffles and bubblers. These are sections of cooling lines that divert the coolant flow into areas that would normally lack cooling, such as cores. A baffle is actually a cooling channel drilled perpendicular to a main cooling line, with a blade that separates one cooling passage into two semi-circular channels, as shown in Figure 8.17. This method provides maximum cross sectional area for the coolant, but it is difficult to mount the divider exactly in the center. The cooling effect and with it the temperature distribution on one side of the core may differ from that on the other side.

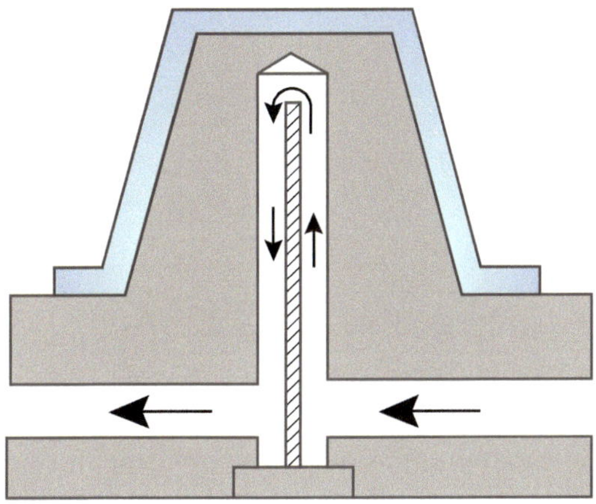

Figure 8.17 Baffle cooling channel

2. Bubbler Cooling Channel

A bubbler is similar to a baffle except that the blade is replaced with a small tube (Figure 8.18). The coolant flows into the bottom of the tube and "bubbles" out of the top, like a fountain. The coolant then flows down around the outside of the tube to continue its flow through the cooling channels. The most effective cooling of slender cores is achieved with bubblers. The diameter of both must be adjusted in such a way that the flow resistance in both cross sections is equal. The condition for this is:

$$\frac{\text{Inner diameter}}{\text{Outer diameter}} = 0.707$$

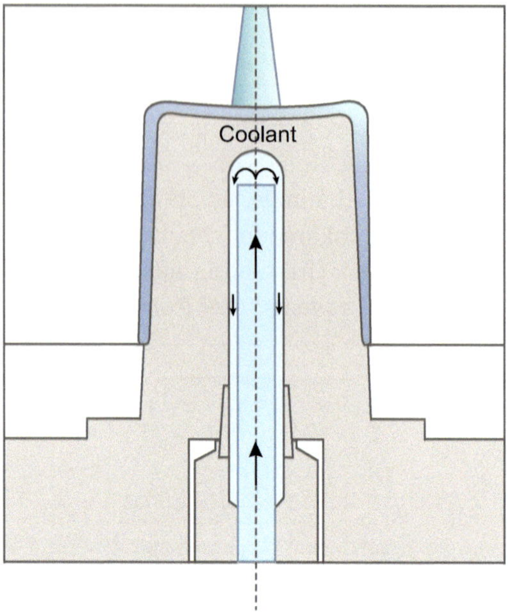

Figure 8.18 Bubbler type cooling channel

3. Indirect Thermal Pin

The use of high heat transfer materials is advantageous for heat transfer to water, which then carries away the heat, as shown in Figure 8.19. The heat is first transferred from the plastic to the upper half of thermal pin, and is dissipated in the lower half of the thermal pin via direct contact with the coolant by convection.

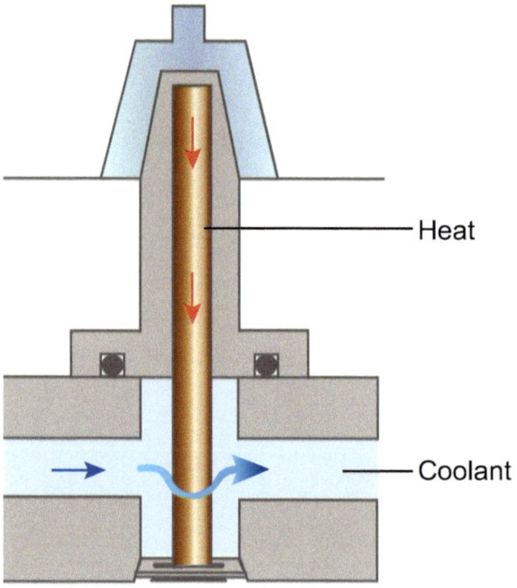

Figure 8.19 Indirect thermal pin

Although a design guide for cooling channels is available, it cannot always be followed, as a result of space limitation of space of the mold base or the geometric design of the product.

8.1.4 Cooling Efficiency: Coolant Flow Consideration

Cooling efficiency is related to the coolant material thermal conductivity, heat capacity, density, viscosity, and coolant flow behavior. To maintain sufficient cooling, the cooling circuit is arranged as far as possible along the contour of the part. The distance from the mold wall should be as short as possible (considering mold material strength).

The Reynolds number (Re) provides an index of flow behavior, defined for a round tube as:

$$\mathrm{Re} \equiv \frac{4Q\rho}{\pi D \eta} \tag{8.1}$$

where ρ is coolant density, η is coolant viscosity, Q is volumetric flow rate, D is diameter, as shown in Figure 8.20.

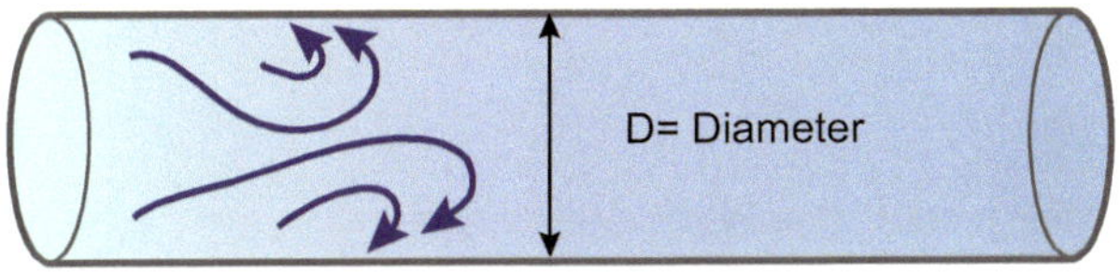

Figure 8.20 Cooling channel characteristic length dimension

For a round tube with laminar flow, Re < 3000; with turbulent flow, Re > 6000, experimentally. The thermal conductivity of water is about five times that of oil. Turbulent flow can create local eddies and provides better heat transfer. It also achieves three to five times as much heat transfer as laminar flow but results in greater pressure drop. Normally, the coolant for the mold is pumped to achieve turbulent flow with Re > 6000. The relationship between pressure drop and flow rate is shown in Figure 8.21.

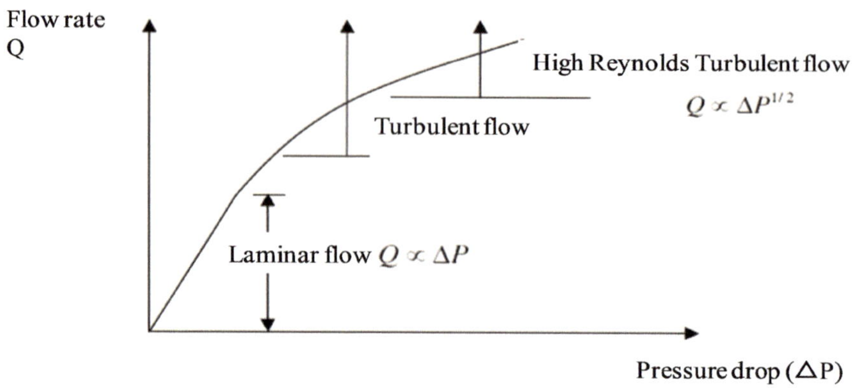

Figure 8.21 Coolant flow rate and pressure drop correlation

Effect of Cooling Channel Design

In this case study, the thickness of a flat plate is 2 mm. Three cooling channel designs were tested with cooling channel diameter d = 3 mm, pitch of each channel P = 10 mm, and fixed distance between cooling channel and part surface D = 6 mm. With these parameters, three cooling channel layouts, serial, parallel, and conformal, are designed, as shown in Figure 8.22.

Figure 8.22 Three common cooling layout design types: (a) serial, (b) parallel, and (c) conformal

The following results are based on a cooling simulation that only simulates cooling. The temperature setting range is 50 to 52 °C (Figure 8.23). It can be seen that the temperature is lowest in the serial cooling channel design.

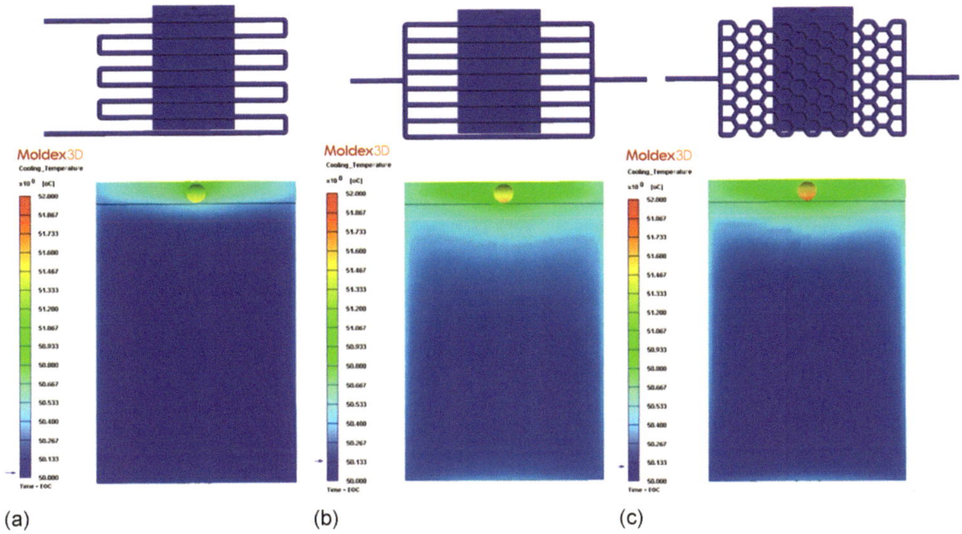

Figure 8.23 Comparison of mold temperature distribution in three common cooling layout types: (a) serial, (b) parallel, and (c) conformal

A comparison of the flow rate for three cooling channel designs is shown in Figure 8.24. It can be seen that in the parallel and conformal cooling designs the coolant flow rate in the region of the projected product is lower.

Based on the energy balance equation, we can obtain the following equation:

$$Q'' = \text{mass flow rate} \times C_p \times \left(T_0 - T_i\right)$$

where T_0 and T_i are temperatures at the outlet and inlet, respectively.

As shown above, the cooling rate (Q'') is proportional to the coolant mass flow rate, the coolant heat capacity (C_p), and the temperature difference between cooling outlet and inlet (ΔT). The serial cooling channel has a higher coolant velocity, and this is consistent with the lower part temperature, as shown in Figure 8.23. The part temperatures of the conformal and parallel cooling channel designs are higher, as the coolant velocities around the part are lower.

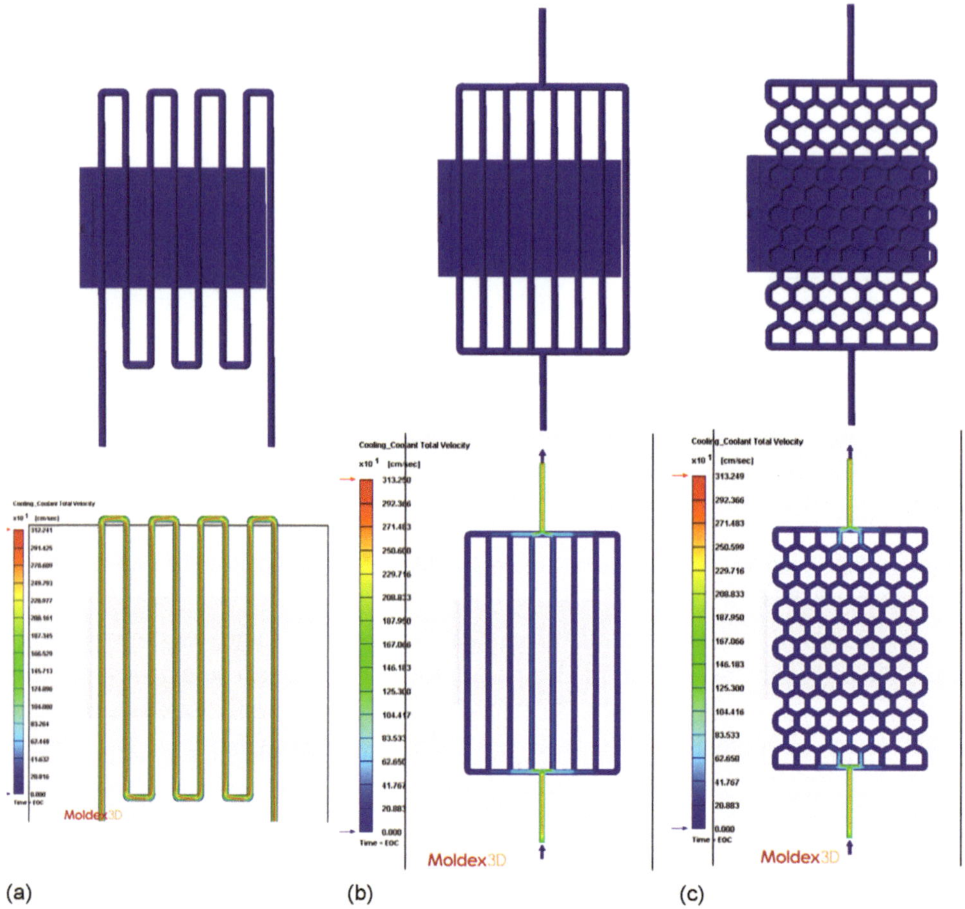

Figure 8.24 Comparison of the coolant flow rate in three common cooling layout types: (a) serial, (b) parallel, and (c) conformal

8.1.5 Cooling Time Estimate

The injection molding cooling time is mainly determined by the thickness of the injection molded parts, but is also associated with the plastic melt temperature, mold temperature, and heat transfer properties of the plastic and mold steel (thermal conductivity, etc.).

Figure 8.25 shows cooling time estimations for several polymers. The main difference is thermal conductivity, heat capacity, and filler type.

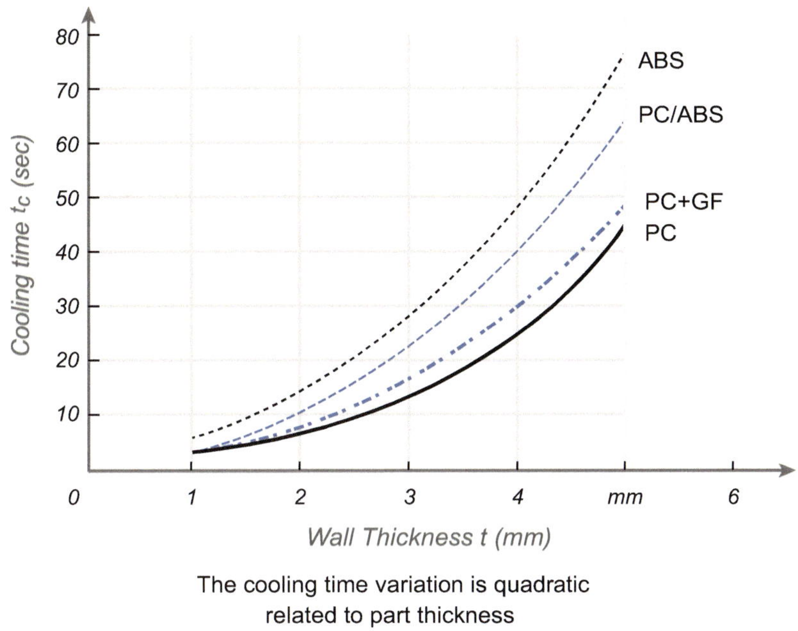

Figure 8.25 Correlation of part thickness and cooling time

There are many ways to estimate the cooling time in practice such as theoretical derivation and several semi-empirical equations and reference formulas which consider the thermal properties of the material:

$$t_C = \frac{H^2}{\pi^2 \alpha} \ln\left(\frac{8}{\pi^2}\frac{T_M - T_C}{T_E - T_C}\right) \tag{8.2}$$

where t_c, the cooling time (s), is proportional to the average cross-sectional thickness of the product that needs to be cooled to the required temperature (T_E); T_M is the injection melt processing temperature; T_c is the mold cooling fluid temperature; H is the product thickness (unit of mm); and α is the plastic thermal diffusivity.

The relationship between the definitions of thermal diffusivity (α) and cooling time is shown in Equation 8.3.

$$\alpha = \frac{k}{\rho C_P} \tag{8.3}$$

$$t_C \propto \frac{h^2}{\alpha} = \frac{h^2 \rho C_P}{k}$$

where k is the thermal conductivity, ρ is the density of the melt polymer, C_p is the heat capacity, h is the maximum part thickness, and t_c is the cooling time.

The cooling time is proportional to the square of the wall thickness. If the thickness is doubled, the cooling time will quadruple.

The prediction of ejection temperature is based on the key parameters stated above. However, using the reference theoretical value, the time will be longer; it is based on the ejecting area in the thickness direction to 70% frozen as cooling time. Figure 8.26 shows the recommended ejection time; note that the cooling time can be reduced significantly with a small increase in de-molding temperature.

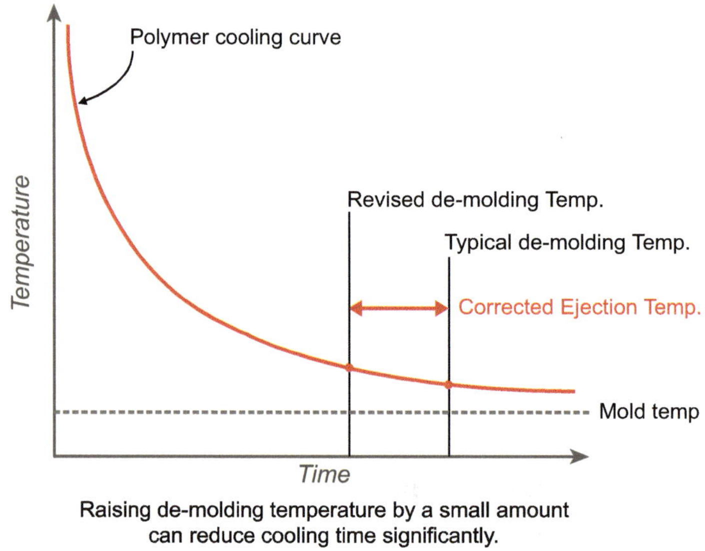

Figure 8.26 Ejection temperature recommendation

8.1.6 Use CAE Cooling Analysis

In this section, some case studies concerning (1) cooling time, (2) hot spots, (3) mold temperature difference, (4) temperature at coolant inlet and outlet, and (5) cooling efficiency are presented to show how CAE can determine proper cooling design and process parameters in a molding process. CAE results listed here ensure the rational design of cooling systems.

1. Cooling Time

Is the cycle time too long? One of the most important tasks in a cooling analysis is to check the cooling time. A shorter cooling time can lead to a shorter molding cycle and hence enhance productivity. Figure 8.27 shows the use of slicing or iso-surface functions to observe the distribution of cooling time within a part. In this example, most of the part is cooled down within 2 s. However, it takes 22 s to cool down to the ejection temperature for the area near a gate. In this way, the impact of the design and process setup on the cooling time is clarified.

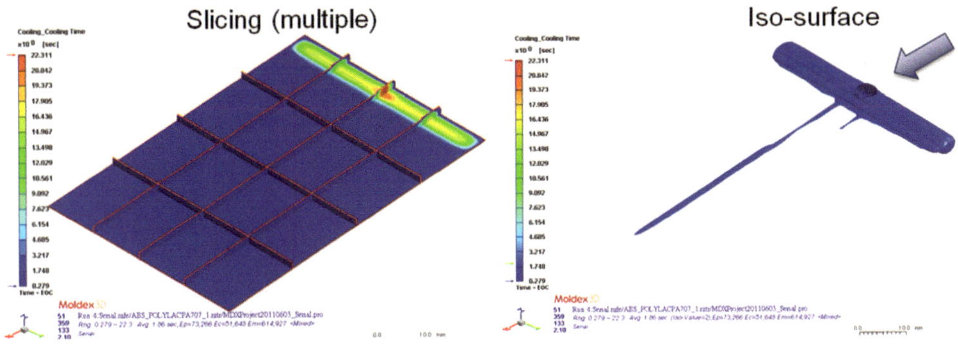

Figure 8.27 CAE helps estimate ejection time

2. Hot Spots

Where is the hot spot? Hot spots occur at places where a local high temperature indicates heat has accumulated there. This is usually related to the part design and cooling system design. A uniform part temperature distribution is always desired. To query the location of hot spots, users can check "temperature" and "cooling time" as shown in Figure 8.28. The green areas are the places where the temperature is still above ejection temperature at the cooling time of 20 s. These are the hot spots.

Figure 8.28 CAE assists in detection of hot spots

3. Mold Temperature Difference

Comparison of the two analyses shown in Figure 8.29 shows that the model with a cooling channel has a narrow mold temperature difference across the part, while the model without a cooling channel has a relatively broad mold temperature difference, especially in the top region. A good temperature distribution is indicated by a narrow temperature difference between the cavity wall and core wall across the part.

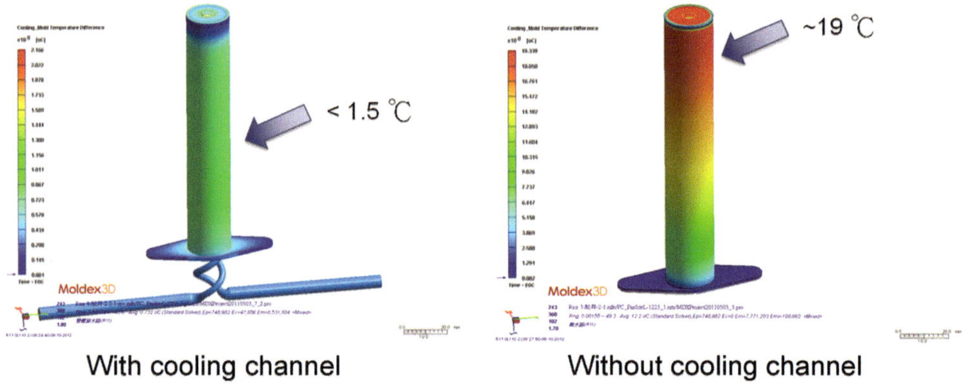

Figure 8.29 CAE results of mold temperature difference

4. Temperature at Coolant Inlet and Outlet

The coolant temperature increases from inlet to outlet due to heat removal from the cavity. If the flow rate is not sufficient, the coolant temperature will rise to such a degree that the cooling effect will decrease. Reduction of the flow rate will increase the difference of coolant temperature between inlet and outlet significantly. In general, the temperature difference should be controlled to be less than 3 °C (Figure 8.30).

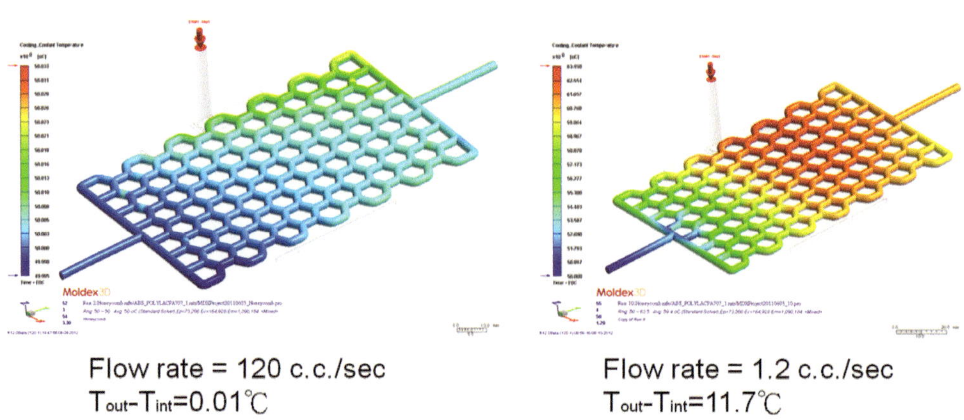

Figure 8.30 CAE results of temperature difference of coolant between inlet and outlet

5. Cooling Efficiency

The cooling efficiency E is defined by the following equation:

$$E = \frac{Q_c}{\Sigma Q_c + Q_m} \times 100\% \tag{8.4}$$

where Q_C is the heat absorbed by an individual cooling channel and Q_m is the heat absorbed by the mold base.

The user can improve a cooling system by checking the results of "channel efficiency". In Figure 8.31, there are three cooling channels: one baffle and two straight channels. Obviously, the baffle absorbs more heat than the other two straight channels. This is an indicator for the cooling channel design for heat removal.

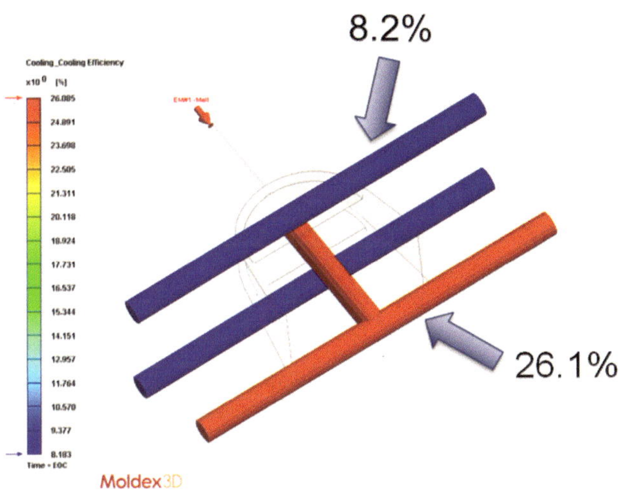

Figure 8.31 CAE results of cooling efficiency of various cooling channels

8.1.7 Conformal Cooling Application

By changing the cooling design to make it closely follow the contours of a part, the temperature distribution will become more uniform and heat/cold spots can be eliminated. This can minimize heat-induced defects such as residual stress, warpage, voids, and sink marks. Product defects are often highly related to its design. For example, if a product has ribs, it is highly possible for sink to occur on the opposite side to the ribs. If the cooling rate is too fast, voids can occur instead of sink marks. Despite the complexity of product defects, which occur for various reasons, conformal cooling can help to minimize the overall warpage - if properly designed. Here we will examine two common defects - sink marks and warpage - and how conformal cooling can solve them. It is recommended to adopt the conformal cooling technique for some products, and Figure 8.32 shows models of several of these.

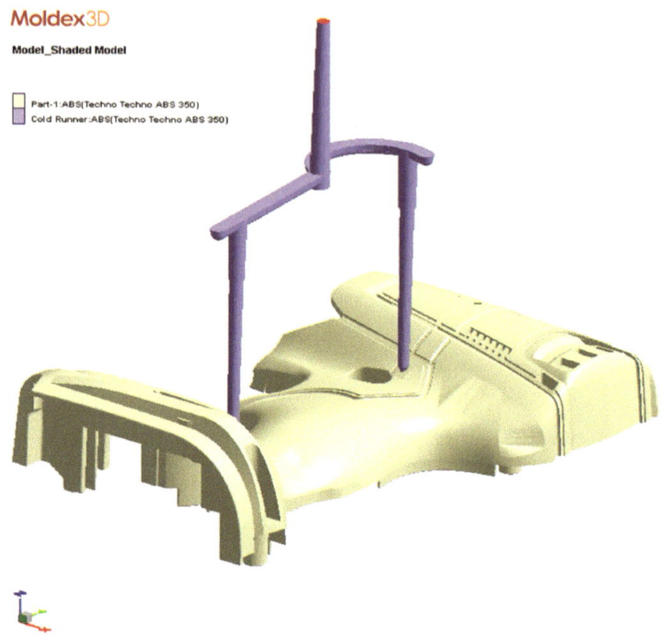

Figure 8.32 The specific product for which a conformal cooling design is recommended

Because the design of cooling channels adopted in conformal cooling applications is usually complicated, CFD simulation can be used to check the detail of coolant flow inside the cooling channel.

3D cooling channel network analysis is applicable for both fast and standard cooling. Additional results are provided by 3D cooling channel network analysis, such as (1) cooling channel Reynolds number, (2) coolant temperature, (3) coolant pressure, (4) coolant velocity, (5) coolant velocity vector, and (6) stream line, as discussed below.

1. Cooling Channel Reynolds Number

Cooling channel analysis shows the Reynolds number distribution inside the cooling channels. For a round cooling line, the value should be greater than 4000 to achieve turbulent flow. Turbulent flow brings better heat transfer and cooling efficiency.

2. Coolant Temperature

Cooling channel network analysis shows the coolant temperature distribution inside the cooling channels, as illustrated in Figure 8.33. A lower temperature can result in higher cooling efficiency with the same flow strength.

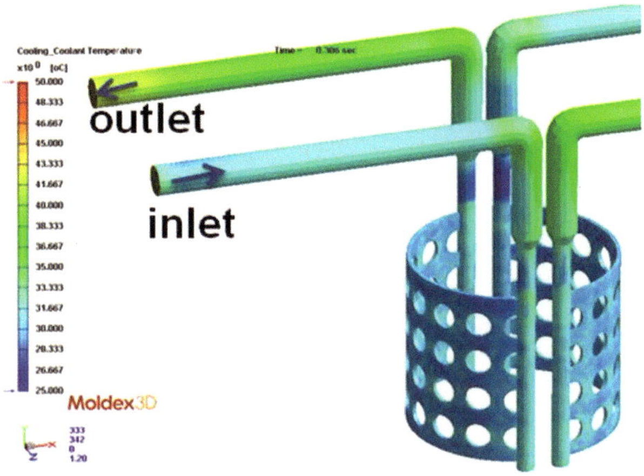

Figure 8.33 Coolant temperature distribution in a conformal cooling channel

3. Coolant Pressure

Cooling channel analysis also shows the pressure distribution inside the cooling channels, as illustrated in Figure 8.34. Higher pressure in the coolant indicates higher energy in pumping to maintain the flow rate and cooling efficiency.

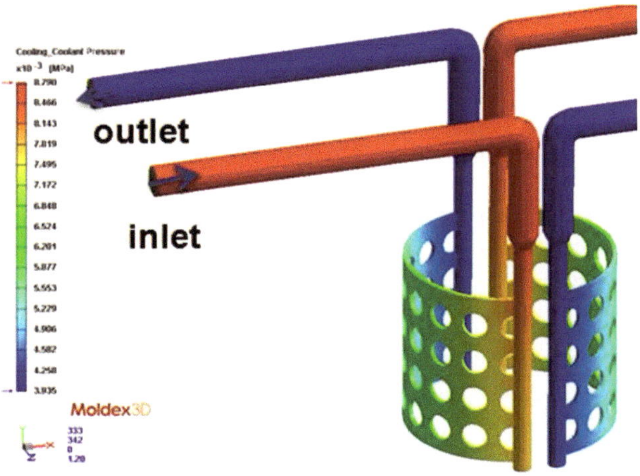

Figure 8.34 Coolant channel pressure distribution in a conformal cooling channel

4. Coolant Velocity

The simulation results show the distribution of different coolant velocity components (X, Y, or Z) inside the cooling channel when running 3D solid cooling channel analysis (solid mesh within cooling channel required). Higher velocity in a cooling channel will lead to greater pressure drop but better heat transfer and cooling efficiency.

5. Coolant Velocity Vector

The simulation results show the speed and direction of the coolant velocity vector inside a cooling channel, as shown in Figure 8.35. Velocity vectors help to understand not only the flow strength but also the flow direction. A higher velocity in a cooling channel will lead to greater pressure drop but better heat transfer and cooling efficiency. Stagnation zones can be identified.

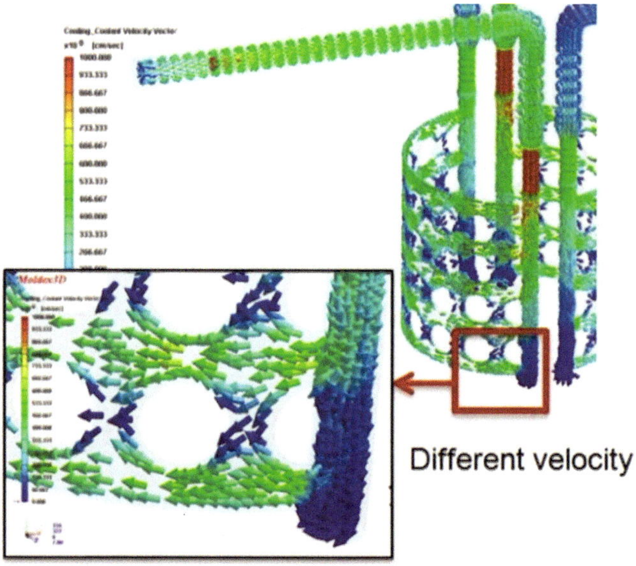

Figure 8.35 Coolant velocity vector in a conformal cooling channel

6. Stream Line

Stream line analysis plots the results along the path of the coolant in the cooling channels from an entrance to end, as shown in Figure 8.36. It can be used to evaluate the cooling efficiency and detect an imbalanced flow of coolant inside a cooling channel.

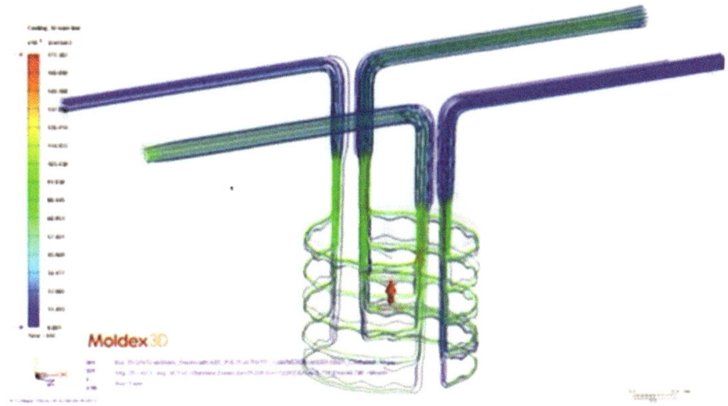

Figure 8.36 Stream line includes flow length and total velocity in a conformal cooling channel

To summarize the benefits of 3D coolant CFD, it can be used to:

1. Estimate properties (temperature, pressure velocity, streamline, etc.) inside the cooling channels and add-ons (e.g., bubbler, baffle, and so on).
2. Support complex cooling design, including multi-inlet and multi-outlet designs.
3. Investigate the influences of various coolant designs on the molding process and product qualities, including shrinkage, warpage, Reynolds number, etc.
4. Revise the existing mold/product design for optimization.
5. Realize the roots of problems and troubleshoot easily.
6. Accumulate knowledge to increase understanding and improve design criteria.

8.2 Practical Applications

8.2.1 Digital Camera Cover

The purpose of this case study is to construct two designs of cooling systems (conventional and conformal) in order to evaluate the cooling efficiency. Different solutions for the digital camera cover product shown in Figure 8.37 were investigated to improve the quality of the molded product through computational analysis. The cycle time and temperature distribution, relating to cooling ability, were compared for two different layouts.

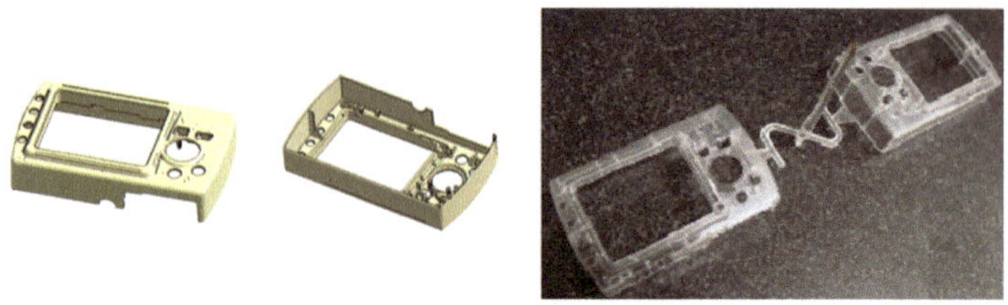

Figure 8.37 CAD and product photo for the case study camera cover

The geometrical model is an electronic device cover which includes a runner system. The dimensions of the model are 270 mm (L) × 60 mm (W) × 21.5 mm (H), and the major thickness is 1.2 mm.

Figure 8.38 shows the conventional and conformal cooling channel designs. There are eight and fifteen cooling channels in the conventional and conformal cooling systems, respectively. For each cavity, the conformal system has two more channels at the core side and one more channel at the cavity side than the conventional system. Also, the conformal system has one additional channel at the sprue. The concept of the conformal cooling design for this model is to increase the cooling channel number, to reduce the cooling channel distances from the part surfaces at the core and cavity sides, and to enhance cooling for the sprue bushing, which is the thickest region in the model.

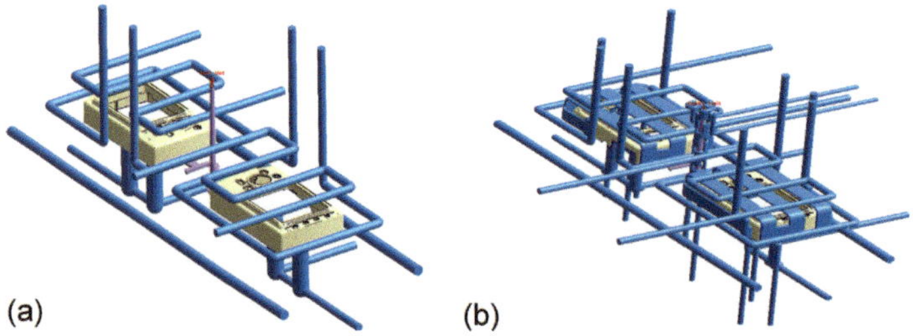

Figure 8.38 The cooling channel designs: (a) conventional and (b) conformal

The material for injection molding was polycarbonate (Mitsubishi NOVAREX 7020), with a working temperature range of 280 to 310 °C and a glass transition temperature of 155 °C. The ejection temperature was 144 °C, the coolant temperature was set at 90 °C, and the flow rate was 120 cm^3/s.

A summary of the set process conditions is given in Table 8.1.

Table 8.1 Molding Conditions

Filling time	0.427 s
Filling pressure	160 MPa (max.)
Packing time	3.15 s
Packing pressure	130 MPa
Cooling time	10.3 s
Mold temperature	90 °C
Material	PC (Mitsubishi NOVAREX 7020)
Melt temperature	280 °C
Ejection temperature	144 °C
Coolant	water
Flow rate	120 cm^3/s
Coolant temperature	90 °C

Figure 8.39 shows the part surface temperature distribution at the end of cooling (EOC) for both designs. The temperature range is set from 90 to 97.3 °C. The two figures reveal that more heat can be removed on the core side cooling channel than the one without it.

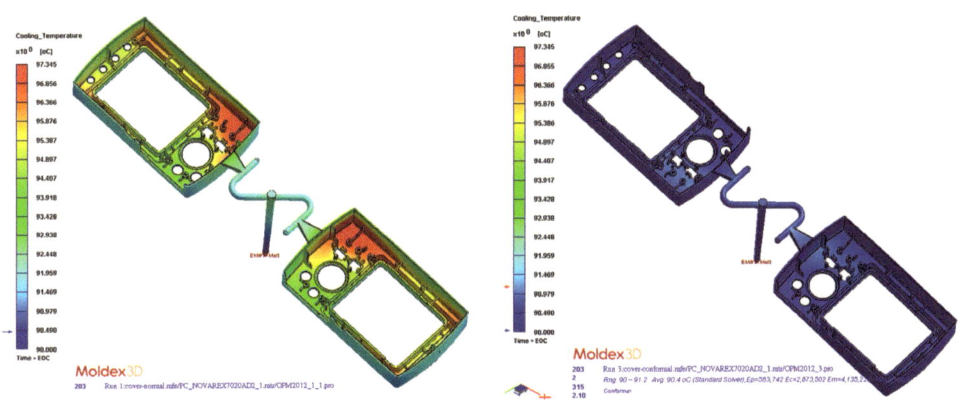

(a) (b)

Figure 8.39 Part surface temperature at end of cooling: (a) conventional and (b) conformal design

Figure 8.40 shows the efficiency of a cooling channel at the end of cooling. In the conformal system, the two channels at the core and the one at the cavity absorb about 45% of the total heat.

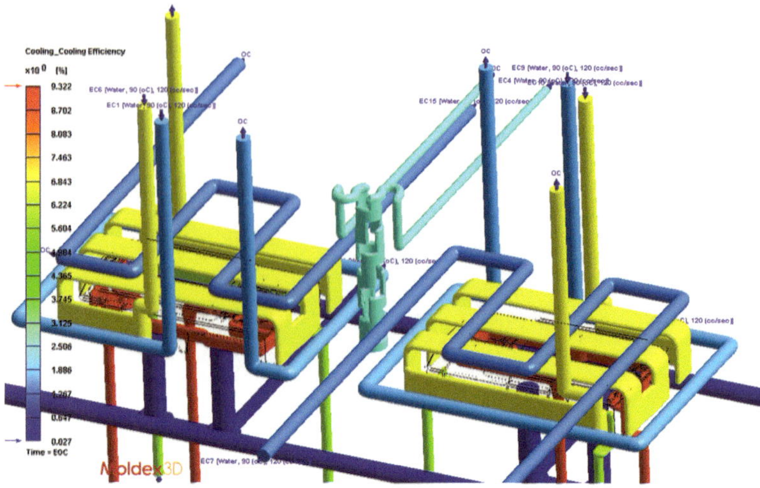

Figure 8.40 Cooling efficiency at end of cooling of the conformal cooling design

Figure 8.41 shows the coolant velocity in the channel. We can see that there are stagnant regions at the corners shown in blue. These regions may induce a relatively low heat transfer rate and easily build up water scale. If the cooling channel were designed with a spiral shape, such regions can be avoided.

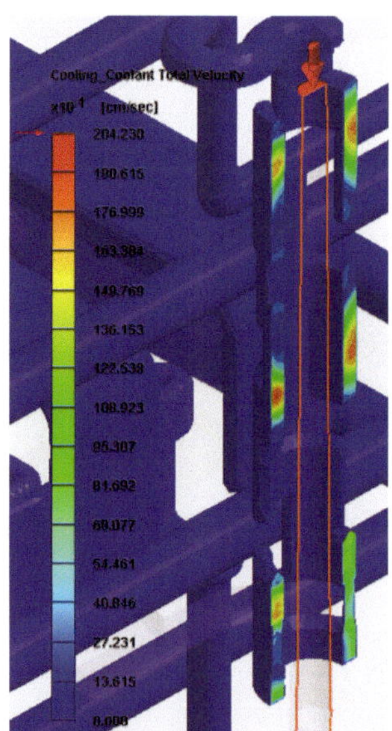

Figure 8.41 Coolant total velocity of sprue cooling channel at end of cooling of conformal cooling design

For the two cooling designs, Figure 8.42 and Figure 8.43 show the temperature distribution of sprue and mold base. The maximum sprue temperature difference between the two cooling designs is less than 1 °C. This result is correspondent with the cooling time result. The mold base temperature difference between the conventional and the conformal cooling systems at the end of cooling is larger than that at the sprue. This is because of the thermal conductivity difference of metal and plastic melt. To sum up, conformal cooling design is much more efficient in removing heat from the system than conventional cooling design.

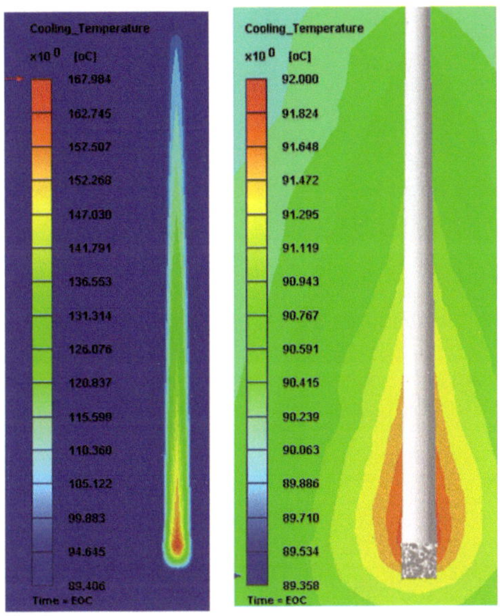

(a) (b)

Figure 8.42 Temperature distribution at end of cooling of the conventional cooling design: (a) sprue and (b) mold

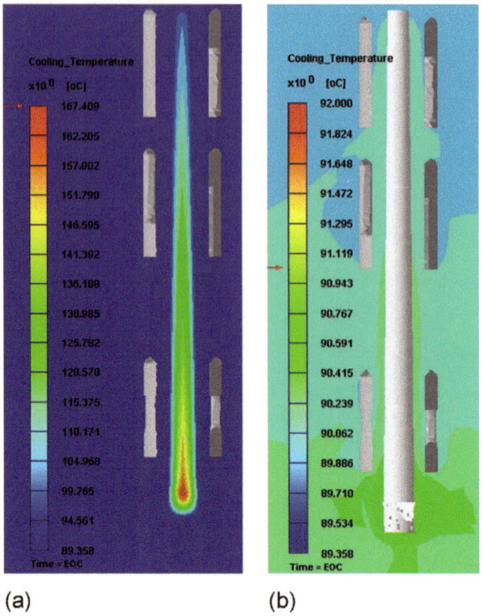

Figure 8.43 Temperature distribution at end of cooling of the conformal cooling design: (a) sprue and (b) mold

8.2.2 Cartridge

In this case, a cartridge model, like the one shown in Figure 8.44, is used for evaluating two different cooling design systems. The hollow interior of this part has a heat accumulation problem if there is no cooling channel inside. This would cause inward warpage after the injection molding process. In this case, we will see the effects of additional cooling channels in the interior area.

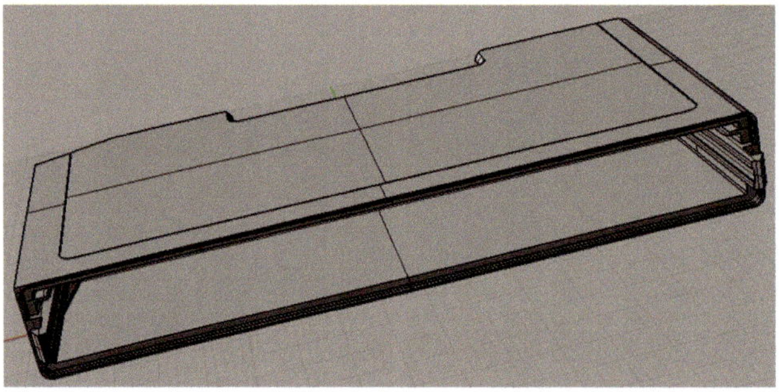

Figure 8.44 Cartridge model

Figure 8.45 shows the conventional and conformal cooling channel designs. The difference is in the interior area.

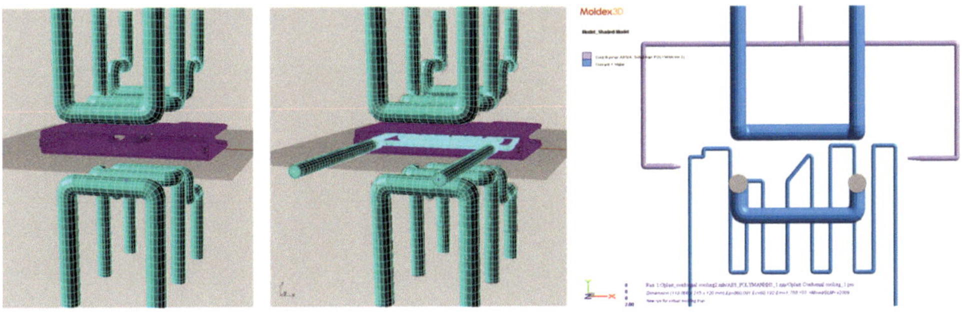

(a) (b) (c)

Figure 8.45 Cooling channel designs: (a) side view of conventional, (b) side view of conformal, and (c) top view of conformal

The dimensions of the part are 90.67 mm (L) × 57.15 mm (W) × 12.85 mm (H), and the dimensions of the mold are 124 mm (L) × 215 mm (W) × 120 mm (H). The plastic material was acrylonitrile butadiene styrene (ABS; A. Schulman, Polyman HH3). The ejection temperature was 106 °C and the freeze temperature was 126 °C.

Table 8.2 shows the detailed molding conditions in this case.

Table 8.2 Molding Conditions

Filling time	0.421 s
Maximum injection pressure	240 MPa
Packing pressure	240 MPa
Packing time	5 s
VP switch-over	98%
Melt temperature	240 °C
Mold temperature	60 °C
Cooling time	20 s

Figure 8.46 shows the cooling temperature from CAE simulation. It seems that the temperature distribution inside the product is primarily shaped by the removal of the accumulated heat by the water coolant. In the conformal cooling channel design, the maximum temperature has dropped significantly from the 115 °C of the conventional cooling design to 69 °C.

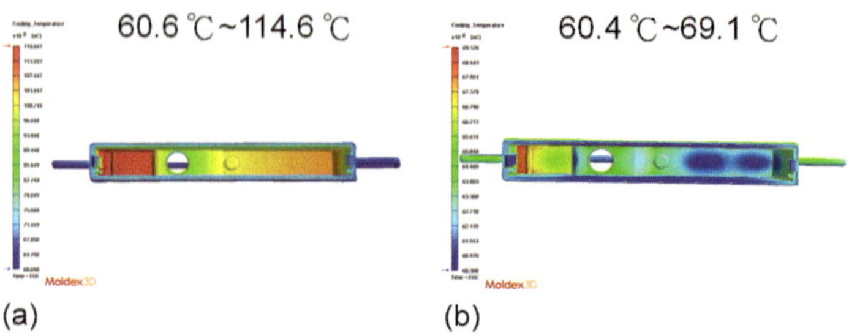

Figure 8.46 Cooling temperature: (a) conventional and (b) conformal cooling channel designs

It is also found that cooling time is decreased from 46.5 to 18 seconds going from the conventional to conformal layouts, as shown in Figure 8.47.

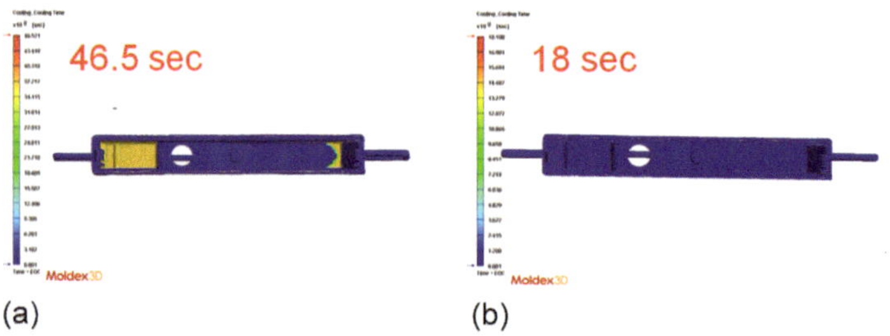

Figure 8.47 Cooling time: (a) conventional and (b) conformal cooling channel designs

The mold temperature difference is an important indicator of thermal displacement magnitude. A conformal cooling channel can reduce the maximum temperature difference from 49.1 °C in the conventional cooling design to 6.5 °C, as shown in Figure 8.48. The temperature difference across the wall thickness is reduced by 85% from 40 to 6 °C.

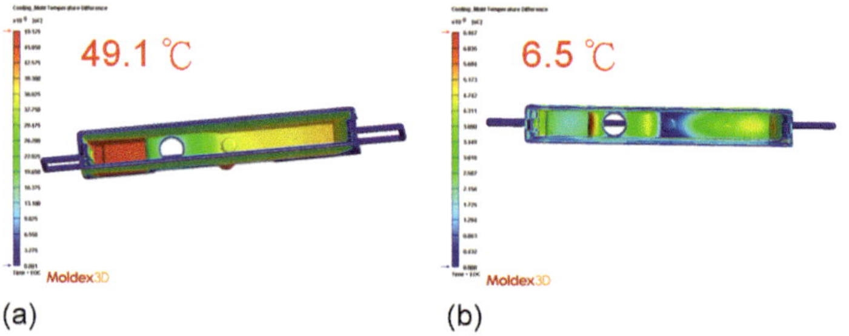

Figure 8.48 Mold temperature difference: (a) conventional and (b) conformal cooling channel designs

In the mold temperature distribution, the maximum temperature drops from 114.7 °C of conventional design to 69.2 °C of conformal cooling channel. The highest temperatures in both designs are located at the lower left corner, indicated by the red circle in Figure 8.49.

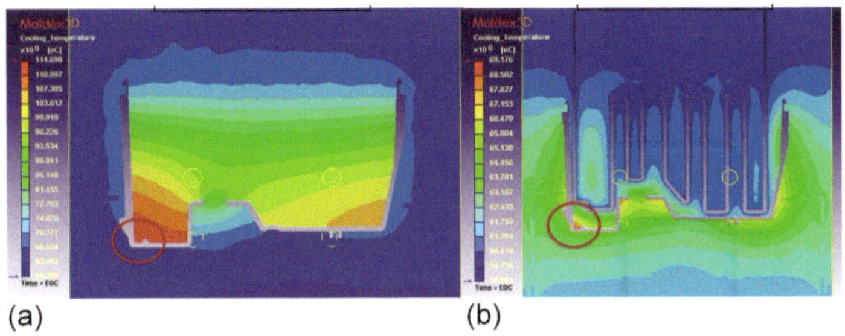

Figure 8.49 Mold temperature: (a) conventional and (b) conformal cooling channel designs

As shown by the total displacement results illustrated in Figure 8.50, the conformal cooling channel design has a significant effect on reducing warpage magnitude.

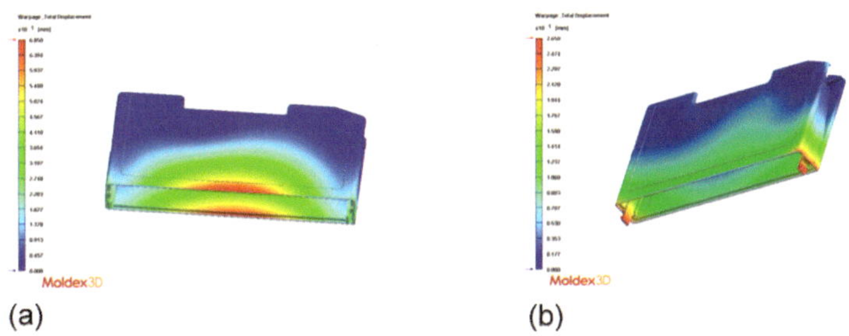

Figure 8.50 Total displacement: (a) 0.001-0.685 mm for conventional and (b) 0.001-0.265 mm for conformal cooling channel designs

Figure 8.51 shows the actual warpage of the two cooling channel designs. We can see that warpage improves significantly in the z-direction and is in good agreement with the simulation result.

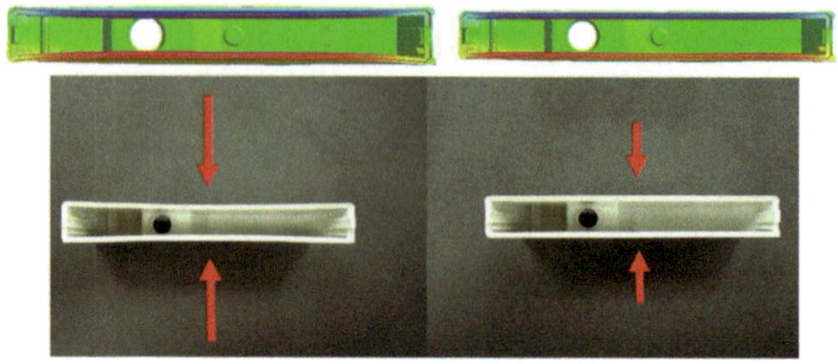

(a) (b)

Figure 8.51 Warpage result comparison: (a) conventional and (b) conformal cooling channel designs

■ 8.3 CAE Case Study

Following the discussions of this chapter and the corresponding CAE verification case, an exercise model, as illustrated in Figure 8.52, is provided in this section for readers to practice CAE application in cooling optimization.

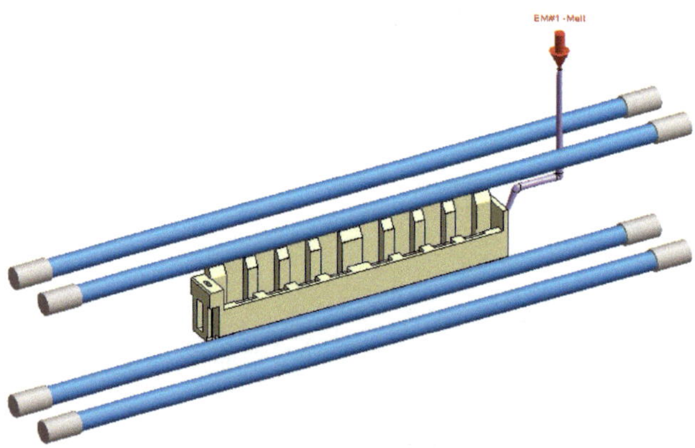

Figure 8.52 Case study sample

Please download the analysis files for CAE Case Study 8.3 from the following website:

https://moldex3d.box.com/s/zr6fvc1vlhbi4ocx111jwd3wmxt4ooif

Then open the .m3j or .mvj file in Moldex3D and answer the questions below according to the analysis results.

Questions:

1. Pre-analysis of the model
   a) Where are the hot spots likely to appear? Possible warping behavior? And explain what factors will cause warpage.
2. Mold flow analysis interpretation
   a) Try to answer the last question from the result of mold flow analysis.
   b) Is the cooling effect of the cooling channel uniform?
3. Effects of molding conditions
   a) Change the coolant of volumetric flow to 1/20 of the default setting (Re = 2946). What is the result?
   b) Change the cooling time to 12 s; what is the result?
   c) Change the coolant temperature to 80 °C; what is the result?
   d) Change the coolant temperature to 100 °C; what is the result?
   e) In this case, which molding parameter has the greatest effect on warpage? Why?
4. Effects of cooling channel design
   a) Change the diameter of cooling channel to 6 mm; what is the result?
   b) Change the number of cooling channels to 6; what is the result?
   c) Change the cooling channel distance to four times the cooling channel diameter; what is the result?
   d) In addition to the original cooling channels, add an additional channel close to the part; what is the result?
   e) As shown above, the cooling channel designs from 4. a)–c) give warpage results similar to the original design, indicating the cooling effects of these

design changes on the product are insignificant. However, comparing design 4. d) with the original one, it is found the effects in the X-direction and Z-direction are contrary. The deformation magnitude in the X-direction is reduced, whereas is the Z-direction it is increased. Explain the cause of this difference.

9 Warpage Control

Shih-Po (Tober) Sun, Wen-Hsin (Debbie) Weng

Warpage is a critical quality issue in injection molded parts. It occurs when there is an uneven distribution of shrinkage throughout the product. It is the most common defect during production, and quite often this defect has to be resolved in haste, relying mostly on past experience of the skilled technician.

In this chapter, causes for warpage (material, part design, and molding conditions) are introduced first, followed by methods to resolve the issue, and lastly a practical example is introduced for better understanding on how to use different molding conditions or utilize molding analysis software to predict and prevent future possible warpage occurrence in the initial stage of part design. It is hoped that, through the prior prediction of warpage occurrence using simulation, manufacturers can improve their product yield and eliminate unnecessary molding trials.

9.1 Basics

While the designer is responsible for the functional and mechanical performance of a part, a molder is required to deliver the part with a good aesthetic quality and meeting the dimensional specifications. When a part fails in these requirements, it is necessary to find the causes and conduct molding process modifications. If simple process modifications cannot resolve the warpage problem, we have to make a mold design change and possibly even extend the modifications to part design. In the following sections, we will discuss types of warpage, possible causes, and solutions.

The most common dimensional defect is in flatness, as it always results in assembly issues. Figure 9.1 shows both a warped product and the product after optimization of flow balance and packing pressure. If edges are warped away from the assembly surface, it is still acceptable because the fixtures on the corner will generate a tight fit between this part and its mating part. However, a warpage in the other direction is considered unwelcome.

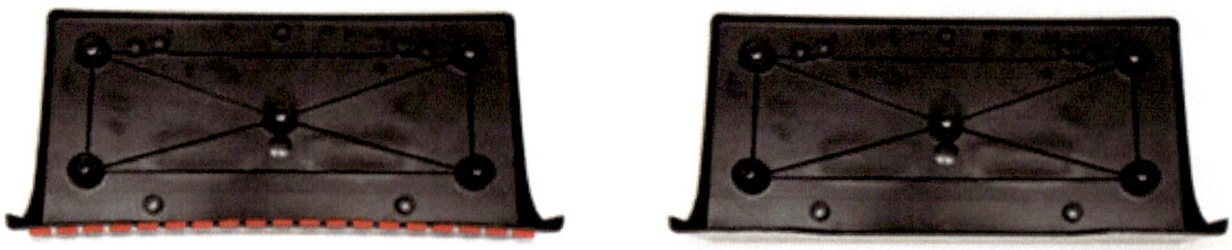

Figure 9.1 A warped electric housing (left) and a flat part after process optimization (right)

Figure 9.2 shows a cooling fan housing suffering from a bulge in the middle of each side of the part. Since there is a moving part, i. e., the fan involved during the assembly, the warpage defect also becomes a functional issue, due to the resulting vibration and noise.

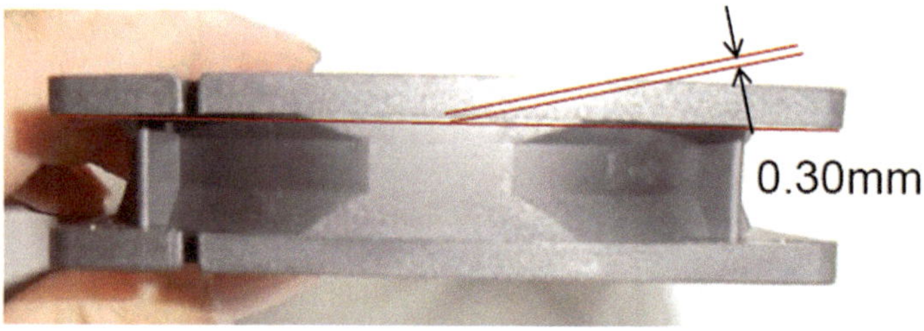

Figure 9.2 A fan housing with a bulge in the middle of each side

The other common type of warpage is a concave shape from a box-like geometry. One example is depicted in the camera housing shown in Figure 9.3. A thickness gauge is used to check the level of warp on the side wall. This deformation is a result of the mold temperature difference between the hot core and cool cavity side of the mold. It is difficult to design enough coolant channels inside the core piece so the part adhering to the core side is subjected to the residual heat and shrinks more than the outside.

Another important type of dimensional specification is roundness, which is relevant in most tight-tolerance joints and fittings. Roundness is easy to measure but difficult to control since the melt flow does not always uniformly advance inside the part and the pressure along the circumference is not equal. Figure 9.4 shows a coordinate measuring system capturing the roundness of a circular part mounted on a fixture.

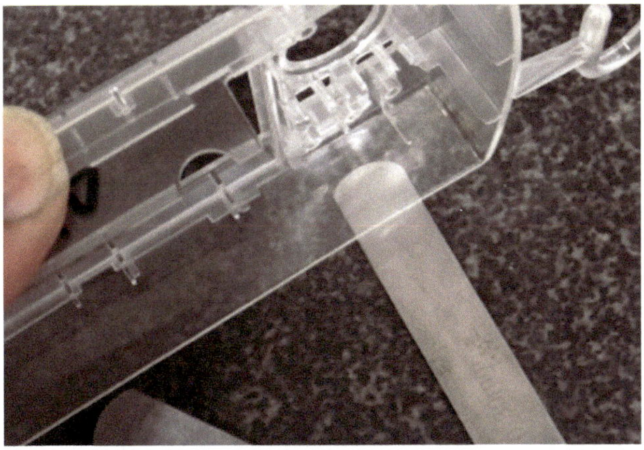

Figure 9.3 Box-shaped product showing a concave shape on the side wall

Figure 9.4 A circular shaft bearing checked on a fixture using a coordinate measurement method

One thing to note is that, in addition to the warpage occurring during molding, there is also warpage that gradually develops after part ejection. Such warpage is quite complex since it depends on a mixture of factors, including the environment and secondary crystallization in a not yet fully crystallized polymer. For example, polyamide (nylon) parts are well-known for their ongoing dimensional changes after being ejected. In this case, we can resort to CAE analysis results to check if relative crystallinity is low so that further crystallization, i.e., secondary crystallization, may occur when the part is exposed to an elevated temperature. In this chapter, however, the focus is on the warpage occurring during the injection molding process only.

9.1.1 The Causes of Warpage

There are various reasons for warpage, including geometric factors such as part design, insert effect, process conditions including pressure and temperature, and material properties such as fibers and Pressure-Specific Volume-Temperature (PVT) behavior. In order to resolve warpage, identifying the cause(s) is critical. Normally a fishbone diagram, as shown in Figure 9.5, is used to list all possible causes for warpage, which then can be compared with the real case to narrow down the source.

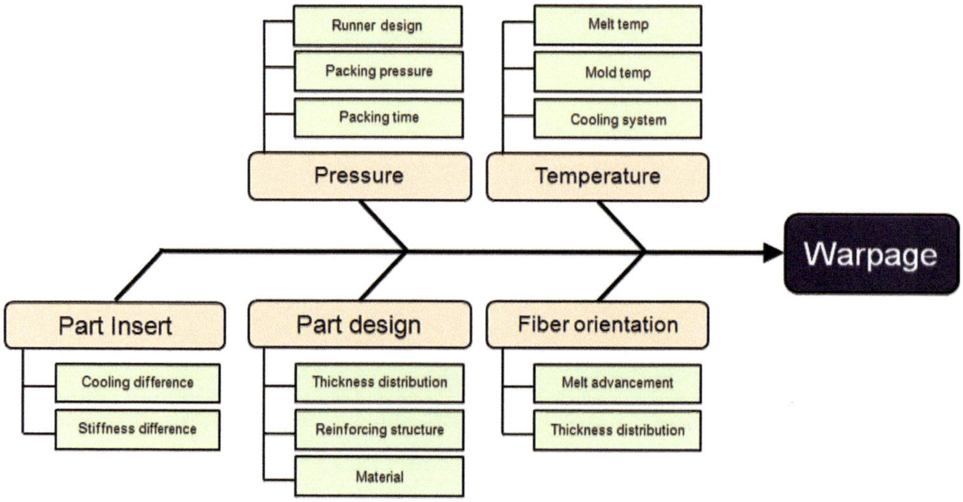

Figure 9.5 Fishbone diagram for warpage troubleshooting

1. Material

The plastic properties affecting shrinkage and warpage include plastic type, degree of crystallization (crystallization rate), and filler properties. Shrinkage behavior varies between different plastic types. For example, polyoxymethylene (POM) is a highly crystallized material and presents significant shrinkage after ejection. Polypropylene (PP), on the other hand, has a relatively minor shrinkage. Amorphous polymers such as poly(methyl methacrylate) (PMMA) have the lowest shrinkage rates. The crystallization rate is related to cooling. Nylon is a material with a low crystallization rate compared to the short injection filling time. The cooling is too fast for a high relative degree of crystallization, which leads to further shrinkage even after mold release. Fillers (inorganic substances) are basically rigid and do not shrink substantially, which makes them ideal for improving the mechanical properties and lowering the shrinkage of the polymer matrix. However, the induced viscosity increase in the melt poses challenges to processing with high filler content.

2. Part Design

Regarding the design of parts and molds, as heat tends to accumulate in the center area of the wall thickness, this area tends to shrink more after the packing pressure is released. The constantly dropping temperature would cause volumetric shrinkage in the original high-temperature areas. A single flat plate does not warp as the shrinkage on both sides is the same; however, warpage may occur if ribs are designed on one side. Usually, the ribs cause more shrinkage due to heat accumulation, such that the plate then bends toward the rib side. In some cases, inclusions such as glass fibers when properly aligned may reduce the rib-side shrinkage so that the warpage direction is opposite to the unfilled one. In the following sections, the mechanisms of how ribs, screw boss, and gussets affect warpage will be introduced.

Inserts are normally made of metal, which is stiffer than plastic, and act as reinforcements. Usually, they decrease the warpage. In some cases, an asymmetric shape of inserts may cause non-uniform shrinkage leading to warpage.

3. Mold Design

As for the mold design features, the gate location affects the packing effect in the part, and an irregular pressure distribution would lead to non-uniform shrinkage and cause warpage. How gates affect pressure distribution will be discussed later.

A good design of cooling channels effectively removes heat away from the part, and also homogenizes the temperature. Warpage is less significant if the temperature becomes uniform. But due to the spatial limitations of the mold (avoiding sliding blocks and ejection pins), the arrangement of cooling channels is often less than ideal. Also, the contact between the mold metal and mold mechanisms is unlike one block of material and thus does not conduct heat so well. On the other hand, some molds utilize high thermal conductivity metals such as beryllium copper to better transfer heat from the hot areas to the cooling system. This allows a shorter cycle time and provides a more uniform temperature distribution.

4. Process

Among all the molding process parameters, temperature and pressure are the most important ones to affect warpage because they relate directly to material shrinkage, i.e., PVT behavior. In addition, the faster the injection speed is, the higher the pressure becomes and so the temperature becomes more uniform, as the time it takes for a plastic material to flow from the gate to the end is shorter.

In the packing stage, the controllable parameters are packing pressure, packing time, multi-stage packing, cooling time, and cooling rate. These are all critical factors that affect shrinkage and warpage. During the filling, packing, and cooling stages of plastic molding, as the temperature and pressure keep changing and are distributed non-uniformly, the density also varies with time and among different areas of the part, leading to non-uniform shrinkage eventually.

The injection molding process can be understood by the changing history of temperature and pressure, the most important causes for part warpage. As shown in Figure 9.6(a), there are five stages for a one-shot cycle process. You can imagine this is the result of any sensor on the runner. The pressure rises when the melt reaches the sensor location in the plastic filling stage (plastic fill), then decreases slightly in the packing stage (injection pressure). After gate sealing (screw offload) and in the cooling stage, the pressure will decrease dramatically until the melt reaches the solid state for mold release.

Figure 9.6(b) and Table 9.1 depict the temperature change of the polymer as it progresses from the inside the barrel to the inside of the cavity. In the feeding stage, from 1 to 2, the polymer enters the barrel and is heated up quickly. Then the polymer melts via heat from shear friction, as shown at 2 to 3, the metering stage. The injection temperature rises in the filling cavity stage from 3 to 4 along with the occurrence of high packing pressure and shear heating. The cooling stage is from 4 to 5, in which the pressure also decreases. A dynamic process like this changes the density throughout the whole process and complicates the material shrinkage behavior.

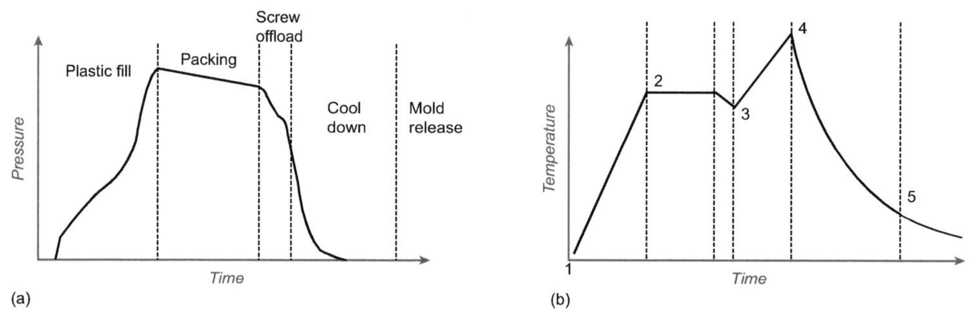

Figure 9.6 (a) Pressure history in mold of an injection molding process and (b) Temperature history of the polymer from inside the barrel until into the cavity

Table 9.1 Injection Process Stages of Polymer Temperature from Inside the Barrel until into the Cavity

Stage (see Figure 9.6)	Description
1→2	Plastic enters the barrel and is heated up quickly
2→3	Plastic absorbs heat until melted fully while maintaining constant temperature
3	Plastic enters the screw to be heated up by a shear friction

Stage (see Figure 9.6)	Description
3→4	Injection molding at high speed. Severe shear friction causes even higher temperature
4→5	Injection completes and the materials are cooled down through mold wall
5	Demolding

9.1.2 Material Effects

In addition to pressure, the molecular chain orientation also contributes to non-uniform shrinkage in the part. Both amorphous and semi-crystalline materials are stretched in the flow field when the temperature becomes higher than the phase-transition temperature including T_g or T_m (Figure 9.7 and Figure 9.8). The stretched chains tend to retract back to their most relaxed state with a globular conformation. If the material is frozen because of the fast cooling near the surface, the strong tendency of the chain to retract causes a larger shrinkage along the flow direction than perpendicular to it. The difference of shear rate leads to a change of the orientation through the thickness. The shrinkage thus becomes anisotropic. If a flow is complex, the shrinkage will not be symmetric across the thickness direction. This will lead to unpredictable warpage for the part; hence, the first step to improving warpage is to simplify the flow. Materials with fibers show even more complex shrinkage behaviors along the direction perpendicular to flow compared to simple polymer chains, as the shrinkage of fibers is different to that of the resin. Warpage is even more complex when the cooling rate of a crystalline material differs, because the induced degree of crystallization and crystalline structure are different so the shrinkage distribution becomes even more unpredictable.

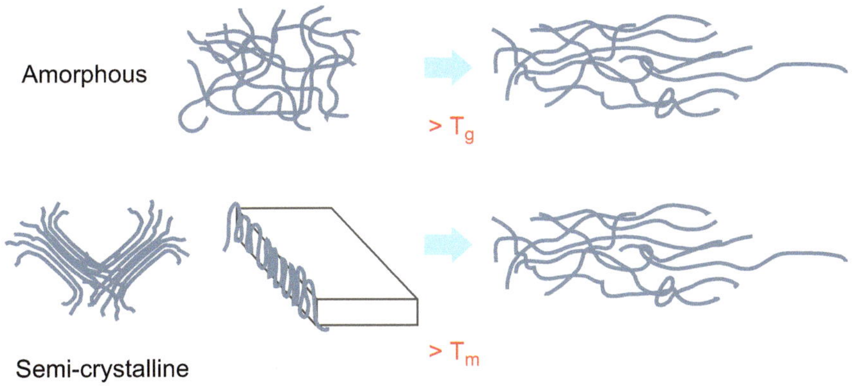

Figure 9.7 Amorphous and semi-crystalline polymers

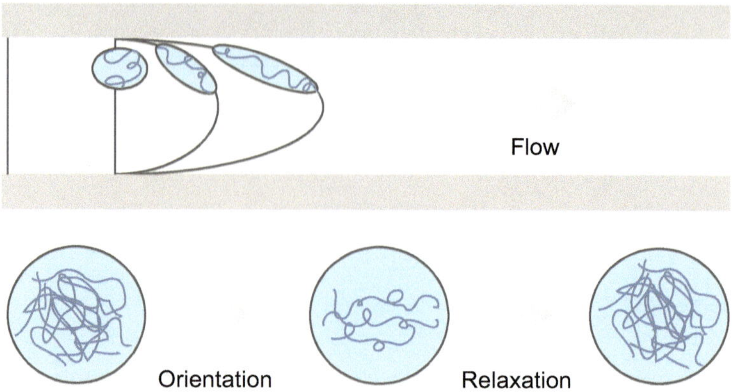

Figure 9.8 Molecular chain orientation caused by a shear flow of fountain flow

Since fibers cannot be stretched, they respond to shear flow by orientation along the flow direction. A simple shear flow results in a uniform orientation, but for fountain flow between the mold walls in injection molding, as shown in Figure 9.9 and Figure 9.10, fibers tend to align along the flow direction near the surface whereas they become perpendicular at the center of thickness area. This anisotropy leads to differential shrinkage between the layers of the part. If the orientation is unsymmetrical between the top and bottom shell layers, non-uniform shrinkage occurs and the part may start to warp.

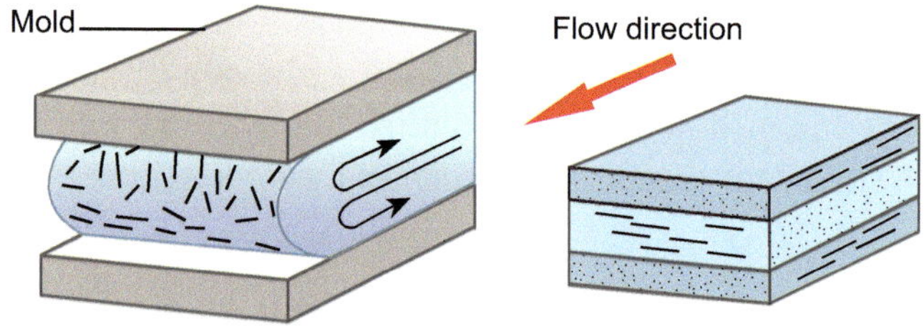

Figure 9.9 Fiber orientation caused by fountain flow

Figure 9.11 shows different warpage behavior of an exemplary ribbed plaque caused by different flow directions for the ribs. When the melt enters the plaque parallel to the rib direction, the ribbed half shrinks less so the part shrinks downward. A flow perpendicular to the rib direction leads to an upward shrinkage instead.

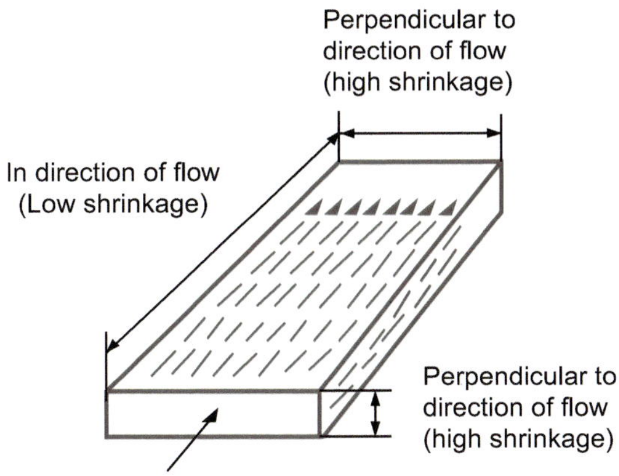

Figure 9.10 Fiber orientation effect on different part shrinkage parallel and perpendicular to the flow direction

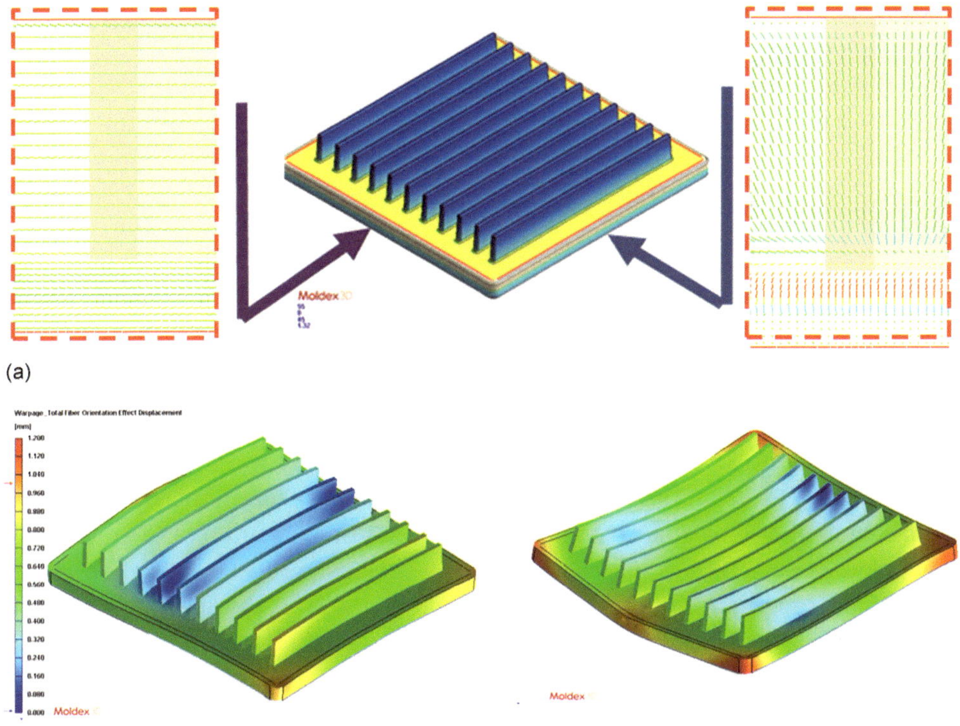

Figure 9.11 (a) Fiber orientation results in the rib section of a plaque with multiple fins. Two gate locations lead to either parallel flow or cross flow inside the fin. (b) The different fiber orientation of fin area leads to different warpage patterns

Apart from the influence of the flow on the orientation of the molecular chains of the plastic material, the occurrence of warpage is also related to shrinkage behavior. A PVT diagram explains the shrinkage properties of a plastic (Figure 9.12; see Section 2.3.3 for more details). The crystalline material has a sharper transition caused by crystallization during cooling and its shrinkage ratio is higher compared to an amorphous material.

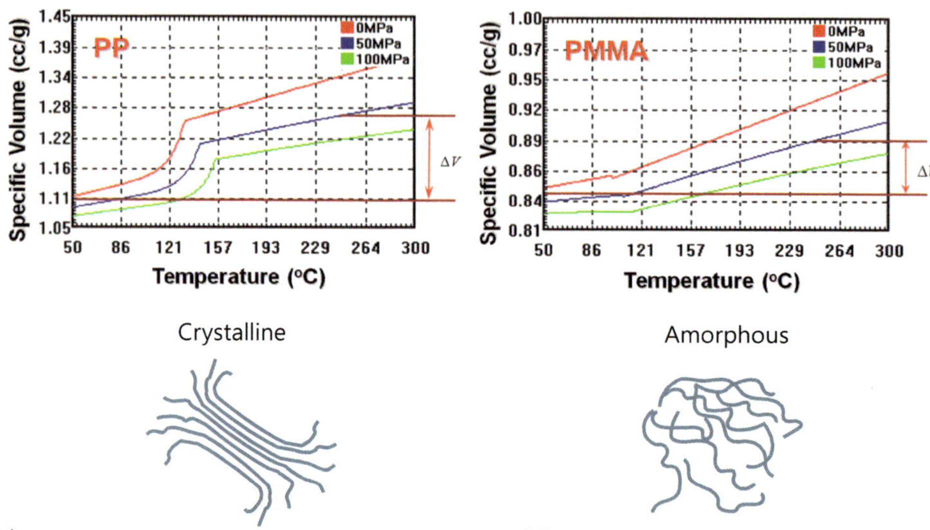

(a) (b)

Figure 9.12 PVT diagrams of (a) crystalline and (b) amorphous plastic materials

9.1.3 Geometrical Effects

In Chapter 3 of this book ("Part and Mold Design"), we have introduced part design principles that focus on the part geometrical effect on the flow of plastic as well as the orientation of molecular chains and fibers, and their result on the anisotropic shrinkage. More details about geometrical effects on warpage are discussed here. Since thickness changes in a part are inevitable, the wall thickness needs to be made as uniform as possible for continuous and gradual change. As shown in Figure 9.13, a smooth transition reduces the internal stress caused by a discontinuity of volume shrinkage.

As cooling is slower for thicker parts, for crystalline plastics, a slower cooling would generate a higher degree of crystallization in a larger area, which results in higher shrinkage. For thin parts, a high flow rate and the shear effect lead to a more significant orientation along the flow direction, which results in a higher internal stress and more severe warpage.

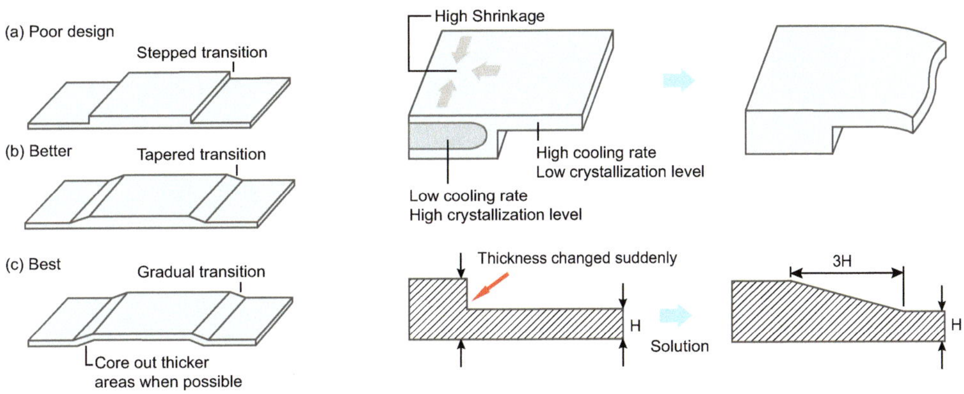

Figure 9.13 Effect of thickness on warpage: a gradual change of thickness is preferred in preventing warpage

For a flat plate or box-shaped product, features such as ribs and bosses are often used to reinforce the part. But, as shown in Figure 9.14, ribs also tend to cause unpredictable results of warpage. If the spacing between the ribs is too small to allow enough cooling channels to pass through, the ribbed area would accumulate heat and induce warpage. For mold makers, it is thus preferred to evaluate the warpage level from a molding simulation based on the materials used, the gating location, and the design of the cooling channels.

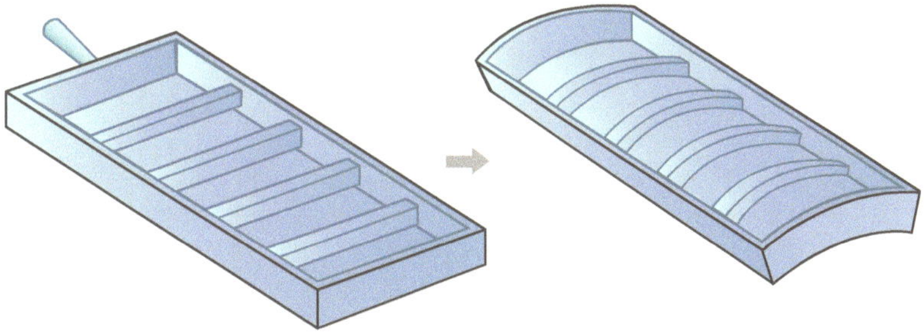

Figure 9.14 Effect of reinforcing rib on warpage: necessary core-out or cooling should be performed to prevent warpage

Cutting ribs into sections is one way of easing warpage. As shown in the top design of Figure 9.15, the bottom of the part is thicker, thus having a higher shrinkage. The thinner ribs do not shrink as much as the bottom, so a concave shape is the result. Below are improvements that can eliminate this warpage, including cutting off the consecutive plane on the upper part, which restrains the shrinkage, or the rib can be designed more symmetrically so that the warpage would not be significant while the stiffness can still be maintained.

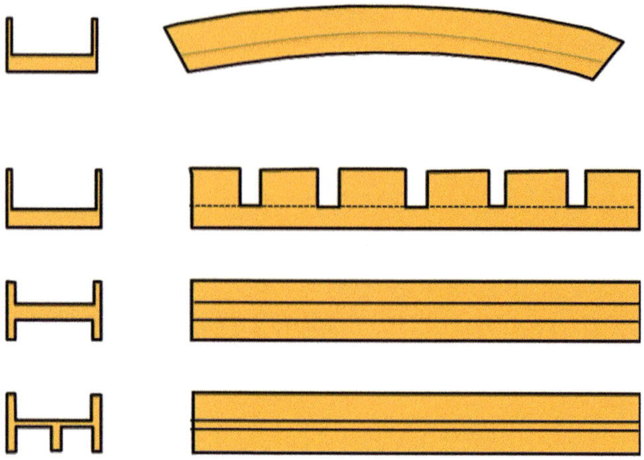

Figure 9.15 Top: uneven thickness distribution leads to warpage; below: improved rib designs to eliminate warpage

When fiber-reinforced plastics encounter a change in thickness, a strong elongational flow orients the fibers as shown in Figure 9.16. The fibers orient better as the flow passes from thick to thin areas as in a converging flow. The effect of this stretching in changing the fiber orientations is stronger compared to shear flow. This is another source of inducing anisotropy into the part apart from shear flow. Parts gated at the center have a strong diverging flow at the beginning of filling. This often leads to a bowl- or dome-shaped warpage.

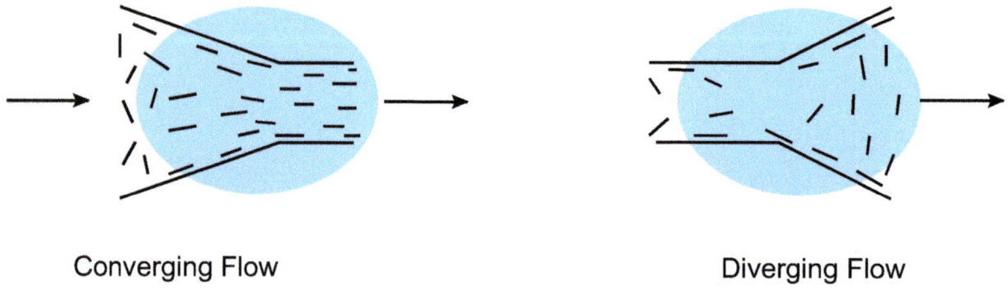

Figure 9.16 Comparison of fiber orientation during converging and diverging flow

9.1.4 Process Condition Effects

Temperature and pressure are the main causes that induce shrinkage. Six factors can be defined in comparison to site conditions: melt temperature, mold temperature, injection speed, packing pressure, packing time, and cooling rate. It is very difficult to test these six factors by trial and error; therefore, we should first understand how they affect shrinkage.

Normally a reduction of shrinkage can be achieved by lowering the melt temperature, increasing the packing pressure, and prolonging the packing time, whereas the mold temperature and injection speed have limited effect on shrinkage. In a cooling process, the mold temperature would not play a critical role unless a material of slow crystallization rate is used, or the mold temperature is too low causing early cooling and insufficient crystallization. The injection speed affects temperature and pressure indirectly. Normally the dynamic range of the injection speed of the machine is not wide, and so is the allowed injection speed range for the part. Therefore, the injection speed, melt temperature, and pressure can all be managed within an adjustable range.

The molding parameters of filling should control the rheological behavior of the plastic material: how much time it needs for filling, and how much pressure should be used for injection. If the injection pressure cannot be evenly distributed in the part, a significant difference in volumetric shrinkage will occur despite applying an excessively high pressure. A localized high injection pressure easily turns into residual stress which deteriorates the part. An ideal injection process ensures that enough material enters the cavity and that the subsequent packing pressure compensates for the shrinkage caused by material cooling. Refer to Chapter 6 for details regarding the effect of pressure on product shrinkage.

Regarding temperature, the shrinkage grows as the temperature of the melt increases. If the melt temperature is too low, the gate will freeze too early rendering the subsequent packing useless. The mold temperature can further be controlled independently via two mold temperature controllers on the two sides of the part separately. This allows warpage to be actively controlled by manipulation of the part shrinkage on both sides. As shown in Figure 9.17, generally the cooling efficiency of the mold core is poor and the shrinkage is high. To solve the problem, the temperature of the mold core can be lowered to ease the shrinkage, while increasing the temperature of the cavity side for better surface quality.

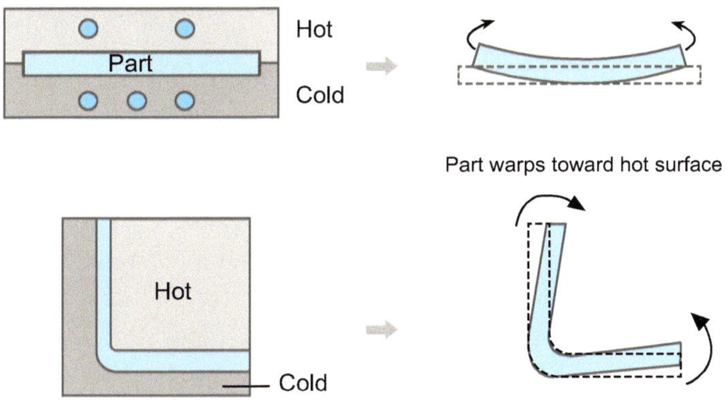

Figure 9.17 Part warpage tendency due to global or localized mold temperature difference

9.1.5 Criteria of CAE Warp Analysis

According to the above principles, CAE can be used for warpage analysis to assess (1) uniform cooling and mold temperature distribution, (2) effective packing before gate freezes, (3) shrinkage due to part geometry, (4) fiber orientation effects on warp, and (5) displacement/flatness prediction:

1. Uniform Cooling and Mold Temperature Distribution

As the cooling channel suitable for both mold cores and mold cavities is different, the induced difference in cooling efficiency would result in distinct shrinkage. It is hard to install a cooling channel at the corner of the mold core, and thus the shrinkage becomes higher. It is thus often seen in box-like parts that the shrinkage at corners is more serious, showing a concave condition around the corners.

In the Moldex3D software, the “Mold temperature difference” function can be used to visualize the difference in mold temperature. Generally, a difference is preferred but which shall maintain a uniform and constant value. The cross-sectional view of the mold temperature could help the user find the problematic hot spot. As shown in Figure 9.18(a), the center of this exemplary fan part shows high temperature even after cooling, which results in notable warpage at the blades (Figure 9.18(b)). If cooling channels can be applied to this side, the warpage can be reduced.

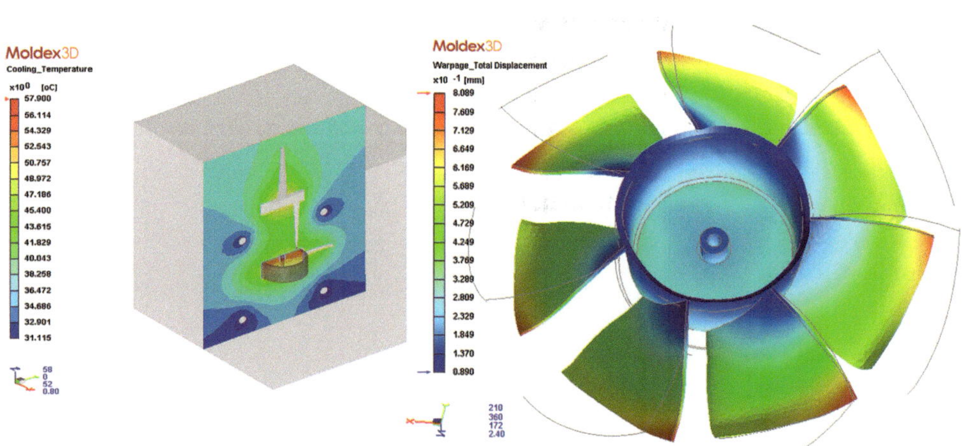

(a) (b)

Figure 9.18 (a) Mold temperature cross-section CAE results showing heat accumulation on the core side, and (b) warpage of the part

2. Effective Packing before Gate Freezes

Gate freezing has a key effect on the success of packing. If the gate seals due to cooling, even a higher packing pressure will not contribute to the effectiveness of packing. Therefore, whether the gate is sealed or not should be particularly observed in a simulation analysis. It can be seen from the temperature distribution chart shown in Figure 9.19 that the high-temperature area (green) locates around the middle of the runner, while the temperature is lower at the gate and inside the part, which means that the pressure cannot be effectively transferred into the part if the packing is done from the right side (Figure 9.19).

Description	
Polymer	PA66
Grade Name	6210GC
Producer	NanYa
Comment	
Last modified date	2010/08/25
Process condition	
Melt temperature (minimum)	270 °C
Melt temperature (normal)	285 °C
Melt temperature (maximum)	300 °C
Mold temperature (minimum)	60 °C
Mold temperature (normal)	80 °C
Mold temperature (maximum)	100 °C
Ejection temperature	210 °C
Freeze temperature	240 °C

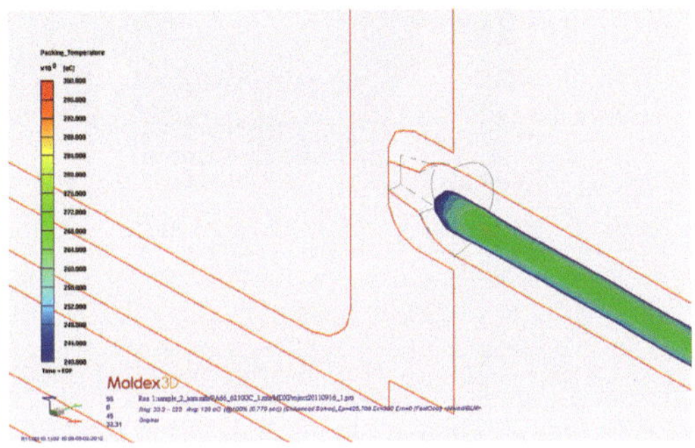

Figure 9.19 Effective packing before gate freezes

3. Shrinkage Due to Part Geometry

Shrinkage difference due to the thickness and geometry characteristics of a part can also affect how a molded part warps. Figure 9.20 shows a slicing of volumetric shrinkage extracted from Moldex3D for a connector case, where its thickness is distributed unevenly. Thicker sections have larger volumetric shrinkage than thinner regions, because heat accumulates more easily in the thicker section, as it is more difficult for cooling to reach the middle area. Because of the uneven thickness distribution, the volumetric shrinkage varies. Therefore, a well-considered part design is an essential point to prevent product warpage.

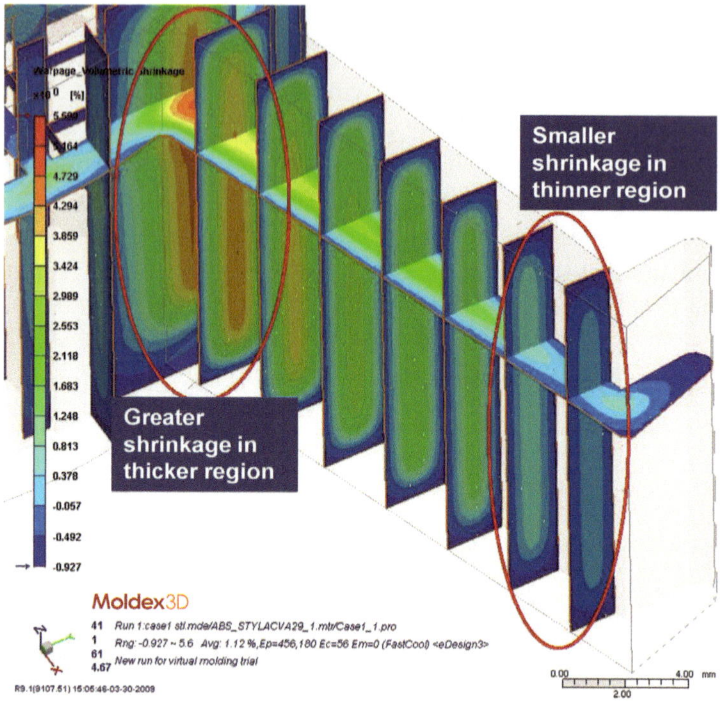

Figure 9.20 Shrinkage difference: thickness and geometry characteristics

4. Fiber Orientation Effects on Warp

As fiber orientation due to flowing creates anisotropic properties in the material, anisotropic shrinkage and warpage are thus generated. The fiber orientation effect can be verified by CAE. Fibers tend to align in the flow (forward) direction, especially in the area close to the mold wall. Shrinkage along the forward direction of fiber orientation is relatively smaller. If fibers are aligned well, the shrinkage along the long axis (forward) becomes smaller; if not, shrinkage would be similar in both directions. A randomized fiber orientation has a more uniform shrinkage along and cross-flow direction compared to the aligned one with an anisotropic shrinkage, as shown in Figure 9.21.

Figure 9.22 shows a poor orientation of the fibers on the top half of the part, leading to isotropic shrinkage, but on the bottom part the fibers are highly oriented giving rise to strong anisotropic shrinkage. However, as the shrinkage along the direction of flow is smaller, we can see as a final result that the shrinkage on the bottom part is small, whereas that on the top half is large, causing the part to warp upward.

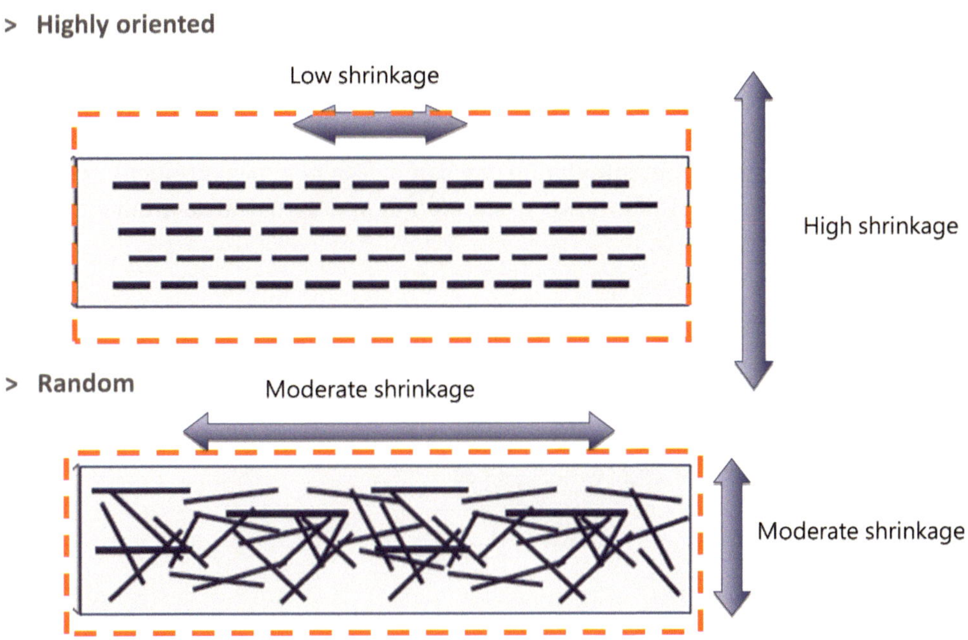

Figure 9.21 Highly-oriented and randomized fiber orientation, and corresponding shrinkage

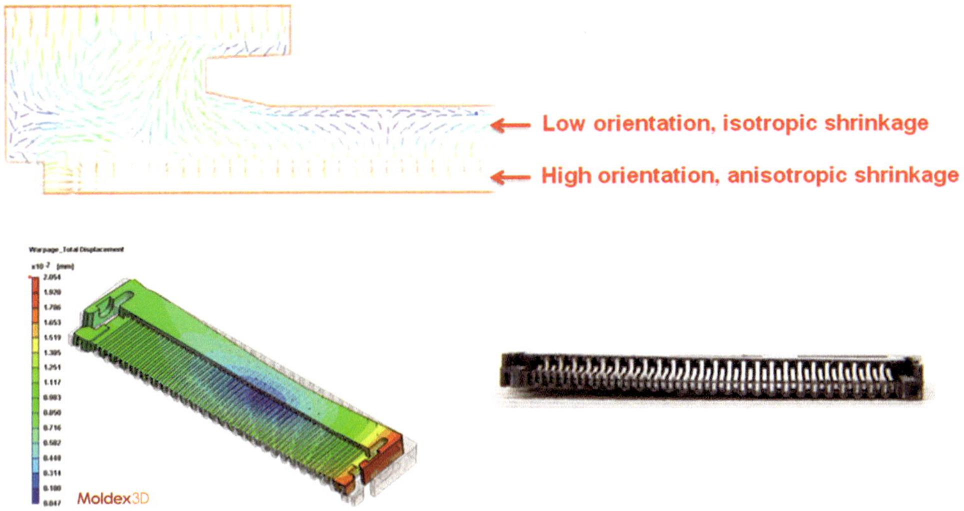

Figure 9.22 Simulation result of fiber orientation of a connector product, and comparison of predicted and actual warpage

5. Displacement/Flatness Prediction

Generally, optical and contact measurements are often used to determine the dimensions of the part. The former uses lights to scan the exterior of the part to determine the dimensions, whereas the latter uses probes to measure the dimensions of specific regions. In simulation, we can define two measurement nodes in order to compare the results between simulation and actual measurements (Figure 9.23).

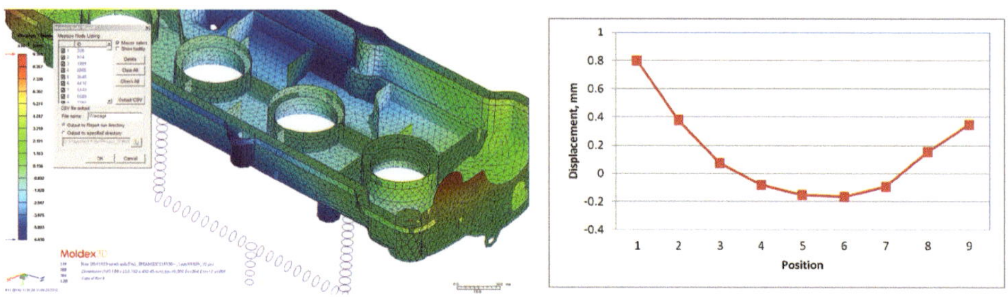

Figure 9.23 Part flatness measurement predicted inside the CAE software

The measure node wizard tracks the coordinate values for each node before and after displacement. The quantified displacement at the targeted region can be recorded and output. The displacement output from all nine measurement nodes shown in Figure 9.23 suggests an upward warpage profile with a distance of 1 mm between the highest and lowest point.

9.1.6 Methods to Minimize Warpage

There are a few practical ways to reduce warpage, including (1) lowering the melt temperature, (2) increasing packing efficacy, (3) using metal inserts or (4) fiber-filled materials, (5) optimizing gate location for higher packing pressure, and (6) using Moldex3D "Design of experiment" module.

1. Lowering Melt Temperature

From the PVT diagram shown in Figure 9.24, we can see that for an initial melt temperature of 184 °C, the volumetric difference generated after returning to room temperature and pressure would be less than that for a high initial melt temperature of 242 °C. Hence, using a low melt temperature for injection would cause a smaller volumetric variation compared to using a high melt temperature.

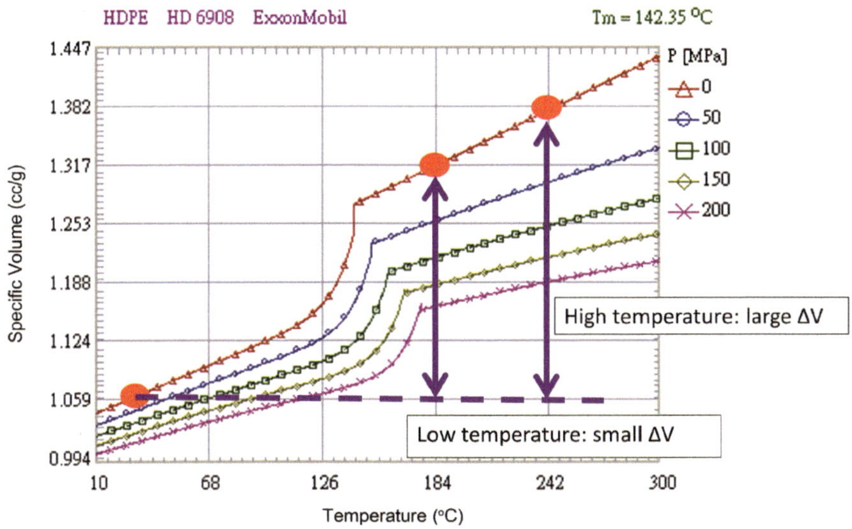

Figure 9.24 Volume difference between molten stage and frozen stage

2. Increasing Packing Efficacy

Using a higher packing pressure would cause a smaller volumetric variation than a lower packing pressure, as shown in Figure 9.25. In the packing stage, when the part starts to cool down, the shrinkage can be compensated by a small amount of the plastic material flowing in slowly from the gate. However, if the pressure is too low to provide sufficient pressure drop to push the plastic forward, the shrinkage can no longer be compensated by additional plastic material. When the gate is sealed, the weight of the part remains unchanged, and as the volume keeps decreasing, shrinkage occurs. Therefore, keeping a high pressure in the packing stage can increase the density, which can compensate for the shrinkage after cooling (Figure 9.25).

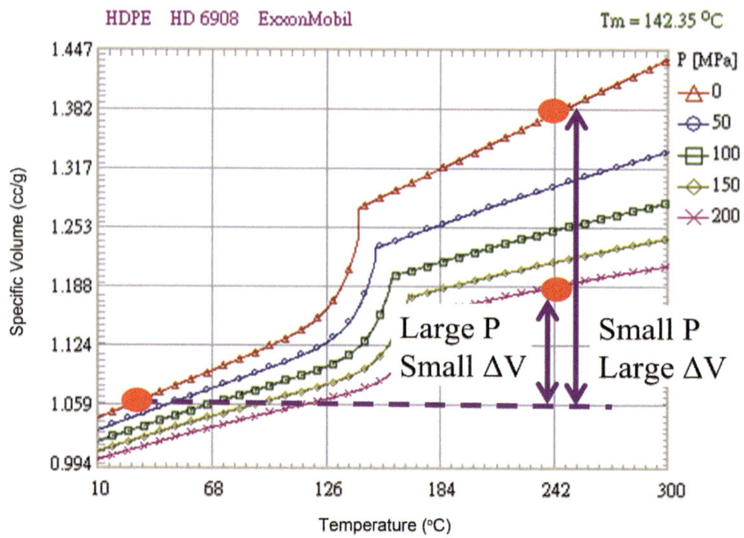

Figure 9.25 Volume difference between molten and frozen states at various packing pressures

In addition, to increase the packing effectiveness, the gate can be enlarged as shown in Figure 9.26 to prevent it from sealing too early, and thus enhance the contribution of packing on preventing the shrinkage of a part.

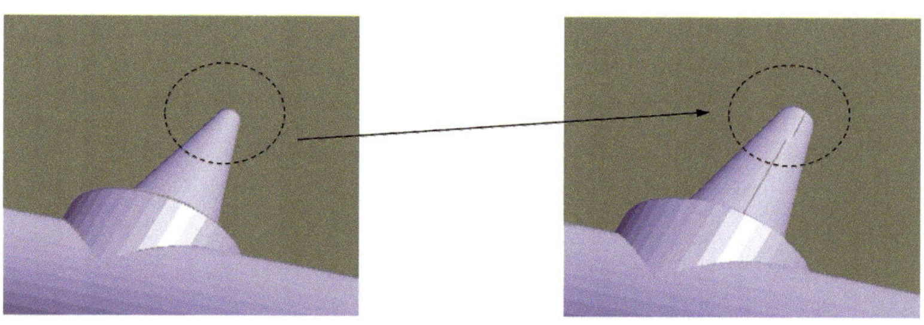

Figure 9.26 Larger gate size for better pressure transfer

Packing is also more effective when the gate size is large enough to easily transfer the pressure to the end of the cavity. An increase in the gate size helps to transfer the pressure and also maintains a higher packing pressure for a longer time before the gate freezes.

3. Using Metal Inserts

Inserts have a higher stiffness to restrain shrinkage and reduce warpage, as shown in Figure 9.27(a); however, they can also create opposite effects. For example, if a part has no warpage initially, after adding inserts to enhance the strength on one side, an unbalanced shrinkage would occur on both sides, and warpage is then induced, as shown in Figure 9.27(b). Thus, warpage may be worsened if the original plastic material shrinks uniformly whereas the insert only counteracts in localized areas.

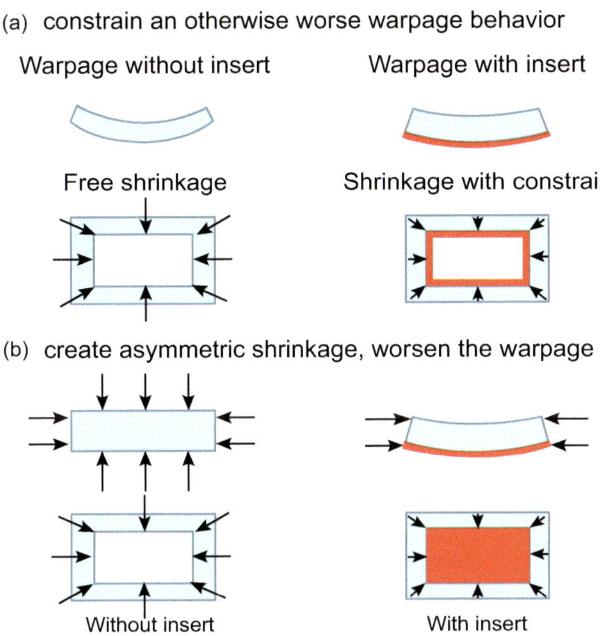

Figure 9.27 Metal inserts: (a) improvement or (b) worsening of warpage

The part shown in Figure 9.28 has a metal reinforcement in the center. It can be seen clearly that the region with reinforcement has a lower shrinkage (blue), whereas in the region without inserts high warpage appears. The metal insert placed in the part reinforces the part by boosting its modulus. Simulation predicts a large displacement at the end of the part where no insert is placed.

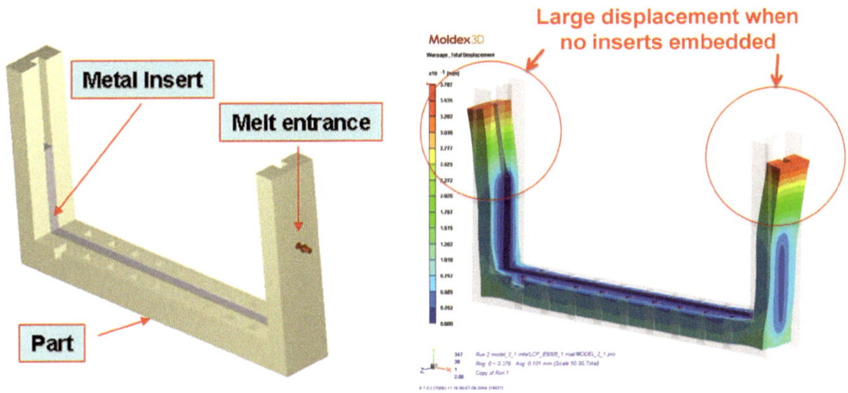

Figure 9.28 Warpage simulation results of regions with and without metal insert

Figure 9.29 presents an example where a metal insert leads to more warpage. Originally, the part shrinks isotropically with small deformation; however, with inserts that restrain the shrinkage on the bottom half, the part warps significantly upward instead.

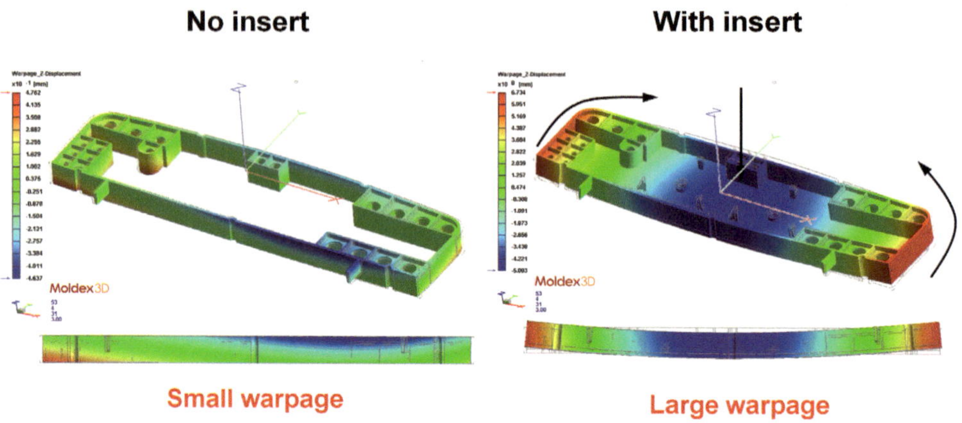

Figure 9.29 Metal insert leads to more warpage for a real part; note that the metal insert is connected to only one side of the part

4. Using Fiber-Filled Materials

Fibers of low thermal shrinkage take up space of within high-shrinkage plastics. Adding fibers increases the part stiffness and thus reduces the warpage.

An example showing the influence of fiber addition on warpage is illustrated in Figure 9.30. The forward direction without fiber has a larger shrinkage, whereas the one with fiber has a smaller shrinkage along the forward direction as fibers shrink less. However, adding fibers would increase anisotropy, and thus generates a larger shrinkage in the vertical direction than in the original.

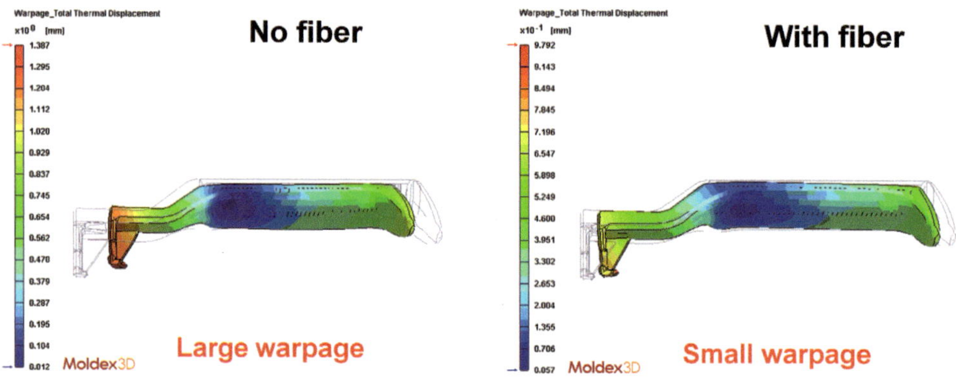

Figure 9.30 Different shrinkages caused by using unfilled or fiber-filled materials (gating from the left side)

5. Optimizing Gate Location for Higher Packing Pressure

The original design shown in Figure 9.31 has four smaller gates, and the geometrical lengths from the sprue to the gates in each are not equal. Because the gates are small, the packing effect generated from each gate is different and leads to serious warpage. The amount of shrinkage can be reduced by enlarging the gate size, adding more gates, and creating runners of equal geometrical length. Thereby, the warpage condition is improved.

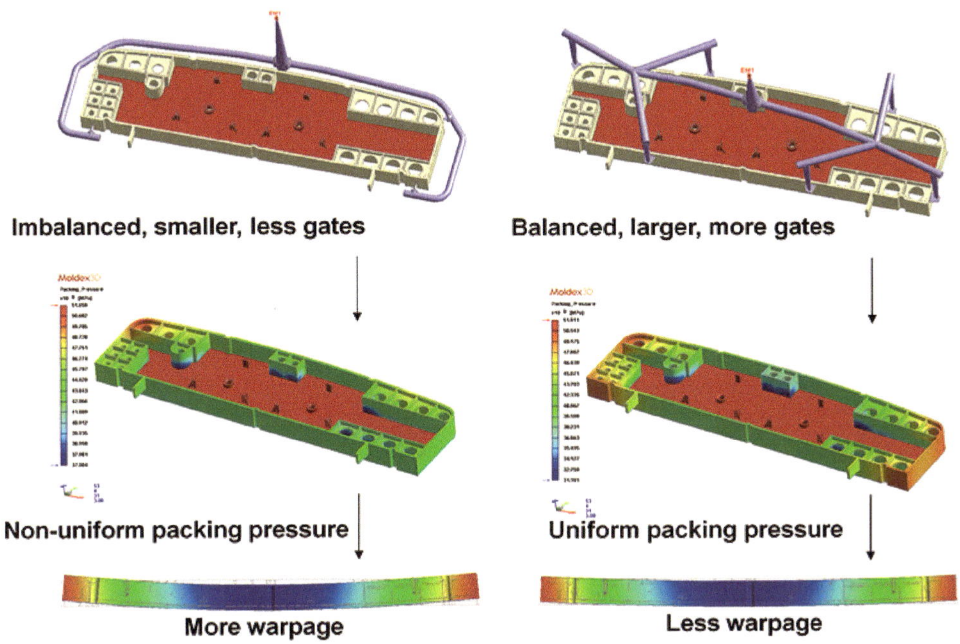

Figure 9.31 Effects of gate location on warpage

6. Using Moldex3D "Design of Experiment" Module

In the case of complicated part dimensions, or if there is no room for part and mold design, and no clear relationship can be identified between process conditions and warpage, CAE software can become useful in determining the optimum molding conditions with virtual experimental designs without conducting actual experiments. The design of experiments (DOE) method can be utilized to find the optimum process conditions with the least amount of warpage (QF, Qualifying Facility: total displacement) The example presented in Table 9.2 shows Run 03 as the most promising process to reduce warpage. The control factors (CF1-4) discussed include melt temperature (CF1), mold temperature (CF2), packing time (CF3), and packing pressure (CF4).

Table 9.2 Utilizing the Design of Experiments (DOE) Method to Find the Optimum Process Conditions with the Least Amount of Warpage (QF Total Disp.)

Run	CF1 [°C]	CF2 [°C]	CF3 [s]	CF4 [MPa]	QF [mm]
01	193.33	26.6/50	17	30.32	6.31
02	193.33	26.6/60	22	40.32	4.11
03	193.33	26.6/70	27	50.32	1.71
08	215.55	26.6/60	17	50.32	2.97
09	215.55	26.6/70	22	30.32	3.15
Best Run	**193.33**	**26.6/70**	**27**	**50.32**	**1.71**

9.2 Practical Applications

CAE Solution on Warpage of Connector

Figure 9.32 shows a high-current connector that is fabricated using polybutylene terephthalate (PBT), a crystalline plastic material. A CAE simulation applied in the engineering development stage finds a serious shrinkage, which cannot meet the customer's demand for high precision in hole dimensions and positions. Such a serious issue has to be fixed in order to avoid affecting the yield of follow-up automatic assembly.

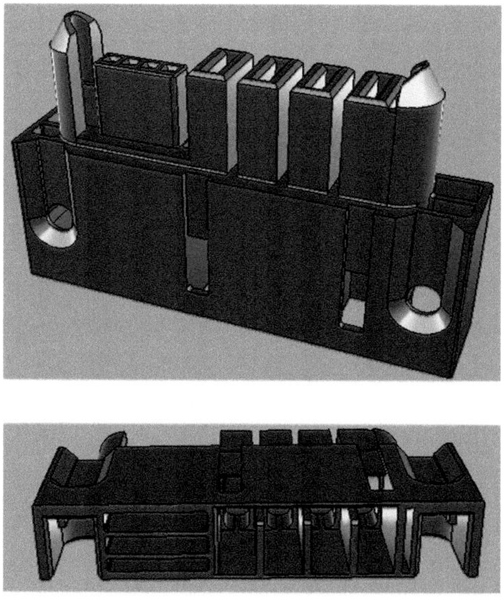

Figure 9.32 Part geometry of a high-current connector with many holes and openings for further assembly

CAE is introduced for (1) design qualification, to provide the optimum design in the shortest possible time and save the time of mold modification; (2) as a communication channel, to provide a base of design change between part design, mold design, and injection molding operators; and (3) for symptomatic treatment, to eliminate the root cause of problem with the visualization function.

The goal of CAE is to improve the shrinkage issue, ensure the part quality, and launch the mass production successfully. Furthermore, any air trap issue in the part should be solved as well in order to meet the customer's demands on the exterior. A quantified target is to improve the shrinkage by over 50%. The critical dimensions should meet the tolerance demands, and as no defects such as air traps and flow marks should affect the appearance.

The molding conditions used in the simulation are as follows: filling time 0.2 s, packing time 2 s, packing pressure 60 MPa, packing switching volume 98%, melt temperature 265 °C, and mold temperature 70 °C. As the original single-gating design shown in Figure 9.33 is suspected of causing a poor packing effect, one more gate is added to increase the packing sources and balance the packing. The additional gate serves as an extra packing source.

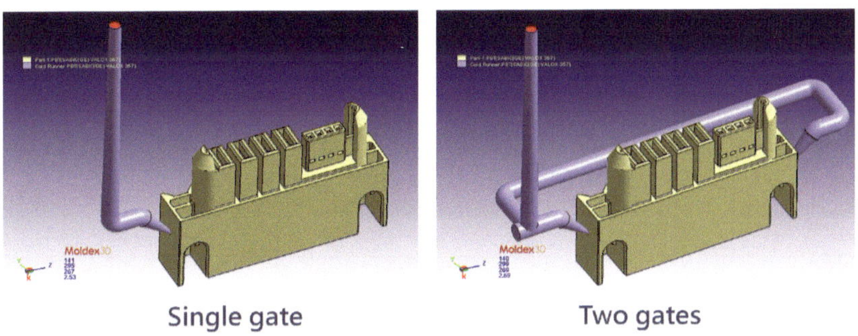

Figure 9.33 Single- and double-gating high-current connector designs

Figure 9.34 shows a volumetric shrinkage simulation result of the original single-gating condition. The red areas represent relatively larger shrinkage. The most significant red area is at the location where the second gate is added, which makes setting up the second gate beneficial to shrinkage reduction.

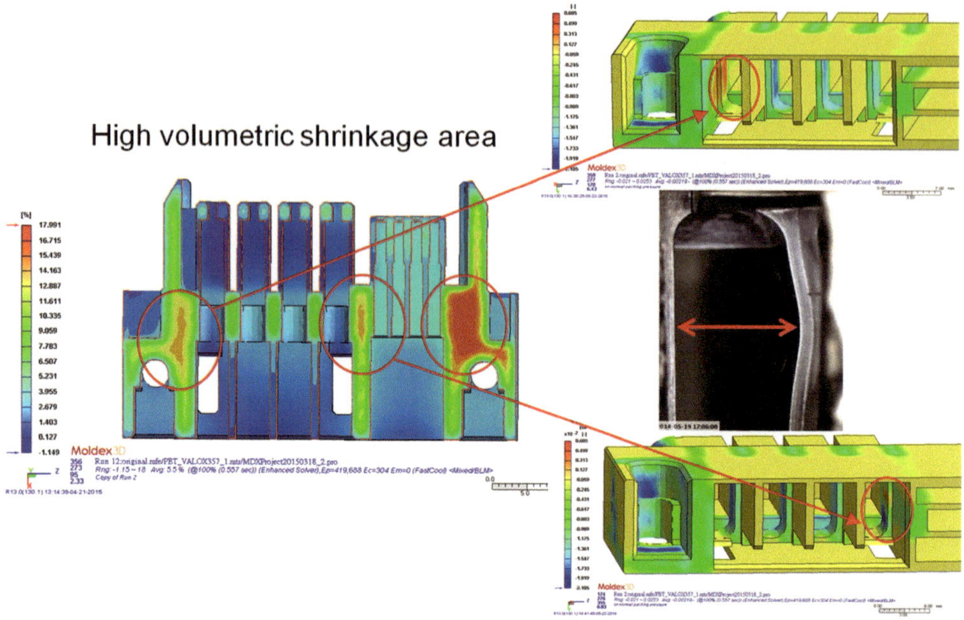

Figure 9.34 High-shrinkage region (marked as red) leaving an enlarged opening of the connector

Figure 9.35 shows the volumetric shrinkage difference between a single- and a double-gating design. The maximum shrinkage is reduced from 17% to 14% after changing to the double gating due to better packing efficiency.

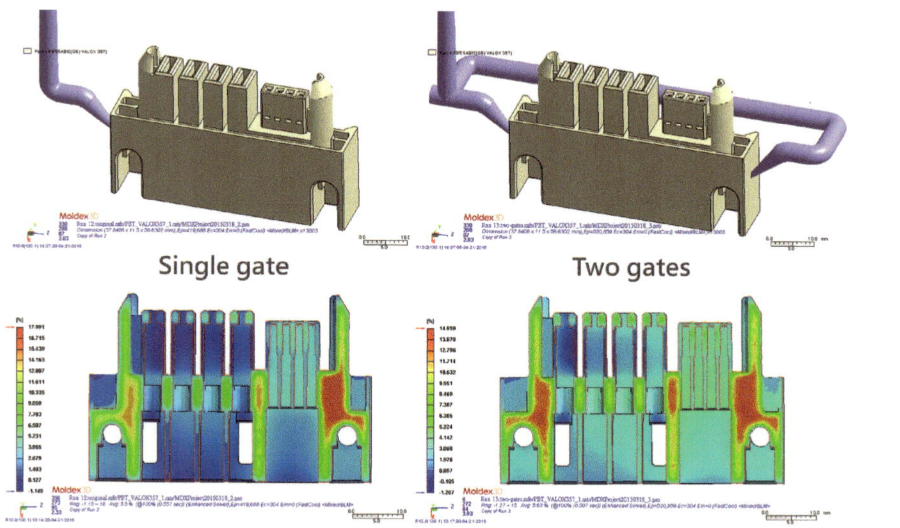

Figure 9.35 Comparison of volumetric shrinkage results between single- and two-gate designs

In addition, the amount of plastic material should be reduced in the part region which has more volumetric shrinkage, in order to reduce the difference of volumetric shrinkage. To avoid affecting the appearance, this reduction can be fulfilled by modifying the mold core and mold cavity, and the effect therefrom is shown in Figure 9.36. As the maximum shrinkage is still nearly 14%, the effectiveness is limited, so the amount of plastic material is further reduced from the slider on the side, which induces the effect shown in Figure 9.37. The region that originally had largest shrinkage is thereby improved significantly.

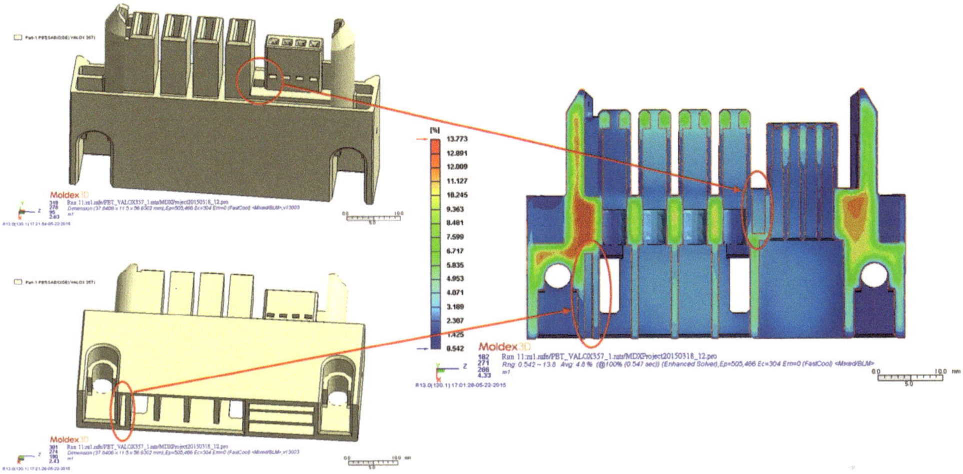

Figure 9.36 Core-out by changing core and cavity side of the mold; the reduction of shrinkage ratio is limited

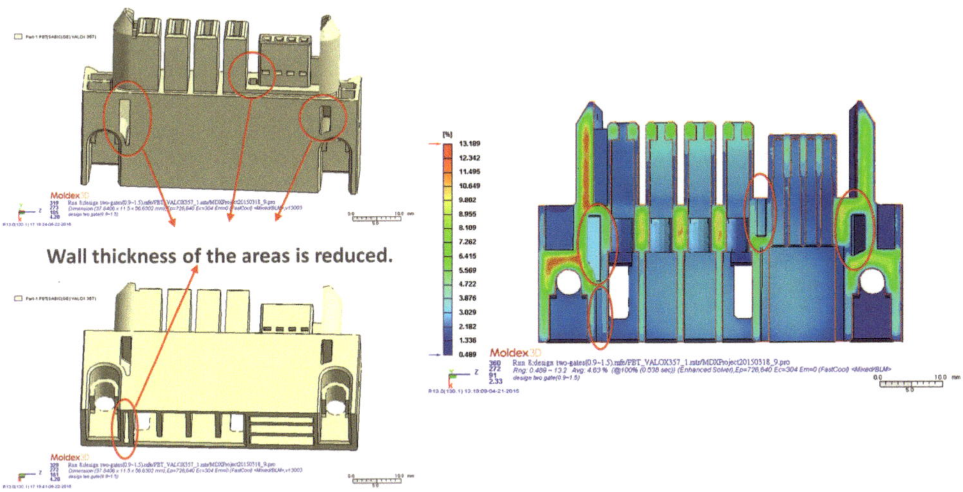

Figure 9.37 Core-out by changing the slider of the mold; the high-shrinkage area of the part is largely reduced

After reduction of the plastic material, the volumetric shrinkage ratio is mitigated successfully, but the problem of air trapping on the surface is induced, as shown in Figure 9.38.

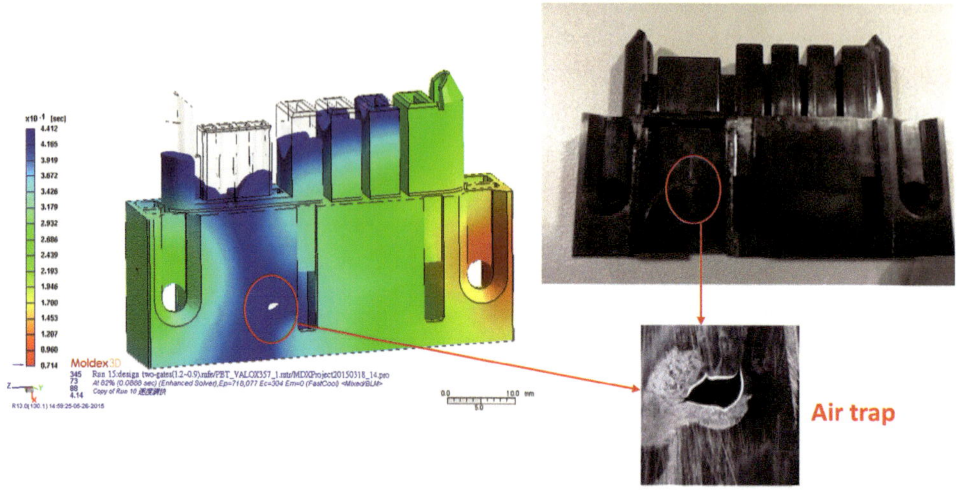

Figure 9.38 Product air trap problem after the core-out design modification

Hence, another design change is required. The air trapping can be solved by adjusting the gate size and change the flowing method. The result in Figure 9.39 shows that the air trap region is moved to the parting line to facilitate venting.

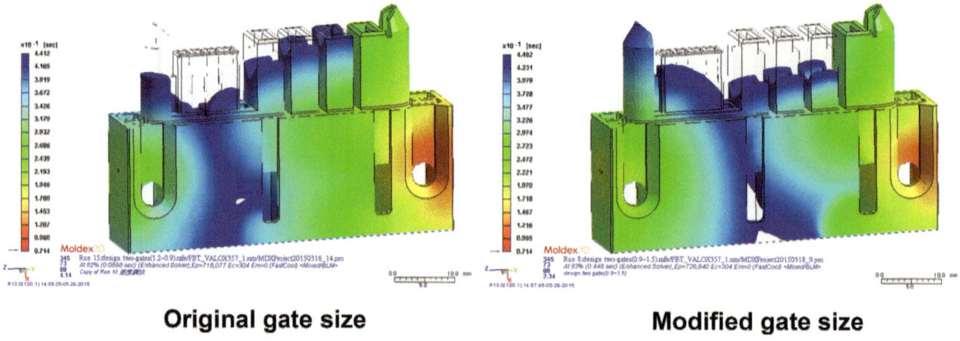

Figure 9.39 Air trap problem resolved by slightly enlarging one selective gate size so that the end of the filling area is moved close to a parting line

The final measurement of the dimensions at different locations as shown in Figure 9.40 (the results are listed in Table 9.3) reveals an improvement of over 70% for the critical dimension of Bottom-1 and Bottom-4 in comparison to the target value of 3.61 mm, and all dimensions can be controlled within 3.61 ± 0.8 mm.

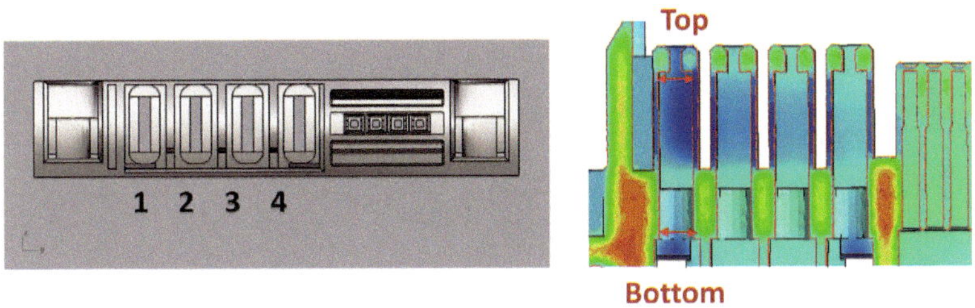

Figure 9.40 Locations of controlled dimensions

Table 9.3 Comparison of Controlled Dimensions of Original and Revised Designs

Locations	Original Design [mm]	Revised Design [mm]	Improved Ratio
Top-1	3.571	3.587	-
Top-2	3.597	3.598	-
Top-3	3.589	3.600	-
Top-4	3.567	3.623	-
Bottom-1	**3.855**	3.683	70%
Bottom-2	3.675	3.678	-
Bottom-3	3.670	3.673	-
Bottom-4	**3.944**	3.684	77%

■ 9.3 CAE Case Study

Following the discussions of this chapter and the corresponding CAE verification case, an exercise model, as illustrated in Figure 9.41, is provided in this section for readers to practice CAE application in warpage control.

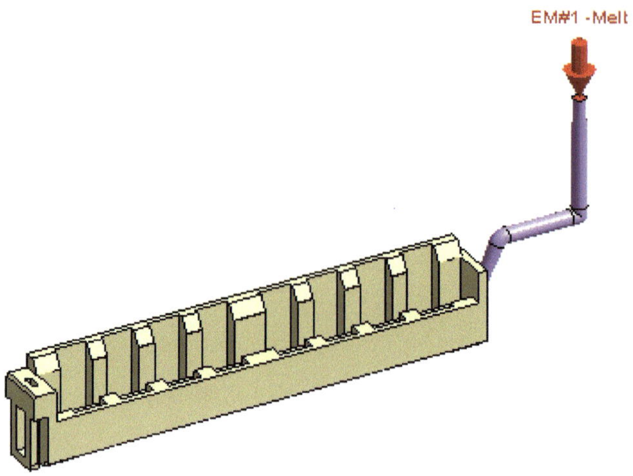

Figure 9.41 Case study sample

Please download the analysis files for CAE Case Study 9.3 from the following website:

https://moldex3d.box.com/s/zr6fvc1vlhbi4ocx111jwd3wmxt4ooif

Then open the .m3j or .mvj file in Moldex3D and answer the questions below according to the analysis results.

Questions:

1. We learned that material selection plays an important role in warpage. Table 9.4 lists four different materials with distinct crystallinity and fiber inclusion.
   a) Provided in this section is a set of simulations using these materials. According to Runs 1 through 4, compare the warpage results and state which material property contributes to a certain warpage result.
   b) How can Moldex3D be used to show the effect of fiber orientation and the results if fibers are randomly aligned instead of oriented?
2. We can make design modifications to ease the warpage of a part. The part shown in Figure 9.42 has been modified to half thickness (0.633 mm to 0.316 mm).
   a) According to Run 5, compare their warpage results.

3. When the mold is changed, e.g. gate enlargement, the packing effect may get better, thus reducing part shrinkage/warpage.
   a) According to Run 6, compare the warpage results after changing the gate location and explain the reason.
   b) According to Runs 7 and 8, show the differences of warpage after enlarging or reducing the gate size.
4. Filling conditions may also play an important role in changing the part warpage.
   a) According to Runs 9 and 10, check the warpage of a highly filled and a not highly filled example and explain the possible reason for the difference.
   b) According to Run 11, show the result of the part if there is only one constant filling speed.
   c) According to Run 12, what is the effect of changing the packing pressure?
   d) According to Run 13, what is the effect of changing the packing time?

Table 9.4 Four Different Materials with Distinct Crystallinity and Fiber Inclusion

Type	Amorphous	Crystalline
Unfilled	PC+ABS CYCOLOY C7410 SABIC(GE)	PA6 Ultramid 8202 BASF
Fiber-filled	PC+ABS CYCOLOY C4210HF SABIC(GE) 10%GF	PA6 Ultramid 8231 G HS BASF 14%GF

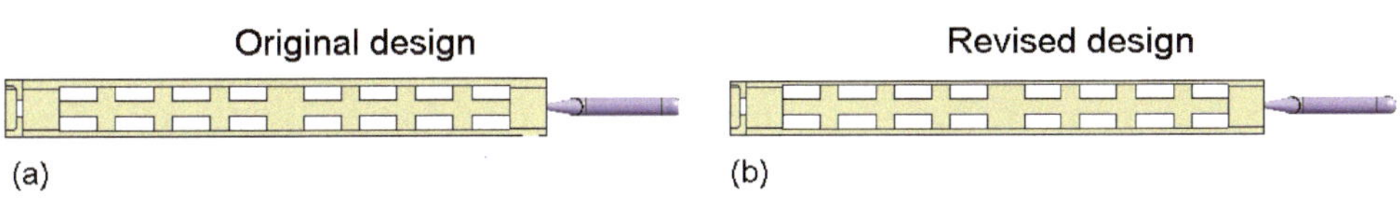

Figure 9.42 The wall thickness of (a) the original design of 0.633 mm, and (b) the revised design of 0.316 mm

10 Fiber Orientation Control

Huan-Chang (Ivor) Tseng

Over the last 30 years, fiber-reinforced thermoplastics (FRT) have increasingly replaced steel-made automotive components as high-performance engineering materials for structural applications in the automotive industry, as shown in Figure 10.1. Apart from their excellent economic benefits and low weight, their big advantage lies in a high level of productivity in processing, which is a result of the short cycle times attainable with thermoplastic matrix systems. Typically these technologies refer to injection or extrusion compression molding of molten glass reinforced plastics within cavities that shape the final part.

Figure 10.1 Automotive components are made by using FRT materials

■ 10.1 Basics

The quality and mechanical properties of FRT materials are defined by the basic materials (the matrix system and the type of fiber), but also to a very large extent by the process-induced flow orientations and the fiber lengths in the finished part. The two categories of fiber length are long fiber (about 10–13 mm) and short fiber (about 0.2–0.4 mm) thermoplastics, referred to as LFRTs (long fiber reinforced thermoplastic) and SFRTs (short fiber reinforced thermoplastic), as shown in Figure 10.2(a) and (b), respectively. The LFRTs can yield continuous fiber reinforcement, but the SFRTs cannot.

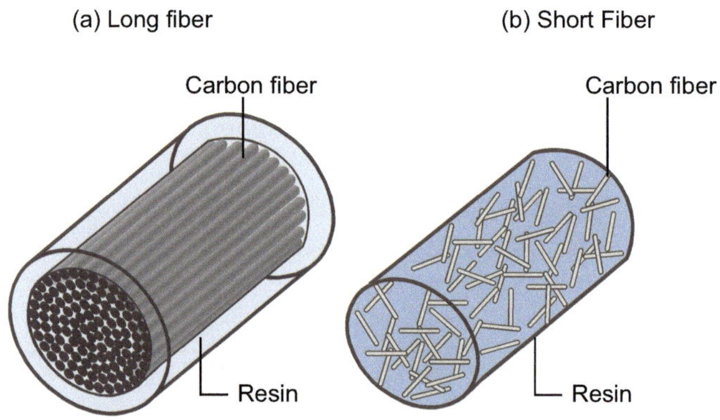

Figure 10.2 Long fiber composites and short fiber composites

The orientation and real fiber length are influenced significantly by the particular processing method and the variation of process parameters. The mechanical properties of real components are broadly scattered as a result of the process-specific plasticizing and flow processes of the melt, the resultant fiber length reduction, and the orientation of the fibers. The mechanical performance potential of long-fiber-reinforced plastics is directly connected to the relative fiber length of the reinforcing fibers used, as shown in Figure 10.3. Therefore, some factors influence mechanical properties; these include fiber length, fiber orientation, fiber content, fiber morphology, fiber chemistry, and the molding process, as shown in Figure 10.4.

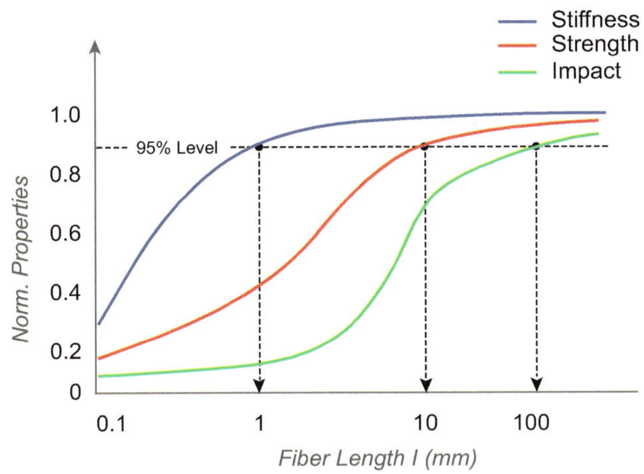

Figure 10.3 Composite properties are shown to increase with fiber length

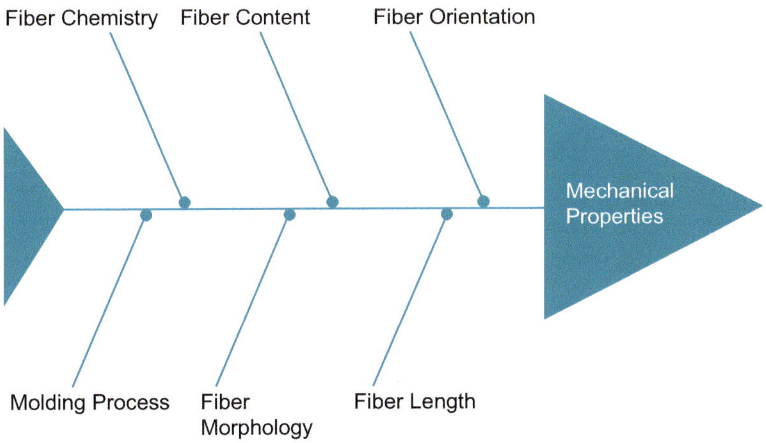

Figure 10.4 Some factors influence mechanical properties

Figure 10.5 shows the flow-direction (x-axis) orientation tensor component A11, the cross-flow-direction (y-axis) orientation component A22, and the thickness-direction (z-axis) orientation component A33 in the central part of the plate. The skin-shell-core-shell-skin structure clearly exists in the injection-molded part of the fiber composites. Fibers in the shell layers are more aligned with the flow direction, which is characterized by large values of A11 and low values of A22, whereas the core layer is quite wide with low values of A11 and high values of A22, representing a rather cross-flow orientation state. Very low values of A33 indicate a nearly planar orientation state.

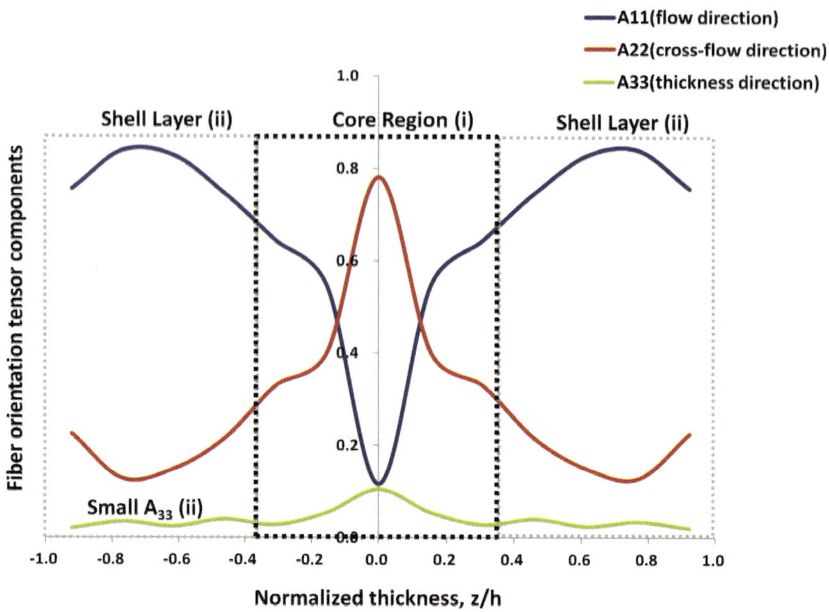

Figure 10.5 Line plot through one slice of the 3D data set, showing the orientation tensor components: flow orientation component A11, cross-flow orientation component A22, and thickness orientation component A33

10.1.1 Process Principle

Production of both SFRT and LFRT materials has been achieved via the injection process, in which the additional fiber composites filled in polymer melts are transported as a suspension into a mold cavity. The effect of divergent and convergent channels of the mold cavity on the alignment of the fibers is clearly illustrated in Figure 10.6.

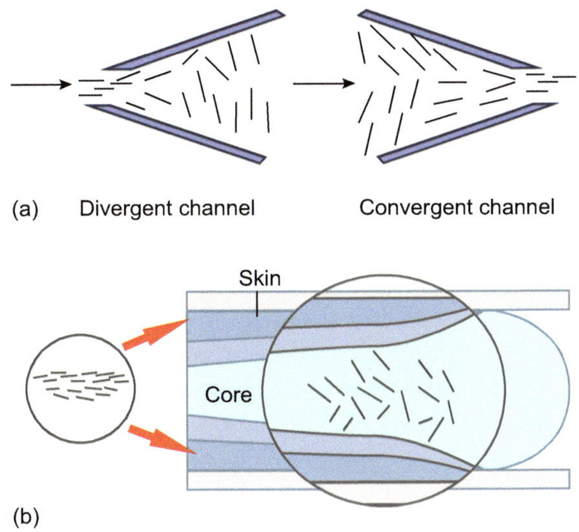

Figure 10.6 Illustration of (a) flow induced orientation of fibers at divergent/convergent channels, and (b) skin-core orientation pattern

10.1.2 Theory Models

A real suspension that is composed of large numbers of fibers is considered collectively with a view of the microstructure. The microscopic detail consists of two components of orientation characterization and modeling that is related to the position and the orientation distribution of fibers. As always, the single rigid fiber is regarded as an axisymmetric rod. The unit vector, **p**, along with the axis direction of the rod, is used to describe fiber orientation. As shown in Figure 10.7, vector **p** can be expressed in surface spherical coordinates, namely, the polar angle, θ, and the azimuthal angle, ϕ:

$$\mathbf{p}(\theta,\phi) = (\sin\theta\cos\phi)\mathbf{e}_x + (\sin\theta\sin\phi)\mathbf{e}_y + (\cos\theta)\mathbf{e}_z \tag{10.1}$$

where $\mathbf{e}_x$, $\mathbf{e}_y$, and $\mathbf{e}_z$ are unit vectors of the Cartesian coordinate axes.

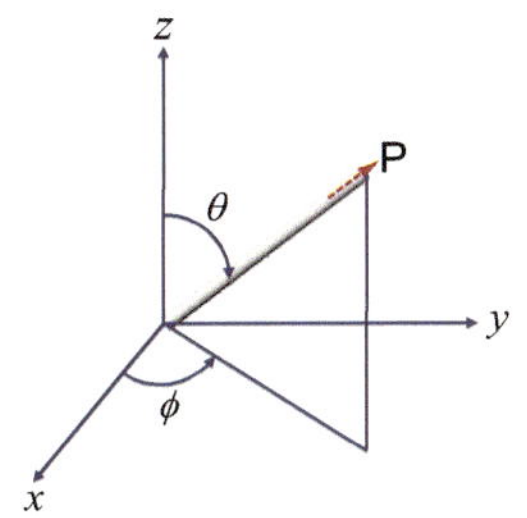

Figure 10.7 Spherical coordinates of the unit vector p(θ,ϕ) defined along with the polar θ and the azimuthal angles ϕ

One approach is to regard the *many* fibers as a sample drawn from an infinite population. The population can be characterized by a probability density distribution function over an orientation space in the surface spherical coordinates, $\psi(\mathbf{p}) = \psi(\theta,\phi)$. This distribution function provides a general statistical description of the orientation state. The probability of finding a certain fiber oriented in a range $\mathbf{p}$ to $\mathbf{p}+d\mathbf{p}$ is given by the product $\psi(\mathbf{p})d\mathbf{p}$, where $d\mathbf{p}$ describes an infinitesimal piece. The integral of this function over the surface spherical coordinates must be equal to a *unit*.

For the sake of completeness, a brief survey of fiber suspension in the rheological area is introduced here, including two representative theoretical models developed over the past three decades: (i) Jeffery's Hydrodynamic Model [1] and (ii) Folgar and Tucker's Isotropic Rotary Diffusion Model [2]. Recently, Phelps and Tucker [3] proposed the Anisotropic Rotary Diffusion Model for long fiber suspension. This is very interesting for the engineering community at present. More importantly, Wang et al. [4] developed the Reduced Strain Closure Model to resolve the quicker response problem of orientation in prediction. More recently, following the aforementioned fiber orientation models, Tseng, Chang, and Hsu [5, 6] developed a new model known as iARD-RPR (improved ARD model and Retarding Principal Rate model).

1. Jeffery's Hydrodynamic Model

The classical analysis by the pioneer Jeffery [1] first concerns itself with the hydrodynamic motion of rigid ellipsoidal particles immersed in a dilute viscous Newtonian fluid, especially when the particle Reynolds number is negligible or the fluid inertia is ignored. Figure 10.8 shows that a swarm of fibers move through a solvent by hydrodynamic motion from their initial random state to a steady alignment state.

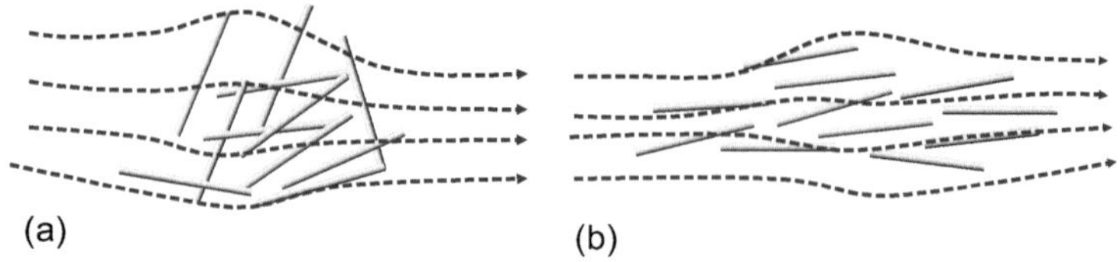

Figure 10.8 Sketch of a swarm of fibers moving through a solvent by hydrodynamics motion: (a) random fibers in an initial state and (b) aligned fibers in steady state

The well-known Jeffery Equation describes a trajectory or an orbit of particles under flows to be written as an expression for the time derivative of the orientation vector, $\mathbf{p}$:

$$\dot{\mathbf{p}} = \mathbf{W}\cdot\mathbf{p} + \xi(\mathbf{D}\cdot\mathbf{p} - \mathbf{D}:\mathbf{ppp}) \tag{10.2}$$

$$\xi = \frac{a_r^2 - 1}{a_r^2 + 1} \tag{10.3}$$

where $\dot{\mathbf{p}}$ represents the material derivative of $\mathbf{p}$ with respect to time. The last term $\mathbf{D}:\mathbf{ppp}$ must be included to stratify the constraint $\mathbf{p}\cdot\mathbf{p} = 1$ in which the fiber's unit length is fixed. $\mathbf{L} = \nabla\mathbf{v} = \mathbf{W} + \mathbf{D}$ is the velocity gradient tensor with its component of $L_{ij} = \nabla_j v_i$, wherein v_i is the component of the velocity in the x_i direction; $\mathbf{W} = \frac{1}{2}\left(\mathbf{L} - \mathbf{L}^{\mathrm{T}}\right)$ is the vorticity tensor; $\mathbf{D} = \frac{1}{2}\left(\mathbf{L} + \mathbf{L}^{\mathrm{T}}\right)$ is the rate-of-deformation tensor. ξ is a shape factor of a particle. The aspect ratio of a shape a_r is the ratio of its longer dimension to its shorter dimension. The fiber's a_r is the ratio of its length l to its diameter d, i.e., l/d. Figure 10.9 illustrates that the values of the parameter ξ equal to 1.0, 0.0, and −1.0, which indicate a thin rod, a spherical molecule, and a thin disc of a particle, respectively.

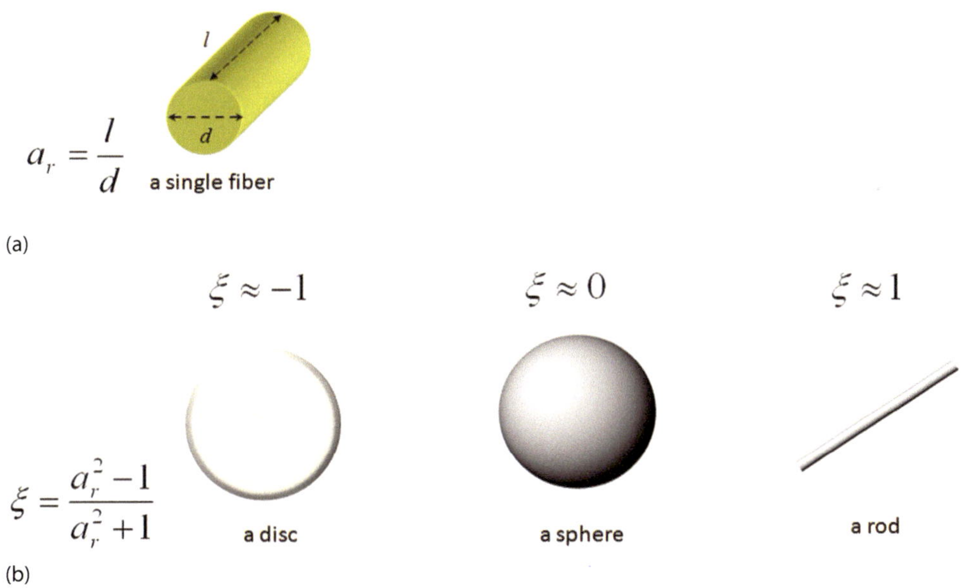

Figure 10.9 Illustration of (a) aspect ratio and (b) shape factor of a particle

The orientation kinetic theory of fiber suspension for Jeffery's hydrodynamics contributor is given by the equation of continuity of the probability density distribution function $\psi(\mathbf{p})$ as a function of the vector $\mathbf{p}$:

$$\frac{d\psi}{dt} = -\nabla_s \cdot (\dot{\mathbf{p}}\psi) \tag{10.4}$$

where $\frac{d}{dt}$ is the material derivative and ∇_s is the surface spherical gradient operator.

2. Isotropic Rotary Diffusion Model

Especially for a non-dilute suspension with a surrounding viscous Newtonian fluid, Folgar and Tucker [2] consider the fiber-fiber interaction induced by rotary Brownian motion. Notably, this interaction is a short-range collision with isothermal fluctuation, but it does not include a long-range interaction of electrostatic forces. In the short fiber suspension, they added the isotropic rotary diffusion term to the Jeffery model. Thus, the equation of continuity with the Jeffery hydrodynamic term and the rotary diffusion term is described as:

$$\frac{d\psi}{dt} = -\nabla_s \cdot \left(\dot{\mathbf{p}}\psi - \dot{\gamma} C_I \nabla_s \psi\right) \tag{10.5}$$

The last term is the so-called Folgar-Tucker Diffusion Model property for isotropic short fibers, wherein $\dot{\gamma}$ is the scalar magnitude of the tensor $\mathbf{D}$, $\dot{\gamma} = (2\mathbf{D} : \mathbf{D})$; C_I is the fiber-fiber interaction coefficient and depends on the volume fraction of fibers and the viscosity and temperature of the suspending fluids. The parameter C_I is usually set between 0.006 and 0.01 [4].

Afterwards, Advani [7] suggested a second moment tensor, **pp**, averaged over probability space to describe the orientation state in surface spherical coordinates:

$$\mathbf{A} = \oint \psi(\mathbf{p})\, \mathbf{pp}\, d\mathbf{p} \tag{10.6}$$

The tensor **A** is symmetric with its trace of unity. $\mathbf{A} = \frac{1}{3}\mathbf{I}$ is in an isotropic random orientation state, where **I** is the identity matrix. Figure 10.10 shows that orientation states correspond to an orientation distribution function and a two-dimensional orientation tensor, examples including isotropic random, partial alignment, and perfect alignment.

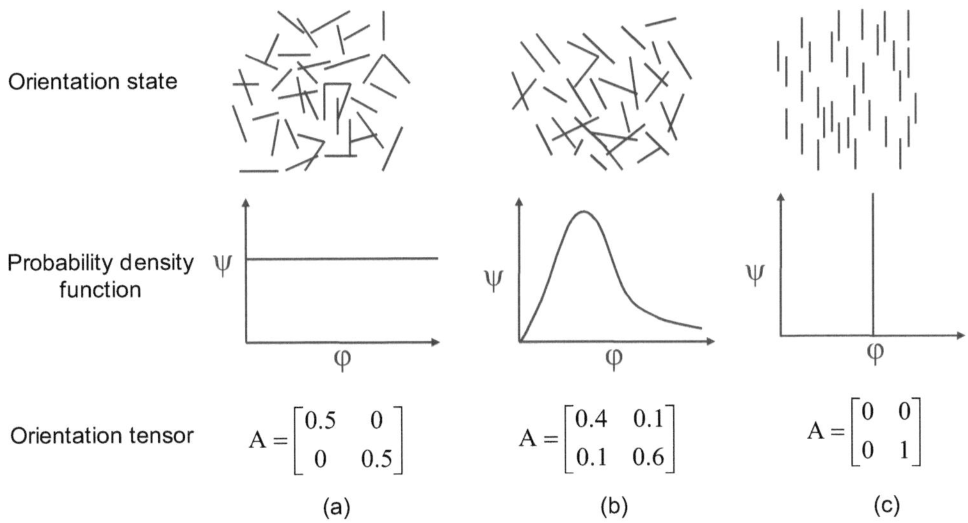

Figure 10.10 Classification of orientation states versus corresponding orientation distributions and two-dimensional orientation tensor: (a) isotropic random, (b) partial alignment, and (c) perfect alignment

Equation 10.5, the standard Folgar-Tucker (FT) orientation equation, therefore, is expressed in the second-order orientation tensor as:

$$\dot{\mathbf{A}} = \dot{\mathbf{A}}_{\mathrm{HD}}^{\mathrm{J}} + \dot{\mathbf{A}}_{\mathrm{DF}}^{\mathrm{FT}} \tag{10.7}$$

$$\dot{\mathbf{A}}_{\mathrm{HD}}^{\mathrm{J}} = (\mathbf{W}\cdot\mathbf{A} - \mathbf{A}\cdot\mathbf{W}) + \xi\left(\mathbf{D}\cdot\mathbf{A} + \mathbf{A}\cdot\mathbf{D} - 2\mathbf{A}_4 : \mathbf{D}\right) \tag{10.8}$$

$$\dot{\mathbf{A}}_{\mathrm{DF}}^{\mathrm{FT}} = 2\dot{\gamma}C_{\mathrm{I}}(\mathbf{I} - 3\mathbf{A}) \tag{10.9}$$

where $\dot{\mathbf{A}}$ represents the material derivative of $\mathbf{A}$ and consists of both Jeffery's hydrodynamic and Folgar-Tucker's diffusive contributors, $\dot{\mathbf{A}}_{\mathrm{HD}}^{\mathrm{J}}$ and $\dot{\mathbf{A}}_{\mathrm{DF}}^{\mathrm{FT}}$, respectively. Notably, $\mathbf{A}_4$ is the *fourth-order orientation tensor*, defined as:

$$\mathbf{A}_4 = \oint \psi(\mathbf{p})\,\mathbf{pppp}\,d\mathbf{p} \tag{10.10}$$

The tensor is also symmetric. The acceptable calculation of such a tensor is obtained through the eigenvalue-based optimal fitting approximation of the orthotropic closure [8] family developed by Tucker and colleagues. In fact, the FT orientation equation has always predicted an initial quicker rate of orientation kinetics at a transient state, as compared to related experiments.

3. Anisotropic Rotary Diffusion Model

The foregoing FT Model, with a fiber-fiber interaction parameter, C_I, describes the isotropic rotary diffusion for short fibers. Differing from the FT Model, Phelps and Tucker [3] further proposed that an *anisotropic rotary diffusion* (ARD) tensor, $\mathbf{D}_r$, in three-dimensional Cartesian coordinates.

Figure 10.11 presents an anisotropic rotary diffusion driven by fiber-fiber collision in a spherical surface coordinate system. The tensor $\mathbf{D}_r$ is properly accounted for in the equation of continuity to obtain:

$$\left[\frac{d\psi}{dt}\right]^{\text{ARD}} = \nabla_s \cdot \left[(\mathbf{I}-\mathbf{pp})\cdot \mathbf{D}_r \cdot (\mathbf{I}-\mathbf{pp})\cdot \nabla_s \psi\right] \tag{10.11}$$

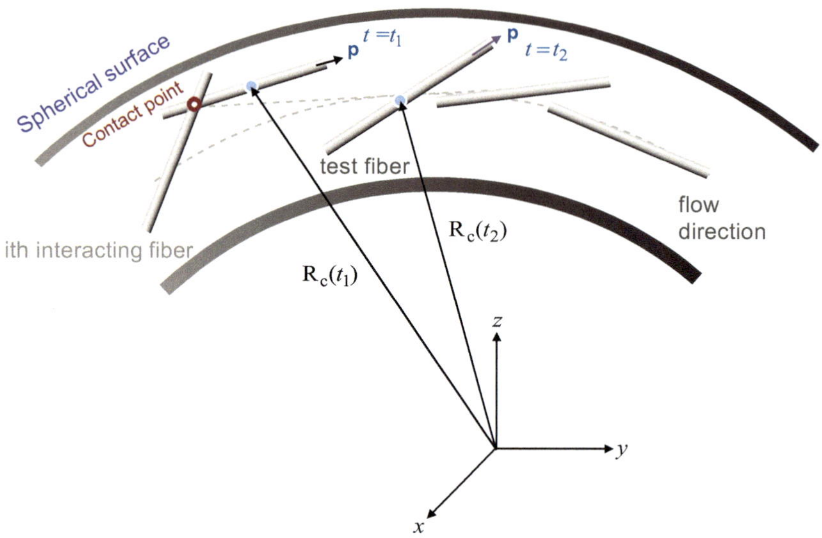

Figure 10.11 Sketch of anisotropic rotary diffusion driven by fiber-fiber collision in spherical surface coordinates

Note that the alternative diffusion tensor, $(\mathbf{I} - \mathbf{pp})\cdot\mathbf{D}_r\cdot(\mathbf{I} - \mathbf{pp})$, indicates a two-dimensional diffusion tensor in the surface spherical coordinate system; $(\mathbf{I} - \mathbf{pp})$ is a projection matrix.

Following Equation 10.11, it can be derived that the ARD orientation equation depends on the tensors $\mathbf{A}$ and $\mathbf{D}_r$:

$$\dot{\mathbf{A}}_{\text{DF}}^{\text{ARD}} = \dot{\gamma}\left[2\mathbf{D}_r - 2\text{tr}(\mathbf{D}_r)\mathbf{A} - 5\mathbf{D}_r\cdot\mathbf{A} - 5\mathbf{A}\cdot\mathbf{D}_r + 10\mathbf{A}_4\mathbf{D}_r\right] \tag{10.12}$$

Phelps and Tucker [3] suggested that the anisotropic diffusion tensor, $\mathbf{D}_r$, approached the Hand tensor [9], which depends on the orientation tensor and the rate-of-deformation tensor:

$$\mathbf{D}_r = b_1\mathbf{I} + b_2\mathbf{A} + b_3\mathbf{A}^2 + \frac{b_4}{\dot{\gamma}}\mathbf{D} + \frac{b_5}{\dot{\gamma}^2}\mathbf{D}^2 \quad (10.13)$$

where the values of b_1, b_2, b_3, b_4, and b_5 are the parameters fit by experimental data. In particular, b_1 corresponds to the fiber-fiber interaction parameter, C_I, of the FT Model. Regrettably, the Jeffery Hydrodynamic Model combined with the ARD Model still results in the initial quicker rate of orientation kinetics in a transient state. This issue is also the same for the standard FT orientation equation.

4. Reduced Strain Closure Model

In fact, with relatively short distance flow and fast filling speed, the transient rate of orientation with time measured in related experiments has always been about 2 to 10 times slower than that predicted by the standard FT model. To treat such a fast response problem of orientation, no matter what isotropic short or anisotropic suspensions, Wang et al. [4] developed a new evolution equation of the second orientation tensor, **A**, called the Reduced Strain Closure (RSC) Model.

Based on eigenvalues, λ_i, and eigenvectors, $\mathbf{e}_i$, of tensor **A**, Tucker et al. applied "the spectral decomposition" of linear algebra:

$$\mathbf{A} = \sum_{3}^{i=1} \lambda_i \mathbf{e}_i \mathbf{e}_i \quad (10.14)$$

Following the standard FT orientation equation, a logical assumption is that the rate of eigenvalues is reduced by a scalar factor κ, while the rate of eigenvectors is unchanged:

$$\dot{\lambda}_i = \kappa \dot{\lambda}_i^{\text{std}} \quad (10.15)$$

$$\dot{\mathbf{e}}_i = \dot{\mathbf{e}}_i^{\text{td}} \quad (10.16)$$

where the factor κ is constrained between 0 and 1.

An improved orientation equation is given by

$$\dot{\mathbf{A}} = \dot{\mathbf{A}}^{\text{std}} - (1-\kappa)\sum_{i=1}^{3} \dot{\lambda}_i^{\text{std}} \mathbf{e}_i \mathbf{e}_i \quad (10.17)$$

The superscript "std" represents the quantity from the *standard* FT orientation equation. Thus, a new expression of the orientation equation, with *one* scalar factor and *two* additional fourth-order tensors, can be obtained:

$$\dot{\mathbf{A}} = \mathbf{W}\cdot\mathbf{A} - \mathbf{A}\cdot\mathbf{W} + \xi\left\{\mathbf{D}\cdot\mathbf{A} + \mathbf{A}\cdot\mathbf{D} - 2\left[\mathbf{A}_4 + (1-\kappa)\left(\boldsymbol{L}_4 - \boldsymbol{M}_4 : \mathbf{A}_4\right)\right] : \mathbf{D}\right\} + 2\kappa\dot{\gamma}C_I(\mathbf{I} - 3\mathbf{A}) \quad (10.18)$$

$$\mathbf{L}_4 = \sum_{i=1}^{3} \lambda_i \mathbf{e}_i \mathbf{e}_i \mathbf{e}_i \mathbf{e}_i \tag{10.19}$$

$$\mathbf{M}_4 = \sum_{i=1}^{3} \mathbf{e}_i \mathbf{e}_i \mathbf{e}_i \mathbf{e}_i \tag{10.20}$$

where the parameter κ is fitted by experimental data; the fourth-order tensors $\mathbf{L}_4$ and $\mathbf{M}_4$ are calculated from the eigenvalues and the eigenvectors. A detailed derivation of the RSC Model is available elsewhere [4]. Note that when $\kappa = 1$ the RSC Equation can recover to the standard FT Equation.

Furthermore, for anisotropic rotary diffusion of long fibers, Phelps and Tucker [3] also *re-derived* their model to obtain the ARD model combined with the RSC Model, namely:

$$\begin{aligned}\dot{\mathbf{A}} &= \mathbf{W}\cdot\mathbf{A} - \mathbf{A}\cdot\mathbf{W} + \xi\left\{\mathbf{D}\cdot\mathbf{A} + \mathbf{A}\cdot\mathbf{D} - 2\left[\mathbf{A}_4 + (1-\kappa)\left(\mathbf{L}_4 - \boldsymbol{M}_4 : \mathbf{A}_4\right)\right] : \mathbf{D}\right\} \\ &+ \dot{\gamma}\left\{2\left[\mathbf{D}_r - (1-\kappa)\boldsymbol{M}_4 : \mathbf{D}_r\right] - 2\kappa\,\mathrm{tr}\left(\mathbf{D}_r\right)\mathbf{A} - 5\left(\mathbf{D}_r\cdot\mathbf{A} + \mathbf{A}\cdot\mathbf{D}_r\right)\right. \\ &\left. + 10\left[\mathbf{A}_4 + (1-\kappa)\left(\mathbf{L}_4 - \boldsymbol{M}_4 : \mathbf{A}_4\right)\right] : \mathbf{D}_r\right\}.\end{aligned} \tag{10.21}$$

This combination equation is called the ARD-RSC Model. Even though the ARD-RSC Equation above seems to be complex with a huge number of tensors compared to the FT Equation, it is suitable for use in the filling process for injection molding simulation.

5. Improved ARD Model and Retarding Principal Rate Model

Following the aforementioned fiber orientation models, Tseng, Chang, and Hsu [1, 2] developed a new model, iARD-RPR (improved ARD model and Retarding Principal Rate model). However, the iARD tensor does not pass the classic rheological rule of Euclidean objectivity, namely, material frame indifference. It is difficult to ignore the non-objective nature due to the fact that different coordinate systems may yield different answers. More recently, Tseng, Chang, and Hsu [5, 6] continued with this effort to propose the objective iARD model. Here, we divide the objective iARD-RPR model into three parts. Firstly, the iARD-RPR equation contains three terms, the Jeffery Hydrodynamics (HD) $\dot{\mathbf{A}}^{\mathrm{HD}}$, the iARD $\dot{\mathbf{A}}^{\mathrm{iARD}}$, and the RPR $\dot{\mathbf{A}}^{\mathrm{RPR}}$, presented as:

$$\dot{\mathbf{A}} = \dot{\mathbf{A}}^{\mathrm{HD}} + \dot{\mathbf{A}}^{\mathrm{iARD}}\left(C_I, C_M\right) + \dot{\mathbf{A}}^{\mathrm{RPR}}(\alpha) \tag{10.22}$$

$$\dot{\mathbf{A}}^{\mathrm{HD}} = (\mathbf{W}\cdot\mathbf{A} - \mathbf{A}\cdot\mathbf{W}) + \xi\left(\mathbf{D}\cdot\mathbf{A} + \mathbf{A}\cdot\mathbf{D} - 2\mathbf{A}_4 : \mathbf{D}\right) \tag{10.23}$$

where $\dot{\mathbf{A}}^{\mathrm{iARD}}$ has two available parameters: the fiber-fiber interaction parameter C_{I} and the fiber-matrix interaction parameter C_{M}; $\dot{\mathbf{A}}^{\mathrm{RPR}}$ has one parameter α, which is to slow down the response rate of fiber orientation; $\mathbf{W}$ is the vorticity tensor and $\mathbf{D}$ is the rate-of-deformation tensor; $\dot{\gamma}$ is the shear rate of $\mathbf{D}$, $\dot{\gamma}=\sqrt{2\mathbf{D}:\mathbf{D}}$; $\xi=\left(a_r^2-1\right)/\left(a_r^2+1\right)$ is the shape factor and a_r is the fiber aspect ratio. However, the fourth-order orientation tensor $\mathbf{A}_4$ is unknown and is not directly obtained. Its calculation, decoupled in terms of the second-order orientation tensor $\mathbf{A}$, is obtained by higher order polynomial closure approximations, such as the orthotropic fitted closure [8] and the invariant-based optimal fitting (IBOF) closure [10].

Secondly, it is significant that the rotary diffusion tensor $\mathbf{D}_r$ depends on the square of the objective rate-of-deformation tensor for defining a new iARD model [5, 6], as shown below:

$$\dot{\mathbf{A}}^{\text{iARD}}=\dot{\gamma}\left[2\mathbf{D}_r-2\text{tr}\left(\mathbf{D}_r\right)\mathbf{A}-5\mathbf{D}_r\cdot\mathbf{A}-5\mathbf{A}\cdot\mathbf{D}_r+10\mathbf{A}_4:\mathbf{D}_r\right] \tag{10.24}$$

$$\mathbf{D}_r=C_I\left(\mathbf{I}-C_M\frac{\mathbf{D}^2}{\|\mathbf{D}^2\|}\right) \tag{10.25}$$

where $\mathbf{D}$ is the symmetric part of the velocity-gradient tensor $\mathbf{L}$, $\mathbf{D}=\frac{1}{2}\left(\mathbf{L}^{\text{T}}+\mathbf{L}\right)$.

The scalar $\|\mathbf{D}^2\|=\sqrt{\frac{1}{2}\mathbf{D}^2:\mathbf{D}^2}$ is the norm of tensor $\mathbf{D}^2$.

Eventually, the RPR model is introduced as:

$$\dot{\mathbf{A}}^{\text{RPR}}=-\mathbf{R}\cdot\dot{\Lambda}^{\text{IOK}}\cdot\mathbf{R}^{\text{T}} \tag{10.26}$$

$$\dot{\Lambda}_{ii}^{\text{IOK}}=\alpha\dot{\lambda}_i,\text{where}\, i,j,k=1,2,3 \tag{10.27}$$

where $\dot{\Lambda}^{\text{IOK}}$ is the material derivative of the diagonal tensor, and its superscript indicates the intrinsic orientation kinetics (IOK) assumption [11]. $\mathbf{R}$ is the rotation matrix; $\mathbf{R}^{\text{T}}$ is the transpose of $\mathbf{R}$, and the superscript T is the transpose operator of a matrix throughout this book. λ_i is eigenvalues of $\mathbf{A}$, $\lambda_1\geq\lambda_2\geq\lambda_3$; $\mathbf{R}=[\mathbf{e}_1, \mathbf{e}_2, \mathbf{e}_3]$ is defined by eigenvector columns of $\mathbf{A}$.

In summary, the iARD-RPR model has just three physical parameters: a fiber-fiber interaction parameter C_{I}, a fiber-matrix interaction parameter C_{M}, and a slowdown parameter α. Foss et al. [12] demonstrated that optimal iARD-RPR parameters are not strongly affected by flow factors such as fill rate and part geometry. The iARD-RPR model has been implemented into a commercial injection molding simulation software, Moldex3D, to allow fiber orientation predictions.

6. The Flow-Fiber Coupling Model

Flow-induced fiber orientation plays an important role in the flow during mold filling, while the flow field is in turn affected by fiber orientation. Previously, Lipscomb et al. [13, 4] developed a constitutive equation for semi-concentrated fiber suspensions using the Dihn-Armstrong flow-fiber coupled approach which can model the effect of fiber orientation on flow stress:

$$\boldsymbol{\tau} = 2\eta_m \mathbf{D} + 2\eta_m \varphi N_p \mathbf{D} : \mathbf{A}_4 \tag{10.28}$$

where $\boldsymbol{\tau}$ is the extra stress tensor; **D** is the rate-of-deformation tensor; η_m is the matrix viscosity; φ is the fiber volume fraction; N_p is a dimensionless parameter. The fourth-order orientation tensor $\mathbf{A}_4$ is determined by using a higher-order polynomial closure approximation [14, 10] in terms of the second-order orientation tensor **A**. Note that $N_p = 0$ indicates a decoupling calculation in the absence of the fiber orientation effect. A decoupled approach has been applied in which the fluid-flow problem is solved first and the resulting velocity field is used to compute the fiber orientation.

Many researchers [15, 16, 17] have done great work on the Dihn-Armstrong coupling model for coupling flow stress and the fiber orientation calculation with the aid of the Hele-Shaw approximation. Previously, VerWeyst and Tucker [16] demonstrated in injection molding simulations of a center-gated disk that the fiber orientation tensor components are independent of the parameter N_p, while the velocity profiles are almost the same. In an end-gated plate made by injection molding, Jørgensen *et al.* [18] recently indicated that the Dihn-Armstrong model cannot describe the result of a broader core induced by plug flow in the case of highly concentrated fiber suspensions. More recently, Tseng and Su [19] achieved state-of-the-art 3D numerical computation technology, not only simulating injection molding flow of a center-gated disk for the long glass fiber-filled polypropylene composite, but also proving that utilizing large N_p values may easily yield divergent numerical results. Solving the constitutive equation of the flow-fiber coupling is truly challenging.

More recently, Favaloro, Tseng, and Pipes [20] developed a new fiber-suspension constitutive equation for the Informed-Isotropic (IISO) viscosity model for concentrated fiber suspensions in non-linear viscous fluids to couple flow field and fiber orientation:

$$\boldsymbol{\tau} = 2\eta_S(\dot{\gamma})\mathbf{D} + 2\eta_S(\dot{\gamma})R_T(\dot{\gamma})\frac{\mathbf{D} : \mathbf{A}_4 : \mathbf{D}}{\mathbf{D} : \mathbf{D}}\mathbf{D} \tag{10.29}$$

where $\eta_S(\dot{\gamma})$ is the shear viscosity of the Cross viscous fluid with fibers; $R_T(\dot{\gamma})$ is the famous Trouton ratio.

Based on the Advani-Tucker fourth-order transversely isotropic viscosity tensor $\boldsymbol{\eta}_4$ [21, 22] depending on orientation tensors, they derived an informed isotropic viscosity scalar [20, 23].

10.1.3 Advantages and Challenges

Using the three-dimensional (3D) mesh ASMI version with RSC, Kleindel et al. [24] performed short glass fiber orientation predictions in an injection molded part with complex geometry. Their conclusions indicated that the orientation in the shell layer was predicted fairly well, while significant deviation was found in the core region. Nguyen et al. [25] demonstrated that the 3D inlet condition can improve the overall prediction of the core region. Furthermore, Wang et al. [4] examined the midplane-mesh ASMI version with ARD-RSC, and set an inlet condition around the gate area to obtain good orientation predictions of long carbon fiber composites. Thus, it is noteworthy that the inlet condition strongly influences the prediction of the RSC Model.

The ARD-RSC Model possesses six parameters wherein the RSC parameter slows down an orientation response; the other parameters are coefficients of the ARD tensor, which is a second-order polynomial tensor-valued function depending on the orientation tensor and the rate-of-deformation tensor. However, excess parameters may cause numerical instability or non-physical behaviors in the computing process if the parameters are not carefully adjusted. Hence, a default parameter set is always used. It appears as though having no criteria is inconvenient for governing these parameters with respect to fiber orientation distribution. To obtain a material-specific parameter set, van Haag et al. [26] performed an automated optimization procedure that provides substantial help for researchers using the ARD-RSC.

Following the state-of-the-art fiber orientation models mentioned above, Tseng, Chang, and Hsu [5, 6] further developed a new fiber orientation model, namely, iARD-RPR (improved ARD model and Retarding Principal Rate model) with just three physical meaning parameters. The iARD-RPR model is able to describe anisotropic fiber orientation. Thus, the iARD-RPR model has been incorporated into a commercial software for injection molding simulation, Moldex3D. Regarding performing the true 3D simulation of injection molding, Moldex3D's Solid Hybrid Model is based on the 3D Finite Volume Method (3D-FVM) technology [27], which can simulate the transient flow field in a complex 3D geometry, thanks to its robustness and efficiency.

In the Manufacturing Systems Research Lab of General Motors (GM) Research and Development, Foss et al. [12] utilized Moldex3D to validate the accuracy of short glass-fiber orientation predictions with two injection molding simulation examples: the end-gated plaque and the center-gated disk. Their predicted orientation distribution exhibited the reliable, classical laminated structure: skin-shell-core-shell-skin. Unlike the midplane computation, the skin-layer orientation prediction was directly calculated using a 3D computation method rather than requiring assumptions to be imposed. Because of the higher numerical resolution of the velocity-gradient tensor, applying the inlet condition was not necessary to enhance the accuracy of orientation prediction. In addition, they demonstrated that the optimal values of the iARD-RPR parameters were not strongly affected by flow factors, such as fill rate and part geometry.

As of today, many researchers still use the midplane mesh model to predict fiber orientation. However, generating a midplane mesh for a complex 3D part is a very time-consuming task, while the midplane calculation based on the Hele-Shaw approximation lacks numerical resolution. For a real complex geometric part, it is indeed a challenge to determine fiber orientation distributions in the requisite designs, including various rib sizes, changes in the flowing direction and changes in thickness. Hence, using the true 3D-mesh computation is a preferable method for analyzing complex 3D parts and the simulation results are generally superior to those of the midplane.

■ 10.2 Practical Applications

10.2.1 Using the iARD-RPR Model for an Injection Molded Center-Gated Disk with Fiber-Reinforced Thermoplastics

Injection molded fiber-reinforced thermoplastics exhibit a distinct shell-core structure. The shell part means that those fibers near the cavity wall are strongly aligned in the flow direction. The core is observed near the cavity center with fiber orientation transverse to the flow. Appling the iARD-RPR model, it is possible to understand how the fiber orientation varies during the mold filling.

Referring to the previous disclosed research report by the Pacific Northwest National Laboratory (PNNL) and the Oak Ridge National Laboratory (ORNL) [28] in the US, as well as a previous study [5, 6], a center-gated disk geometry is shown in Figure 10.12. The disk is about 90 mm in radius with a gap thickness of 2 h = 3 mm. One measurement point is located at 34 mm in the radial direction. The mold and melting temperatures were set at 70 °C and 238 °C, respectively. The filling time and volume flow rate were 0.59 s and 130 cc/s, respectively. The packing time of 10 s and packing pressure of 50 MPa were given, and all of the processing conditions are addressed in Table 10.1.

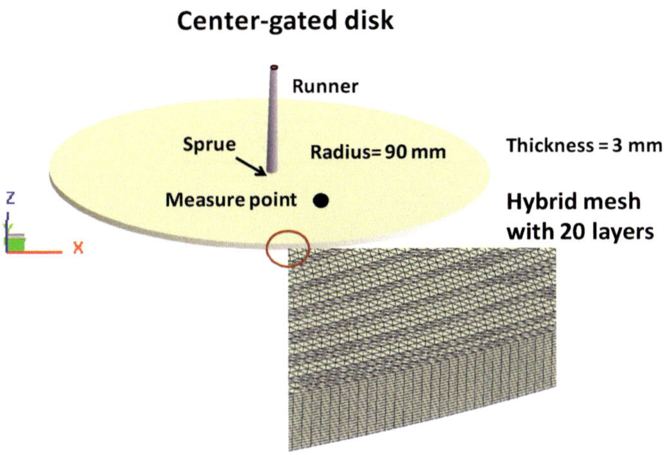

Figure 10.12 The injection-molding geometry of the center-gated disk (90-mm radius and 3-mm thickness) and its hybrid mesh (20 layers) for 3D numerical computation

Table 10.1 Processing Conditions for the 40 wt.% LGF/PP Composite During the Injection Mold Filling

Fill time [s]	0.59
Fill rate [cc/s]	130
Mold temperature [°C]	70
Melt temperature [°C]	238
Pack time [s]	10
Pack pressure [MPa]	50

The researched material is the 40 wt.% long glass fiber immersed in polypropylene matrix (40 wt.% LGF/PP) with a density of 1.22 g/cm^3. The initial fiber diameter and length are 17.4 µm and 13 mm, respectively. The post-extruded average fiber length is about 2 mm.

The 3D model was first developed by one of the CAD (Computer Aided Design) programs and then meshed by creating hexagonal (HEXA) elements with 20 layers, as shown in Figure 10.13. In this study, the Moldex3D simulation program was used in the disk during the injection molding process. Figure 10.13 shows the sprue pressure history from the filling stage (0.59 s) to the packing stage (10 s).

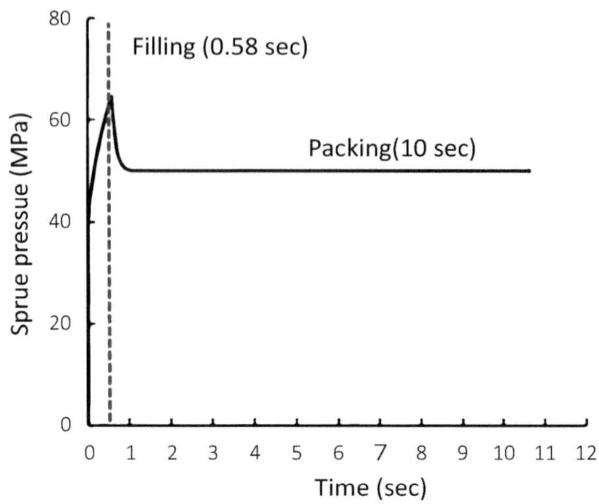

Figure 10.13 Sprue pressure history from filling stage to packing stage

For fiber orientation prediction, the iARD-RPR model parameters are listed in Table 10.2. Additionally, the initial condition of the fiber orientation tensor at the runner entrance is set as the isotropic state. Due of the higher 3D numerical resolution and the runner system consideration in the computation, it was unnecessary to set an experimental inlet condition around a gate for enhancing the accuracy in determining the fiber orientation.

Table 10.2 Optimal Parameters of the iARD-RPR Fiber Orientation Model for the 40 wt.% LGF/PP Composite

CI	0.01
CM	1.0
α	0.1

Both fiber orientation components, A11 and A22, correspond to the flow and the cross-flow directions, respectively. As a result, Figure 10.14 and Figure 10.15 show the predicted gap-wise distribution of fiber orientation tensor components, A11 and A22, at the measurement point located at 34 mm in the radial direction. The solid and dash lines are at the end of the filling and packing stages, respectively. For the experimental data on fiber orientation, reference can be made to the PNNL-ORNL report [3]; these predicted distributions exhibit a distinct core-shell structure. The fibers found in the shell region near the cavity are strongly aligned with the flow direction, as opposed to the others found in the core region near the cavity center, which are transverse to the flow. In particular, the core width of the orientation distribution for the filling is obviously broader than that for the packing. Significantly, this prediction is in qualitative agreement with the earlier experimental results on long fibers in the study by Bailey and Rzepka [29], who concluded that the core thickness is reduced during packing. As a whole, the iARD-RPR predictions for the core region and shell layers agree fairly well with the PNNL-ORNL results of the fiber orientation data.

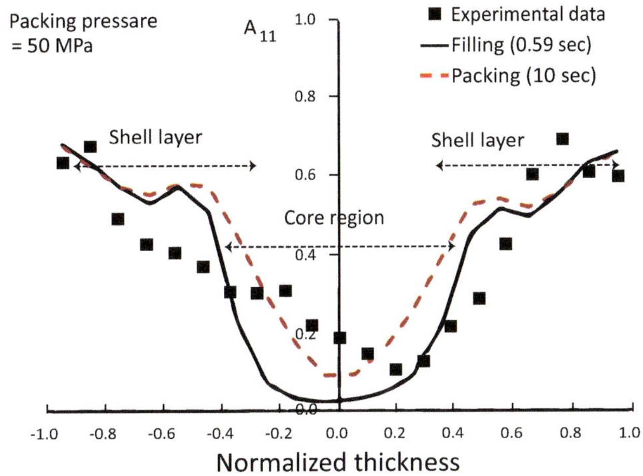

Figure 10.14 Comparison of the predicted orientation distributions along the flow directions (A11) through normalized thickness at the measure point for the filling and packing stages with experimental validation for the 40 wt.% LGF/PP composite

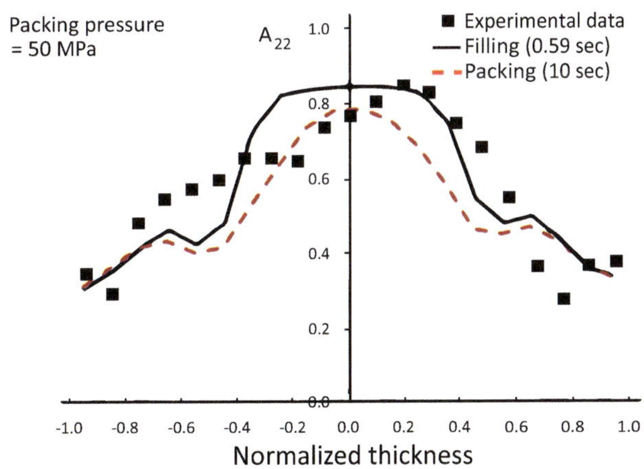

Figure 10.15 Comparison of the predicted orientation distributions along the cross-flow directions (A22) through normalized thickness for the filling and packing stages with experimental validation for the 40 wt.% LGF/PP composite

After a cavity has been completely filled, the density and viscosity of a compressible polymeric melt affect the state of fiber orientation during the packing stage. Bailey and Rzepka [29] proposed the mechanism of the large viscosity gradient from the lower viscosity of the core region to the higher viscosity of the shell layer during the packing stage.

10.2.2 Comparison of iARD-RPR Models under GNF Decoupling and IISO Coupling

Based on the previous studies of Wonisch and Wüst [30] and Tseng et al. [31], commercial BASF composites were used for this comparison: 35 wt.% short glass fiber-reinforced polyamide 66 (35 wt.% SGF/PA66), sold under the name Ultramid A3WG7. The geometry of the injection molding parts was designed as a 150 × 150 × 3 mm^3 plate. For the BASF plate, one central location of the gap-wise fiber orientation distribution across the thickness was measured by means of micro-computed tomography (micro-CT) scanning [32]. In reference to the work of Wonisch and Wüst [30], the micro-CT images were evaluated in order that a digital continuous curve of the fiber orientation tensor components as a function of the thickness could be obtained.

For the comparison, we used Moldex3D commercial injection molding simulation software to perform the 3D injection molding simulation and fiber orientation predictions. The initial orientation condition at the runner entrance was set as an isotropic state. It was unnecessary to provide an inlet condition for the experimental fiber orientation distribution set around a gate, due to the higher 3D numerical resolution and solid runner system's considerations in the computation. The primary aim here was to compare the iARD-RPR models in respect of GNF decoupling and IISO coupling.

The molten plastic injected from the injector nozzle traverses a runner and a gate, and then fills up the cavity. As shown in Figure 10.16, the iso-contour of melt time is displayed in the flow front pattern. The initial melt front is obviously convex parabolic. After approaching a roughly 70%-filled cavity, the flat front is maintained under GNF (generalized Newtonian fluids) decoupling, while the IISO coupling has a "concave parabolic" free surface. It appears that the IISO coupling is able to simulate some peculiar, irregular filling patterns for fiber-reinforced melts at high fiber concentrations: the free surface advances faster along the cavity side walls.

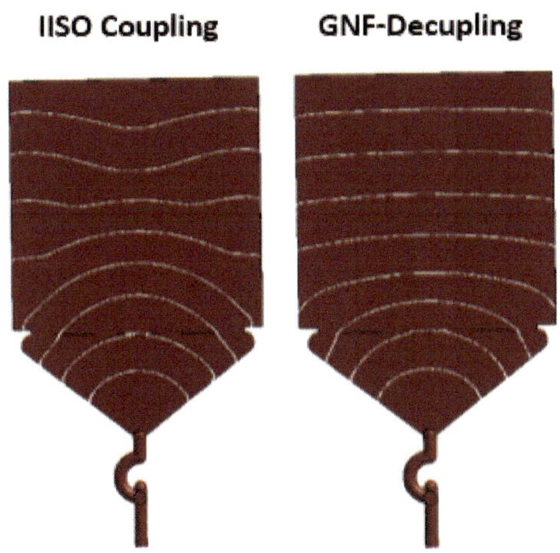

Figure 10.16 Simulated melt front of fiber composite in an iARD-RPR model for comparing GNF decoupling and IISO coupling

Figure 10.17 shows the predicted flow-direction orientation component A11 distribution. The Folgar-Tucker model with GNF decoupling yielded satisfactory results in the shell layer, but significant deviation still occurred in the core region, shell-core interface and skin boundary. Besides, IISO coupling can yield a broader core region, which is in good agreement with the experimental data. However, the standard Folgar-Tucker equation cannot completely dominate the shell-core interface and skin boundary.

The significant results obtained with the iARD-RPR fiber orientation model are shown in Figure 10.18. Clearly, the skin boundary, shell height and shell-core interface are completely predicted by the iARD-RPR model. However, the core width is not good enough for GNF decoupling. In addition, Figure 10.19 shows the average velocity profile through the thickness at the center of the plate. The velocity profile for the IISO coupling is correspondingly close to being plug flow at the midplane, while the narrow bell-shaped flow profile is common for GNF decoupling. In summary, the iARD-RPR model attached to the IISO coupling is able not only to predict the accurate fiber orientation of the skin-shell core structure, but also to simulate the concave parabolic melt front and the plug-flow in reality.

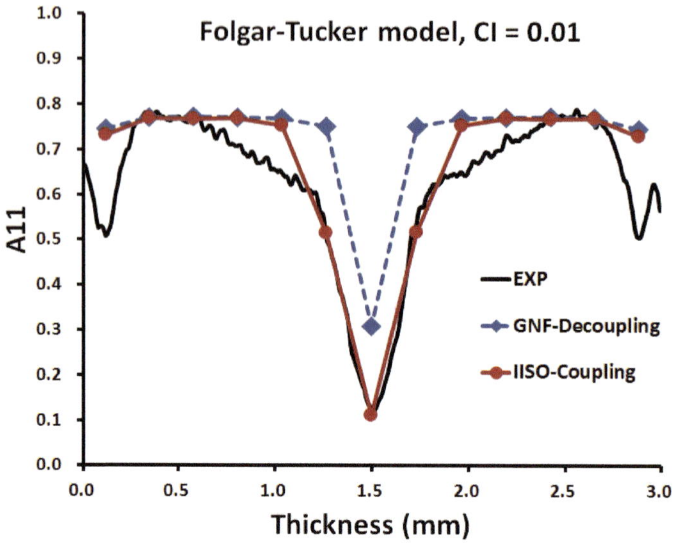

Figure 10.17 Predicted flow orientation component A11 distribution for the Folgar-Tucker model with GNF decoupling and IISO coupling

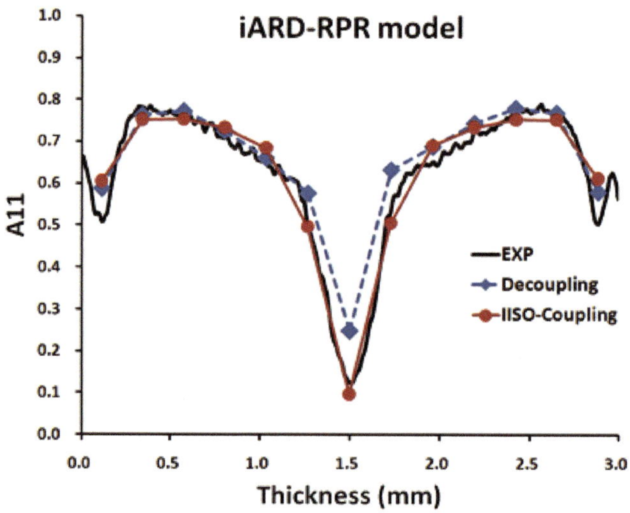

Figure 10.18 Predicted flow orientation component A11 distribution for the iARD-RPR model with GNF decoupling and IISO coupling

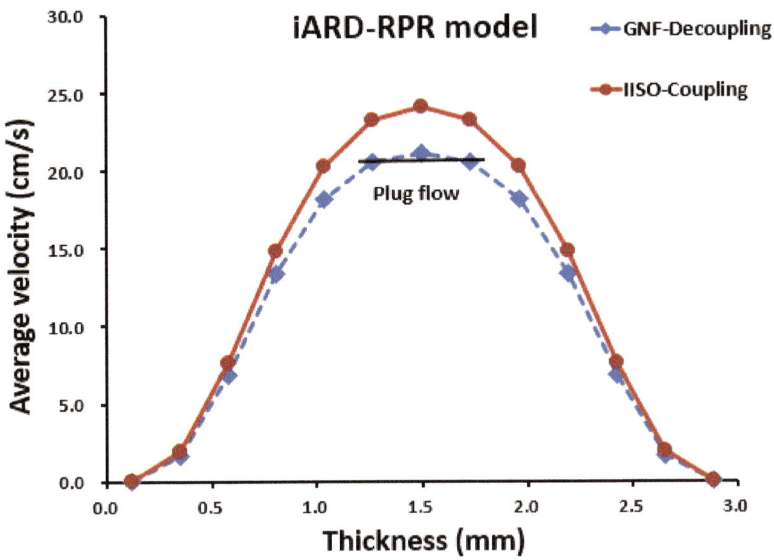

Figure 10.19 Average velocity profile for the iARD-RPR model with GNF decoupling and IISO coupling

Previously, there was an urgent need to improve the numerically stable and convergent results of Dihn-Armstrong flow-fiber coupling in fiber suspension rheology. Here, we have demonstrated the use of the informed isotropic (IISO) constitutive equation. For injection-molded fiber composites, the simulated melt front primarily exhibits a free surface advancing faster along the side walls in reality, as well as a plug flow velocity profile. Furthermore, the predicted fiber orientation distributions with a broader core are in excellent agreement with experimental data.

10.2.3 The Influences of Material Flow and Fiber Interaction on Fiber Orientation and Product Quality during Injection Molding

In order to understand the changes in the fiber microstructure (including fiber orientation, fiber length, fiber concentration) of the composite material used in injection molding, and the impact of this microstructure on the injection product, Moldex3D was used on an actual product for in-depth theoretical discussion and micro-computed tomography (micro-CT) and image processing analysis to verify the microstructure characteristics [33]. A tensile test of the mechanical properties was also carried out and the effect of flow on the microstructure characteristics of the fiber was initially evaluated and quantified so as to permit a study of deformation by warpage and of the mechanical properties of the product. This example is a composite geometric design having three different types of gating for ASTM D638 standard test pieces, as shown in Figure 10.20.

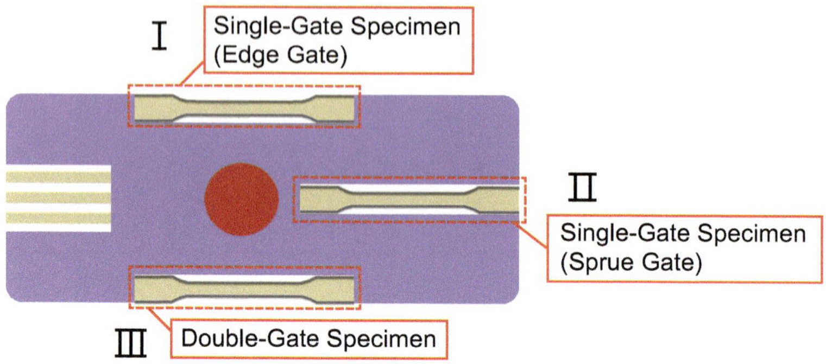

Figure 10.20 Case model with three types of different gating for ASTM D638 standard test pieces [33]

Moldex3D was used to observe the warpage from several different perspectives, as shown in Figure 10.21, and then physical experiments were performed. Table 10.3 compares the simulation data with the experimental data.

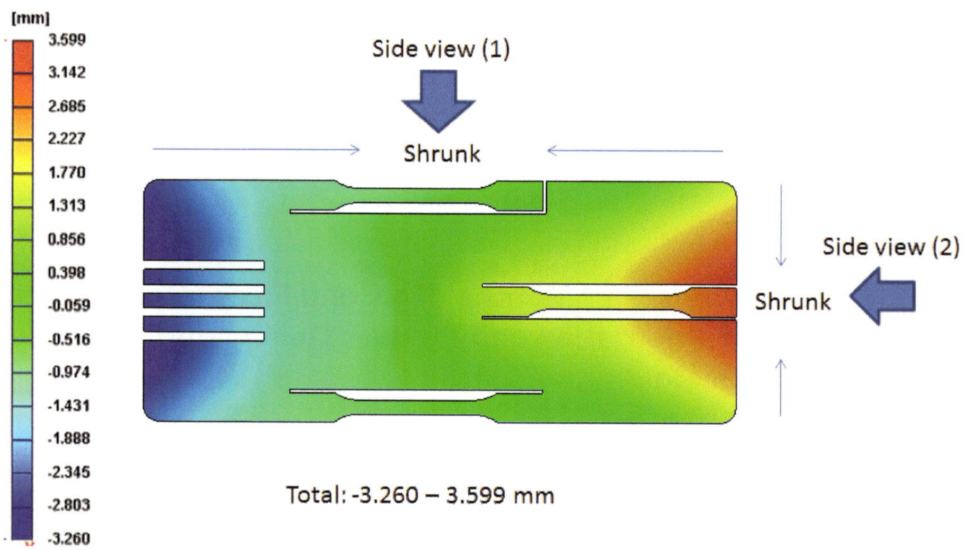

Figure 10.21 Observation of warpage from different perspectives

Table 10.3 Comparison of Warpage in Simulation and Experiment

Model	Trend	Simulation	Experiment	Error
I	Warp downward	2.25	2.44	7.7%
	Bent inward	2.05	2.19	6%
II	Warp upward	1.76	1.93	9%
	Gate shrink	28.25	28.58	3%
III	Warp upward	0.85	1.01	16%
	Bent inward	8.6	9.51	1%

The results show that the simulation is in good agreement with the experiment. To understand the influence of different fiber lengths on the mechanical properties, the tensile strength of three kinds of test pieces was measured. The results are summarized and shown in Figure 10.22.

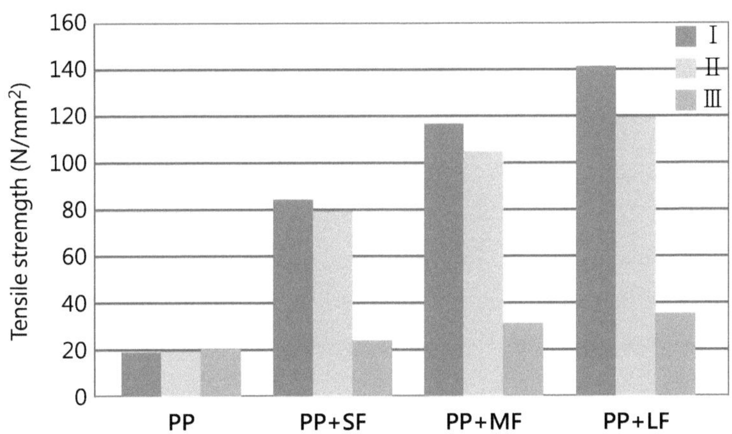

Figure 10.22 Tensile strength results for three test pieces

The results show that the tensile strength of the same kind of test piece will increase with increase in fiber length. In Model I, for example, the strength of pure PP is about 20 N/mm^2, while the strength of PP with long fiber is about 140 N/mm^2. A similar trend can be found for Model II in that the strength has increased from 20 N/mm^2 (pure PP material) to 120 N/mm^2 (long fiber material). However, in Model III, weld lines occurred due to the double-sided gating, so the increase in strength is relatively insignificant, only increasing from 20 N/mm^2 to about 40 N/mm^2. In addition, further comparison of the strength difference between Model I and Model II shows that, when fiber-containing composite materials are used, the strength of Model I is always greater than that of Model II. The strength of medium-fiber materials is 118 N/mm^2 for Model I and 105 N/mm^2 for Model II. Similarly, for long-fiber composite materials, Model I possesses a strength of 140 N/mm^2 and Model II, 120 N/mm^2. The reason for this phenomenon may be the entrance effect caused by the side gating, which creates a relatively strong fiber orientation effect. In summary, the simulation analysis and the prediction of macrophysical properties are in good agreement with the experimental results.

An in-depth study of the microscopic fiber orientation corresponding to the aforementioned macrophysical characteristics is discussed and shown in Figure 10.23, where the trends of the overall fiber orientation are quite consistent between the simulation analysis and experimental results. However, there are still differences in the amount of change in A11 and A22.

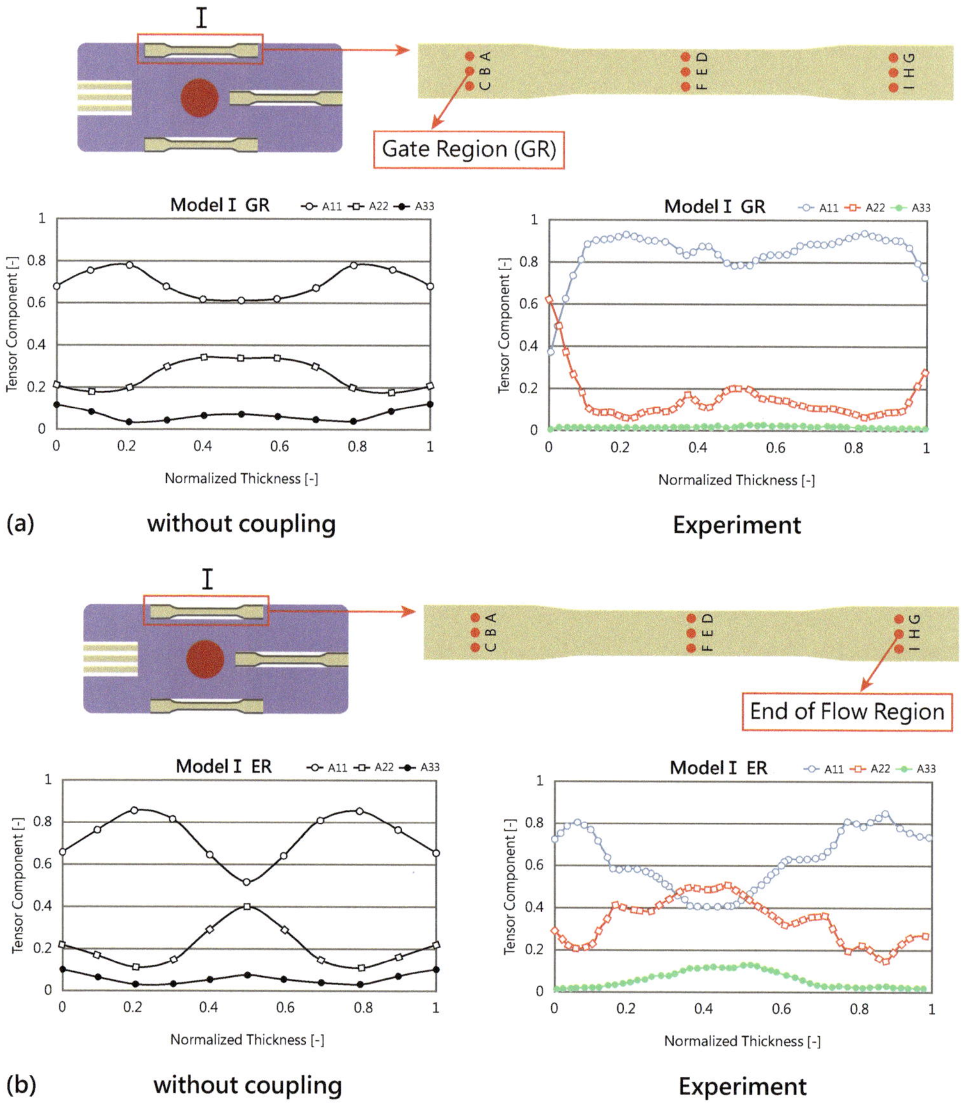

Figure 10.23 Comparison of fiber orientation in the simulation and experiments for Model I: (a) near gate region, NGR, and (b) end of filling region, EFR

Favaloro et al. [20] proposed the IISO viscosity model as a way to estimate the viscosity change under the influence of polymer melt flow and fiber interaction, and then Tseng and Favaloro [34] revised this IISO model to produce the Revised IISO model and added it to Moldex3D. Thus, the coupling effect of flow and fiber can be considered in the above study of the three ASTM D638 standard test piece systems with different flow fields. As shown in Figure 10.24, the A11 and A22 of the simulation analysis are quite close to the image analysis experiments, where the fiber orientation (A22) perpendicular to the flow direction will obviously dominate near the end of the flow in Model I (Figure 10.24(a)) and Model II (Figure 10.24(b)).

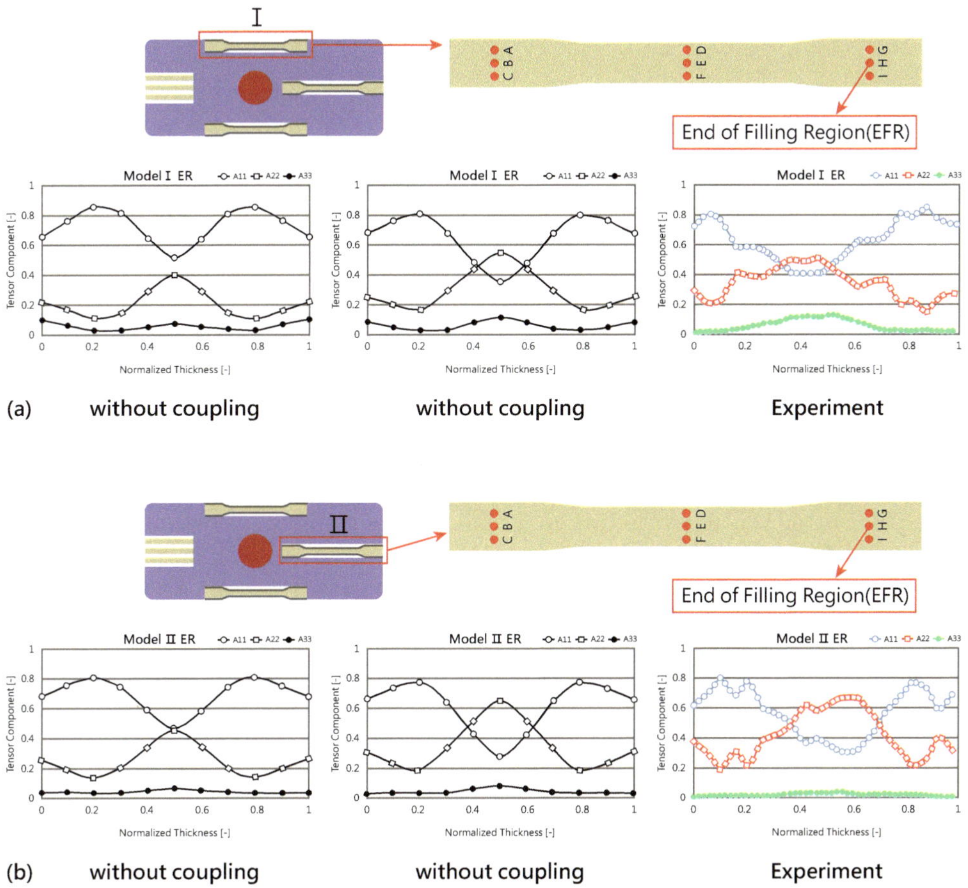

Figure 10.24 EFR fiber orientation results of simulations and experiments, where the simulations are considered with and without flow-fiber-coupling effects: (a) Model I and (b) Model II

■ 10.3 CAE Case Study

Following the discussions of this chapter and the corresponding CAE verification case, an exercise model, as illustrated in Figure 10.25, is provided in this section for the readers to practice CAE application in fiber orientation control.

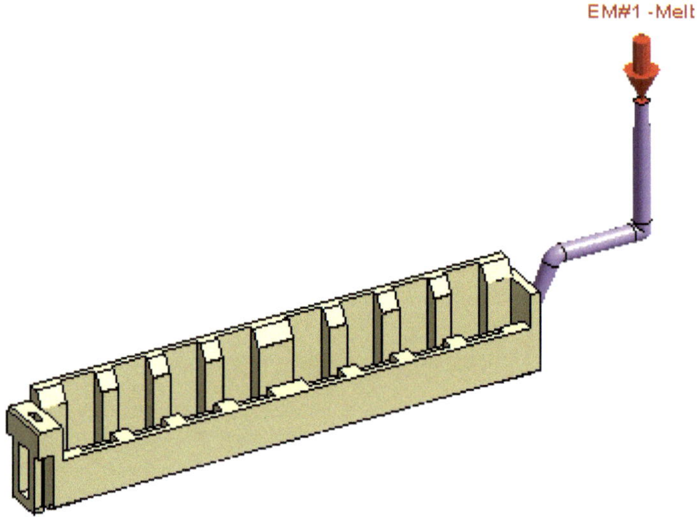

Figure 10.25 Case study sample

Please download the analysis files for CAE Case Study 10.3 from the following website:

https://moldex3d.box.com/s/zr6fvc1vlhbi4ocx11jwd3wmxt4ooif

Then open the .m3j or .mvj file in Moldex3D and answer the questions below according to the analysis results.

Questions:

1. Comparison for different fiber content (see Table 10.4)
   a) Compare the warpage for different fiber contexts and explain the possible reasons.
   b) For fiber-filled thermoplastics, how do you understand the effect of fiber orientation on warpage?
2. Changing the design of products
   a) How does the warpage change for the 14% fiber content if the holes are filled at the bottom as shown in Figure 10.26?
3. Design of mold
   a) Check the effect of different gate locations: origin design vs. center-gated.
   b) Check the effect of different gate sizes: origin design vs. smaller gate size.
4. Condition of processing
   a) Check the effect of changing filling time: 0.04 s vs. 0.2 s.
   b) Check the effect of changing flow rate profile: three-section vs. one-section.

Table 10.4 Fiber Content of Some PA6 Resins

Fiber content	Grade Name
0%	PA6 Ultramid 8202 BASF
14%	PA6 Ultramid 8231 G HS BASF
33%	PA6 Ultramid 8233 G HS BASF
44%	PA6 Ultramid 8234 G HS BASF

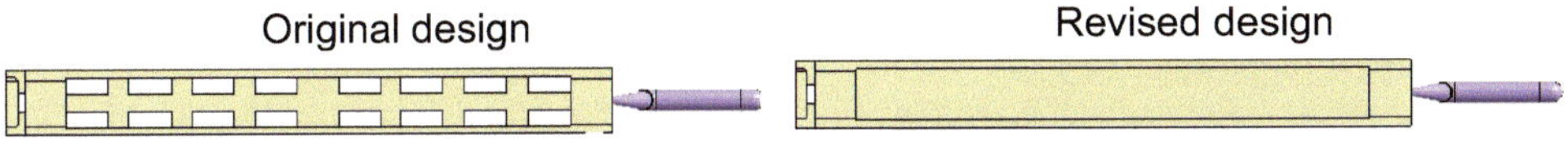

Figure 10.26 Design change in a part: (a) original design and (b) revised design

References

[1] Jeffery, G.: The Motion of Ellipsoidal Particles Immersed in a Viscous Fluid, *Proceedings of the Royal Society of London* A, vol. 102, 1922, p. 161

[2] Folgar, F.; Tucker III, C.: Orientation Behavior of Fibers in Concentrated Suspensions, *Journal of Reinforced Plastics and Composites*, vol. 3, 1984, p. 98

[3] Phelps, J.H.; Tucker III, C.: An Anisotropic Rotary Diffusion Model for Fiber Orientation in Short- and Long-Fiber Thermoplastics, *Journal of Non-Newtonian Fluid Mechanics*, vol. 156, 2009, p. 165

[4] Wang, J.; O'Gara, J.F.; Tucker III, C.: An Objective Model for Slow Orientation Kinetics in Concentrated Fiber Suspensions: Theory and Rheological Evidence, *Journal of Rheology*, vol. 52, 2008, pp. 1179–1200

[5] Tseng, H.-C.; Chang, R.-Y.; Hsu, C.-H.: Phenomenological Improvements to Predictive Models of Fiber Orientation in Concentrated Suspensions, *Journal of Rheology*, vol. 57, 2013, p. 1597

[6] Tseng, H.-C; Chang, R.-Y.; Hsu, C.-H.: An Objective Tensor to Predict Anisotropic Fiber Orientation in Concentrated Suspensions, *Journal of Rheology*, vol. 60, 2016, p. 215

[7] Advani, S.: "Prediction of Fiber Orientation During Processing of Short Fiber Composites", Ph.D. Thesis, University of Illinois at Urbana-Champaign, 1987

[8] Cintra Jr, J.S.; Tucker III, C.: Orthotropic Closure Approximations for Flow-Induced Fiber Orientation, *Journal of Rheology*, vol. 39, 1995, p. 1095

[9] Hand, G.: A Theory of Anisotropic Fluids, *Journal of Fluid Mechanics*, vol. 13, 1962, p. 33

[10] Chung, D.H.; Kwon, T.H.: Invariant-Based Optimal Fitting Closure Approximation for the Numerical Prediction of Flow-Induced Fiber Orientation, *Journal of Rheology*, vol. 46, 2002, pp. 169–194

[11] Tseng, H.-C.; Chang, R.-Y.; Hsu, C.-H.: Method and Computer Readable Media for Determining Orientation of Fibers in a Fluid. Patent US 8,571,828 B2, 2013

[12] Foss, P.; Tseng, H.-C.; Snawerdt, J.; Chang, Y.-J.; Yang, W.-H.; Hsu, C.-H.: Prediction of Fiber Orientation Distribution in Injection Molded Parts Using Moldex3D Simulation, *Polymer Composites*, vol. 35, 2014, p. 671

[13] Lipscomb II, G.G.; Denn, M.M.; Hur, D.U.; Boger, D.: The Flow of Fiber Suspensions in Complex Geometries, *Journal of Non-Newtonian Fluid Mechanics*, vol. 26, 1988, pp. 297–325

[14] VerWeyst, B.E.: "Numerical Predictions of Flow-Induced Fiber Orientation in Three-Dimensional Geometries", Ph.D. Thesis, University of Illinois, 1998

[15] Chung, S.T.; Kwon, T.H.: Coupled Analysis of Injection Molding Filling and Fiber Orientation, Including in-Plane Velocity Gradient Effect, *Polymer Composites*, vol. 17, 1996, pp. 859–872

[16] VerWeyst, B.E.; Tucker III, C.L.: Fiber Suspensions in Complex Geometries: Flow/Orientation Coupling, *The Canadian Journal of Chemical Engineering*, vol. 80, 2008, pp. 1093–1106

[17] Mazahir, S.; Vélez-García, G.; Wapperom, P.; Baird, D.: Evolution of Fibre Orientation in Radial Direction in a Center-Gated Disk: Experiments and Simulation, *Composites Part A: Applied Science and Manufacturing*, vol. 51, 2013, pp. 108–117

[18] Jørgensen, J.K.; Andreassen, E.; Salaberger, D.: The Effect of Fiber Concentration on Fiber Orientation in Injection Molded Film Gated Rectangular Plates, *Polymer Composites*, vol. 40, 2019, pp. 615–629

[19] Tseng, H.-C.; Su, T.-H.: "Coupled Flow and Fiber Orientation Analysis for 3D Injection Molding Simulations of Fiber Composites", in: Polymer Processing Society (PPS-34), 2018

[20] Favaloro, A.J.; Tseng, H.-C.; Pipes, R.B.: A New Anisotropic Viscous Constitutive Model for Composites Molding Simulation, *Composites Part A: Applied Science and Manufacturing*, vol. 115, 2018, pp. 112–122

[21] Advani, S. G; Tucker III, C.L.: Closure Approximations for Three-Dimensional Structure Tensors, *Journal of Rheology*, vol. 34, 1990, pp. 367–386

[22] Advani, S.G.; Tucker III, C.L.: The Use of Tensors to Describe and Predict Fiber Orientation in Short Fiber Composites, *Journal of Rheology*, vol. 31, 1987, pp. 751–784

[23] Favaloro, A.J.: "Rheological Behavior and Manufacturing Simulation of Prepreg Platelet Molding Systems", Ph.D. Thesis, Purdue University, 2017

[24] Kleindel, S.; Salaberger, D.; Eder, R.; Schretter, H.: Prediction and Validation of Short Fiber Orientation in a Complex Injection Molded Part with Chunky Geometry, *International Polymer Processing*, vol. 30, 2015, p. 366

[25] Nguyen, B.N.; Fifield, L.S.; Yocum, A.; Schu, C.: "Predictive Engineering Tools for Injection-Molded Long-Carbon-Fiber Thermoplastic Composites–Pacific Northwest National Laboratory (PNNL)", FY 2014 Annual Progress Report-Lightweight Materials R&D, U.S. Department of Energy, 2014

[26] van Haag, J.; Bontenackels, C.; Hopmann, C.: "Fiber Orientation Prediction of Long Fiber-reinforced Thermoplastics: Optimization Of Model Parameters", in SPE ANTEC Conference, Orlando, 2015

[27] Chang, R.-Y.; Yang, W.-H.: Numerical Simulation of Mold Filling in Injection Molding Using a Three-dimensional Finite Volume Approach, *International Journal for Numerical Methods in Fluids*, vol. 37, 2001, p. 125

[28] Nguyen, B.N.; Jin, X.; Wang, J.; Phelps, J.H.; Tucker III, C.L.; Kunc, V.; Bapanapalli, S.K.; Smith, M.T.: "Implementation of New Process Models for Tailored Polymer Composite Structures into Processing Software Packages", U.S. Department of Energy, Pacific Northwest National Laboratory, 2010

[29] Bailey, R.; Rzepka, B.: Fibre Orientation Mechanisms for Injection Molding of Long Fibre Composites, *International Polymer Processing*, vol. 6, 1991, p. 35

[30] Wonisch, A.; Wüst, A.: More Precise Part Design: Accurate Simulation of Fiber Orientation of Glass Fiber-Reinforced Plastics, *Kunststoffe international*, no. 9, 2014, pp. 80–83

[31] Tseng, H.-C.; Chang, R.-Y.; Hsu, C.-H.: Numerical Predictions of Fiber Orientation for Injection Molded Rectangle Plate and Tensile Bar with Experimental Validations, *International Polymer Processing*, vol. 33, 2018, pp. 96–105

[32] Gandhi, U.; Sebastian, D.B.; Kunc, V.; Song, Y.: Method to Measure Orientation of Discontinuous Fiber Embedded in the Polymer Matrix from Computerized Tomography Scan Data, *Journal of Thermoplastic Composite Materials*, vol. 29, 2016, pp. 1696–1709

[33] Huang, C.-T.; Chu, J.-H.; Fu, W.-W.; Hsu, C.; Hwang, S.-J.: Flow-induced Orientations of Fibers and Their Influences on Warpage and Mechanical Property in Injection Fiber Reinforced Plastic (FRP) Parts, *International Journal of Precision Engineering and Manufacturing-Green Technology*, vol. 8, 2021, p. 917

[34] Tseng, H.-C.; Favaloro, A.J.: The Use of Informed Isotropic Constitutive Equation to Simulate Anisotropic Rheological Behaviors in Fiber Suspensions, *Journal of Rheology*, vol. 63, 2019, p. 263–274

11 Hot Runner Optimization

Tsai-Hsin (Sam) Hsieh

In hot runner system applications, apart from the determination of the basic dimensions of the hot runner and the location of gate, issues that need to be considered include whether the heating is uniform across all regions in the hot runner system, whether the temperature control of the heater is precise, and the impact of heat dissipation from the hot runner to mold base on the part appearance, molding cycle, shrinkage, and warpage.

With the progress of computer technology, using CAE software to assist with part and mold development has become a standard operation procedure in the industry. In this chapter, the simulation techniques of hot runner molds and their industrial applications will be introduced along with the sharing of a few actual successful industrial case studies.

11.1 Basics

Before exploring the hot runner, let us examine what are the problems encountered in molds that use conventional cold runner systems.

First, excessive injection pressure and clamping force resulting from increased pressure loss make it more likely to generate the short shot, which requires a machine of larger capacity for molding and thus increases the production cost. Also, cold runners become waste after the plastic molding and result in a follow-up recycling and crushing issue. Furthermore, as the cold runner has a larger dimension which requires a longer cooling time, the molding cycle time is thereby increased and causes the production capacity to be decreased. Moreover, using a cold runner system may generate the problem of premature cooling of melt before entering the mold cavity, which would lead to poor gloss, weld line, and other exterior issues. In summary, poor part quality might be foreseen, such as with unbalanced shrinkage and deformation due to a poor packing effect, sink marks, residual stress, and other issues.

In the optical and medical product industries, which require high product quality with higher cost in material usage, the hot runner system has advantages to help achieve the goal of quality assurance and manufacturing cost reduction. Regarding optical products, the most common issue is that the stress induced during the injection process remains in the plastic parts. Such residual stress would affect material properties and fails in the original goal of the optical design. For this, the characteristic of low pressure loss in the hot runner system provides a good way to avoid excessive residual stress induced from high injection pressures. As for medical products, using hot runner systems is still necessary due to the strict demand in dimensional precision and product stability. In addition, the plastic materials used for the above-mentioned products are more expensive; producing no waste that saves material cost is another advantage of using hot runner systems. For common consumer goods such as plastic bottle caps, food containers, and other mass-produced products, the use of hot runner systems provides a method to boost production capacity.

11.1.1 Process Principle

The hot runner system is very complicated (Figure 11.1). Apart from the determination of the basic dimensions of hot runner and the location of gate, consideration also needs to be given to issues such as whether the heating is uniform across all regions in the hot runner system, whether the temperature control of the heater is precise, and the impact of heat dissipation from the hot runner to mold base on the part appearance, molding cycle, shrinkage and warpage.

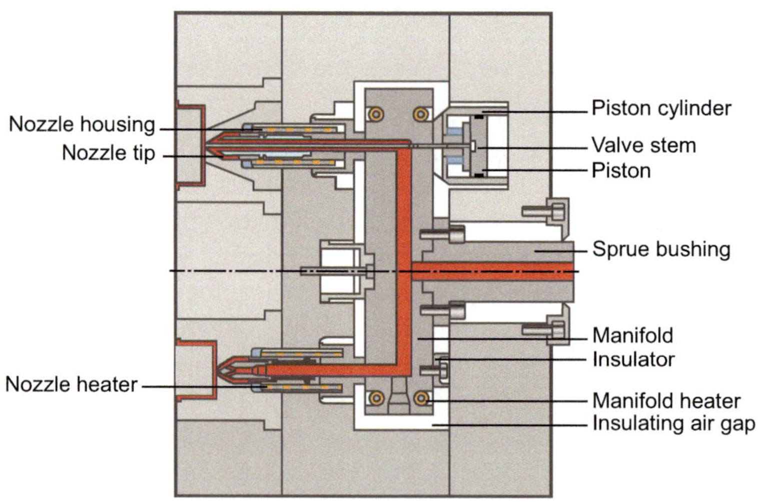

Figure 11.1 Hot runner system components

Based to the design of nozzles, there are two types of gates, according to how they are controlled: the open type (also called as the hot gate) as shown in Figure 11.2(a)-(c), and the valve type as shown in Figure 11.2(d).

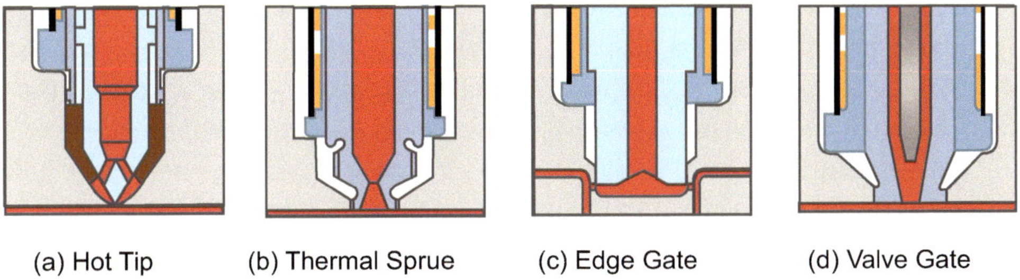

Figure 11.2 Hot runner gate types

1. Hot-tip gate

   The hot-tip gate is applicable for the part design that allows tiny gate residue on the surface or bottom of the plastic part. The length of gate residue is affected by the diameter and land of the gate, the cooling around the gate, and the material used. Hot-tip gate can be used for various plastic materials, and its maximum gate diameter is 3 mm normally. For the injection molding of tiny parts, multi-tip gating can be used, which has up to four heads for a single nozzle and can mold four parts at a time. If a larger gate diameter is required, the valve gate is recommended as the cycle time is bound to be affected.

2. Thermal sprue gate

   Thermal sprue gate is suitable for cold runners which introduce an extended nozzle tip as a part of the molded plastic part. The reversed taper inside the nozzle is able to break the runner waste effectively. The length of sprue waste is determined by the gate diameter and the material used. The thermal sprue gate is applicable for both amorphous and crystalline plastics.

3. Edge gate

   If the trace of the gate is not allowed to appear at the top or on the inside of a plastic part, an edge gate is preferred for injecting plastic from the side of the plastic part. The gate is cut upon the opening of the mold, and the trace of the gate is similar to that of a cold runner. Such a gating method is under many constraints and should be considered thoroughly upon application.

4. Valve gate

   The valve gate, which is a trend of development of hot runner systems, can be divided into the spring valve and the hydraulic/pneumatic control valve. The former utilizes the injection pressure to push the spring piston, and as the pressurized plastic moves to the gate, the spring piston moves backward automatically to allow the plastic to pass smoothly; when the pressure drops, the spring piston then returns to its original position to close the gate. Though such systems requires no external driving force, as time goes by, the spring undergoes fatigue and deterioration, and the gate system eventually fails to work when the elasticity of spring is gone.

Another design is the hydraulic/pneumatic controlled nozzle, whose working principle is to drive the valve up and down via a pneumatic and hydraulic system. During injection, the valve is pulled backward so as to let the plastic flow out of the nozzle freely; after the injection, the valve is pushed forward to close the gate completely to prevent further flowing of the plastic. Such a system is suitable for parts with a strict requirement in appearance; it has a big gate diameter so can reduce the residual stress in the plastic part effectively, and can achieve a more effective filling at a lower injection pressure.

Upstream to the nozzle is the manifold (Figure 11.3), whose main purpose is to distribute the melt. In a system with multiple nozzles, the manifold is able to distribute the melt injected from the machine nozzle to the location where each nozzle lies. Therefore, it is critical to ensure the flow balance in a hot runner system by considering the equivalence of the following factors: runner length, runner size, corner count, corner angle, and melt head position. Thus, the melt distributed to each gate should be of the same properties, such as flowing through the same flow length, experiencing the same times of directional change, and having the same temperature and pressure, so as to achieve identical part properties in each mold in a multi-cavity system.

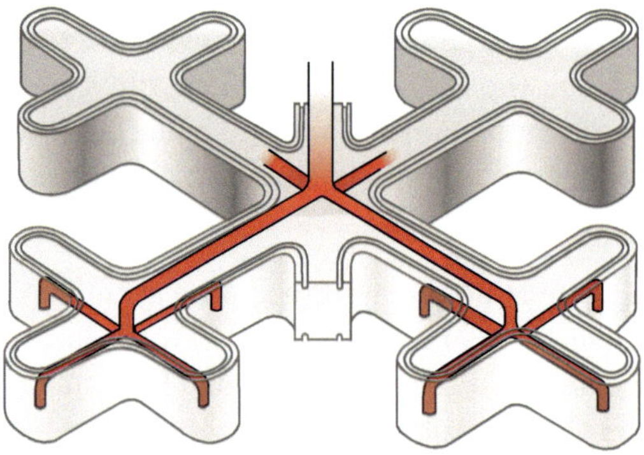

Figure 11.3 Hot runner manifold

Normally the design of a manifold has the following guidelines:

1. Flow balance

   In order to pursue the identical part properties of each mold in a multi-cavity system, the melt distributed to each gate should be of same properties, such as flowing through the same flow length, experiencing the same times of directional change, and having the same temperature and pressure.

2. Optimization of runner size

   The size determination of a hot runner should consider maintaining consistent shear stress in different sections as well as a lower pressure loss. A larger-sized hot runner has characteristics of low shear, long hesitation, and low-pressure loss, while a smaller-sized hot runner has a higher flow rate and shear and is often used in a process requiring frequent color change and short hesitation, so is able to complete a color change in a shorter time and avoid an overlong hesitation. Therefore, the size of hot runner would affect process conditions.

3. Runner surface smoothness

   After machining, dirt must be cleaned away from the runner to ensure the plastic flows smoothly at around corners, while eliminating the possibility of blocking of the plastic. A polishing process is performed at the end.

4. Uniform heating

   The heater itself should have good temperature uniformity and guarantee an even temperature distribution along the runner.

5. Minimum heat mass

   Narrow down and shorten the design of manifold to reduce its heat mass and achieve a more precise control on the temperature. Use the air layer for heat insulation to reduce heat loss.

6. Robust structure

   Injection molding is a process performed at high temperature (over 300 °C) and high pressure (over 200 MPa). The manifold needs to be designed to withstand the stress during the injection process under a high process temperature with sufficient structural strength to avoid being destroyed.

The development flow of a general hot runner system design is shown in Figure 11.4, and includes the following steps:

1. Customer provides the design drawings of plastic parts, designated material types and models, and part specifications such as a location of the weld line, deformation allowance, etc.
2. Evaluate according to the above information provided by the customer and mainly determine the nozzle quantity, locations, arrangement of heating coils

and heaters, the size of melt channel, and other relevant design aspects by referring to part dimensions, material properties, and injection positions.

3. Carry out a CAE simulation analysis to verify the design of the hot runner system, and obtain the final confirmation with the customer before actual manufacturing.
4. Start the manufacture of relevant components of the hot runner system. Provide recommendations on molding conditions under certain situations.
5. After delivery of the hot runner system to the customer for mold assembly, make a molding trial and obtain feedback information from the customer.

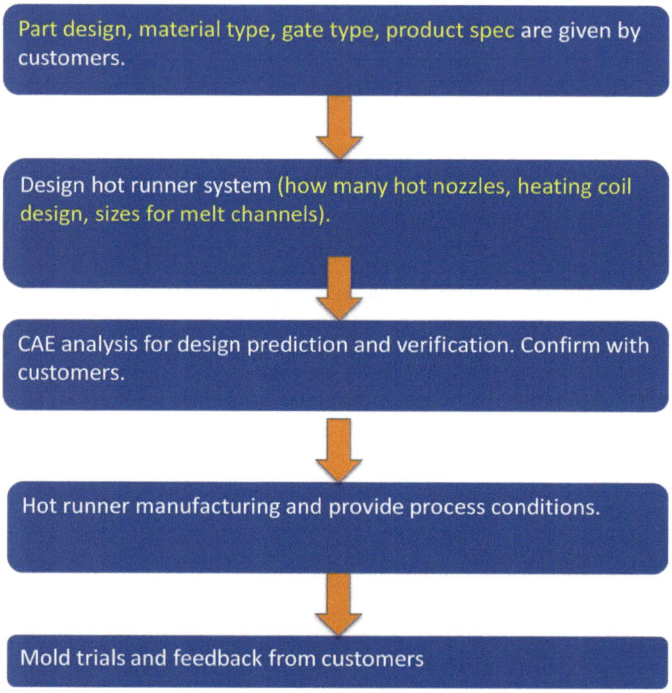

Figure 11.4 General development procedure of a hot runner system

11.1.2 Temperature Control in a Hot Runner System

Normally, a heating system consists of heating coils (bars) and sensors (Figure 11.5). The temperature control of hot runner metal and melt is achieved by the heating system. Once the heating coil is turned on, the temperature starts to rise; it is monitored by sensor nodes. Generally, the design is for one heating coil with one sensor for temperature control.

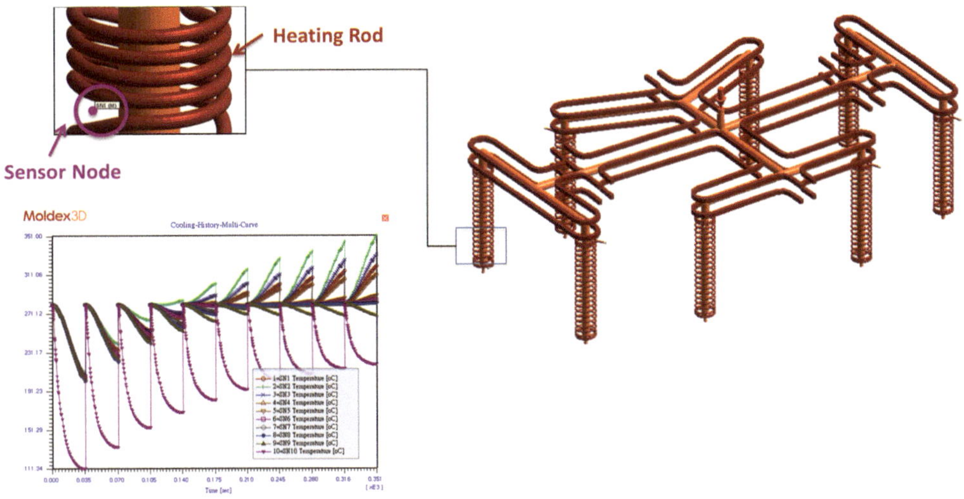

Figure 11.5 Hot runner heating system

Temperature Control Method

A common heating temperature control is the ON/OFF control (Figure 11.6). There are only two statuses of the heating coil (rod): ON and OFF. ON means that the heating is at 100% power, while OFF means a complete shutdown with zero power output. The status of the heating coil (rod) is determined by the temperature detected by sensor nodes. On a controller, the upper and lower limits of temperature can be set, and when the temperature feedback from the sensor reaches the upper limit, the heating coil would be turned off, and vice versa.

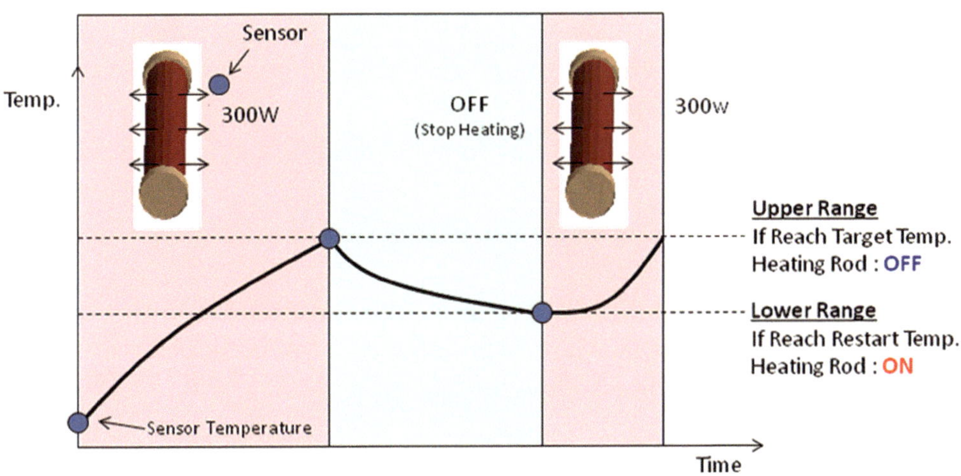

Figure 11.6 Heating rod control method: on/off control

Another common heating temperature control is the PID control (Figure 11.7). In contrast to the ON/OFF control, the PID control has a power output between 0 and 100%. Similarly, the status of the heating coil (rod) is determined by the temperature detected by sensors, and after a formula calculation, the power level is determined. The feature of this control is to control the temperature to the target value and maintain it in a stable manner; in the case of any variation, the adjustment in power level would be done according to the mechanism mentioned above.

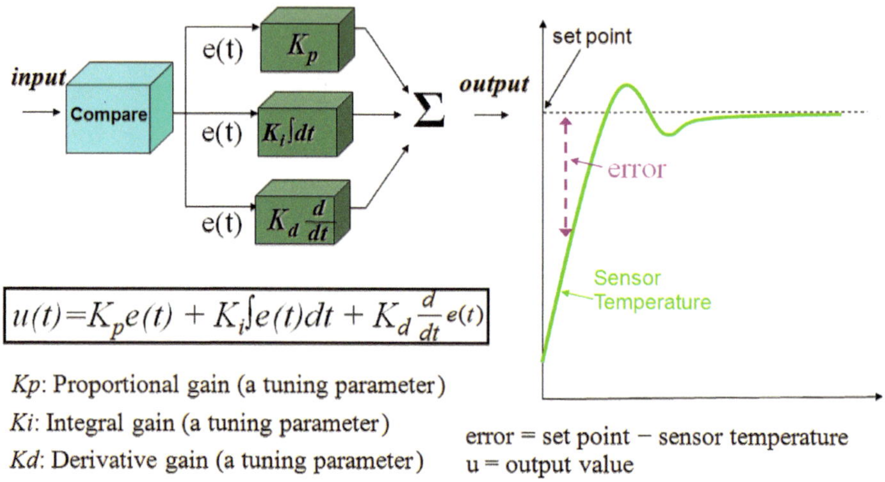

Figure 11.7 Heating rod control method: PID control

11.1.3 Advantages and Challenges

The main advantages of a hot runner system over its cold runner counterpart can be summarized by the following:

- Elimination or reduction of regrind from the mold, resulting in material savings
- Energy savings and faster cycles times, since cooling and injection cycles can be reduced
- Although tooling can be costly, it offers the opportunity for thinner, more economical wall sections based on design rather than flow constraints
- Increased design freedom concerning gate location
- Reduced injection pressure which enables easier cavity filling and reduced molded in stress, resulting in superior part quality
- Manual sorting of the molded product and runner is eliminated

Disadvantages of using hot runner systems:

- Complicated system structures, high machining precision is required, high cost in mold manufacturing and maintenance, easy decomposition for thermally sensitive materials, low flexibility for the design of gating location, difficult for temperature control.

Challenges of using hot runner systems:

- Hot runner system requires better temperature control to maintain melt temperature.
- In general, choosing external heating method can improve the distribution of heat, but high mold temperature during the injection process increases the cycle time.
- To understand the importance of hot runner controls, we need to examine how they actually control temperature, processing implications, and mold design considerations.
- The development of the hot runner system is very challenging. We can ask ourselves a few questions:
  - The hot runner system is very complicated. How to design/optimize?
  - The mechanism behind hot runners in injection molding is still not clear. How to investigate?
  - Material issue. How is material type related to hot runner design?
  - Process condition control issue. How to increase process window?
  - Product quality issue. How hot runner design affects product quality?

In the past, the product development cycle started from part design, and moved from mold design, mold construction, through molding trials, to production (Figure 11.8). Without design verification the production process usually requires more time and investment due to countless molding trials.

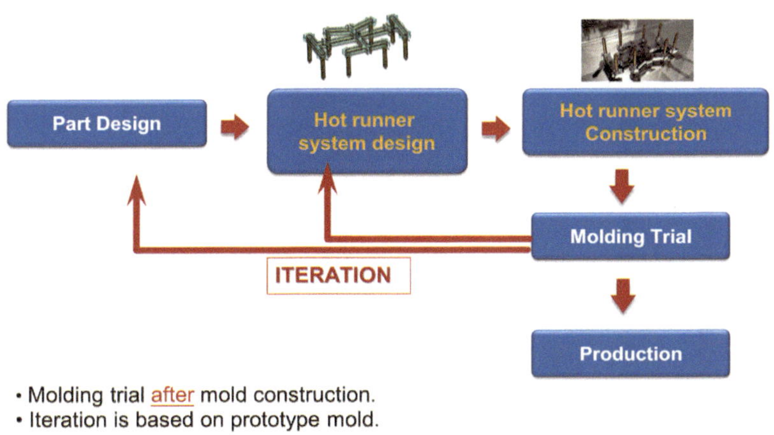

Figure 11.8 Traditional approach: trial and error

CAE is a process simulator that evaluates the effect of material properties, process conditions, and part/mold design on the process dynamics and part quality. During the product development, CAE technology not only helps product and mold verification, but also eliminates mold trials and shortens time to market (Figure 11.9).

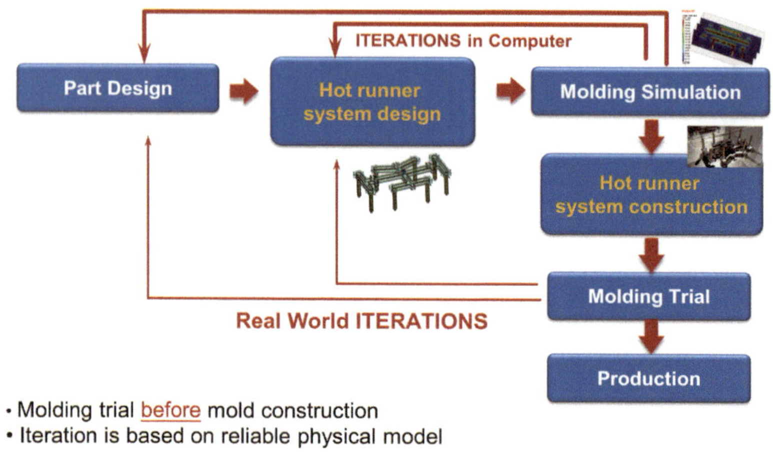

Figure 11.9 Computer-aided engineering (CAE)

Why is there a need for CAE in a hot runner system?

The mechanism behind the hot runner system is too complicated to be fully understood. In addition, there are some critical issues in the current hot runner technology, such as temperature control issues, flow imbalance, and material degradation. Therefore, simulation technology is highly needed for hot runner designers and makers to examine their designs before the real manufacturing. CAE helps us simulate the real hot runner system and provide the temperature analysis profile/history in the runners and mold base.

Due to the complexity of a hot runner system and its high cost, it is necessary to incorporate CAE in the up-front analysis. The features of using CAE for the analysis of a hot runner system are as follows:

1. Simulate an actual hot runner system and provide the melt temperature distribution and history in the runner and mold base.
2. Examine the hot runner system design for the design improvement and optimization of internal components, such as heating coils, manifolds, and nozzles.
3. Deeply understand the impact of hot runner design on the molding process and part quality, such as on injection pressure, clamping force, weld lines, shrinkage, warpage, and other problems.
4. Assist with the design change of molds and parts.

5. Help diagnose the cause of the problem and verify the solution.
6. Help accumulate enormous professional knowledge to establish the know-how and design guidelines for hot runner systems.

In general molding analysis, the melt temperature is deemed as in an ideal evenly distributed condition. However, such an assumption may be different from actual conditions. To do a complete analysis, the CAE Advanced Hot Runner (AHR) Analysis can be introduced. In contrast to general molding analysis, the AHR analysis is able to consider all components in a hot runner system, which are divided into three major categories (Figure 11.10): hot runner metal, heating system, and heat conduction boundary conditions.

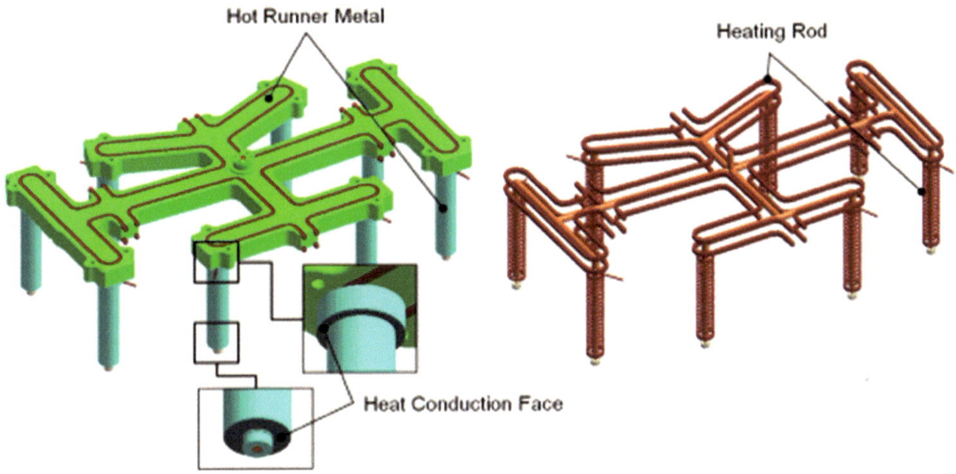

Figure 11.10 Hot runner mesh model components

For example, a hot runner system consists of many metal components, which are given different names and material properties according to their functionalities, such as sprue bushings, manifolds, nozzles, valve pins, etc. In AHR analysis, each component can be set for its heat conductivity, specific heat, density, and other thermal properties (Figure 11.11).

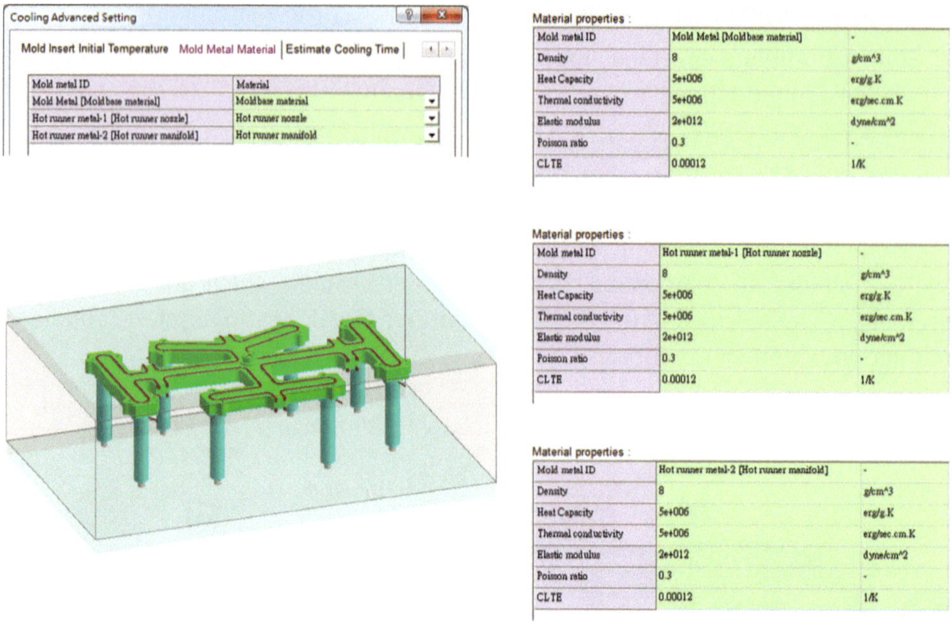

Mold metal ID	Material
Mold Metal [Moldbase material]	Moldbase material
Hot runner metal-1 [Hot runner nozzle]	Hot runner nozzle
Hot runner metal-2 [Hot runner manifold]	Hot runner manifold

Material properties :

Mold metal ID	Mold Metal [Moldbase material]	-
Density	8	g/cm^3
Heat Capacity	5e+006	erg/g.K
Thermal conductivity	5e+006	erg/sec.cm.K
Elastic modulus	2e+012	dyne/cm^2
Poisson ratio	0.3	-
CLTE	0.00012	1/K

Material properties :

Mold metal ID	Hot runner metal-1 [Hot runner nozzle]	-
Density	8	g/cm^3
Heat Capacity	5e+006	erg/g.K
Thermal conductivity	5e+006	erg/sec.cm.K
Elastic modulus	2e+012	dyne/cm^2
Poisson ratio	0.3	-
CLTE	0.00012	1/K

Material properties :

Mold metal ID	Hot runner metal-2 [Hot runner manifold]	-
Density	8	g/cm^3
Heat Capacity	5e+006	erg/g.K
Thermal conductivity	5e+006	erg/sec.cm.K
Elastic modulus	2e+012	dyne/cm^2
Poisson ratio	0.3	-
CLTE	0.00012	1/K

Figure 11.11 Hot runner metal properties

The last of the three major categories of hot runner system components is the setting of thermal conductivity boundary condition (Figure 11.12). As the main purpose of a hot runner system is to maintain the melt temperature with a heating system, the operating temperature is much higher than that of the mold base.

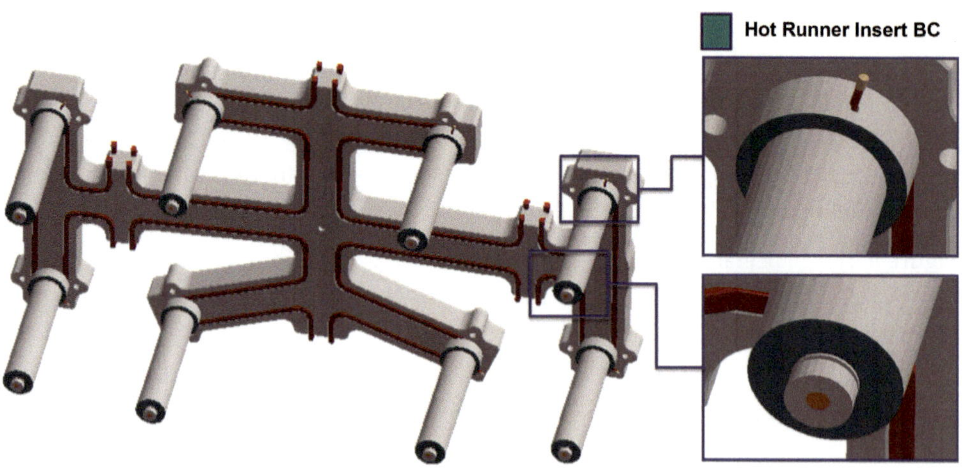

Figure 11.12 Heat conduction faces

Substantially, AHR is able to simulate an actual hot runner system, including nozzles, heating coils, bushings, manifolds, and other components, as well as the melt temperature analysis in depth, pressure drop, shear-heating, and other physical phenomena in the melt channel, and provide the dynamic temperature distribution of the mold. Review the following key analytical points to assess the benefit of developing a hot runner system.

1. Prediction of Temperature Distribution of Hot Runner Metal

As shown in Figure 11.13, from the prediction result of temperature distribution of the hot runner metal, we can know whether the temperature is evenly distributed in different metal components of each region and find out potential issues. Accordingly, if necessary, all aspects of the design can be reviewed, such as the power and arrangement of heaters or heating coils, location of sensing nodes, thermal conduction contact faces, etc., so as to achieve the goal of even temperature distribution.

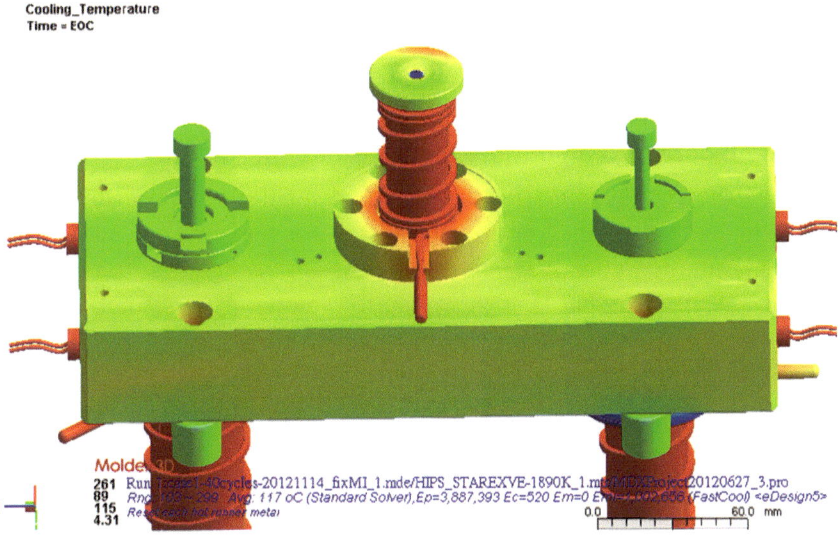

Figure 11.13 Temperature distribution of hot runner metal

2. Prediction of Mold Base Temperature Distribution

As shown in Figure 11.14, the actual temperature of the region where the mold base contacts the hot runner system would be higher than the set temperature of the mold base due to the high temperature of the hot runner system. Such a phenomenon would lower the cooling efficiency of the mold base and lead to a longer cooling time and molding cycle, which increases the manufacturing cost. In addition, plastic parts are likely to warp or deform due to the uneven shrinkage resulting from the temperature difference between the mold core and mold cavity, thus leading to a poor yield of finished parts.

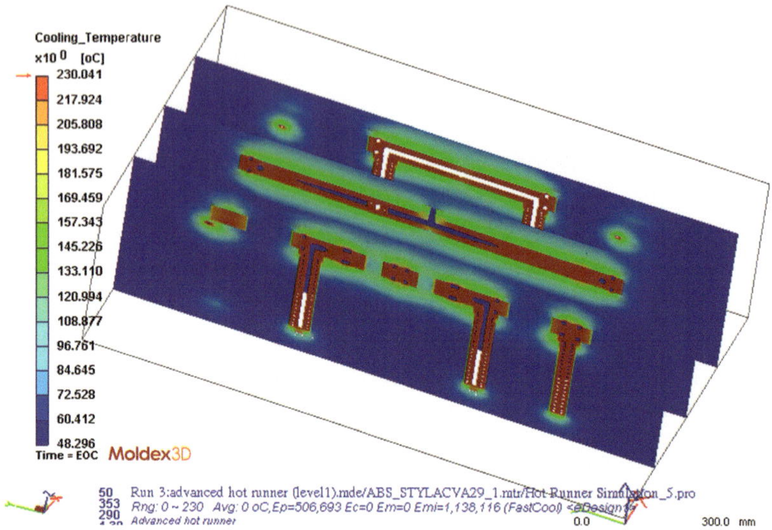

Figure 11.14 Mold base temperature distribution prediction

3. Verification of Melt Temperature in the Hot Runner

Figure 11.15 shows the simulation result which verifies whether the heating system is able to heat up the melt evenly. If the power is too low, the melt temperature cannot be maintained effectively; however, if it is too high, an unnecessary cost would be induced and, what is even worse, the temperature cannot be controlled properly, which causes material decomposition and affects the quality of output plastic parts.

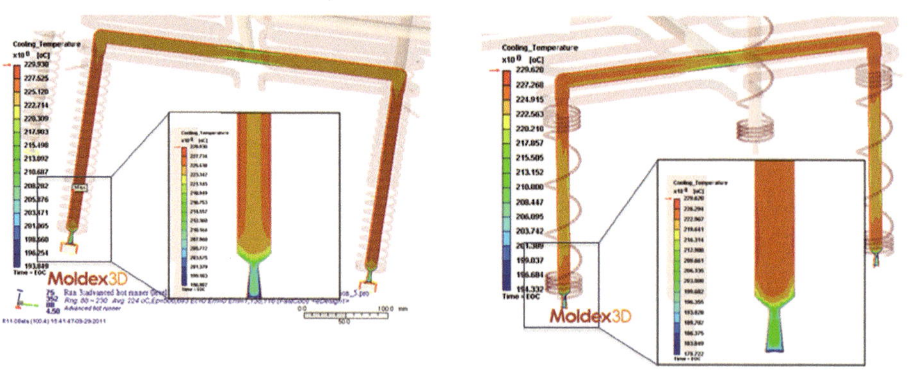

More uniform temperature along hot runner nozzle   Non-uniform temperature along hot runner nozzle

Figure 11.15 Heating rod power and size verification

Figure 11.16 shows an example of overheating. In the beginning, a heating coil with a uniform density is designed, but through the simulation, an overheating issue at the nozzle region is found which might lead to a material decomposition problem.

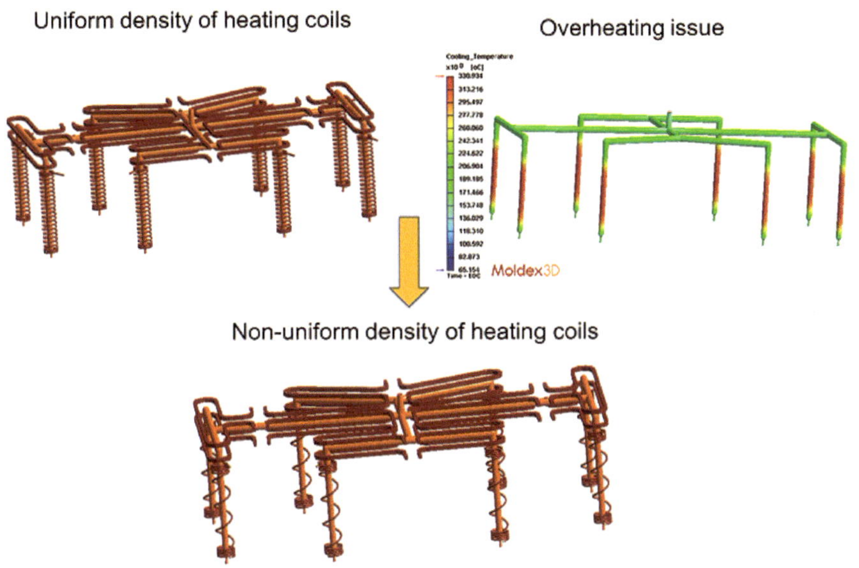

Figure 11.16 Overheating problem example

4. Verification of Heater or Heating Coil Arrangement

There are two points to verify the uniformity of internal temperature of a hot runner system: one is to determine whether the heating coiling heats up the hot runner uniformly, and the other is to check whether the location of the sensing node is determined properly. The goal is a uniform melt temperature, which is able to ensure the quality of the plastic part, whereas with a too-high or too-low temperature there is a potential liability to cause poor part quality.

Figure 11.17 shows a case study which verifies different arrangements of heaters or heating coils through the simulation. According to the analysis result, it can be seen that the heating coil design on the right side displays a more uniform temperature distribution than that on the left side.

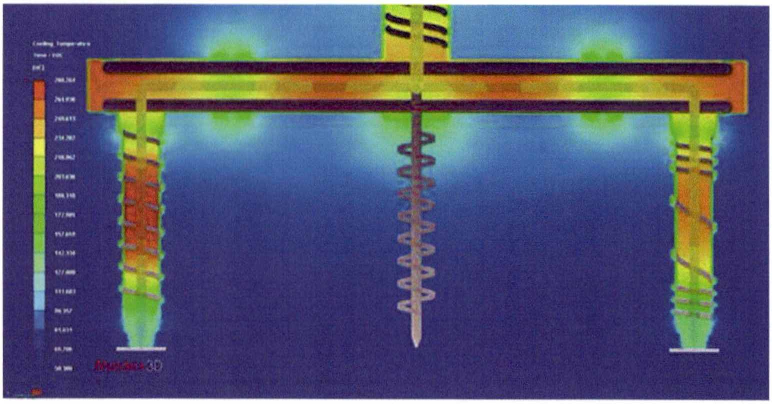

Figure 11.17 Heating rod layout verification

5. Effect of Filling and Molding under Different Temperature Control Conditions

In order to understand the effect of temperature control of each nozzle on the internal melt temperature, we can analyze and verify the filling behavior resulting from different nozzle temperatures, such as the variation in contribution of each gate, the location of weld lines, etc.

For example, in the simple system shown in Figure 11.18, a frame, there are two drops and two heating coils for the hot runner system. If the heating is not controlled properly, such as if the heating power is larger on the right side, we will see flow imbalance. The flow on the right side is at a high temperature and hence flowing faster than on the left side. Therefore, we can predict the welding lines do not occur in the middle, but rather on the left.

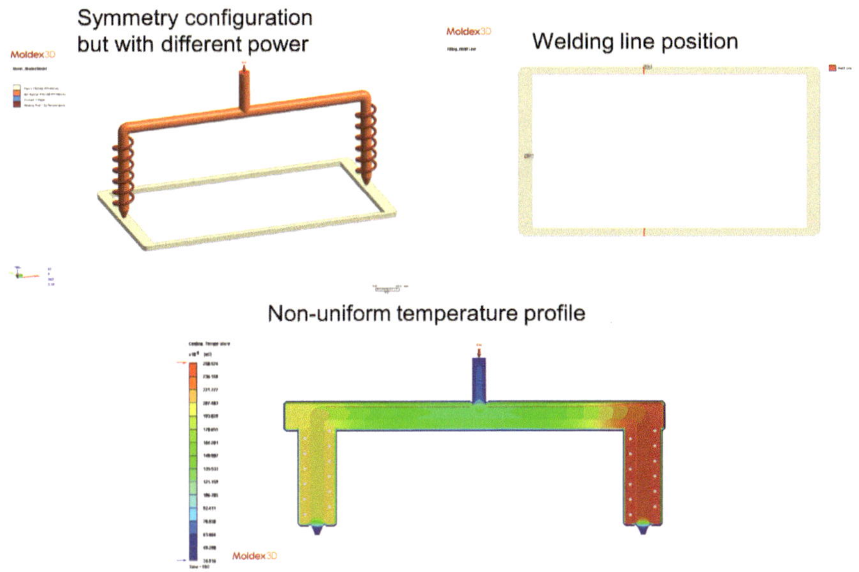

Figure 11.18 Flow imbalance example

The governing equations used to simulate a hot runner system are described in Section 5.2.1.

■ 11.2 Practical Applications

11.2.1 CAE Verification on a Single-Gate Hot Runner System

In this study, this numerical simulation technology is applied to a real hot runner system manufactured by ANNTONG IND. CO. LTD. In the hot runner experiment, the injection molding process is conducted cycle by cycle. The melt temperature is measured at the same locations as the positions of sensor nodes 1 to 10 in the simulation. There are seven sensor lines arranged in the hot runner system, which is represented by the sensor nodes (Figure 11.21) in the simulation. Figure 11.19 shows the hot runner system and the metal mold used in the experiment. The heating coils are connected to the power supplier. The size of the mold base is 250.00 × 250.00 × 320.00 mm.

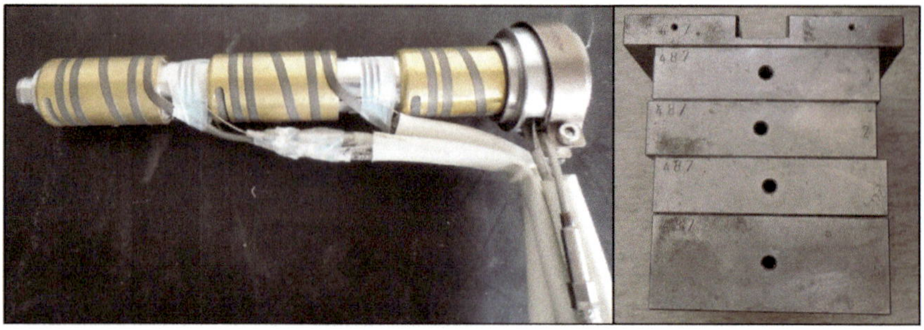

Figure 11.19 The hot runner system and metal mold used in the experiment

This system includes the part, hot runner nozzle, three sets of heating coils, three brass bushings which the heating coils are wrapped around, a metal mold, and other bushing components. The material for the metal mold is P20. The thermoplastic material for the melt is PC (Panlite L-1225Y). The constructed CAE model for this hot runner system is shown in Figure 11.20, which also indicates the materials used for each hot runner component. Components shown in the same color are made of the same material.

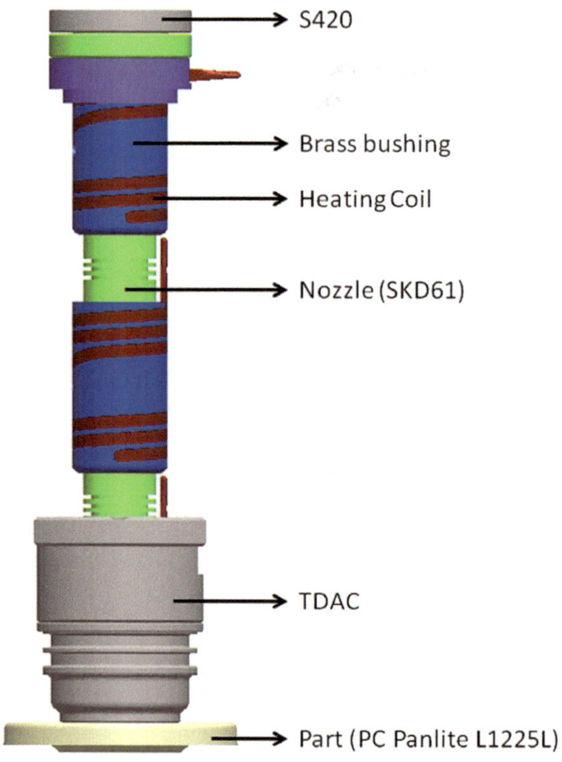

Figure 11.20 The CAE model for the hot runner system

Table 11.1 lists the material properties, including density, heat capacity, and thermal conductivity, for each hot runner components in this system. These parameters are the inputs in the analysis. This is to consider the effects of different thermal properties.

Table 11.1 Material Properties for Hot Runner Components

	P20 (Moldbase)	SKD61	Brass	TDAC	S420
Density [g/cm^3]	7.75	7.81	8.50	7.81	7.72
Heat capacity [$\times 10^6$ erg/g·K]	4.62	4.60	3.85	4.62	4.62
Thermal conductivity [$\times 10^6$ erg/s·cm·K]	2.90	2.47	16	2.60	2.43

Figure 11.21 shows the boundary conditions information and sensor node locations, which are defined according to the real experiments. The surfaces highlighted in a dark green color are defined as the contact areas with the metal mold. The undefined surfaces are automatically assumed as heat insulation due to the air layer around these areas. The air is assumed as a very poor heat conduction material. The number 1 to 10 sensor nodes are used to measure melt temperature function of time. These nodes are placed along the central line of the hot runner nozzle. The number 1 to 7 sensor nodes (or 1A, 2A, 3A, 1B, 2B, 3B, 1C) are used to control on/off for the heating coils. For example, the sensor nodes SN4, SN3, and SN2 (or 1A, 2A, 3A) are assigned to each heating coil, respectively. The target temperature is 280 °C and the range is ±2 °C. The power for each heating coil is 400 W or 610 W, as indicated in Figure 11.22. If the sensor node 1A reaches the temperature of 282 °C, the corresponding heating coil is switched off. In contrast, if the temperature goes down to 278 °C, the heating coil is turned on. This is the same for the sensor nodes 2A and 2B with their corresponding heating coils.

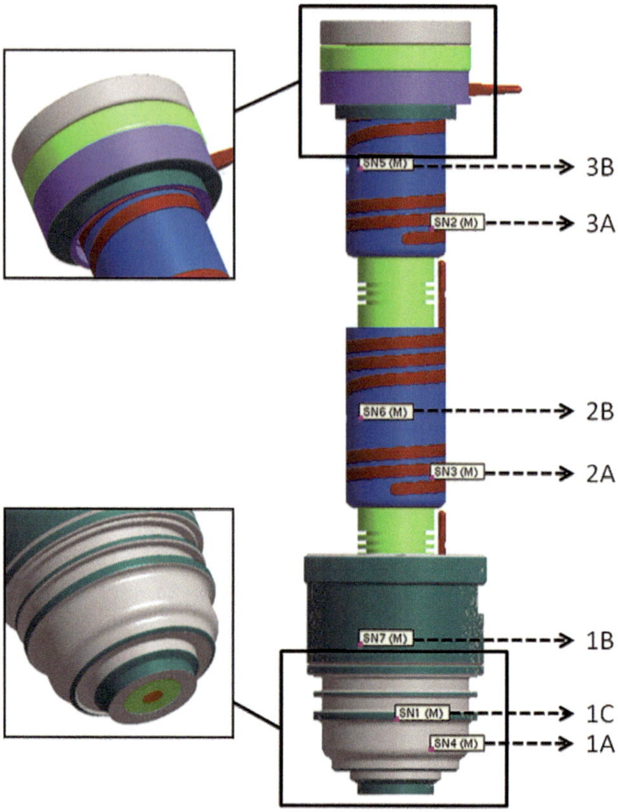

Figure 11.21 Boundary conditions and sensor node positions

The filling time and packing time are both 5 seconds. The cooling channel layout and coolant (water) temperature are also shown in Figure 11.22. The cooling time and opening time are 20 s and 5 s, respectively. They are determined from the process conditions in the real injection molding experiments. They are counted in one injection molding cycle for transient cooling analysis. Therefore, the cycle time is 35 seconds. The melt temperature on the sensor nodes 1 to 10 can be predicted and compared with the experimental measurements.

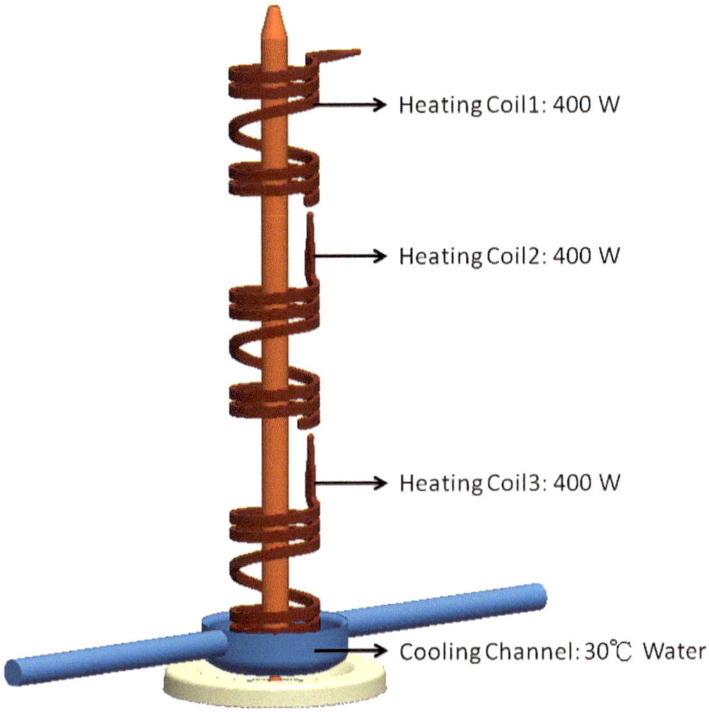

Figure 11.22 Heating coils and cooling channel

Table 11.2 shows the molding parameters. The temperature measuring will begin after 10 shots to ensure the melt temperature is stable.

Table 11.2 Injection Molding Process Parameters

Filling speed	40 mm/s
Melt temperature	280 °C
Mold temperature	80 °C
Packing pressure	10 MPa
Injection time	5 s
Packing time	5 s
Cooling time	20 s
Mold open time	5 s
Cycle time	35 s

Figure 11.23 displays the simulated melt temperature field inside the runner or melt channel at the end of 10 injection molding cycles. The sensor node 2 detects higher melt temperature in the runner due to the higher power from the heating coil around that area. The sensor node 10 detects much lower melt temperature because of the cooling channel and heat conduction contact with the metal mold in the hot runner system. The melt temperature field at different times can be shown and observed also.

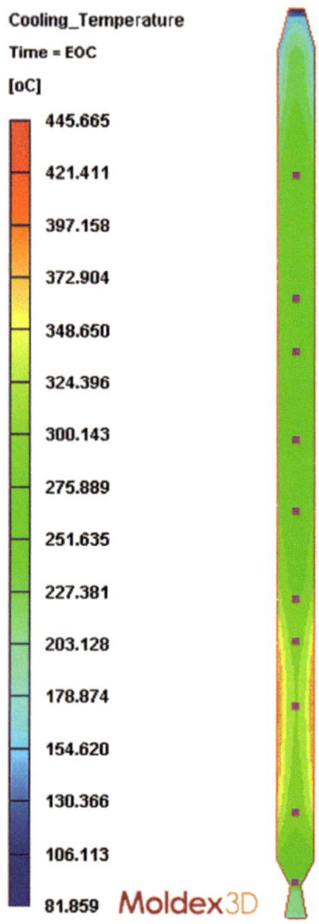

Figure 11.23 Location of sensor nodes and temperature distribution inside the runner after 10 cycles

Figure 11.24 demonstrates the simulated temperature field inside the mold base after 10 injection molding cycles. The mold base near the cooling channel has lower temperature as indicated in dark blue color.

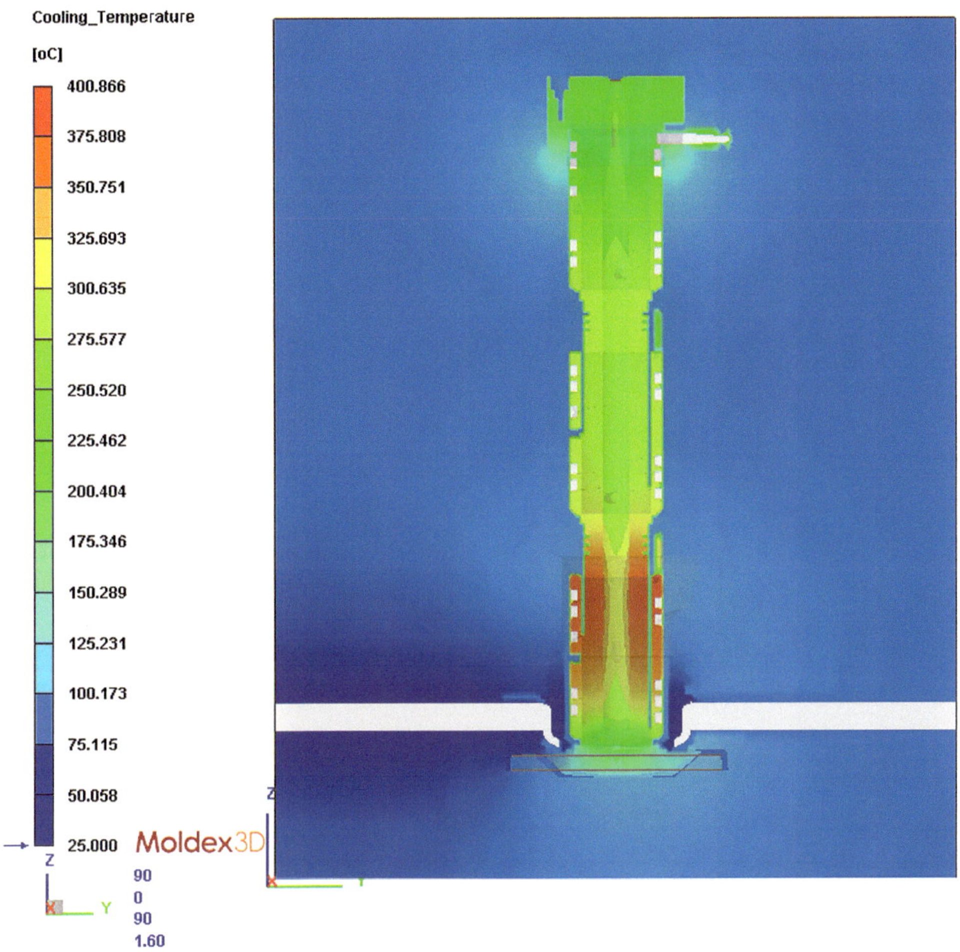

Figure 11.24 Temperature distribution inside of mold base after 10 injection cycles

Figure 11.25 displays the simulated melt temperature history curves for the sensor nodes 1 to 10. This records the dynamic response of melt temperature cycle by cycle and allows for melt temperature investigation at the end of the specific cycle by extracting the data. It can be observed on a monitor that the melt temperature from each sensor node goes up cycle by cycle.

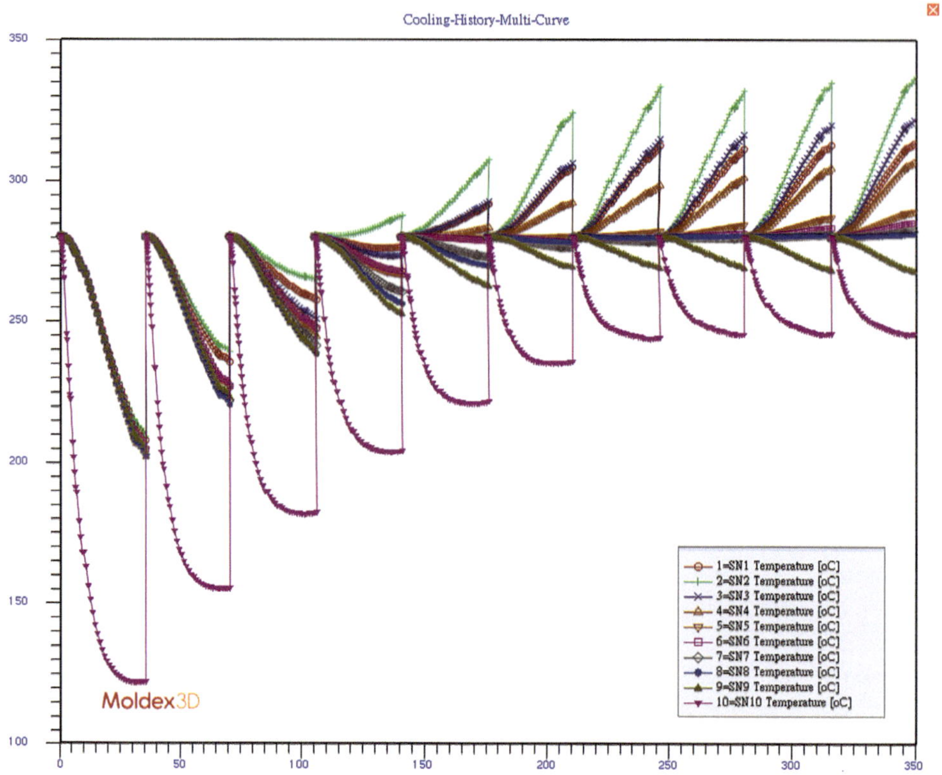

Figure 11.25 Melt temperature history curves for the sensor nodes 1 to 10

Validation from the Experiment

Figure 11.26 shows the experimental melt temperature distributions as a function of location using controlling sensor nodes 1A, 2A, and 3A. Position zero means the location of the gate. The temperature was measured after several injection molding cycles and assumed to reach steady state. The simulation result is also plotted together and compared with the experimental one.

Figure 11.27 shows the experimental melt temperature distributions as function of a location using controlling sensor nodes 1C, 2A, and 3A. Both the experimental and simulation results are similar in the case of controlling sensor nodes 1A, 2A, and 3A.

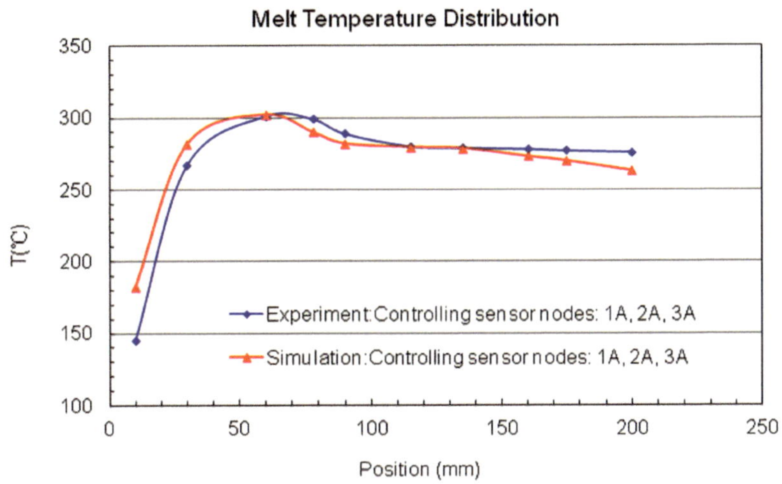

Figure 11.26 Melt temperature distribution at the controlling nodes 1A, 2A, and 3A

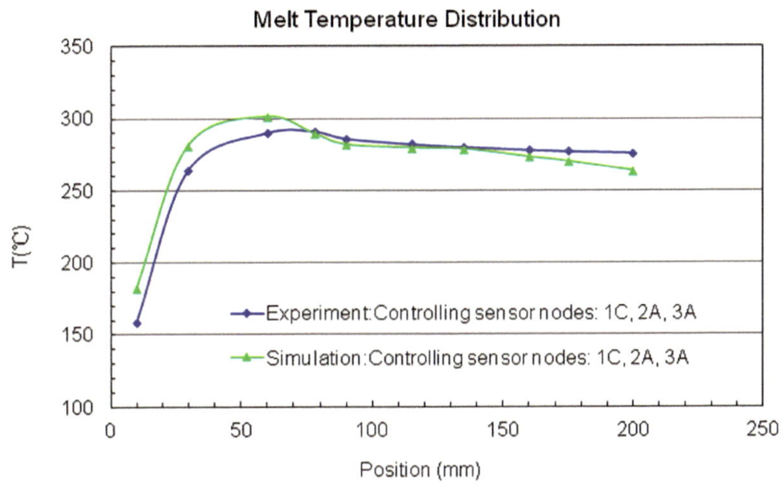

Figure 11.27 Melt temperature distribution at the controlling nodes 1C, 2A, and 3A

Figure 11.28 shows the experimental melt temperature distributions as a function of location using controlling sensor nodes 1B, 2B, and 3B. In the experimental result, there is no obvious higher temperature region at the position of 60 mm. The simulation result also indicates this phenomenon. This is different to the previous two cases.

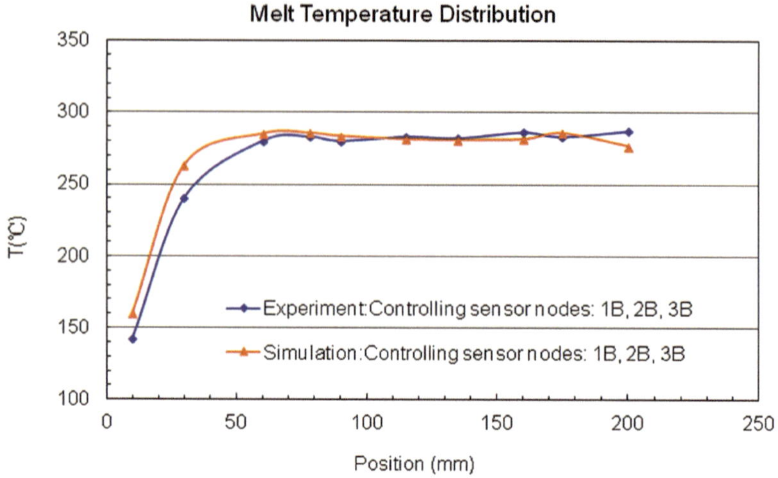

Figure 11.28 Melt temperature distribution at the controlling nodes 1B, 2B, and 3B

11.2.2 CAE Pin Movement Control of Valve Gate

In order to assist the user of the hot runner in maximizing the benefit of the sequential control technique and achieving the goal of quality improvement in the plastic parts, a hot runner supplier proposed a new solution which is able to control the speed of a valve gate. The CAE molding analysis software also adapts to this trend and provides the "simulation technology of valve pin movement speed control" as another option of advanced valve gate control for engineers to use apart from the general gate control (Figure 11.29).

In an advanced valve gate simulation analysis, the CAE engineer can use a mesh editor (Figure 11.30) to create the model based on the geometric shape of the actual valve gate during the mesh preprocess stage, including hot runner geometry, valve pin geometry, stroke, gate control number, etc. As the product designer does not need to simplify the geometry and sacrifice important geometrical information, the real three-dimensional solid meshing technology considers the most realistic valve gate design, which can drastically improve the precision of simulation analysis.

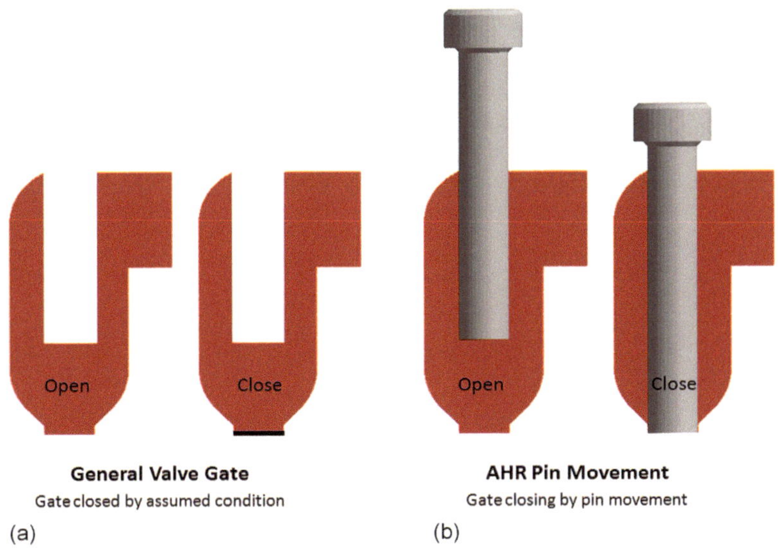

Figure 11.29 Different control methods in simulation for a valve gate system: (a) general and (b) pin movement

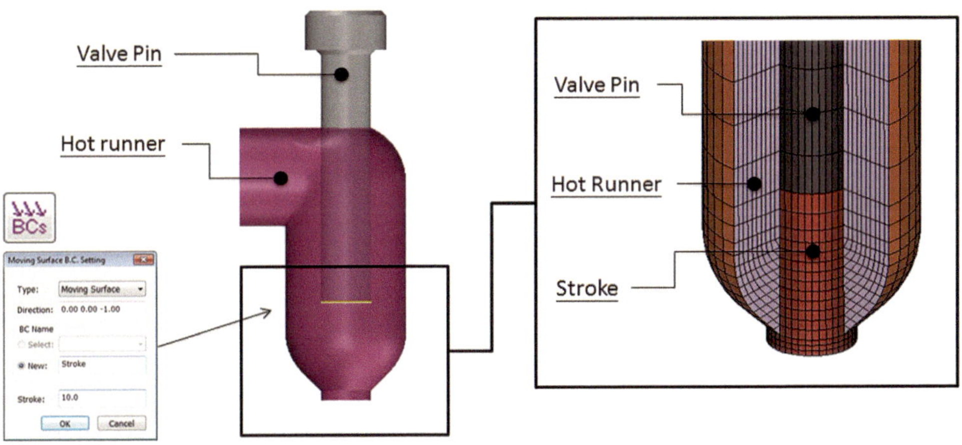

Figure 11.30 User-friendly interface to build up the mesh of valve gate

In the stage of molding condition setting, the control of each group of the gate can be configured as well, such as the initial state of the gate (opened or closed), pin movement speed, and levels of speed; this is an indispensable tool for analytical engineers with advanced valve gate simulation requirements. As shown in Figure 11.31, the following configurations can be achieved via the interface of pin movement speed setting.

1. Valve pin: According to the attributes of the mesh model, the pin movement speed of the valve gate can be configured respectively or by group.

2. Initial status: Set the initial status of a gate to open or close according to the simulation requirement.
3. Maximum velocity: Set the actual maximum movement speed of the valve pin.
4. Valve pin speed diagram: Set the movement timing and speed of the valve pin according to the simulation requirement.

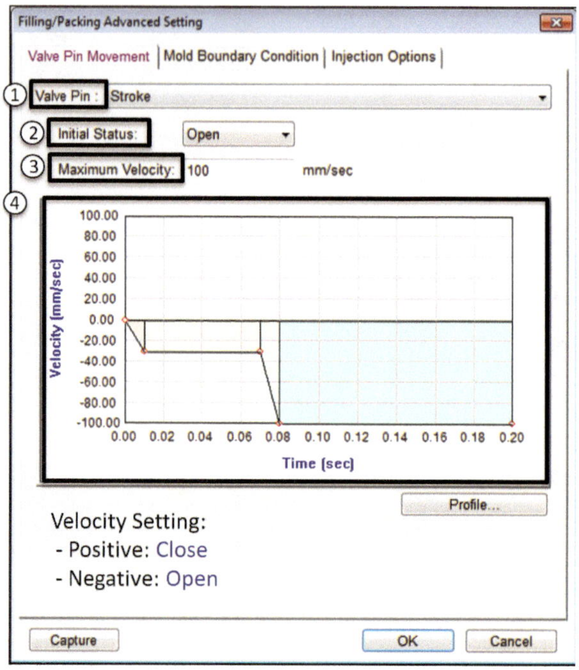

Figure 11.31 Process conditions setting interface of pin movement

From Figure 11.32, we can see that through a real 3D simulation control, the velocity field at the gate region varies as the pin position in the hot runner changes. The complete analysis data in the CAE molding analysis enables CAE engineers to configure and verify the pin movement speed control parameters with sufficient information, and if necessary, the geometry of the hot runner gate can be explored to further optimize the design.

For multi-point sequential gating cases, the influence of pin speed control on the flowing melt front as well as the improvement result can be verified via the flow analysis of filling. As shown in Figure 11.33, this is a case of three gates with sequential control; the three gates G1, G2, and G3 open in sequence. Figure 11.33(a) illustrates the simulation result of a one-level valve pin speed control, which shows a significant change in melt front speed when the gates G2 and G3 are opened to release the pressure instantly. Such a phenomenon could further lead to surface flow marks and other potential exterior defects. Figure 11.33(b) illustrates the simulation result after optimizing the speed control of the valve pin, which shows that the melt front speed does not change obviously while the gates G2 and G3 are opened. From these analyses, the influence of pin speed control on the flowing melt front can be verified. Furthermore, the effectiveness of the process can be increased, so as to ensure the surface flow mark issue is actually improved or even eliminated.

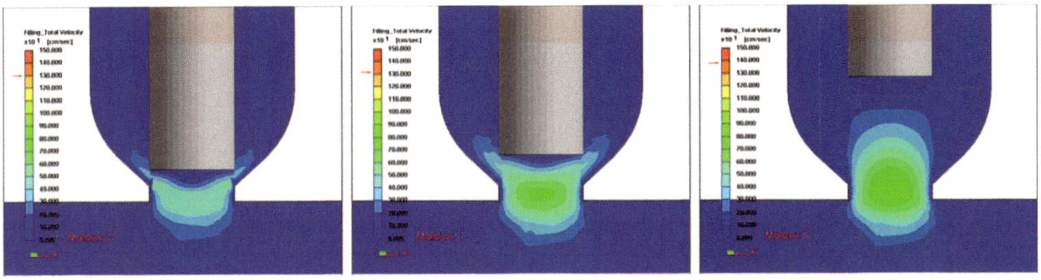

Figure 11.32 The melt velocity distribution near the gate is affected by the movement of the pin

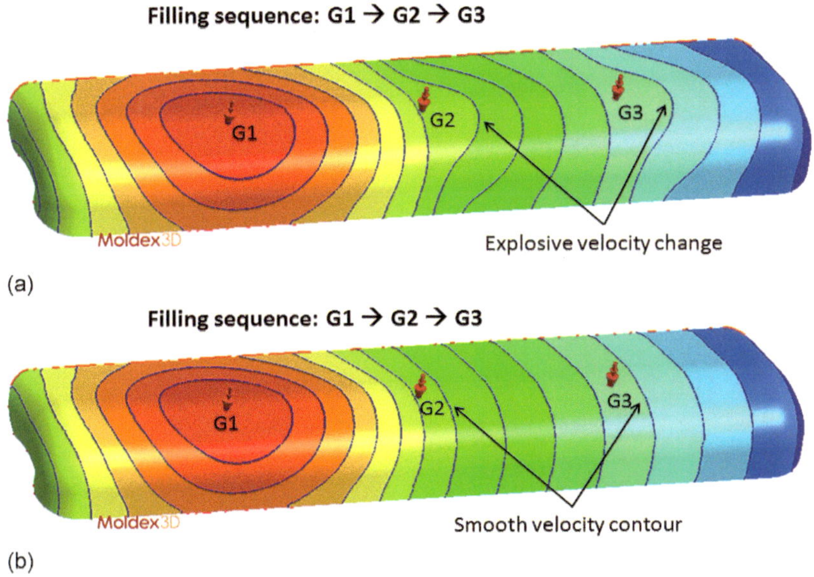

Figure 11.33 Melt filling velocity contours of valve gate systems for (a) one- and (b) several-segment velocity settings in pin movement

The hot runner system has become one of the important designs in the plastic injection molding process, and the industry also continues to invest research resources for developing more advanced and wide-ranged applications. By using

CAE simulations and the pin speed control technology, the user of a hot runner can verify and optimize the design before actual development (Figure 11.34), and reduce the chances of surface defects occurring. The part quality can be managed with the maximum benefit that the sequential control technology can deliver.

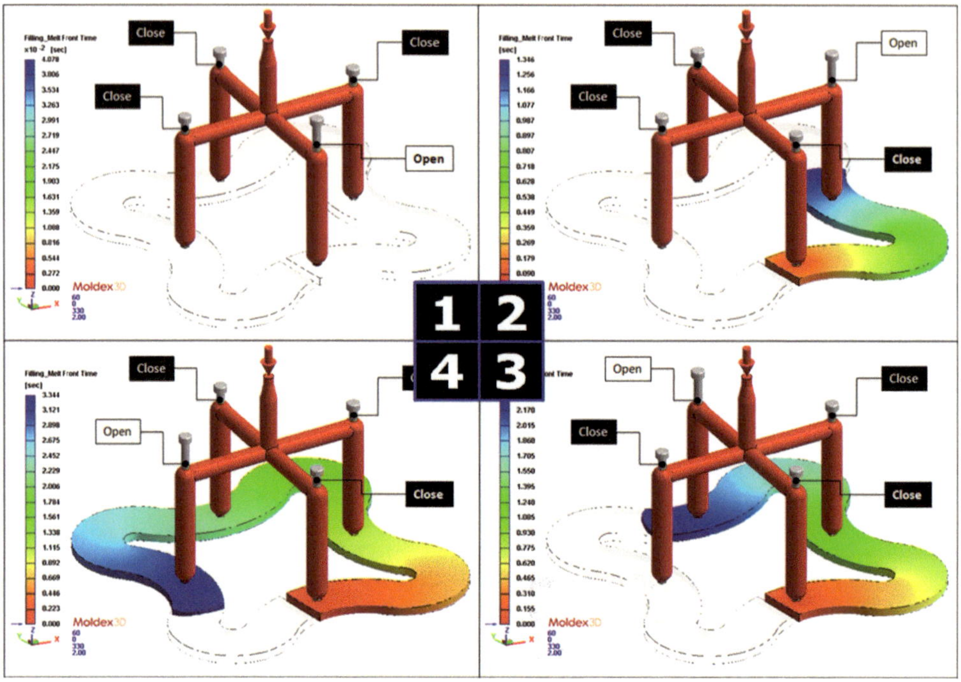

Figure 11.34 The application of combined technologies of pin movement and sequential control for valve gate system

■ 11.3 CAE Case Study

Following the discussions of this chapter and the corresponding CAE verification case, an exercise model, as illustrated in Figure 11.35, is provided in this section for readers to practice CAE application in hot runner optimization.

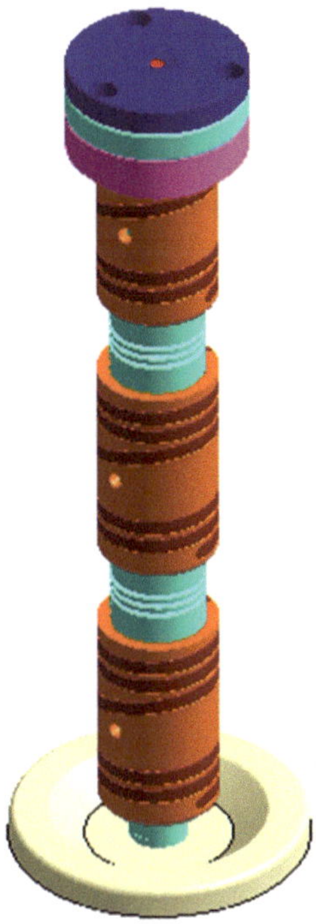

Figure 11.35 Case study sample

Please download the analysis files for CAE Case Study 11.3 from the following website:

https://moldex3d.box.com/s/zr6fvc1vlhbi4ocx111jwd3wmxt4ooif

Then open the .m3j or .mvj file in Moldex3D and answer the questions below according to the analysis results.

Questions:

1. Analysis model information
   a) Find the value or amount of:
      i. part thickness
      ii. hot runner length
      iii. gate diameter

        iv. heating rod
        v. sensor node
2. Analysis results with default setting
    a) Capture the temperature distribution image of:
        i. melt in hot runner
        ii. hot runner metal
        iii. mold base
    b) Capture the history curve plot of sensor nodes
3. Find the influence by adjusting the heating rod parameter of:
    a) lower heating power
    b) higher heating power
4. Figure out the heating rod behavior when the control method is ON/OFF mode
5. Figure out the heating rod behavior when the control method is PID mode
6. Figure out the heating rod behavior when the heating method is by Temperature mode

12 Co-/Bi-Injection Molding

Chih-Chung (Jim) Hsu, Yu-Sheng (Tim) Chou

There are many terminologies to define different types of multi-component molding (MCM) processes. To get a better understanding, we have proposed the definition for MCM processes as shown in Figure 12.1. Specifically, MCM processes can be divided into two categories. The first type is producing with a distinct interface. For example, insert molding, over-molding, and sequential multiple-shot molding belong to this type. During the molding process, each component has a distinct interface and they will not influence each other. On the other hand, for the processes on the right-hand side of Figure 12.1, the interface between two materials is uncertain due to the injection process sequences. The process type-uncertain interface is the topic that this chapter will discuss.

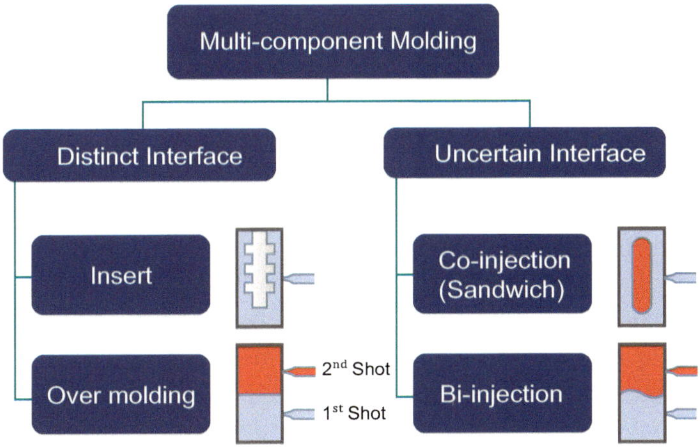

Figure 12.1 Multi-component molding processes can be divided into two categories: (left) with distinct interface system, and (right) with uncertain interface system

■ 12.1 Basics

12.1.1 Process Principle

1. Co-Injection Molding

Co-injection molding, also known as sandwich injection molding, was developed in the 1980s, and is a molding technology using two identical or different but compatible polymer melts to form a part of a core wrapped by a skin. Such a process technique can be divided into two types, one-channel and two-channel, according to the design of the nozzle, as shown in Figure 12.2.

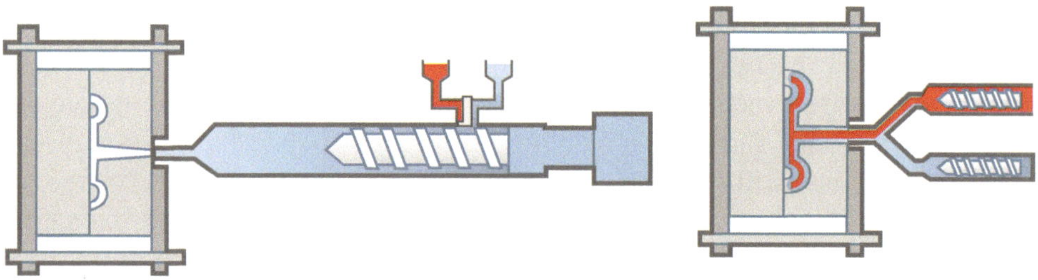

A simultaneous injection which uses a single barrel

(a)

A control valve that let the melts enter the cavity in a sequential order

(b)

Figure 12.2 (a) One-channel and (b) two-channel technique for co-injection molding

We can take a one-channel process as an example, whose process flow is shown in Figure 12.3. Figure 12.3 shows the melt injected in order by a single runner system. The melt injected first is called the skin layer, while the one injected afterwards is called the core layer. Step (1) shows the first injection of the skin layer whose filling amount should be adjusted in advance so that the second injection of the core layer can be done effectively. Step (2) shows the filling of the second injection. In order to improve the part appearance, normally the second injection of melt does not penetrate the first one. Of course, if color mixing is intended for the part, the core layer would be allowed to penetrate the skin layer. Step (3) shows that some parts would be completed with the skin material being used for filling once more to wrap up the gate. In this way, the core layer would not be exposed, affecting the appearance, and the parts can be ejected once the melt is cooled and cured.

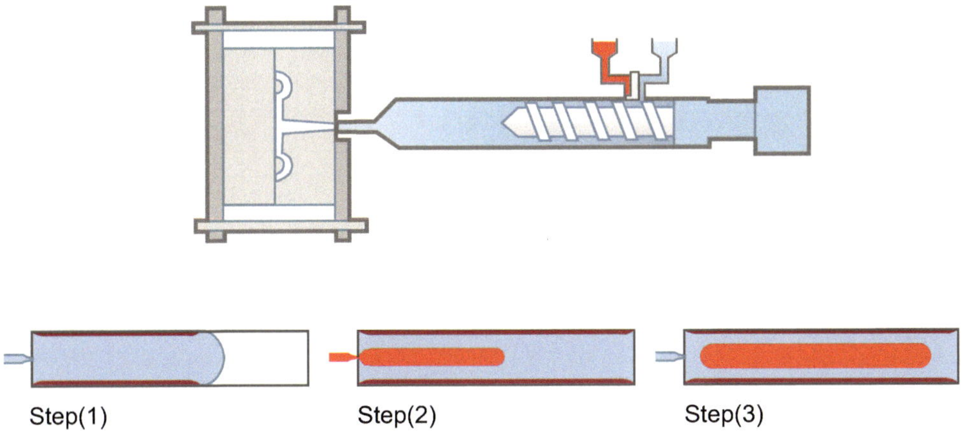

Figure 12.3 Process sequence of co-injection molding

Fragile materials can be wrapped inside soft and elastic plastics in order to achieve buffering against impact and protect the structures within. Therefore, the compatibility between materials is critical for co-injection molding. A material compatibility chart is shown in Figure 12.4.

2. Bi-Injection Molding

In comparison with co-injection molding, the bi-injection molding process is to inject two different materials A/B into the mold cavity simultaneously via two independent injection systems, and the finished part would possess weld lines obviously. As shown in Figure 12.5, such a molding process is similar to conventional injection molding. As co-injection molding injects melt into the same gate sequentially, it is easy to tell the bi-injection molding process apart from the co-injection molding process by checking the position of weld line on the finished part.

	Material Combinations	ABS	ASA	CA	EVA	PA 6	PA6.6	PBT	PC	PE	PET	PMMA	POM	PP	PPO Mod.	PS	PSU	Plasticsed PVC	SAN	TPE-A	TPE-E	TPE-S		TPE-U	TPE-V	EPDM	NR/SBR	SBR	LSR
Thermoplastics	ABS	🔵	🔵	🔵				🔵	🔵		🔵	🔵			🟠	🟠		🔴	🔵	🔴	🔴	🔴	🔵Ⓜ	🟠	🔴				
	ABS / PC																			🔴	🔴	🔴	🔵Ⓜ	🟠					
	ASA	🔵	🔵	🔵	🔵				🟠			🟠			🟠	🟠			🔵					🔵					
	CA	🔵	🔵	🔵	🔴										🟠	🟠			🔵					🔵					
	EVA		🔵	🔴	🔵					🔵				🔵		🔵													
	PA 6					🔵	🔵			🔴			🔴							🔴	🔴	🔴	🔵Ⓜ	🟠	🔴	🔴	🔴	🔴	🔴Ⓜ
	PA 6 (Mod.+ 25%GF)																			🟠	🟠	🟠	🔵Ⓜ	🟠	🟠				
	PA 6.6					🟠	🔵	🔵	🟠	🔴			🔴			🟠				🔴	🔴	🔴	🟠Ⓜ	🟠	🔴	🔴	🔴	🔴	🔴Ⓜ
	PA 6.6 (Mod. + 25%GF)																			🟠	🟠	🟠	🔵Ⓜ	🟠	🟠				
	PA 6.12																												
	PA 12 (Mod. + 25%GF)																			🟠	🟠	🟠	🔵Ⓜ	🟠	🟠				
	PBT						🔵	🔵	🔵							🔴				🔴	🟠	🔴	🔵Ⓜ	🟠	🔴	🔴	🔴	🔴	🔵Ⓜ
	PC	🟠	🟠				🔴	🟠	🔵	🟠						🔴	🟠			🔴	🔵	🔴	🔵Ⓜ	🟠	🔴				🟠Ⓜ
	PC / PBT																			🔴	🔵	🔴	🔵Ⓜ	🟠	🔴				🔵Ⓜ
	PE	🟠	🟠	🟠	🔵	🔴	🔴		🟠	🟠		🔴	🔴	🟠		🟠				🔴	🔴	🟠	🔵Ⓜ	🔴					
	PET	🔵									🟠		🟠						🔵	🔴	🔴	🔴	🔵Ⓜ	🔴	🔴				
	PMMA	🟠	🟠							🔴		🔵				🔴		🔵	🟠			🔴	🟠Ⓜ	🟠					
	POM	🟠	🟠	🟠		🔴	🔴			🔴			🔵	🔴		🟠			🟠	🔴	🔴	🔴	🔴	🟠					
	PP	🟠	🟠	🟠	🔵	🔴	🔴					🔴	🔴		🔴	🟠			🟠	🔴	🔴	🔵	🔵	🔴	🔵				🟠Ⓜ
	PPO Mod.	🟠	🟠	🟠										🔴	🔵	🔵				🔴	🔴	🔴	🟠Ⓜ	🔴	🔴				
	PPE Mod.																									🔵Ⓟ	🔵Ⓢ	🔵Ⓢ	
	PS	🟠	🟠	🟠	🔵	🟠	🟠		🟠						🔵	🔵			🟠	🔴	🔴	🔴	🟠Ⓜ	🔴	🔴				
	PSU								🔵								🔵												
	PCV-hart																			🔴	🔴	🔴	🔴	🔵					
	SAN	🔵	🔵	🔵				🔵	🔵		🔵	🔵	🟠		🟠	🟠		🔵	🔵	🔴	🔴	🔴	🔵Ⓜ	🔵	🔴				
TPE	TPE-E							🟠	🔵		🟠																		
	TPE-U	🔵	🔵	🔵		🔵	🔵		🔵						🟠	🟠			🔵					🔵					
D	BMC																												
Elastomers	EPDM							🔵																		🔵			
	NR																										🔵		
	SBR																											🔵	
	LSR																												🔵
		Thermoplastics																		TPE						Elastomers			

🔵 Good adhesion 🟠 Poor adhesion 🔴 No adhesion

Ⓜ Adhesion modified Ⓟ Peroxide cross-linkage Ⓢ Sulphur-linkage

Figure 12.4 Compatibility table for general polymers [1]

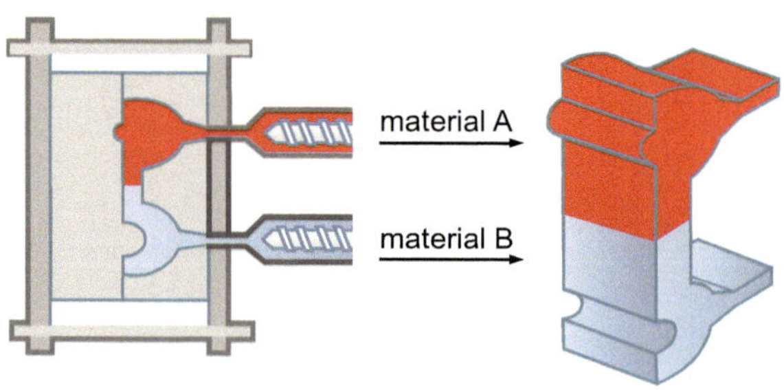

Figure 12.5 Process sequences of bi-injection molding

12.1.2 Advantages and Challenges

The advantages of co-injection molding are as follows:

1. Materials with low viscosity can be used for the core in order to reduce the injection pressure.
2. Concerning eco-friendliness, recycled secondary materials can be used for the core.
3. Flexible for different usage characteristics. For example, soft materials can be used as the skin layer for thick parts, while hard materials can be used for the core, or plastic foam can be used for the core to reduce the weight.
4. Core materials of lower quality can be used to reduce the cost.
5. Expensive materials with particular surface properties, such as electromagnetic insulation or high electrical conductivity, can be used for the skin or core layer in order to enhance the part performance.
6. A proper combination of skin and core material is able to reduce the residual stress of finished part and enhance mechanical strength or surface properties.

Refer to Table 12.1 for the combinations and applications of different materials, which clearly utilize the advantages of co-injection molding in various parts.

Table 12.1 Skin/Core Combinations [2]

Combination	Properties	Application
Soft feel skin/hard core	High strength core with soft feel skin	Door handles, gear levers
Unfilled skin/core with conductive filler	Electromagnetic interference shielding	Computer housing
Virgin skin/recycled core	Environmentally friendly production, cost saving	Garden furniture, automotive bumpers
Unfilled skin/reinforced core	High surface finish, structural performance	Automotive door handles
In mold paint/variable core	No finishing of product required after molding	Wheel trims
Unfilled skin/foamed core	Good surface finish, low density, high rigidity	Automotive body panels
Pigmented skin/uncolored core (or reverse)	Reduced pigment cost, aesthetics	Yogurt pots

The bi-injection molding process uses two independent gates to inject melt into the mold cavity. Such a process comes with the following advantages:

1. Straightforward process engineering.
2. High efficiency on production, finishing dual-color products in one time.

3. Short cycle time and low cost for mold.
4. Identical material brings better structural strength.
5. Satisfaction of consumers' requirements for abundant colored plastic items.

However, although there are many advantages of co-/bi-injection molding, the main challenge faced in today's manufacturing is to control the material spatial distribution inside the cavity for these molding processes. This is especially important for structural applications as product stiffness depends largely on the skin/core distribution of two different materials. For the co-injection molding process, unlike sequential over-molding, which has a distinct interface, the skin/core interfacial flow front cannot be controlled with ease. If mold design modification is allowed, manipulating the gate position effectively changes the skin/core distribution inside the cavity. As in Figure 12.6(a), the fountain flow behavior of the skin leaves a frozen layer on the cavity wall, while the sequential shot penetrates into the melted core through the path of least resistance to displacing the first material. Because of this fountain flow behavior, the first material is driven by the melt front, forming an additional frozen layer all the way forward. Hence, an ideal co-injection molded part exhibits a core completely encased by the skin except for the regions near the gate. The possibility of breakthrough behavior, as shown in Figure 12.6(b), is also important to determine the skin/core filling ratio and melt strength.

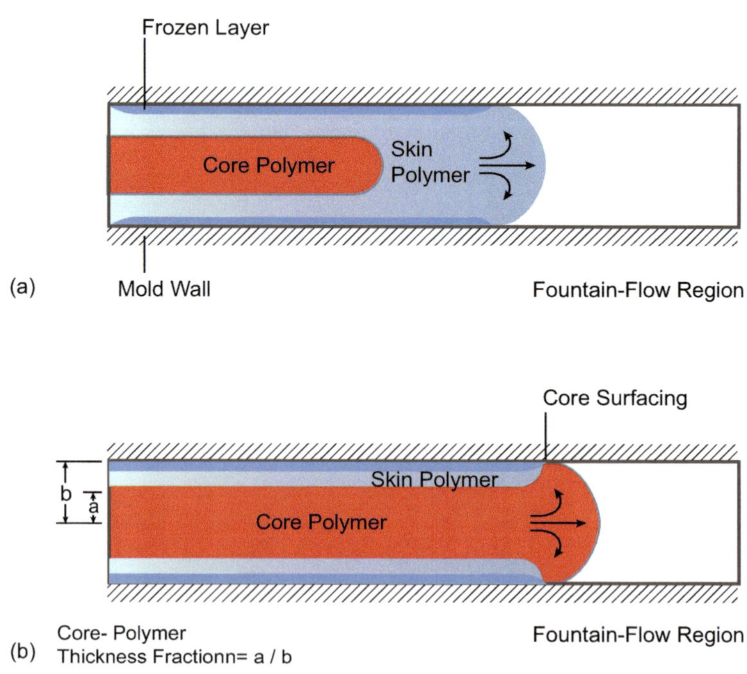

Figure 12.6 (a) The penetration behavior of skin/core melt, and (b) melt breakthrough behavior

Besides the challenges in co-injection molding mentioned above, there is one common topic concerning this process that should be addressed. This topic is how the simultaneous material combination of co-/bi-injection achieves high compatibility and improved structural strength. The interface strength compared to distinct interface methods of multi-component molding (MCM) can be seen in Figure 12.7, where it is shown that a liquid-to-liquid combination (co-/bi-injection) will be better than solid-to-liquid combination (MCM) effect. This interfacial strength depends on molecular diffusion, similar to judging the strength of a weld line.

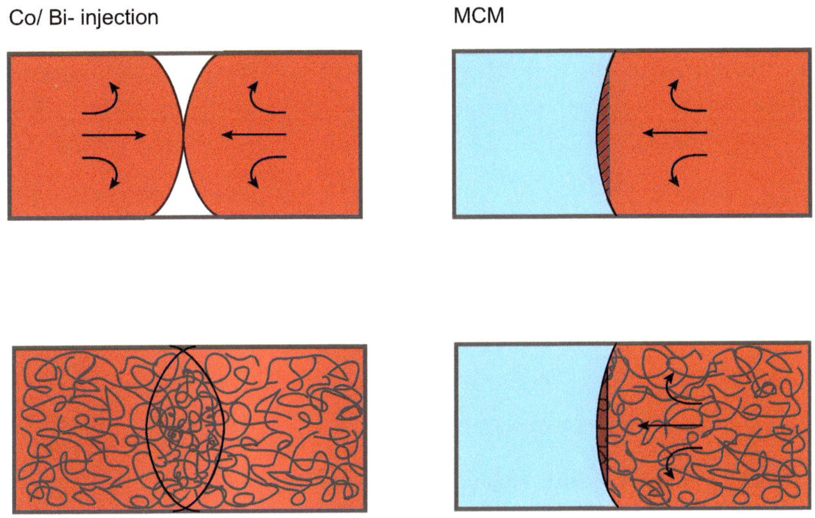

Figure 12.7 Interfacial strength comparison [3]

In summary, due to the injection of two dissimilar materials, the primary simulation challenges of co-/bi-injection include:

1. Optimization of the combination of two material properties
2. Maximization of overall flow performance ratio of different materials
3. Avoiding blow through and maintaining quality surface finish
4. Predicting weld line defects for plastic intensity and cornering effect

12.1.3 Theory Models

Assuming that the materials for the skin layer and the core layer are compressive Newtonian fluids, and neglecting the influence of surface tension, the velocity field, pressure field, and temperature field, a melt flow can be represented by the governing equations mentioned in Section 5.2.1.

To track the advancement of the interface position over the molding process, a volume fractional function f_i is defined to describe the filling status of each cell. The fractional volume function is governed by the following transport equation:

$$\frac{\partial f_i}{\partial t} + \nabla \cdot (\mathbf{u} f_i) = 0 \qquad \text{Equation 12.1}$$

Two transport equations for polymer and gas are solved to determine the advancement of interfaces and subscript i denotes for different transported material. For $i = 1$, the transported material is skin polymer, where $f_1 = 1$ is defined as the cell filled by the skin polymer melt, $f_1 = 0$ as the empty cell enclosed by air or the cell occupied by the core polymer; For $i = 2$, the transported material is core polymer, where $f_2 = 1$ defines as the cell filled by the core polymer, $f_2 = 0$ as the empty cell enclosed by air or core polymer melt. Hence, the interfaces (skin polymer/air, skin/core polymer, core polymer/air) are located within the cells with $0 < f_i < 1$.

In order to know the position of the polymer-air and polymer-polymer interfaces, we introduce f_c which represents the position of the interface for different materials. A front tracking value greater than f_c denotes a region filled by the polymer, while a smaller than f_c value corresponds to an empty region (filled with the other material, respectively).

12.2 Practical Applications

12.2.1 Co-Injection Molding of Fork Model

In this case, the multi-cavity co-injection molding model in reference [4] will be rebuilt using Moldex3D, whose molding conditions will be used for the molding analysis to examine the flow behavior of the experimental materials under different injection speeds, and the physical phenomena of the multi-cavity co-injection molding system will be explored in detail using CAE tools.

Such a co-injection molding experiment uses a mold designed with three cavities. It has a one-to-three trapezoidal runner and three disk-shaped cavities. The mesh model built for the simulation analysis is shown in Figure 12.8; there are 20 layers of mesh along the thickness direction. The cold runner of each mold cavity is numbered sequentially as runner 1, runner 2, and runner 3 for the convenience of follow-up discussion.

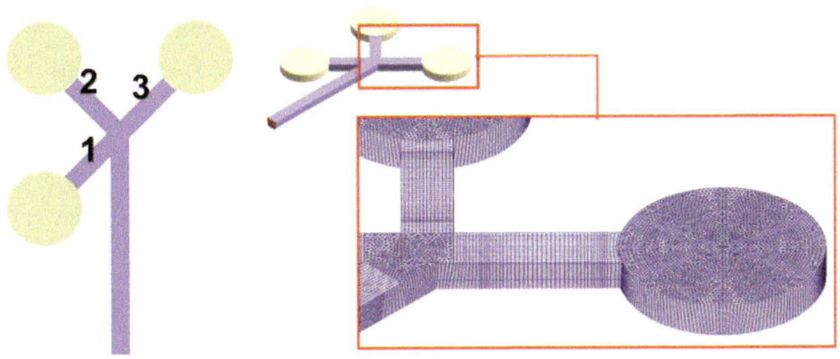

Figure 12.8 The 3D mesh used in this case

The material used is GPPS685, and its corresponding viscosity and PVT can be found in the Moldex3D material databank. The mold temperature is 60 °C, the melt temperature is 220 °C, and the three speeds are high (51 cm^3/s), medium (35.7 cm^3/s), and low (10.2 cm^3/s). The filling behavior simulated under different injection speeds is shown in Figure 12.9; from left to right are the simulation results under low, medium, and high injection speeds respectively. The material used for the skin layer and the core layer is the same, with only the core material dyed for a clear observation. It is found in the experiment [5] that the filling behavior of the material of the core layer is similar at the low injection speed, and the filling distance of the core layer in the secondary runner closest to the primary runner is the longest. A totally different behavior is seen under a high injection speed, such that the runner which had the longest filling distance at the low injection speed now have the shortest.

Figure 12.9 Simulation results of flow behavior in different molding conditions (from left to right, low, medium, and high injection speeds)

In order to understand the simulation difference between the different injection speeds, the temperature distribution of the center slicing plane from the simulation result is explored and shown in Figure 12.10. According to the distance between each runner to the primary, the flow length order is runner 1 > runner 2 > runner 3. At a high injection speed, runner 1 will be filled up by the skin material and is unfavorable for the follow-up filling of the core material. The flow length of core layer differs under various injection speeds: runner 1 has the longest flow length at the low injection speed, while runner 3 has the longest flow length at the high injection speed. This result is explained by the temperature difference at each injection speed. The shear-heating effect becomes more significant at a high injection speed, which leads to runner 3 taking up the largest portion of shear heating comparing to the other two, so the flow length of the core layer of runner 3 therefore becomes the longest.

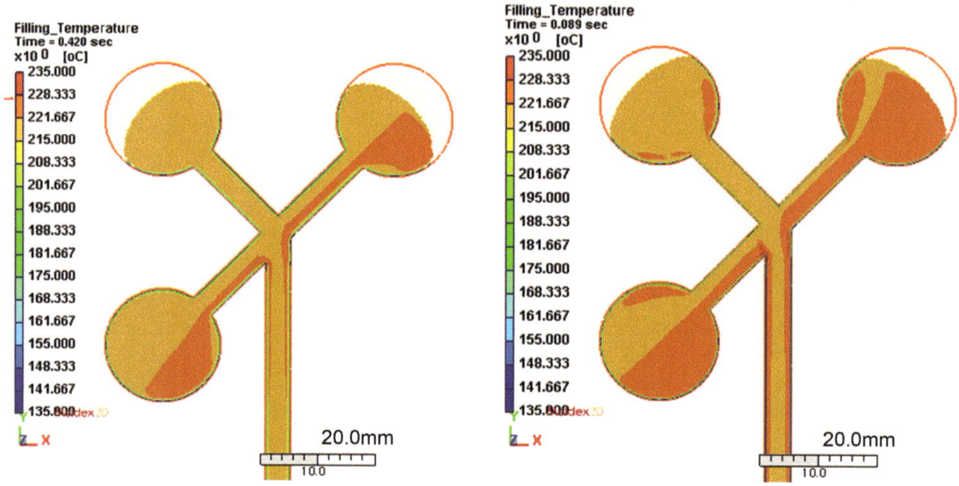

Figure 12.10 The temperature distribution of center slicing plane at (a) lower and (b) higher injection speeds

12.2.2 Co-Injection Molding: Core Breakthrough and Flow Imbalance

We have presented numerical simulations concerning two important defects encountered in the sandwich molding process: the core breakthrough in a thin-walled part and the flow imbalance in a multi-cavity mold. The part thickness largely determines how far the skin melt front advances before breakthrough happens. The thickness effect on core breakthrough was studied by using the spiral molds with the thicknesses of 1, 2, 3, and 5 mm. The spiral length was 900 mm. A similar mold scheduled for further verification studies is shown in Figure 12.11. The sandwich injection molding machine was a twin-barrel, 85-ton, maximum 130 MPa device. The injection rates were all 50 cc/s in this study. The skin and core material used was PMMA (Chi-Mei Acryrex CM 211). The melt temperature and mold temperatures were 205 °C and 70 °C, respectively.

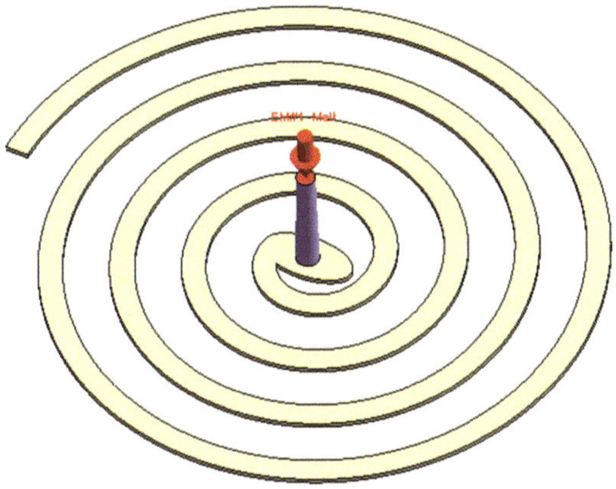

Figure 12.11 The mold to be used in the breakthrough study showing its spiral-shaped cavity

An example of the predicted skin ratio distribution for the 2-mm cavity is shown in Figure 12.12(a), where the red color suggests a skin-rich region on both sides of the mold and the blue region at the flow front shows the core penetration. The breakthrough region is clearly indicated as the blue-colored area. Figure 12.12(b) exhibits the cross section of the sandwich part. The core is encased in the skin layer with a rounded edge despite the square cavity shape.

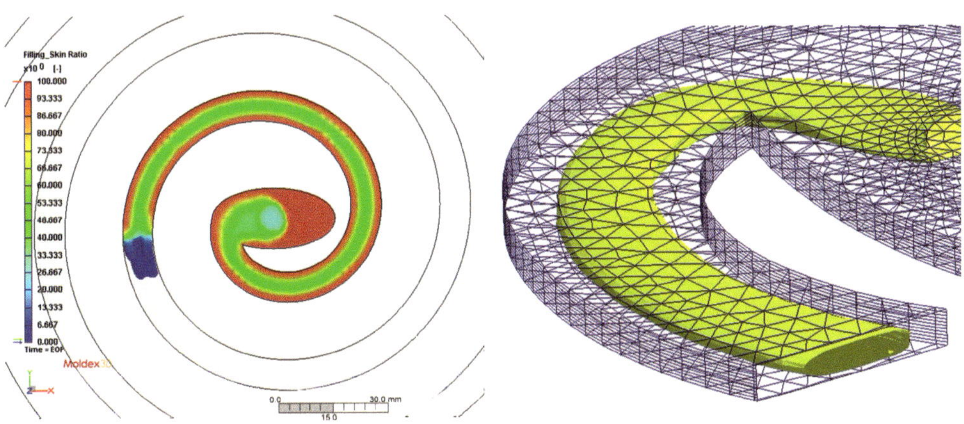

Figure 12.12 The skin ratio distribution along the spiral mold when breakthrough occurs: (a) top view and (b) side view

On increasing the core filling ratio (i. e., the core entered at an earlier time), the core penetration occurred in an earlier stage and the flow length decreased, as shown in Figure 12.13. According to an apparent linear interpolation with the desired core ratio, the flow length can be predicted roughly.

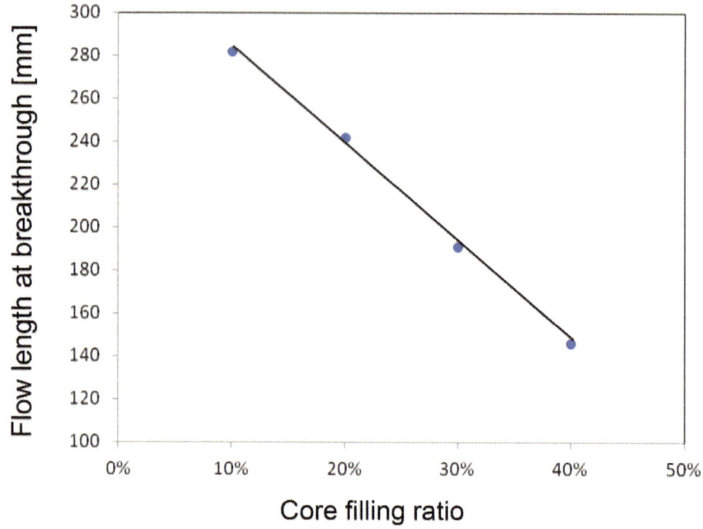

Figure 12.13 Maximum flow length decreased as the core filling ratio increased for the 2-mm thickness mold

Under the same condition of filling ratio, the flow length depended on the cavity thickness. While the frozen skin layer thickness remained the same, larger thickness means the amount of skin material could extend to a longer distance when the ratio remained the same. The breakthrough flow length at a thickness of 1, 2, 3, and 5 mm were tested, and the results are presented in Figure 12.14. It is apparent that, for the thickness of 1 mm or less, it would be difficult to make large enough parts, due to early core breakthrough.

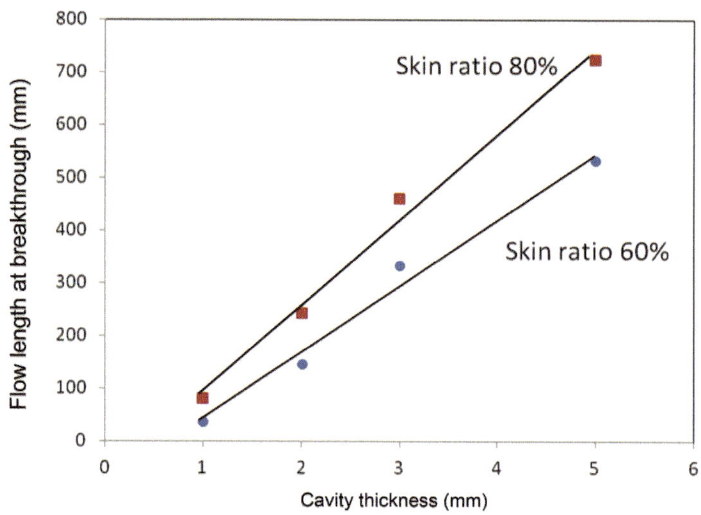

Figure 12.14 Maximum flow length increased as the cavity thickness increased

Regarding the individual advancement of skin or core, flow length versus injection time graphs are shown in Figure 12.15. The slopes can be seen as the melt front speed. As shown in Figure 12.15(a), the core traveled at a faster speed after it entered the cavity and before it penetrated the skin layer. The core volumetric flow rate remained the same as the skin. However, due to the shrunken passage caused by the frozen skin layer, the core layer traveled at a higher speed. Once it penetrated the skin, the passage opened and the core slowed down. Moreover, note that for the 2-mm cavity in Figure 12.15(b), the slope in the skin-advancing region is not a straight line since the injection pressure had already reached the maximum pressure so the target flow rate could not be reached.

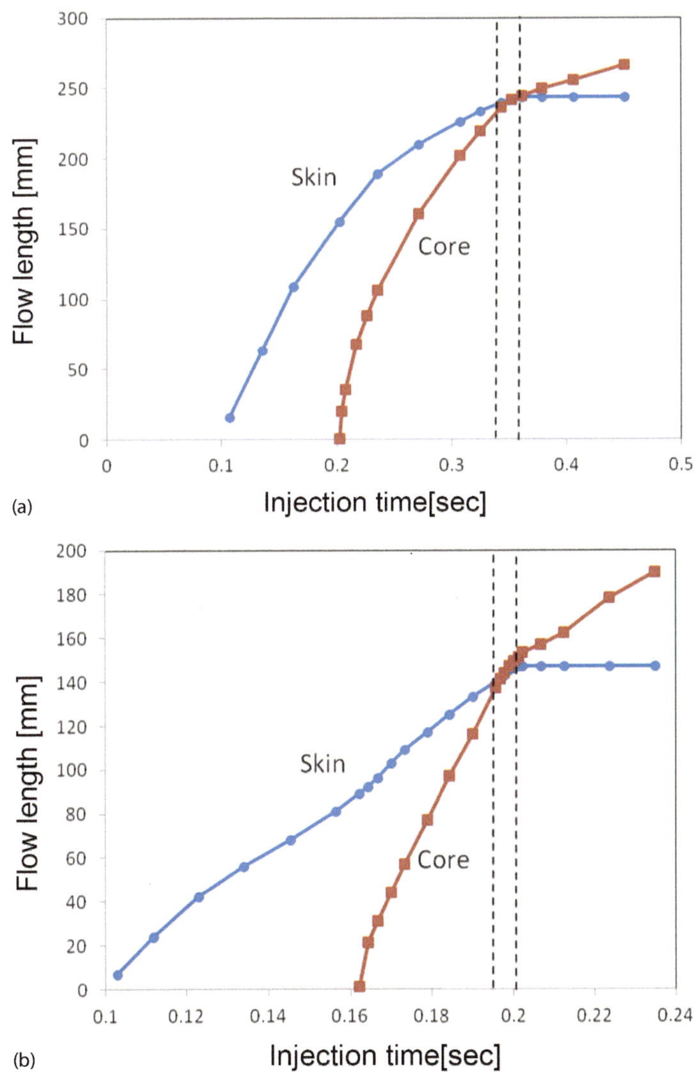

Figure 12.15 The melt front evolution with time of (a) 5-mm and (b) 2-mm thickness cavities, where the stage of core expansion is indicated between the dotted lines

The difference of pressure required to expel the skin at the front also affected how the core advanced. Being able to predict these important aspects regarding skin/core movement in test models is essential for application in parts with more complex geometry. There is a transition zone, i.e. a "core expansion region" as proposed by Watanabe [5], who observed the skin/core movement near the breakthrough area. The core slows down to be in sync with the skin rate until the skin is depleted. The core melt front then expands in the thickness direction due to the drop of pressure from the insufficient skin layer. This expansion of core layer was also found in the simulation, as shown in Figure 12.16.

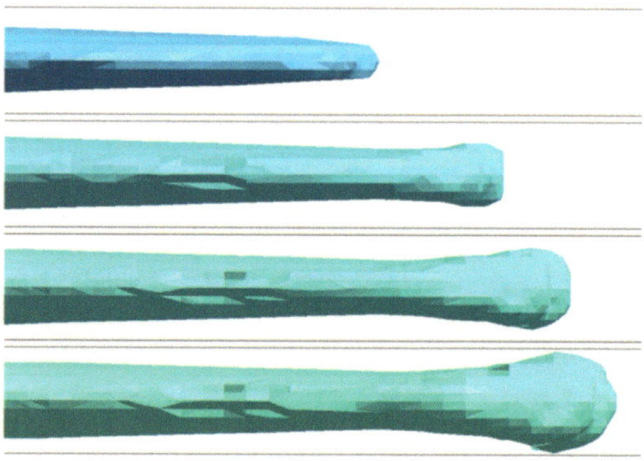

Figure 12.16 Side view showing that the core expanded in thickness direction as the melt front advanced right before breakthrough

12.2.3 Co-Injection Molding: Fiber Orientation Predictions

The fiber orientation will affect the mechanical properties of the material and is discussed in this section with the simulation of a co-injection part. The case geometry is ASTM D638 Type V standard test piece, as shown in Figure 12.17(a), and the material used is PP Globalene SF7351. Figure 12.17(b) shows the measure node position.

The simulation result is shown in Figure 12.18(a), where A11 in flow direction shows a higher value on the part surface, rising from 0.75 to 0.85 on the surface, and then falling to 0.25; contrary to the flow direction, A22, perpendicular to the flow direction, first drops from 0.2 to 0.15 and then rises to 0.5, while A33 along the thickness direction is the lowest. The same trend can be found in the experimental data scanned by µ-CT, as shown in Figure 12.18(b). The experimental and simulation results are slightly different in the regions from 0 to 0.2 in the thickness direction. This difference is due to the shrinkage of the product thus affects the fiber orientation of the solid layer. Overall, the results of experimental measurements and simulation predictions are consistent.

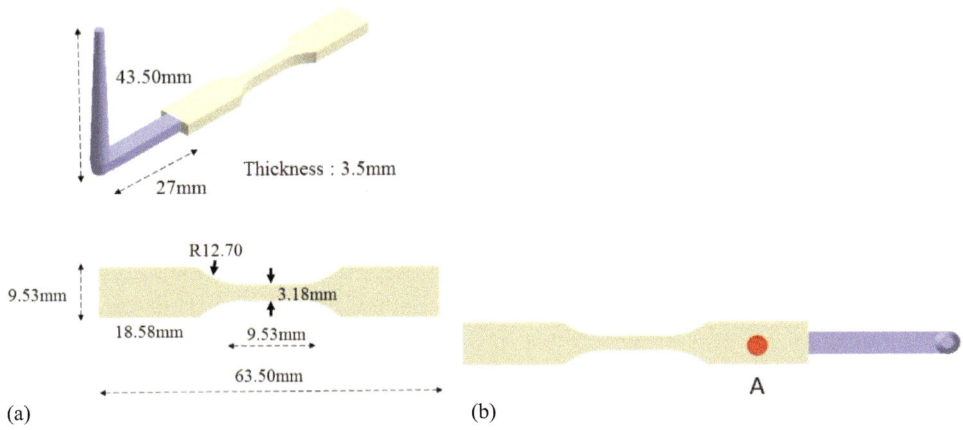

Figure 12.17 (a) Model Geometry and dimensions and (b) the measure node of point A

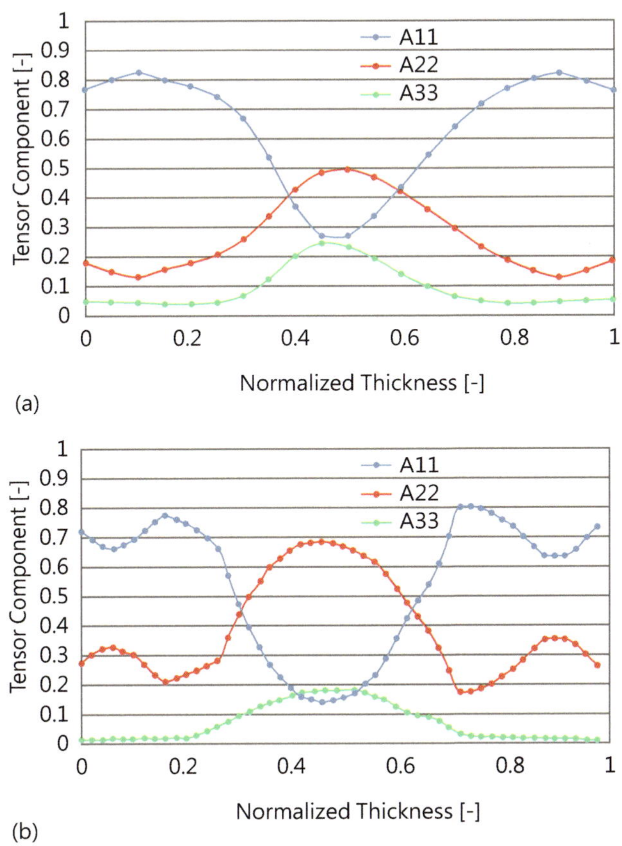

Figure 12.18 Comparison of the fiber orientation distribution between (a) simulation and (b) experiment [6]

12.2.4 CAE Case of Bi-Injection Molding

As the bi-injection process is to inject different materials from different gates, the focus of the simulation lies on the merging position of these two different materials in the part. Hence, in the following case the feeding ratio of two different materials from different feed inlets is explored. As shown in Figure 12.19, the geometry has a length of 100 mm, width of 30 mm, and thickness of 5 mm. In order to simplify the exploration of different feeding ratios, the viscosity of a Newtonian fluid is adopted. We fix the flow rate from EM#2 as Q_2 but change the flow rate from EM#1 as Q_1, and then the effect of difference in flow rate can be investigated.

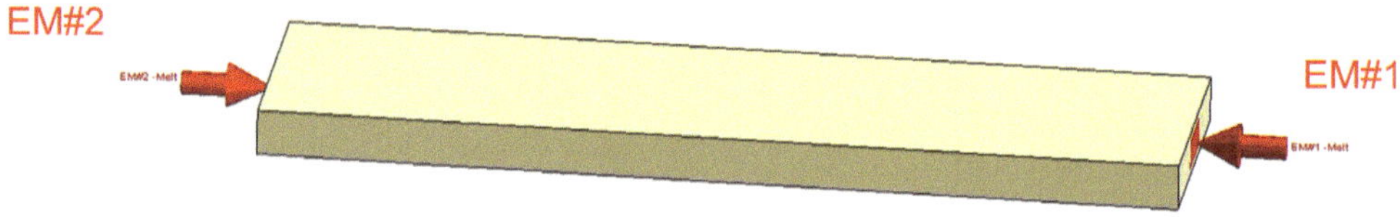

Figure 12.19 The 3D geometry of part

Figure 12.20 shows the filling behaviors under different feeding ratios. When $Q_1/Q_2 = 1$, as the switching ratio is 50%, which represents an identical flow rate in both inlets, the merging point appears in the middle of the part. When $Q_1/Q_2 = 2$, which means the flow rate ratio is 2, the merging point appears at 1/3 of the total length of the part. And when the flow rate ratio increases to 3, that is, $Q_1/Q_2 = 3$, it can be seen obviously that the merging point appears at the 1/4 of the total length of the part. The above three results effectively demonstrate the difference in merging point position under different flow rate ratios based on a fundamental phenomenon verification of the bi-injection process.

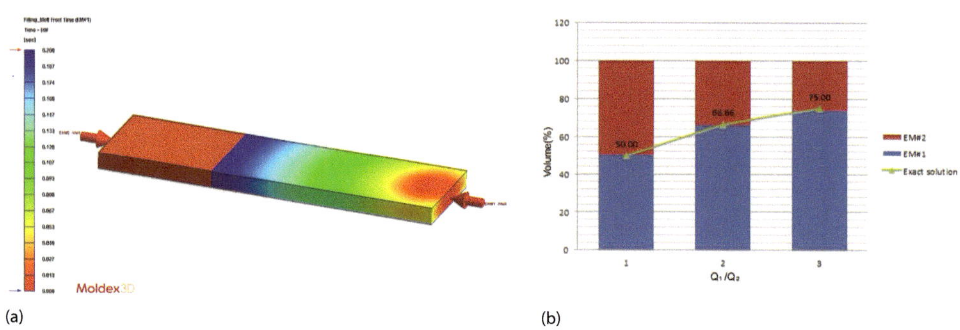

Figure 12.20 Simulation results for flow rate ratio of (a) $Q_1/Q_2 = 2$, and (b) comparisons of different flow rate ratios

In the part shown in Figure 12.21, there are two inlets for materials of different colors to be injected. EM#1 locates on the top cover, while EM#2 locates on the base. The material used here in both cases is PP. The simulation is to check the location of the weld line, by which the molding conditions are adjusted.

Figure 12.21 The 3D geometry of part

As the mold cavity of inlet 1 has a more complicated geometric feature, it is anticipated that the left side of the cavity would require a higher injection speed during filling. Therefore, the same flow rate of injection speed (20 cc/s) is given to the both sides first to understand the computed result of injection pressure. It is found just as expected at the design phase that the feature on the left mold cavity results in a higher injection pressure at the filling end, as shown in Figure 12.22.

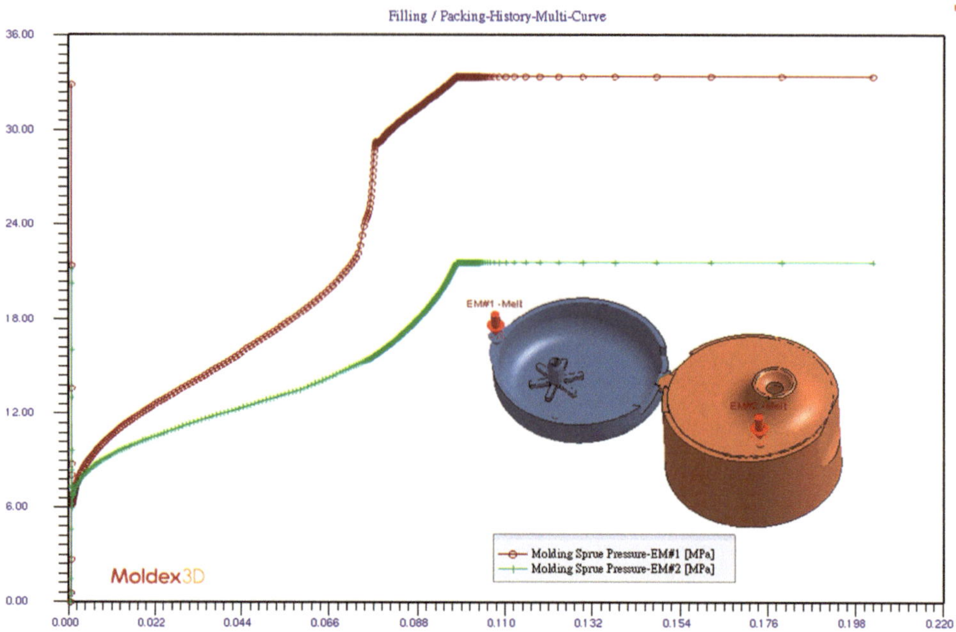

Figure 12.22 Simulation result of sprue pressure comparison for same flow rate ratio

As the volume of both cavities is similar, the location of the weld line appears at the joint of the cap, as shown in Figure 12.23.

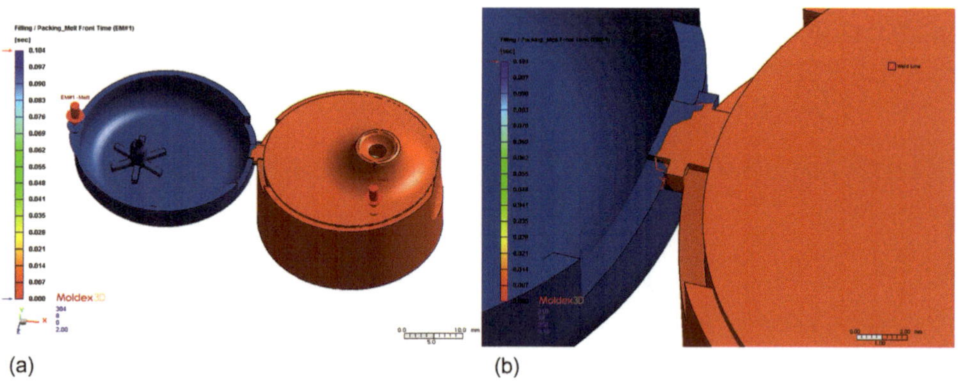

Figure 12.23 Simulation results for same flow rate ratio: (a) melt front distribution and (b) weld line position

In order to prevent fatigue failure at this joint due to repeated opening and closing, the flow rate at inlet 2 is further adjusted to observe how the position of the weld line changes. Figure 12.24 shows an analysis result of gradually raising the flow rate at inlet 2. It can be seen that the location of the flow front gradually moves toward the left cavity while the weld line is moved away from the joint of the cap. Figure 12.25 shows a zoom-in of this region for clear observation.

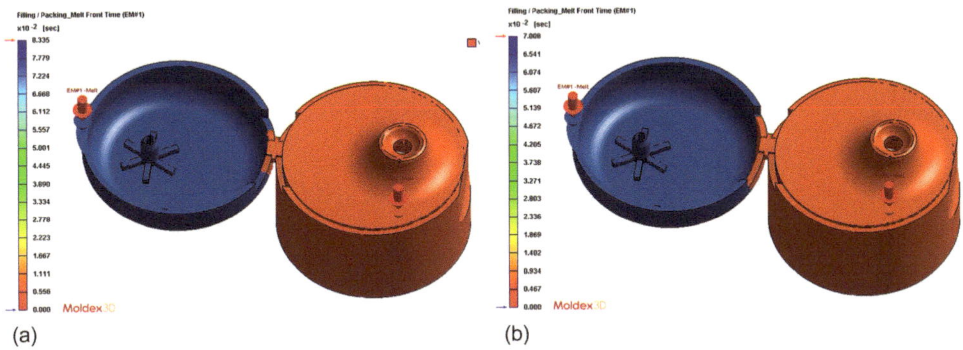

Figure 12.24 Simulation results for different flow rate ratios: (a) flow rate of EM#2 = 30 cc/s and (b) flow rate of EM#2 = 40 cc/s

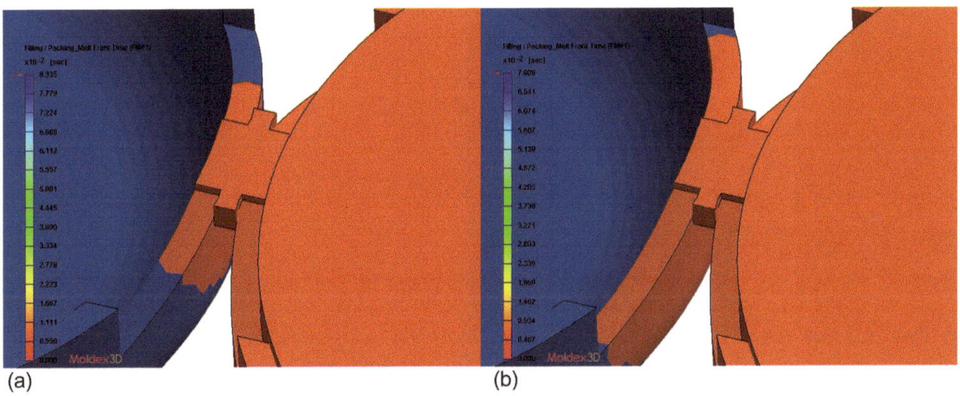

Figure 12.25 Weld line positions for different flow rate ratios: (a) flow rate of EM#2 = 30 cc/s and (b) flow rate of EM#2 = 40 cc/s

We can also compare the clamping force computational results, as shown in Figure 12.26 under three different designs. It is found that the injection pressure of inlet 1 can be reduced while the flow rate at inlet 2 is raised, and the weld line can be controlled to the expected location as well. When the flow rate of inlet 2 is 40 cc/s, the overall clamping force required is relatively small comparing to the other two flow rate designs. Therefore, through an analysis, the optimum process conditions can be further obtained.

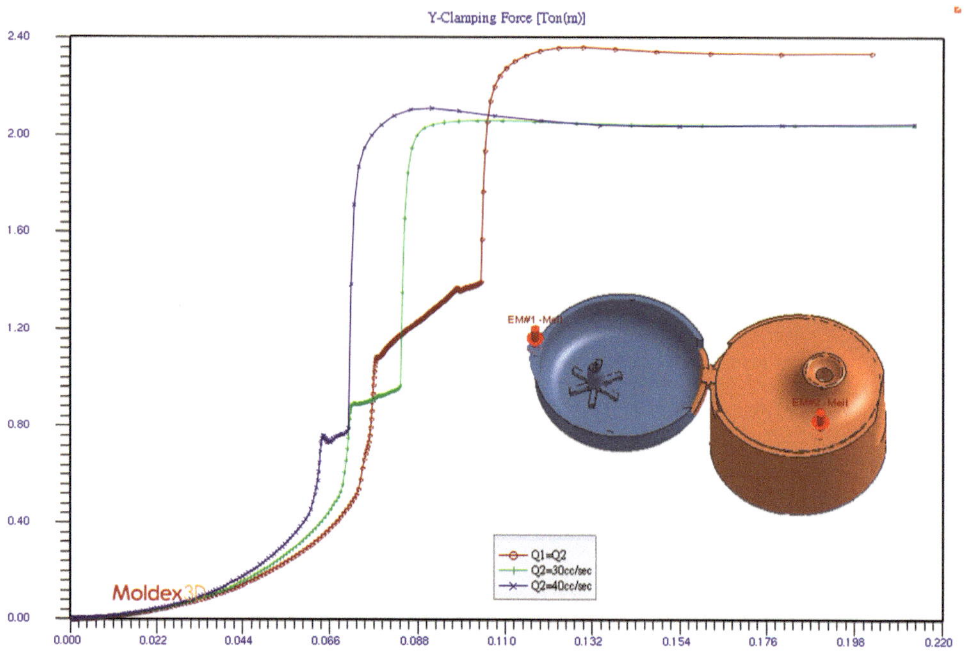

Figure 12.26 The comparisons of clamping force for different process conditions

■ 12.3 CAE Case Study

Following the discussions of this chapter and the corresponding CAE verification case, an exercise model, as illustrated in Figure 12.27, is provided in this section for readers to practice CAE application in co-/bi-injection molding.

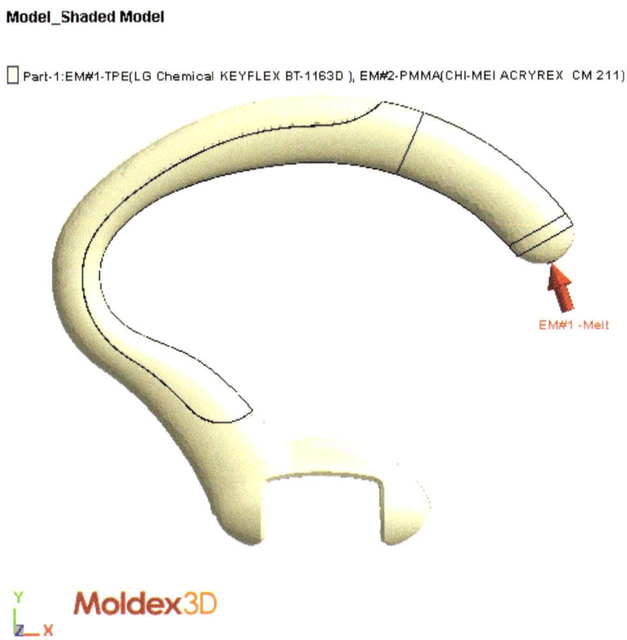

Figure 12.27 Case study sample

Please download the analysis files for CAE Case Study 12.3 from the following website:

https://moldex3d.box.com/s/zr6fvc1vlhbi4ocx111jwd3wmxt4ooif

Then open the .m3j or .mvj file in Moldex3D and answer the questions below according to the analysis results.

Questions:

1. Pre-analysis of the model
   a) What is the max L/t of the case? If taking TPE as the material, does short shot occur? Why?
   b) Where will the filling ends and weld lines be?
2. What are the results of default process conditions for melt front time (skin/core), history of sprue pressure, and warpage behavior?

3. If we switch to a different entry time for the core material, what results will be influenced?
4. Based on Run 1, if we change the process conditions such as melt temperature of skin material/core material and mold temperature, what will be the results?

References

[1] Sumitomo (SHI) Demag Plastics Machinery GmbH: "Multi-Component Technology" [Online]. *https://www.interempresas.net/FeriaVirtual/Catalogos_y_documentos/197307/TP_Multi_gb.pdf*

[2] Goodship, V.; Love, J.: *Multi Material Injection Moulding*, Rapra Review Reports, Smithers Rapra Press, 2002

[3] Huang, K.-C.: "Study of Flow Length Breakthrough and Warpage in Co-Injection Molding", Master Thesis, Department of Mechanical Engineering, National Yunlin University of Science and Technology, 2014

[4] Yang, W.; Yokoi, H.: Visual Analysis of the Flow Behavior of Core Material in a Fork Portion of Plastic Sandwich Injection Molding, *Polymer Testing*, vol. 22, 2003, p. 37

[5] Watanabe, D.; Ishiaku, U.S.; Nagaoka, T.; Tomari, K.; Hamada, H.: The Flow Behavior of Core Material and Breakthrough Phenomenon in Sandwich Injection Molding: Part I: Dependence on Viscosity and Injection Speed of Skin/Core Materials, *International Polymer Processing*, vol. 18, 2003, p. 398

[6] Huang, C.-T.; Chen, X.-W.; Fu, W.-W.: Investigation on the Fiber Orientation Distributions and Their Influence on the Mechanical Property of the Co-Injection Molding Products, *Polymers*, vol. 12, 2020, p. 24

13 Gas-/Water-Assisted Injection Molding

Chih-Chung (Jim) Hsu, Yu-Sheng (Tim) Chou

Gas-assisted injection molding (GAIM) is a revolutionary plastics molding technology developed during the past decades. GAIM has been promoted to overcome some inherent weaknesses of conventional injection molding processes. By extending the concept of GAIM, with the substitution of water for gas as the assisting fluid, the concept of water-assisted injection molding (WAIM) is formed. This latter method has the advantages of a shorter product cycle, lower cost, and good packing capability, and has been gradually introduced to make various products.

13.1 Basics

The main concept of GAIM, or WAIM, is to inject a fixed amount of melted resin as in the conventional injection molding process, followed by an injection of a controllable amount of fluid from the gate or a preinstalled gas/water channel into the core of mold cavity with the flowing molten plastic, as shown in Figure 13.1. As fluid, nitrogen (N_2) is usually used for gas assisting, while water at room temperature is used for water assisting. Hereinafter the gas-/water-assisted process will be commonly referred to as fluid assisted.

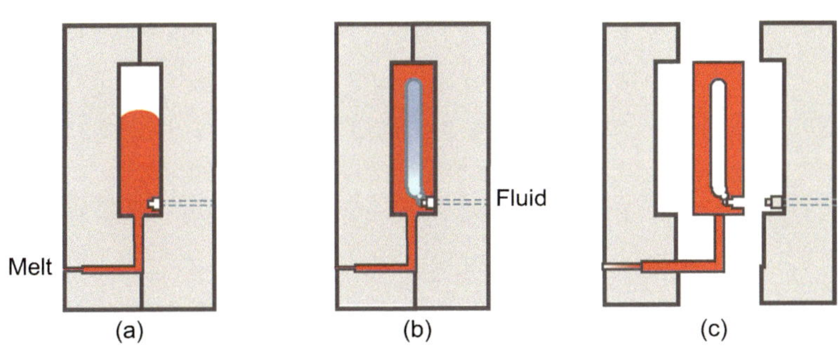

Figure 13.1 Schematic diagram of the fluid-assisted injection molding process: (a) melt filling, (b) fluid injection and packing, and (c) part demolding

13.1.1 Process Principle

According to the amount of plastic melt injected into the mold cavity, the process can be divided into two categories:

1. Short-shot: The amount of resin injected is smaller than the volume of the mold cavity, and with the assistant fluid the resin is expanded and shaped, which is suitable for making stick-shaped or large-sized hollow parts. The hollowed core ratio of a finished part is around 10–35%.
2. Full-shot: The amount of resin injected is the same as the volume of the mold cavity. For the time being, the assistant fluid does not facilitate the filling but is for the packing during cooling in order to compensate for the shrinkage of melt, which is suitable for the molding of thin-plate parts. The hollowed core ratio is 3~15% and the method is very effective for improving sink marks and warpage. There are new improvements in the full-shot molding process, which will be mentioned later.

13.1.1.1 Short-Shot Process

1. Filling Stage

The short-shot molding process of fluid-assisted molding is shown in Figure 13.2. As shown in Figure 13.2(a), as for general injection molding, a certain amount of melt is injected by the injection machine, starting from the barrel, through the nozzle, runner, gate, and into the mold cavity. The biggest difference from general injection molding is that the melt injected is not required to fill up the whole mold cavity, but just to fill the mold cavity partially.

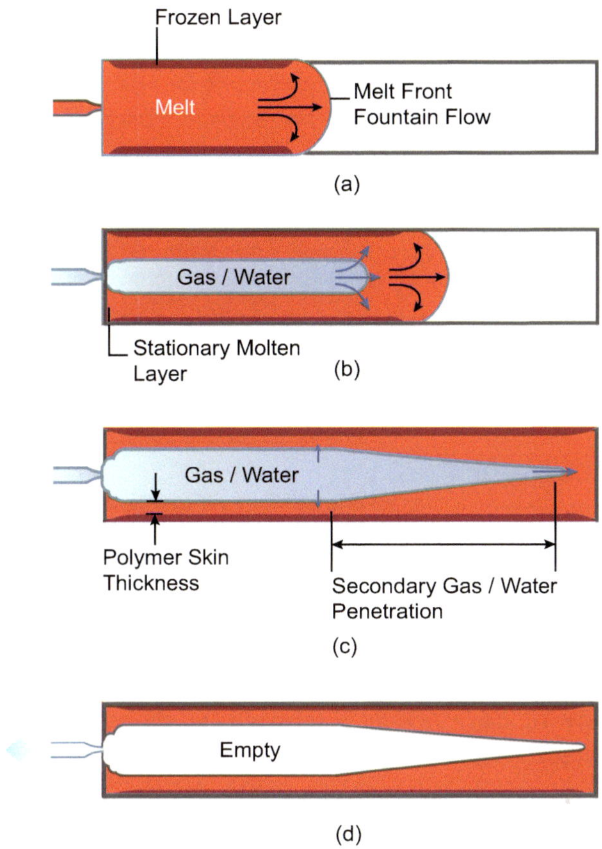

Figure 13.2 The molding process of fluid-assisted molding

2. Fluid-Assisted Filling Stage

As shown in Figure 13.2(b), after a certain amount of melt is injected into the mold cavity, the nozzle is closed and the fluid starts to be pressurized and pushed into the mold cavity after a time delay. The fluid drives the melt to fill up the whole mold cavity along the path with minimum flow resistance. Such filling done by the fluid is called the primary fluid penetration.

3. Fluid Packing Stage

As shown in Figure 13.2(c), the fluid pressure still remains for a certain period in the packing process, and the fluid is left in the mold cavity, while a certain pressure is kept for the packing of the mold cavity in order to compensate for the shrinkage of melt induced by cooling. By referring to Figure 13.3, it can be seen that the filling done by the fluid is completely to fill up the space induced from the shrinkage of melt after cooling, which is called the secondary fluid penetration, and such behavior differs among plastics of different shrinkage ratios.

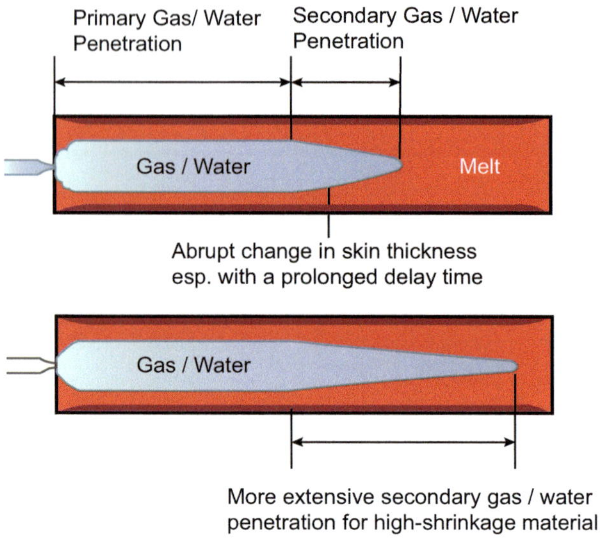

Figure 13.3 Gas/water penetration in GAIM/WAIM; definition of primary and secondary penetration

4. Fluid Release and Cooling Stage

As shown in Figure 13.2(d), the fluid is released and the opening is filled with plastic after the packing ends and the part is cured completely. The mold is then opened to eject the finished part. In the case of a liquid-assisted process where the liquid is injected through a pinhole into the mold cavity, there is no need for plastic to close up the hole as the liquid needs to come out therefrom. If necessary, the liquid inside can be removed with gas.

13.1.1.2 Full-Shot Process

Full Shot Overflow Method

When using this method, as shown in Figure 13.4, the molding die should have two cavities: a molding cavity and an overflow cavity. A valve which separates the two cavities is also needed. After the molding cavity has been completely filled with molten material, the valve is opened and highly pressurized fluid is introduced into the cavity. The molten material is forced to flow into the overflow cavity. This method can be used to manufacture parts with a high-quality surface. The required pressure of the fluid is much lower. However, the method may result in a waste of the polymer material.

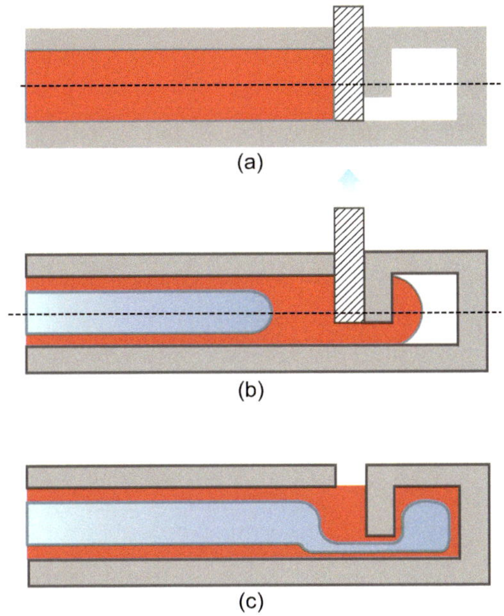

Figure 13.4 Full shot overflow method

Full Shot Pushback Method

As shown in Figure 13.5, the sprue and the fluid inlet are positioned at the two ends of the molding cavity. The injected fluid will force the plastic melt to flow back to the barrel. Hesitation marks can be eliminated by this method, and the part of material which flows back to the barrel can be reused. However, the nozzle needs to be specifically designed to prevent the backflow of the melt at the material injection stage; a fluid transfer system should also be set up.

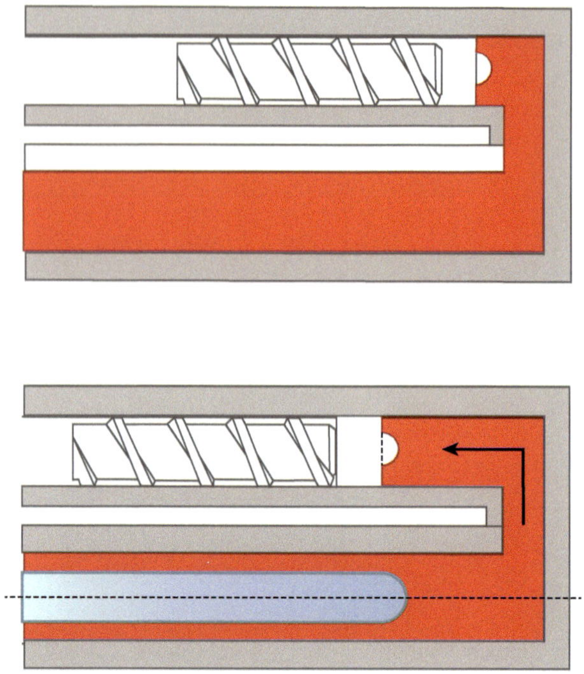

Figure 13.5 Full shot pushback method

13.1.2 Advantages and Challenges

Comparing to conventional injection, as the fluid keeps packing inside the plastic so that the pressure can be applied to the surface of finished part effectively and evenly from the core of thickness, the injection pressure required for the injection machine can be drastically reduced. Such process is also called "low-pressure hollow injection molding", and has great benefit on the improvement of shrinkage, sink marks, dimensional precision, and residual stress of finished parts. The advantages of the fluid-assisted injection molding are summarized below:

1. Reduced Injection Pressure of Machine

The maximum injection pressure required for the machine in a fluid-assisted injection molding process is only 30–50% of that required in a conventional injection molding, as shown in Figure 13.6.

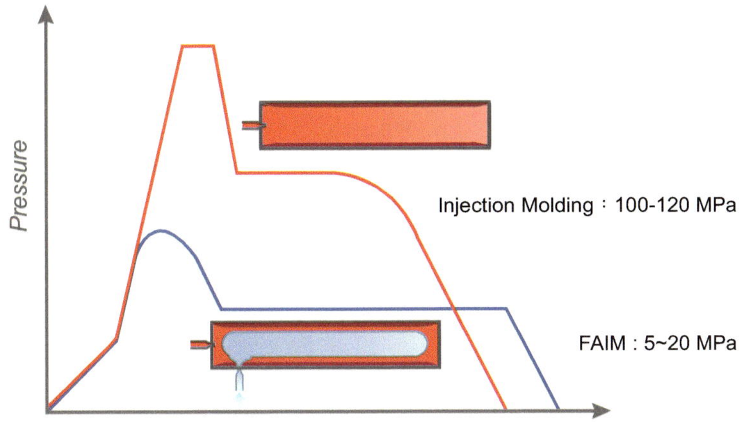

Figure 13.6 The difference in pressure between conventional and fluid-assisted injection molding processes

2. Improvement of Unbalanced Packing Issues

An unbalanced packing in the cavity is easily seen at the packing stage of conventional injection molding, as it varies according to the distance from the gate: the closer to the gate, the better the packing effectiveness, and vice versa. Such issue often causes over-packing or under-packing locally. By using assistant fluids for packing, the unbalanced packing pressure issue can be improved as fluids have better pressure transmission capability (due to the poorer compressibility of gases and liquids comparing to melts), as shown in Figure 13.7.

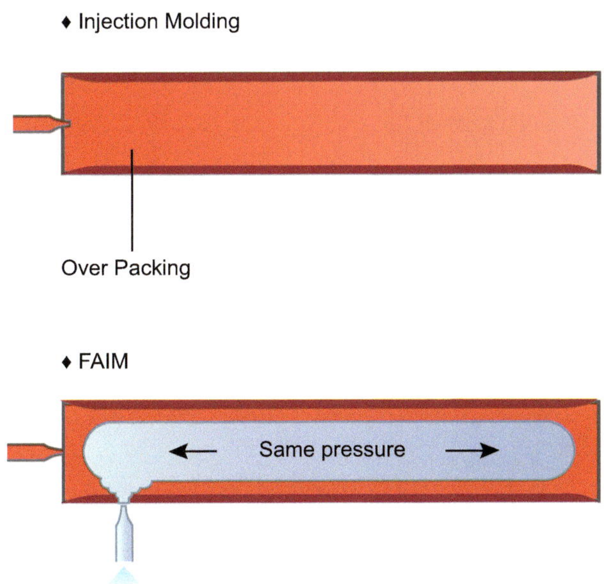

Figure 13.7 The packing phase difference between conventional and fluid-assisted injection molding processes

3. Light Weight and Costs Down

As a result of the core-out of melt, fluid-assisted injection molding is able to reduce the weight of part drastically, reaching over 50% of weight reduction for a stick-shaped part with thick cross-section, while for a part with thick ribs on a thin wall the weight can be reduced up to 35%, as shown in Figure 13.8.

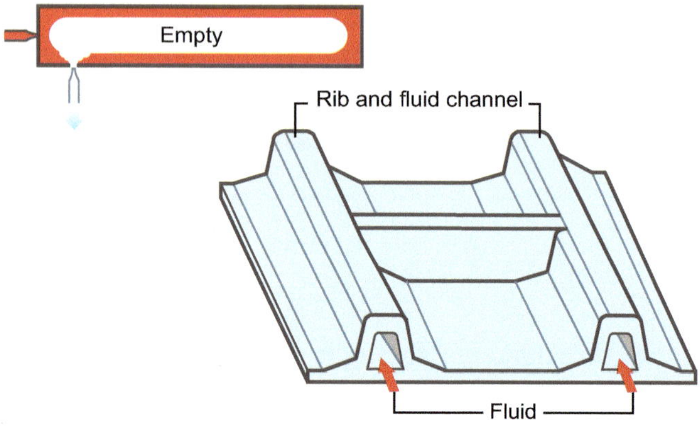

Figure 13.8 Core-out behaviors in the fluid-assisted injection molding process

4. Reduced Manufacturing Time

The cooling time in the production process almost dominates the whole cycle time of the part. Generally speaking, the cooling time is in proportion to the square of the thickness of melt. As the fluid-assisted injection molding (FAIM) greatly cores out the melt from the inside of the part and reduces the thickness of melt, the cooling time is then shortened drastically, as shown in Figure 13.9.

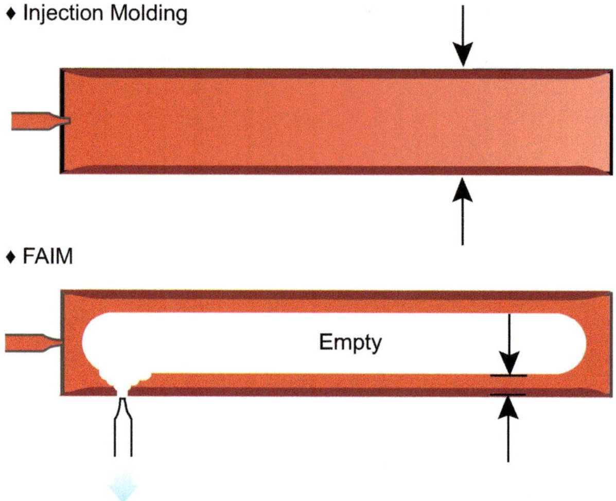

Figure 13.9 The thickness difference between conventional and fluid-assisted injection molding processes

5. Reduced Sink Marks

Using fluids to assist the filling can maintain a better packing inside the mold cavity, and with a smaller melt ratio (reduced shrinkage), the generation of sink marks can be relieved effectively, as shown in Figure 13.10.

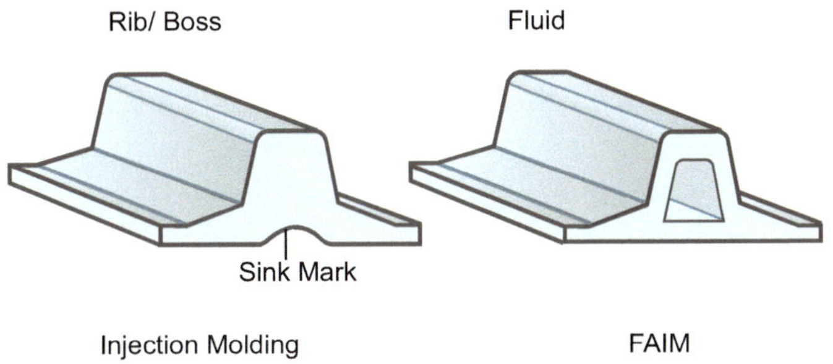

Figure 13.10 The sink mark difference between conventional and fluid-assisted injection molding processes

6. Reduced Shrinkage Caused by Thickness

The volume of a plastic part shrinks as it cools down. The thicker the region is, the higher the shrinkage becomes. Using fluid-assisted injection molding can result in effective packing with coring out of the internal melt, with reduced shrinkage at the thicker regions, as shown in Figure 13.11.

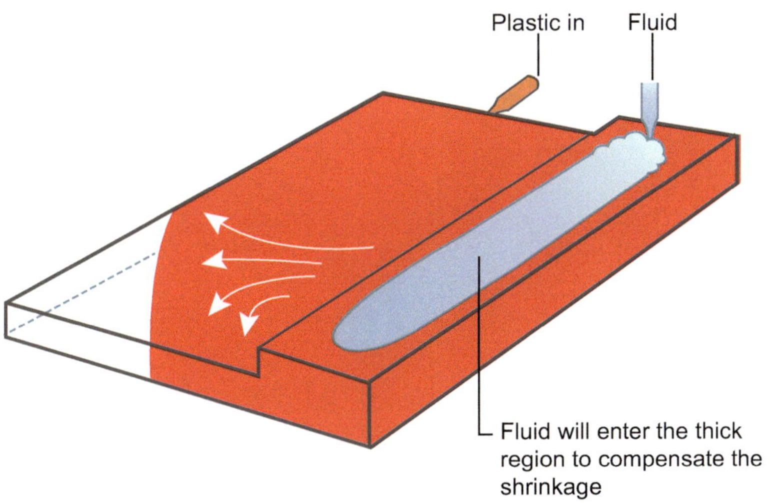

Figure 13.11 The application of fluid-assisted injection molding process in the thicker regions

7. Reduced Cooling Time

This advantage is exclusively owned by water-assisted injection molding. Water at room temperature is often used as the assistant fluid in liquid-assisted injection molding. The thermal conductivity of water is 40 times higher than that of gas, while the thermal capacity of water is four times larger than that of gas, making liquid effective in cooling molded parts directly from the inside. Using water can result in a 25% reduction of the cycle time of molding one plastic part compared to using a gas as the assisting medium. Also, a suitable liquid is able to facilitate the molding under low-pressure conditions. This advantage is the most specific and outstanding benefit possessed by liquid-assisted molding.

However, GAIM/WAIM also have common issues in the process, which are described as follows:

1. Fluid Blow-Through

In short-shot molding, as the injected fluid pushes the melt front along the path of the least resistance, a gas blow-through phenomenon may occur easily if the pre-filling is insufficient or the injected gas pressure is too high, as shown in Figure 13.12. However, in case of a too-small pressure for the assistant fluid, insufficient core-out would occur.

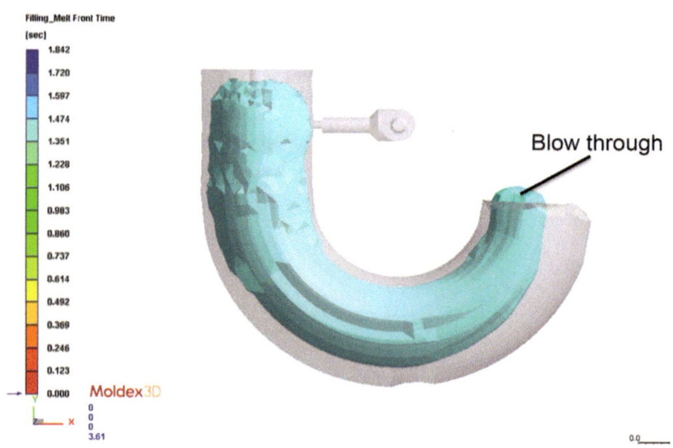

Figure 13.12 Gas blow-through behavior

2. Hesitation Marks

In the fluid-assisted injection molding process, a delay time (to avoid the backflow of melt) is required when it is switched to fluid filling after the melt filling is completed. However, if the delay time is too long, the melt front will be cooled down to form a frozen layer and thus hesitation marks are generated, as shown in Figure 13.13.

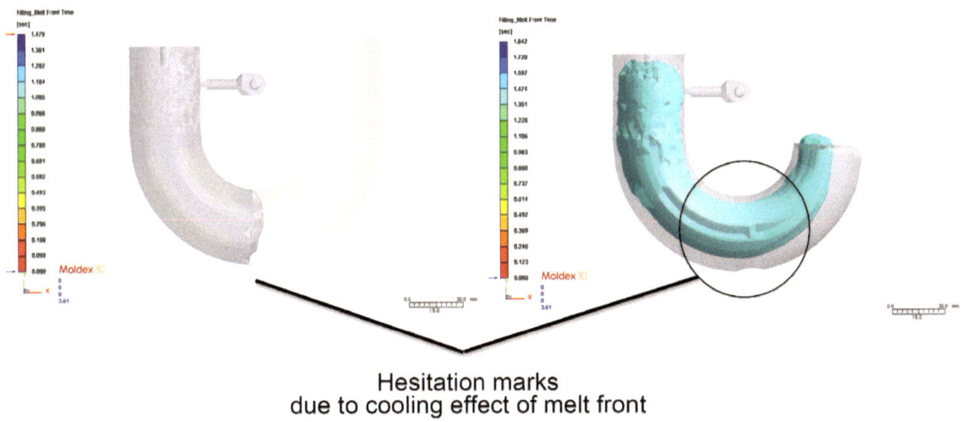

Figure 13.13 Hesitation marks due to short shot method

3. Corner Effect

When a fluid is passing through a corner, it tends to flow along the inner radius due to the smaller resistance. This can mean that the melt becomes too thin, leading to insufficient part strength or even defects, as shown in Figure 13.14.

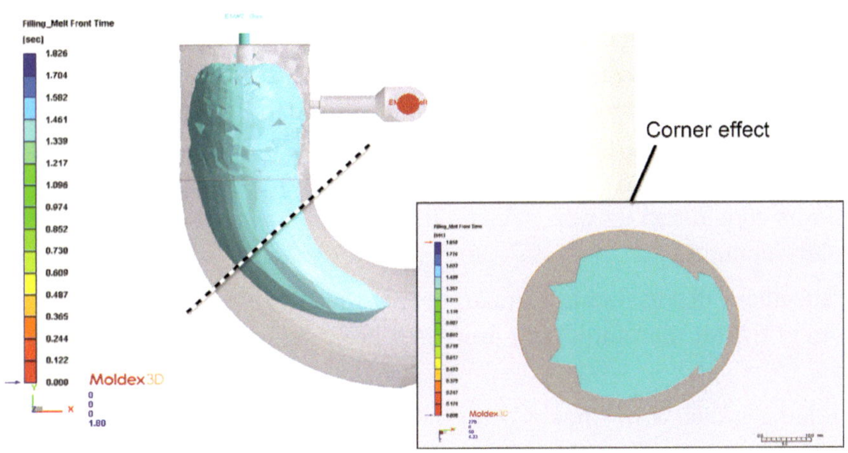

Figure 13.14 Corner effect

4. Fingering Effect

During the injection process, the assistant fluid can easily penetrate through the defined gas (liquid) channel into the thin-walled area of a part due to an overpressure. Such behavior would leave fingerprint-like flow marks on the surface, as shown in Figure 13.15.

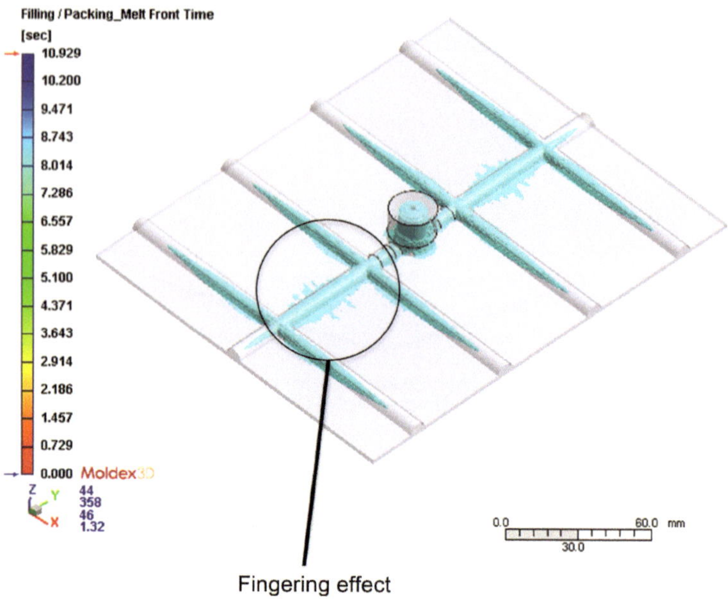

Figure 13.15 Fingering effect

5. Process Control Effect

Due to the good thermal conductivity and low cost of water, WAIM helps a lot in the improvement of molding quality and brings a breakthrough development for fluid-assisted injection molding. However, as the controllability of water is poor, apart from the consideration on how to inject the liquid into the mold cavity, the removal, recycling, and route planning of residual liquid after the part is molded need to be considered as well, which often causes a great increase in the cost of mold development.

From the above discussion it is apparent that the advantages of WAIM are similar to those of GAIM; therefore, both processes are compared again in more detail as follows:

1. Molding cycle: WAIM has a better cooling effect, so, in comparison to GAIM, has a shorter molding cycle and a higher efficiency.
2. Finished part variety: WAIM is able to produce larger-sized plastic parts while achieving a thinner residual wall thickness; however, GAIM is limited by part dimensions and geometric shapes.
3. Production equipment: WAIM comes with a lower equipment setup cost, while GAIM is often protected by patents and thus the equipment cost is higher.
4. The medium used: WAIM is capable of using low-cost media which are easily obtainable for operation (i.e., water), while GAIM requires compressive nitrogen as the medium, which costs more.

Refer to Section 5.2.1 and Section 12.1.3 for the theoretical equations used for WAIM and GAIM, among which the biggest difference is that the melt tracking equation should be specially modified for assistant fluids, and is arranged as follows.

To track the advancement of the interface position over the molding process, a volume fractional function f_i is defined to describe the filling status of each cell. The fractional volume function is governed by the following transport equation:

$$\frac{\partial}{\partial t} f_i + \mathbf{u} \cdot \nabla f_i = 0 \qquad \text{Equation 13.1}$$

Two transport equations for polymer and gas are solved to determine the advancement of interfaces and subscript i denotes for different transported materials. For $i = 1$, the transported material is polymer, where $f_1 = 1$ is defined as the cell filled with the polymer melt, and $f_1 = 0$ as the empty cell enclosed by air or the cell occupied by the injected gas. For $i = 2$, the transported material is injected gas, where $f_2 = 1$ defines the cell filled with the injected gas, and $f_2 = 0$ the empty cell enclosed by air or polymer melt. Hence, the interfaces (polymer/air, polymer/gas, gas/air) are located within the cells with $0 < f_i < 1$. In order to realize the position of the polymer-air and gas-polymer interfaces, we introduce f_c, which represents the position of the interface of different fluids. A front tracking value greater than f_c denotes a region filled by the polymer, while a smaller than f_c value corresponds to an empty region (filled with air or gas, respectively).

Regarding the core-out region for the assistant fluid, the quality of finished part is determined by the hollowed core ratio, which is the average radius of the gas core-out area in a cross section of the finished part divided by the radius of the finished part, as shown in Figure 13.16. The finished part is dissected along the gas penetration direction, and the ratio of gas core-out at each cross section is recorded.

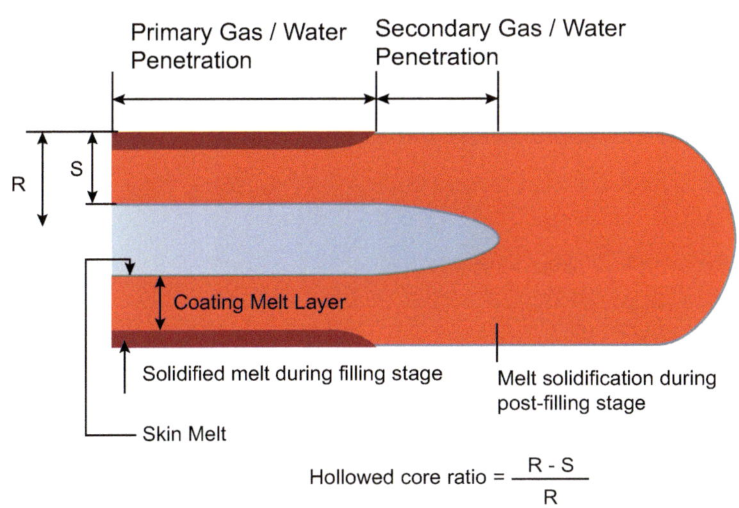

Figure 13.16 The definition of hollowed core ratio

■ 13.2 Practical Applications

As the interaction between fluid and melt in a process is fairly complicated, there are still many questions waiting to be explored, such as fingering effect, blow-through effect, gloss difference, hesitation sink mark, transition mark, floating fiber, and other commonly seen problems. These issues along with the fluid penetration effect require a good auxiliary tool to help with deeper study. In this section, we will use a few case studies to explain the above issues sequentially from GAIM to WAIM.

13.2.1 CAE Verification on GAIM

Figure 13.17 shows the geometry of a paper clip mold used in this study. The diameter of the circular pipe is a uniform 10 mm. The material used is the non-crystalline transparent PS, and the material parameters and process conditions are listed in Table 13.1 [1].

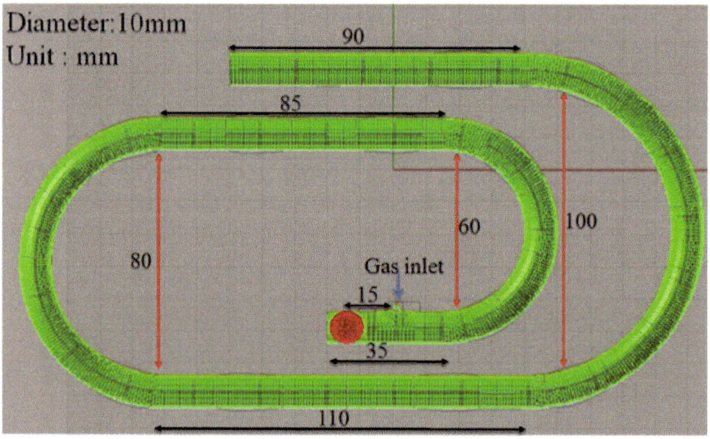

Figure 13.17 The 3D mesh and geometry used in this case [1]

Table 13.1 The Process Conditions Used in this Case

Melt temperature [°C]	230
Mold temperature [°C]	60
Filling time [s]	1.4
Prefill [%]	81
Delay time [s]	0
Gas injection time [s]	5.0
Gas pressure [MPa]	10

First, we compare the result of gas penetrating the melt front between the experiment and the simulation. Figure 13.18 shows a high compliance in trend between the simulation and experiment regarding the gas penetration condition. The gas penetration condition can be divided into primary penetration and secondary penetration. In the primary penetration region, the hollowed core ratio is evenly distributed, while in the secondary penetration region, as the melt starts to cool down and shrink to create space for the gas to penetrate, the hollowed core ratio varies much along the penetration direction. Such a result is also seen in the hollowed core ratio distribution, as shown in Figure 13.19.

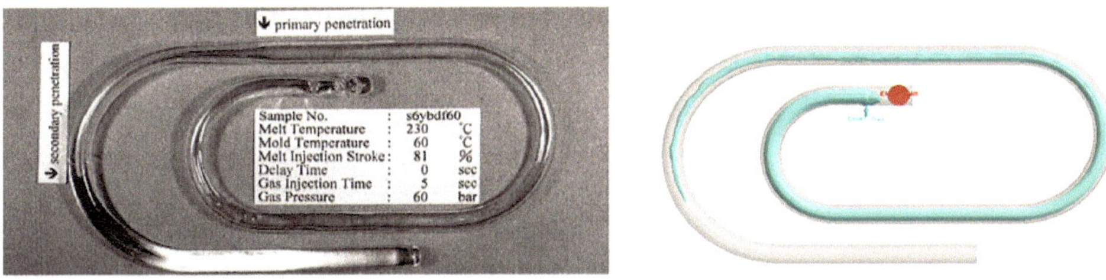

(a) (b)

Figure 13.18 The comparison between (a) experiment and (b) simulation

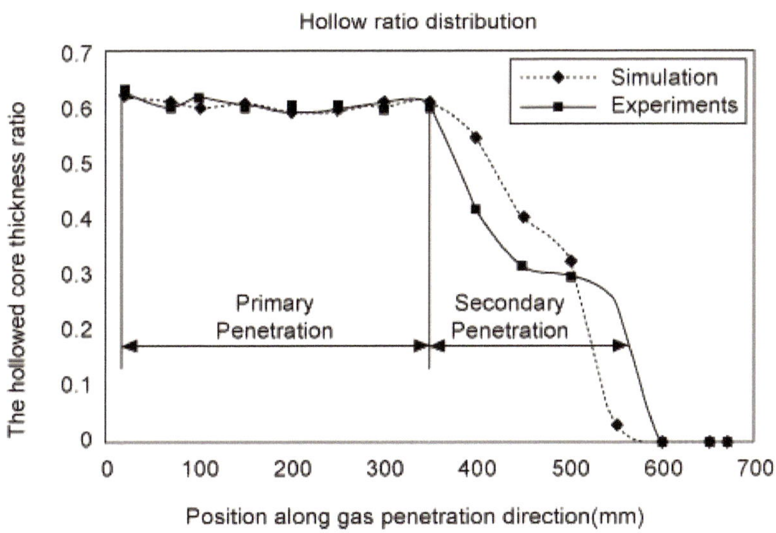

Figure 13.19 Hollowed core ratio along gas penetration direction [1]

Figure 13.20 indicates that the average hollowed core ratio of finished part decreases as the delay time increases, which is because the increase in delay time leads to a decrease in melt temperature in the mold cavity, which increases the viscosity and thickens the frozen layer of finished part, resulting in a smaller average hollowed core ratio in the penetration zone.

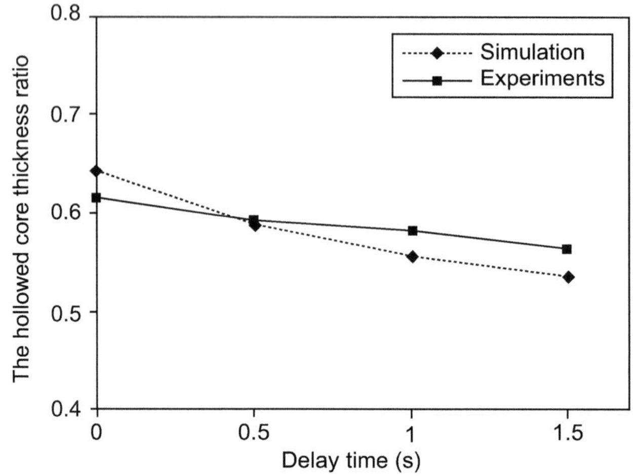

Figure 13.20 The influence of delay time on hollowed core ratio

By increasing the gas pressure, as shown in Figure 13.21, a thinner skin layer is generated, which increases the average hollowed core ratio of the finished part. The hollowed core ratio approaches a constant value as a certain high gas pressure is reached.

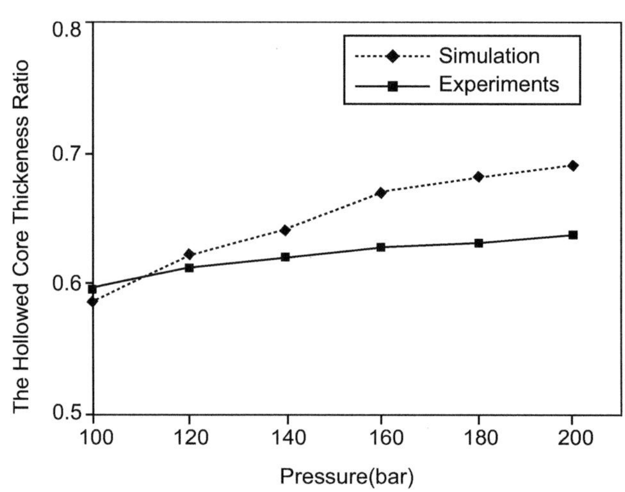

Figure 13.21 The influence of gas pressure on hollowed core ratio

If the melt temperature is higher, as shown in Figure 13.22, the melt viscosity becomes lower and the flowability increases, which leads to a thinner frozen layer, and results in an increase in average hollowed core ratio of the finished part.

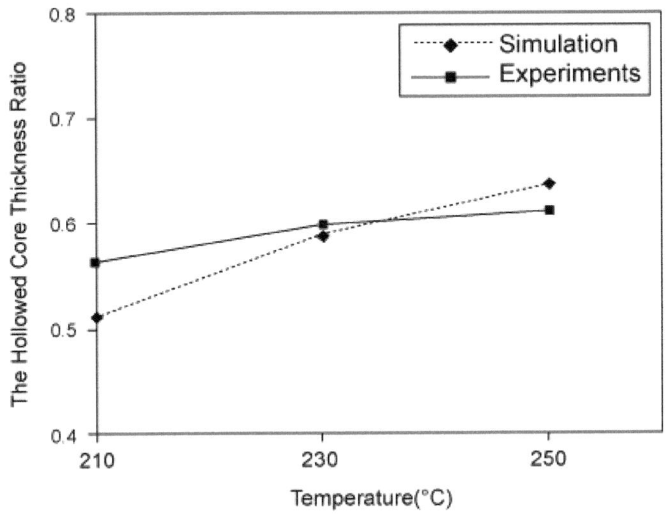

Figure 13.22 The influence of melt temperature on hollowed core ratio

This case study demonstrates that the prediction obtained from the simulation complies well with the experimental results. The hollowed core ratio of the finished part can be increased by increasing either the gas pressure or the melt temperature. Also, the primary and secondary gas penetration are both seen in the experiment and the simulation.

13.2.2 CAE Verification on WAIM

The current study uses a three-dimensional flow simulator (Moldex3D R10) to investigate the flow behavior with three different models under different processing conditions. Three different geometry models were investigated in the current study (Figure 13.23). Figure 13.23(a) shows the flat plate model, Figure 13.23(b) shows the two-thickness model, and Figure 13.23(c) shows the symmetric model. The dimensions of the flat plate are 35 mm (W) × 100 mm (L) × 5 mm (H). The water pin is located near the entrance of the cavity with dimensions of 1 mm × 1 mm. Model 2 has a thickness variation from 5 to 3 mm at half of the plate length. Model 3 has a U-shaped symmetric geometry above the flat plate.

(a) Model 1 (b) Model 2 (c) Model 3

Figure 13.23 The 3D geometry for different models used in this study

The material used for the simulation is semi-crystalline polypropylene. The modified Cross model is used for modeling the viscosity of polymer melt as a function of pressure, temperature, and shear rate. The fill time is specified to be 0.95 s. The melt temperature at the sprue entrance is from 180 to 195 °C. The maximum injection pressure is specified as 262 MPa. The mold temperature is from 50 to 65 °C. The pre-filled percentage of melt short-shot is from 80 to 90%. The water injection time is specified as 5 seconds, with pressure from 8–12 MPa and temperature from 75 to 82 °C. Water pressure fluctuation is ignored. The delay time is specified from 0–0.5 seconds.

Figure 13.24 shows the melt and water front at the end of filling for each model. Compared with experimental results, the flow behavior is pretty closed [2]. The flat plate model shows a generally uniform penetration behavior in Figure 13.24(a). For the model with two different thicknesses in Figure 13.24(b), the water front reaches thickness transition point with a bubble width expansion. The bubble width then shrinks due to the resistance of the thinner cavity. Figure 13.24(c) shows that symmetric model has a good stability when water penetrates through the two channels.

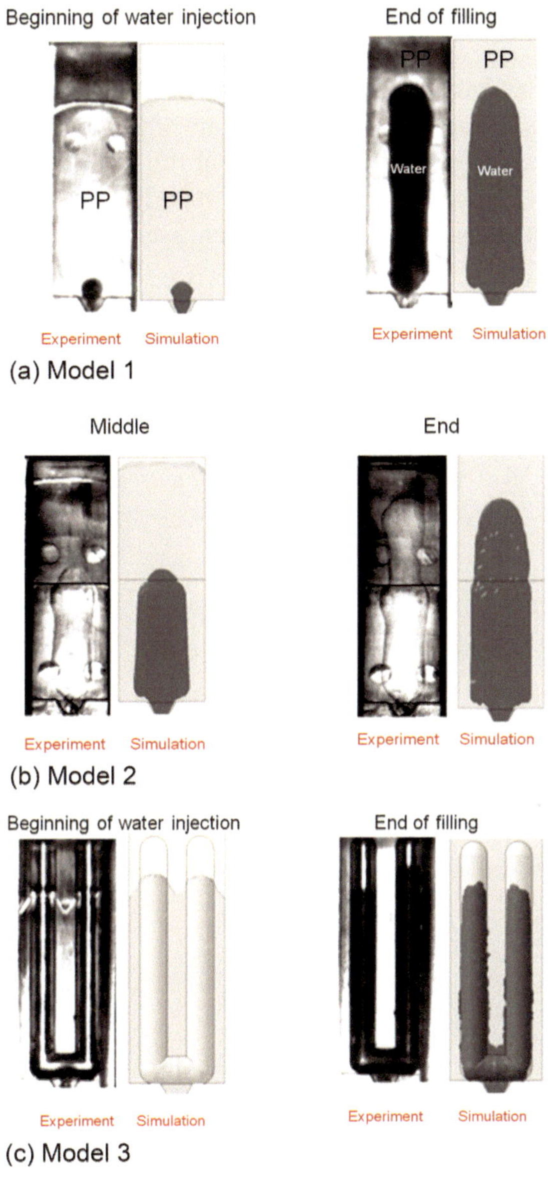

Figure 13.24 Comparisons between experiment and simulation

In this study, the numerical simulations present the three-dimensional flow behavior of water-assisted injection molding. Three different models were studied, with good agreement with experimental results. Water penetrates symmetrically in the U-shape model and hesitates at the thickness transition point in the two-thickness model. The results are helpful to implement a more sophisticated part fabrication in a WAIM process.

13.2.3 CAE Verification on GAIM: Fingering Effect

To distinguish the driving force of fingering in GAIM, this section uses a simple part with a center sprue gated rectangular plate of uniform thickness and a fish-bone layout of gas channels. The dimensions of the plate are referenced from [3] and described as in Figure 13.25. The plastic material used is polycarbonate (GE/Lexan141). The corresponding processing conditions are listed in Table 13.2.

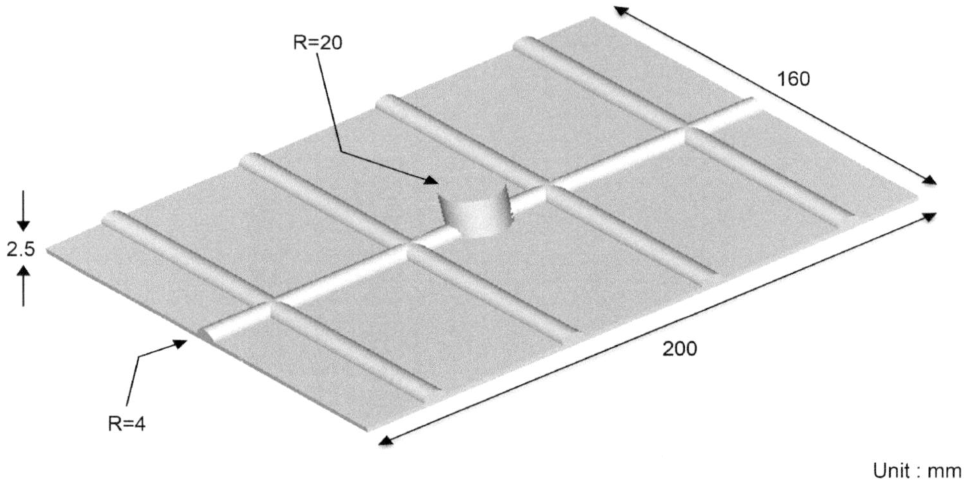

Figure 13.25 Geometrical parameters of a fish-bone gas channel plate

Table 13.2 Process Conditions

Melt/mold temperature [°C]	300/90
Filling time [s]	1.0
Prefill [%]	95
Delay time [s]	0.1
Gas injection time [s]	10
Gas pressure [MPa]	5.0

Figure 13.26 shows the results of the gas fingering patterns by the numerical simulation. First, simulation results on the gas penetration are quite consistent with the experimental study, both in penetration patterns and length [3]. Secondly, it is observed on the main channel between Branch 1 and 2, and Branch 3 and 4, that the gas bubbles penetrated outside the designed gas channels and into the thin sections, forming finger-shaped patterns. The numerical prediction shows a similar trend as well.

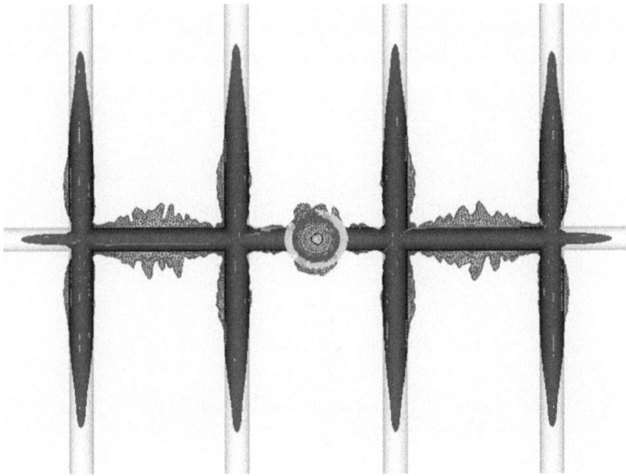

Figure 13.26 Simulation fingering patterns of gas fronts for a fish-bone gas channel plate

We further investigate the mechanism of the finger-shaped forming phenomenon by different time steps of the filling stage as shown in Figure 13.27(a) to Figure 13.27(c). In this case, when the gas is introduced as partially melt filled (the so-called short shot type), at the filling time of 5.31 sec, both primary and secondary gas penetration behaviors can be observed as in Figure 13.27(a). At this stage, no visible fingering behavior is observed in the main channel between Branch 1 and 2 or Branch 3 and 4. However, when more gas is pushed to enter the cavity as shown in Figure 13.27(b) and Figure 13.27(c), the gas bubbles encounter more resistance to flow forward since the melt has become gradually frozen at this stage; thus, the gas will backtrack when encountering resistance and start to penetrate into the thin plate sections. At this time, it will start to form tiny fingering. After the cavity is completely filled, the gas bubbles will occupy the main gas channel, and the continuously entering gas flow will then penetrate all possible regions, which ultimately leads to a significant fingering effect as shown in Figure 13.27(c). The simulation results are quite consistent with those of the experimental study.

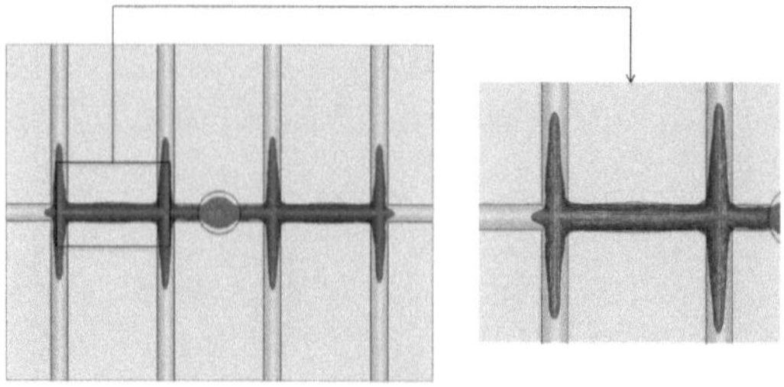

(a)

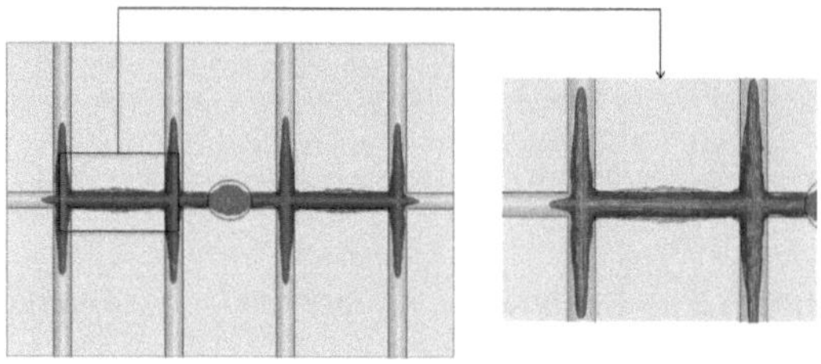

(b)

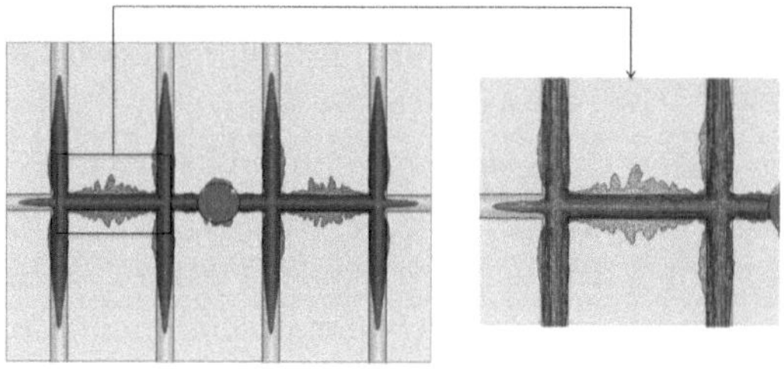

(c)

Figure 13.27 Gas fingering behavior for a fish-bone gas channel plate: (a) time = 5.31 s, (b) time = 6.51 s, and (c) time = end of filling

■ 13.3 CAE Case Study

Following the discussions of this chapter and the corresponding CAE verification case, an exercise model same, as illustrated in Figure 13.28, is provided in this section for readers to practice CAE application in gas-/water-assisted injection molding.

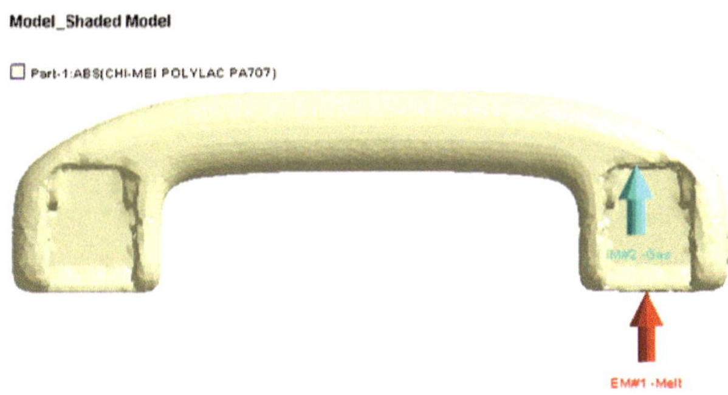

Figure 13.28 Case study sample

Please download the analysis files for CAE Case Study 13.3 from the following website:

https://moldex3d.box.com/s/zr6fvc1vlhbi4ocx111jwd3wmxt4ooif

Then open the .m3j or .mvj file in Moldex3D and answer the questions below according to the analysis results.

Questions:

1. Pre-analysis of the model
    a) What is the max L/t of the case? If ABS is taken as the material, does short shot occur? Why?
    b) Where will be the filling ends and weld lines?
2. What are the results of default process conditions for melt front time (melt/gas), history of sprue pressure, warpage behaviors?
3. If we switch to a different enter time for core material, what results will be influenced?
4. According to Run 1, if we change the process conditions such as melt temperature, mold temperature, delay time, duration time, gas pressure, what will be the results?

References

[1] Lu, B. H.: "Study of Gas Penetration Characteristics and Formation of Hollowed Core Thickness in Gas-Assisted Injection Molding Process", Master thesis, Department of Mechanical Engineering, Chung Yuan Christian University, Taiwan, 1996

[2] Wu, Y.-C-; Liu, S.-J.: Flow Visualisation of Water Assisted Injection Moulding Process, *Plastics, Rubber and Composites: Macromolecular Engineering*, 34, 2005, pp. 227–231

[3] Lu, X.; Chiang, H.; Fong, L.; Zhao, J.; Chen, S.: Study of "Gas Fingering" Behavior in Gas-Assisted Injection Molding, *Polymer Engineering and Science*, 39, 1999, pp. 62–77

14 Foam Injection Molding

Yuan-Jung (Dan) Chang, Li-Yang (Robert) Chang, Chih-Wei (Joe) Wang

The motivation of microcellular injection molding is to enhance the product structure strength using less plastic material, by application of the foaming technique in the injection molding process, where the cell size is intended to be smaller than the defect dimension of the polymer chains [1]. First, this chapter introduces the basic process principles, advantages, challenges, and theory models of microcellular injection molding. Second, a real case of an oil pan demonstrates how to use CAE in assisting innovative process development [2]. Finally, a case study provides the reader with some exercises to practice CAE settings and interpret analysis of microcellular injection molding.

14.1 Basics

The foaming process applies blowing agents and forms the polymer into a two-phase structure of solid and gas. Based on different types of blowing agents, the process can be classified as physical, chemical, or reactive foaming, as shown in Table 14.1.

Table 14.1 Classification of Foaming Systems

Foaming Type	Polymer	Blowing Agents
Physical foaming	Polystyrene (PS), polypropylene (PP)	Nitrogen (N_2), carbon dioxide (CO_2)
Chemical foaming	Polystyrene (PS), polypropylene (PP)	Azodicarbonamide (AC), sodium bicarbonate ($NaHCO_3$)
Reactive foaming	Polyurethane (PU)	Water (H_2O)

Physical blowing agents (PBAs) are gases dissolved into a polymer mixture under high pressure. The gases will then release when foaming has started. Chemical blowing agents (CBAs), as their name implies, apply chemicals to produce gas in a thermal decomposition process. The gases generated by CBAs are mostly nitrogen (N_2) and carbon dioxide (CO_2), which behave much like a PBA. Reactive foaming utilizes mixed reactive components to produce gas and cross-linked polymer at the same time via chemical reactions, such as polyurethane foaming.

Microcellular technology has been developed enormously in foaming processes since it was introduced by Dr. Nam Suh and co-workers at MIT in the early 1980s [3]. Subsequently, the microcellular technique was executed in reciprocating screw injection machines from Trexel and Engel in 1998 [4]; MuCell® is the trademark of the microcellular injection molding technology owned by Trexel, Inc.

14.1.1 Microcellular Process Principle

Basic ideas of microcellular technology are briefly illustrated in Figure 14.1. Firstly, molten plastic in the injection barrel is mixed with a supercritical fluid (SCF), either nitrogen (N_2) or carbon dioxide (CO_2). Through the specialized screw, the plastic melt theoretically can dissolve the maximum amount of SCF to become a well-mixed single-phase mixture. During the injection, a large amount of cell nucleation appears due to pressure drop, which causes the formation of bubbles at the same time. Then the products are shaped inside the mold with a lot of microcellular bubbles. For typical microcellular foam, the cell size is around 3 to 100 mm and the cell density is around 10^6 to 10^9 cells/cm^3 [5].

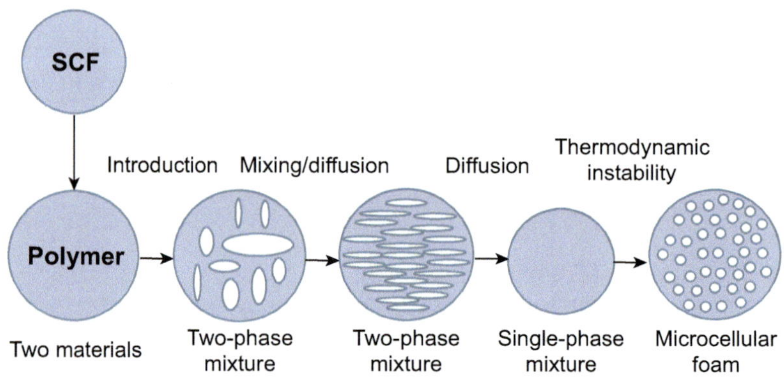

Figure 14.1 Illustration of the essential steps in microcellular injection molding

Technically speaking, the key point of MuCell® is to control cell growth, which involves four steps:

1. Gas dissolution:

   Introduction of a SCF into the barrel of a plastic melt forms a high-pressure single-phase mixture.

2. Nucleation:

   When the melt mixture is injected into the cavity, a large number of nucleation sites form due to pressure drop.

3. Cell growth:

   Cell growth and coalescence occur during molding.

4. Shaping:

   After solidification of the plastic, the part is shaped in the mold.

Therefore, to control the cell growth mechanism precisely, the behavior of the SCF phase transition must be understood. As the phase diagram in Figure 14.2 shows, the material will transfer to a gas- or liquid-like fluid, namely a supercritical fluid, when it reaches a state with a pressure higher than the critical pressure (P_c) and a temperature higher than the critical temperature (T_c). SCFs possess gas-like diffusivity (high diffusion rate), liquid-like density (high density), gas-like viscosity (low viscosity), and low surface tension so that they are especially suitable for the MuCell® molding technology providing good homogeneity in plastic melts [6].

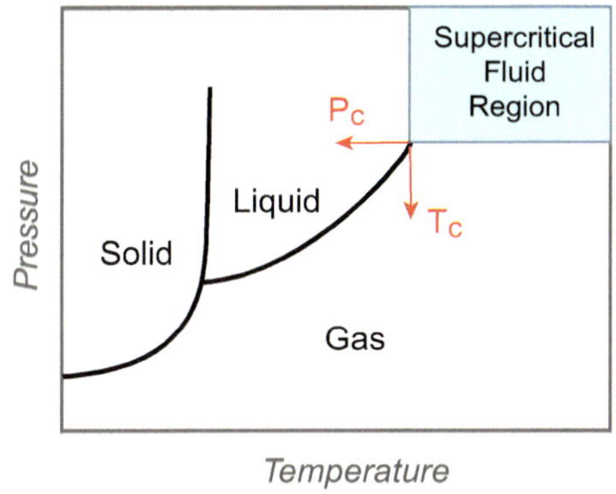

Figure 14.2 Phase diagram

The most commonly used SCFs are carbon dioxide (CO_2) and nitrogen (N_2) and their critical states are listed in Table 14.2. Carbon dioxide has better solubility to plastics than N_2 [7], but the process and cell quality can be more easily controlled with nitrogen. Hence, nitrogen is more popular than carbon dioxide for this purpose.

Table 14.2 Critical Points of Carbon Dioxide and Nitrogen [8]

Gas	Critical Point	
	P_c [kg/cm²]	T_c [°C]
CO_2	72.8	31.1
N_2	3.5	−147

Higher weight reduction by less injected volume can provide a lower pressure in the cavity favoring cell growth. A high pressure drop rate is helpful for the nucleation process and leads to high cell density with small bubble size. Regarding the different cavity pressures, there are three methods of microcellular injection molding:

1. Low-Pressure Foaming

Usually, the low-pressure foaming method applies to the short-shot injection. The low-pressure state will induce the dissolved SCF to nucleate from the melt mixture. Then the volume of the mixture increases to occupy the whole cavity. The mechanism is illustrated in Figure 14.3. However, the bubbles broken on the surface will lead to surface defects like high roughness and swirl marks, and also tend to induce a too-large cell size, and the uneven distribution of cell sizes makes the part rupture eventually.

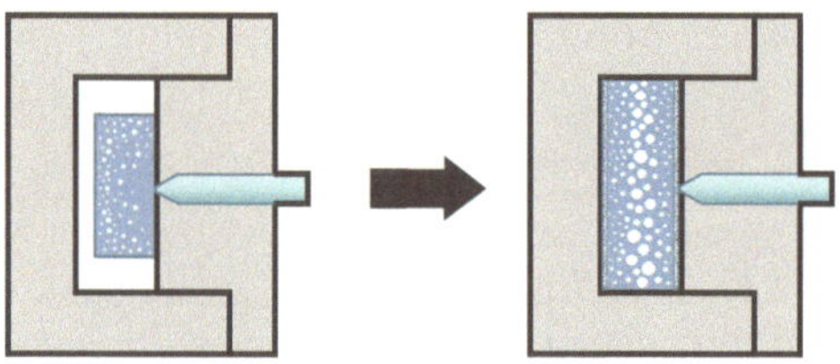

Figure 14.3 Illustration of low-pressure foaming

2. High-Pressure Foaming (Core-Back)

To obtain a better surface quality and low density of the products, the high-pressure foaming method, also called the core-back process, has been developed with a characteristic higher than 10 MPa in the fully filled cavity. The mechanism is illustrated in Figure 14.4. The cavity pressure is released to be low enough for cell growth to take place, but only in the core layer region of the part so a better surface quality with fewer swirl marks can be achieved. The key point is that the frozen skin layer is hard enough before the foaming.

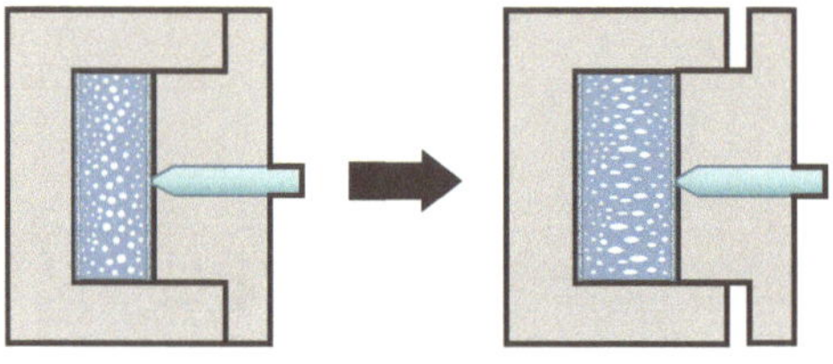

Figure 14.4 Illustration of high-pressure foaming (core-back)

The obvious difference between core-back technology and traditional microcellular foam injection molding is that there is a dwelling (packing) stage once the filling stage is complete. During this process, the pressure is maintained and the temperature decreased at the same time, so that the bubbles are re-dissolved and the temperature decreases to a level suitable for foaming, so that the core-back part can be given a better quality foam. As shown in Figure 14.5, the main processes of core-back are as follows:

1. Fill the melt into the mold.
2. Delay the process to cause the bubbles to re-dissolve in the melt. This process is needed to prepare for a uniform cell structure.
3. Release the pressure in the mold through core retraction.
4. The bubbles are nucleated, grown, and finally shaped, and the part is demolded.

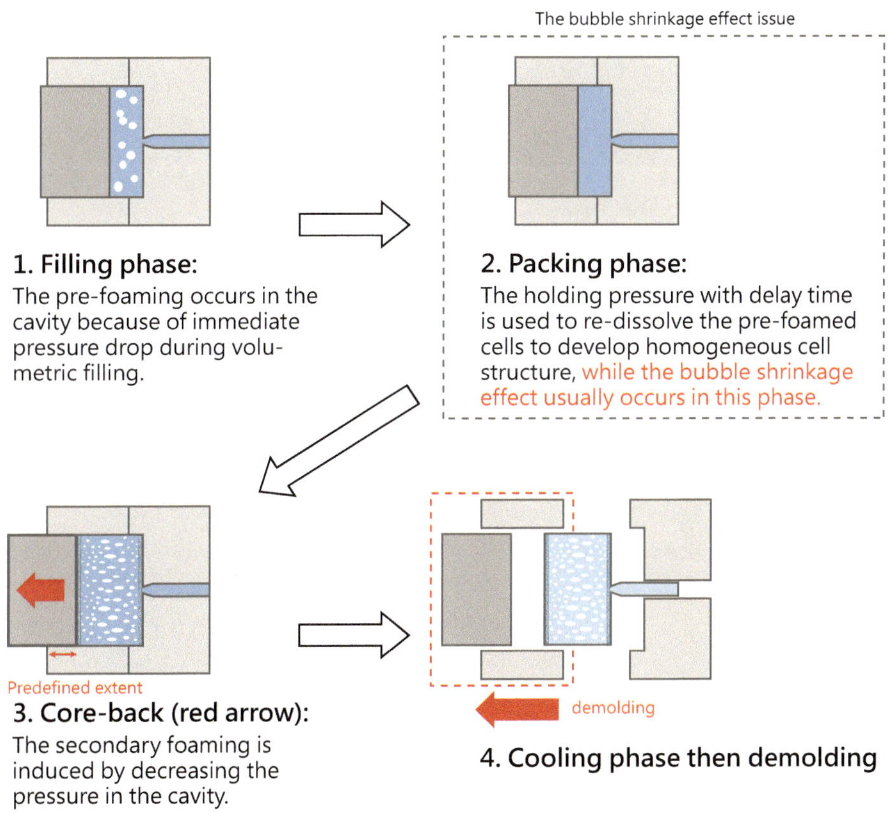

Figure 14.5 Mechanism of the core-back process

The challenges of the core-back process are not only to control the molding parameters such as temperature or pressure, but also closely relate to the dwell time, core-back speed and core-back distance. In general, there will be an unfoamed skin layer on the surface of the part, and the cell size of the core layer will be larger than the surface.

3. Foaming with Counter Pressure

The counter pressure method supplies a gas pressure against the melt front during the filling stage [9]. The counter pressure can be controlled by the venting mechanism and provides a flow resistance to the melt, which results in a smoother surface of the melt front (Figure 14.6). Due to the high pressure against to the melt front, the cells will not grow up so the cell size in the melt front zone is smaller compared to conventional foaming. It is also called non-swirl mark molding technology.

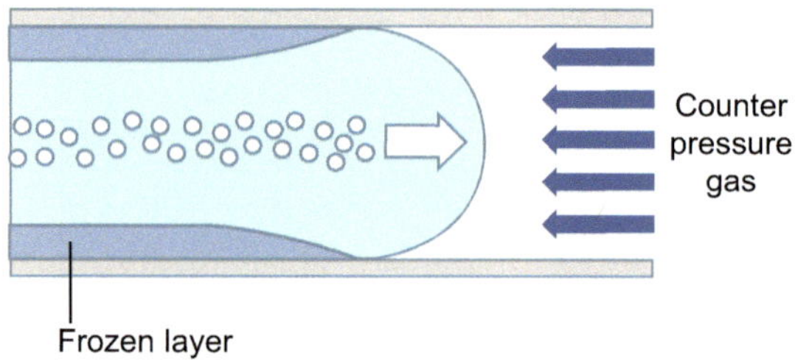

Figure 14.6 Illustration of foaming with counter pressure

14.1.2 Advantages and Challenges

The foaming plastic consists of two phases: gas and solid. The gas phase is shaped either as closed bubbles isolated from each other or open bubbles with interconnective channels. These bubble structures give the products useful properties such as light weight, heat insulation, sound absorption, vibration reduction, and so on [9]. Microcellular foaming plastic provides the following advantages in injection molding:

1. Reduced manufacturing cost:

   The mixture of a SCF and a plastic melt has a lower viscosity than the pure plastic and the low-pressure foaming implemented by short-shot injection brings about the non-fully filled cavity. Both the lower viscosity and the non-fully filled cavity lead to a decreased molding pressure and clamping force.

2. Reduced amount of plastic:

   The specific volume of a gas is much higher than that of a plastic. When a gas is used in the molding process, the weight of the product can be reduced by 6 to 12%.

3. Improved product qualities:

   The lower molding pressure leads to lower residual stress of the molded part. The melt volume expansion from cell growth furnishes lower shrinkage and higher dimensional precision, which provide higher freedom in product design. A dimensional accuracy of 0.01 mm and even 0.001 mm can be achieved. Furthermore, no sink marks will appear at thin-wall and thick-rib locations.

4. Available for thin products:

   The cell size is small enough for the production of thin parts to which the conventional foaming process cannot be applied.

5. Low impact on the environment:

   Foaming agents that are harmful to the environment, such as hydrochlorofluorocarbons (HCFC) and hydrocarbons, can be replaced by carbon dioxide and nitrogen, which are natural components of the atmosphere.

6. Low transport cost per unit weight:

   Less energy is needed to transport the lower-density products.

Microcellular injection molding has many advantages but there are some application limitations for high-quality appearance products because of the cell structure. Besides the product limitations, additional equipment is needed, such as a specialized screw and gas metering system. As shown in Figure 14.7, the SCF is added in the mixing stage of the screw and precise control of the SCF dosage is very important. Therefore, the challenges of potentially poor surface quality and extra cost of molding equipment and SCF dosage control system should be taken well into account before adoption of a microcellular injection molding process.

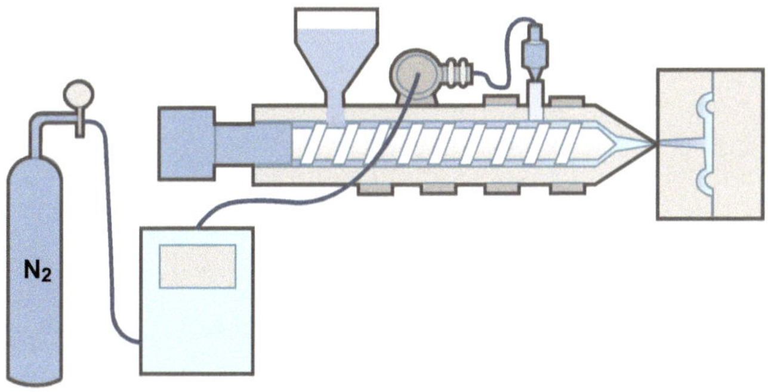

Figure 14.7 MuCell® molding equipment

14.1.3 Theory Models

The mechanisms controlling the processes of cell nucleation and cell growth as illustrated in Figure 14.8 are key to understanding microcellular injection molding. Theory models of (1) cell growth and (2) cell nucleation which are also used for molding simulation are introduced in this section.

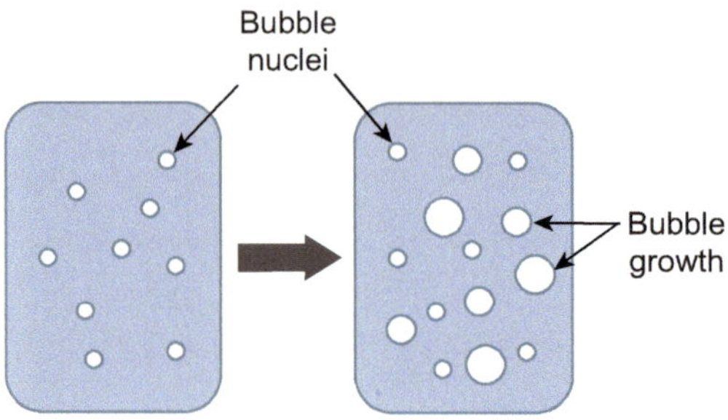

Figure 14.8 Illustration of cell nucleation and cell growth

1. Cell Growth Models

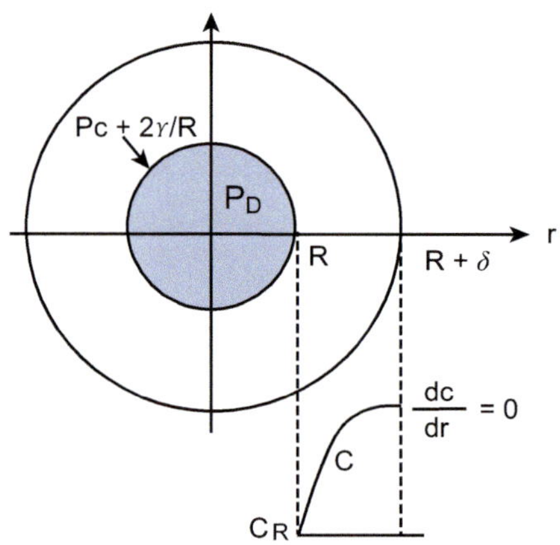

Figure 14.9 The cell growth model

Microcellular foaming occurs after injection of the melt into the mold cavity. The dynamic behavior of cell growth within a macroscopic polymer melt can be illustrated as in Figure 14.9 [10], where the cell radius R is described by:

$$\frac{dR}{dt} = \frac{R}{4\eta}\left(P_D - P_C - \frac{2\gamma}{R}\right) \tag{14.1}$$

where η is the melt viscosity, P_D the bubble pressure, P_C the ambient pressure, and γ the surface tension.

The gas mass transfer at the cell surface (gas-polymer interface) is:

$$\frac{d}{dt}\left(\frac{4\pi}{3}\frac{R^3 P_D}{R_g T}\right) = 4\pi R^2 D \left.\frac{\partial c}{\partial r}\right|_{r=R} \quad (14.2)$$

where R_g is the gas constant, T the temperature, c the dissolved gas concentration, and D the diffusion coefficient.

During cell growth, the dissolved gas concentration near the cell is described as a diffusion equation:

$$\frac{\partial c}{\partial t} = D\left[\frac{1}{r^2}\frac{\partial}{\partial r}\left(r^2 \frac{\partial c}{\partial r}\right)\right] \quad (14.3)$$

The dynamic cell growth behavior is described by the gas mass transfer at the cell surface proposed by Han and Yoo [11]:

$$\frac{d}{dt}\left(P_D R^3\right) = \frac{6D\left(R_g T\right)\left(c_\infty - c_R\right)R}{-1+\left\{1+\frac{2/R^3}{R_g T}\left(\frac{P_D R^3 - P_{D0} R_0^3}{c_\infty - c_R}\right)\right\}^{1/2}} \quad (14.4)$$

where the concentration has the following relation:

$$\frac{c_\infty - c}{c_\infty - c_R} = \left(1 - \frac{r-R}{\delta}\right)^2 \quad (14.5)$$

and P_{D0} is the initial bubble pressure, R_0 the initial bubble radius, c_∞ and c_R the dissolved gas concentration in liquid bulk and at bubble surface, respectively, and d the concentration boundary thickness.

2. Cell Nucleation Models

Cell nucleation occurs because the pressure of the mixture drops from the sprue to mold cavity. The cell nucleation rate $J(t)$ can be described as a classical exponential function [12]:

$$J(t) = f_0\left(\frac{2\gamma}{\pi M_w / N_A}\right)^{1/2} \exp\left(-\frac{16\pi\gamma^3 F}{3k_B T\left(\bar{c}(t)/k_H - P_C(t)\right)^2}\right) N_A \bar{c}(t) \quad (14.6)$$

where f_0 and F are fitting parameters, M_w is the gas molecular weight, N_A the Avogadro number, k_B the Boltzmann constant, k_H the solubility parameter, and $\bar{c}$ the average concentration of dissolved gas.

Cell nucleation competes with cell growth and can be described as coupled form with the mass conservation of the dissolved gas $\bar{c}(t)V_{L0}$ [10]:

$$\bar{c}(t)V_{L0} = c_0 V_{L0} - \int_0^t \frac{4\pi}{3} R^3(t-t',t') \frac{P_D(t-t',t')}{R_g T} J(t') V_{L0} dt' \quad (14.7)$$

where V_{L0} is the volume of the polymer matrix.

■ 14.2 Practical Applications

The application of CAE in MuCell® can be demonstrated with the case of an oil pan [13]. Several molding parameters can be set and each of them has different effects on the product. In this section, these molding effects are studied and discussed both by experiments and CAE analysis. The consistency between CAE results and the experiments is then verified.

14.2.1 CAE Verification on Microcellular Injection Molding: Case 1

1. Case Information

The experiments and CAE analysis were executed by Trexel, Inc., and Moldex3D, respectively. In this case, the effects of SCF dosage, weight reduction, and injection speed on cell density and part transparency are examined. First, case information is provided and then the three molding parameters gas dosage, weight reduction, and injection speed are sequentially discussed.

The CAE model and the part thickness distribution are shown in Figure 14.10 and Figure 14.11, respectively. The dimensions of the part are 762 × 267.4 × 101.7 mm³. The thickness ranges from 0.9 to 3.2 mm and the average is around 2.7 mm. The rib heights range from 2 to 4 mm, and the maximum boss height is around 9.5 mm. The volume of the part is 971 cm³. Polypropylene (PP) and nitrogen are used as the matrix and SCF, respectively. The mold temperature is 40 °C and the melt temperature is 220 °C.

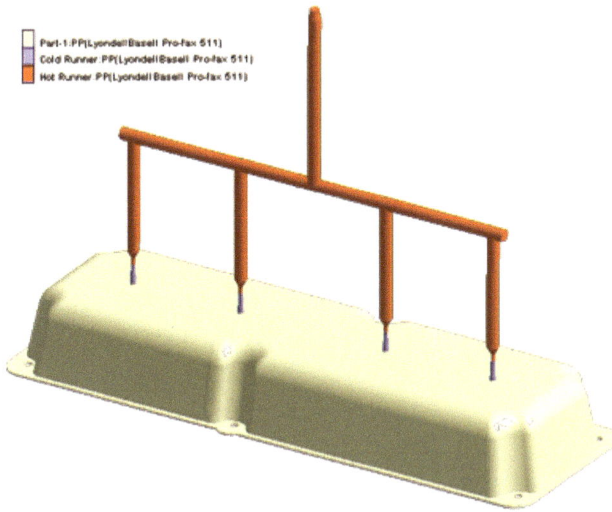

Figure 14.10 CAE model including part, cold runner, and hot runner

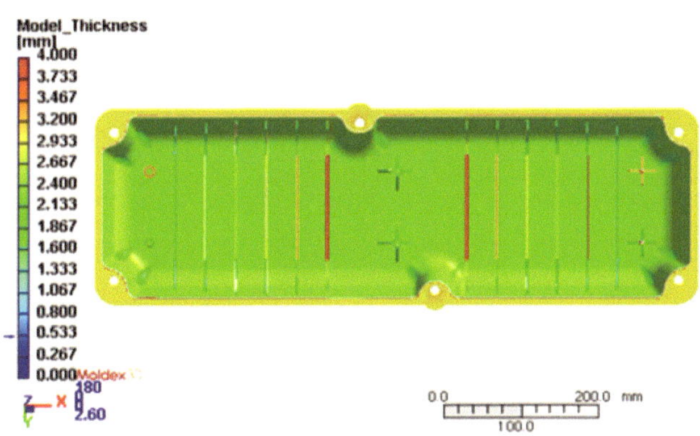

Figure 14.11 The thickness distribution of the part

2. Effects of Gas Dosage

To understand the molding effects induced by the amount of the SCF, the variable condition of gas dosage amount of samples A to C is set to 0.4, 0.6, and 0.8 wt.%, respectively, as listed in Table 14.3; the injection speed is 6 in/s and the weight reduction 12%.

Table 14.3 Process Conditions for Evaluation of Gas Dosage Effects

Sample	A	B	C
Injection speed [in/s]	6		
Weight reduction [%]	12		
Gas dosage amount [wt.%]	0.4	0.6	0.8

The CAE cell density and the actual appearance of samples A to C are shown in Figure 14.12(a) and Figure 14.12(b), respectively. The cell density ranges from 0 to $4 \times 10^6/\text{cm}^3$ and the CAE analyses show that the higher the gas dosage used the higher the cell density formed. Regarding the appearance of the real product, a higher cell density would provide a less transparent product. For sample A, the blue color shown in Figure 14.12(a) represents the low cell density distribution, which corresponds to the obvious transparency of the whole product in Figure 14.12(b). As for sample C, the light green color in Figure 14.12(a) corresponds to the white, non-transparent product in Figure 14.12(b). Thus, the gas dosage effect on the appearance quality is clarified, which also implies that CAE can help find the proper amount of the used SCF for the microcellular injection molding process.

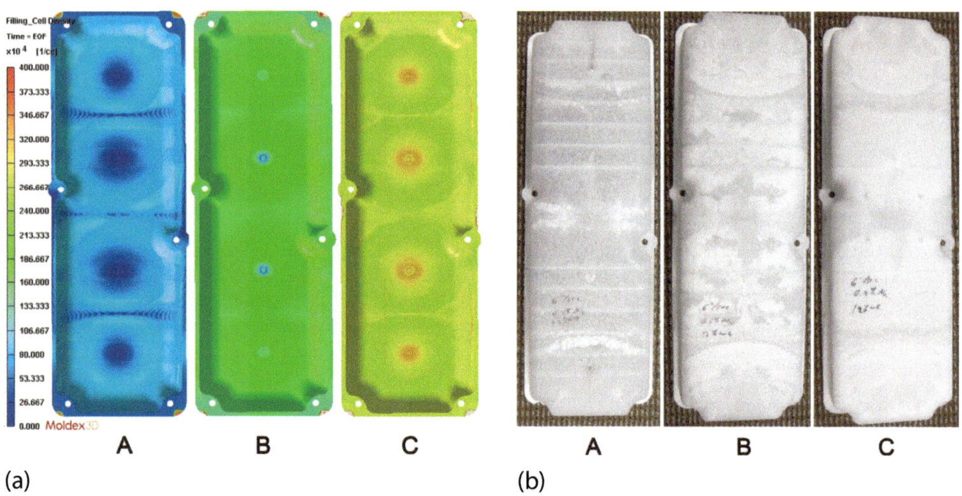

Figure 14.12 Results of samples A, B, and C: (a) simulation of cell density and (b) actual appearance

Furthermore, local features of samples A and B are shown in Figure 14.13, where Sample B of Figure 14.13(b) shows more details for the correlation between CAE cell density and actual appearance. The ① arrows point to the high-density area (red color) corresponding to the white, non-transparent portion of the real part. The ② arrows point to the low-density area (green color), also the weld line location, corresponding to the transparent portion. Moreover, the large-area transparent appearance of sample A can be demonstrated in Figure 14.13(a).

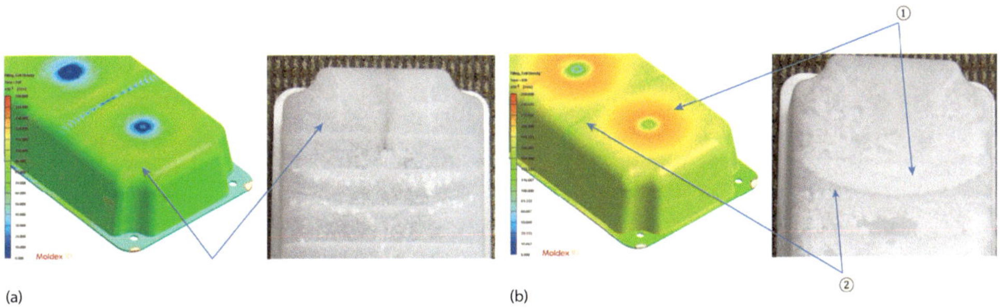

Figure 14.13 Local features of (a) sample A and (b) sample B

Real and CAE results of part weight and weight reduction experiments and analyses of samples B and C are compared in Table 14.4 and illustrated in Figure 14.14. The predicted values are slightly overestimated but the pattern is the same as in the real experiments. With increasing gas dosage, cell size decreases while cell density increases.

Table 14.4 Quantitative Comparison of Experimental and CAE Results of Samples B and C

Sample (gas dosage)		B (0.6 wt.%)	C (0.8 wt.%)
Experiments	Part weight/weight reduction	712 g/12.32%	717 g/11.71%
CAE	Part weight/weight reduction	723 g/12.95%	728 g/12.35%
	Avg./std. cell size [mm]	43.22/10.11	40.69/6.13
	Avg./std. cell density [10^6/cm^3]	1.8/0.17	2.7/0.13

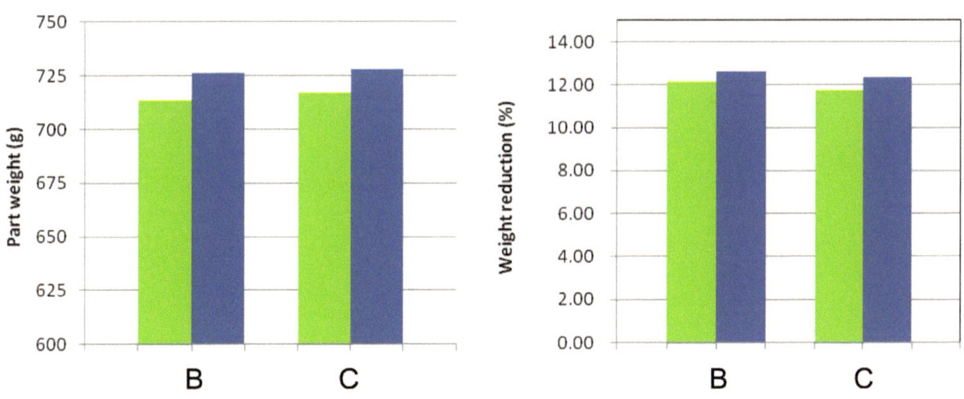

Figure 14.14 Part weight and weight reduction of samples B and C: green color represents real experiments and blue color for CAE results

3. Effects of Weight Reduction

The setting condition of a higher weight reduction could provide a lower pressure environment which favors cell growth. The variable process condition of weight reduction of samples D, E, and A is set to 4, 8, and 12%, respectively, as shown in Table 14.5; the injection speed is 6 in/s and the gas dosage amount 0.4 wt.%.

Table 14.5 Process Conditions for Evaluation of Weight Reduction Effects

Sample	D	E	A
Injection speed [in/s]	6		
Weight reduction [%]	4	8	12
Gas dosage amount [wt.%]	0.4		

The edge features of sample A and sample D are shown in Figure 14.15(a) and Figure 14.15(b), respectively. Sample A shows that the higher weight reduction of 12% leads to non-transparency at the corners of the part, whereas the lower weight reduction of 4% of sample D displays more transparency at the same location. However, the cell density in the main area of the product remains almost unaffected by the weight reduction.

Real and CAE results of part weight and weight reduction experiments and analyses of samples D and E are compared in Table 14.6 and illustrated in Figure 14.16. The predicted values show the same trend as the experimental results. The cell size increases with weight reduction.

Table 14.6 Quantitative Comparison of Experimental and CAE Results of Samples D and E

Sample (weight reduction)		D (4%)	E (8%)
Experiments	Part weight/weight reduction	776 g/4.37%	749.5 g/7.70%
CAE	Part weight/weight reduction	791 g/4.77%	760 g/8.50%
	Avg./std. cell size [mm]	43.19/11.15	47.91/11.85
	Avg./std. cell density [10^6/cm^3]	0.77/0.17	0.77/0.20

(a)

(b)

Figure 14.15 Edge features of (a) sample A and (b) sample D

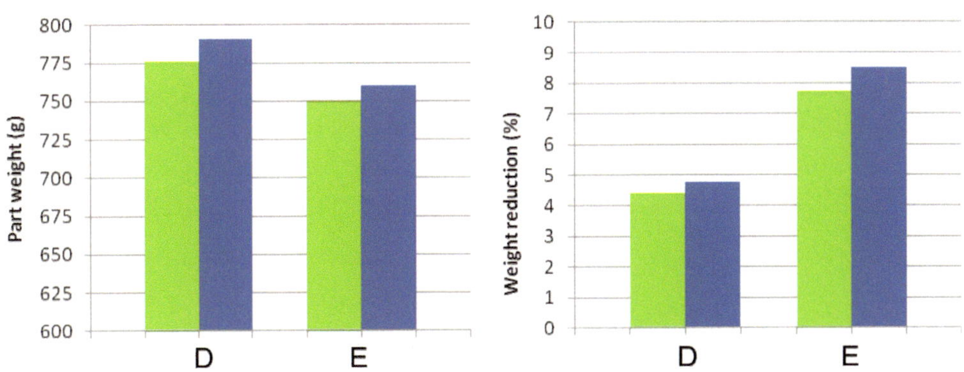

Figure 14.16 Part weight and weight reduction of samples D and E: green color represents real experiments and blue color for CAE results

4. Effects of Injection Speed

The injection speed is also an important process condition because it influences the melt flowing pattern during molding. The injection speed of samples F, G, and C is set to 2, 4, and 6 in/s, respectively, as shown in Table 14.7; the weight reduction is 12% and the gas dosage amount 0.8 wt.%.

Table 14.7 Process Conditions for Evaluation of Injection Speed Effects

Sample	F	G	C
Injection speed [in/s]	2	4	6
Weight reduction [%]	12		
Gas dosage amount [wt.%]	0.8		

CAE cell density results and the actual appearance of samples F, G, and C are shown in Figure 14.17(a) and Figure 14.17(b), respectively. The cell density ranges from 0 to 4 × 10^6/cm^3 and the three samples show the same order of magnitude of the cell density. Under the low injection speed, the gas is more easily released from the melt front, which causes the short shot defect, as shown at the corners of sample F shown in Figure 14.17(a).

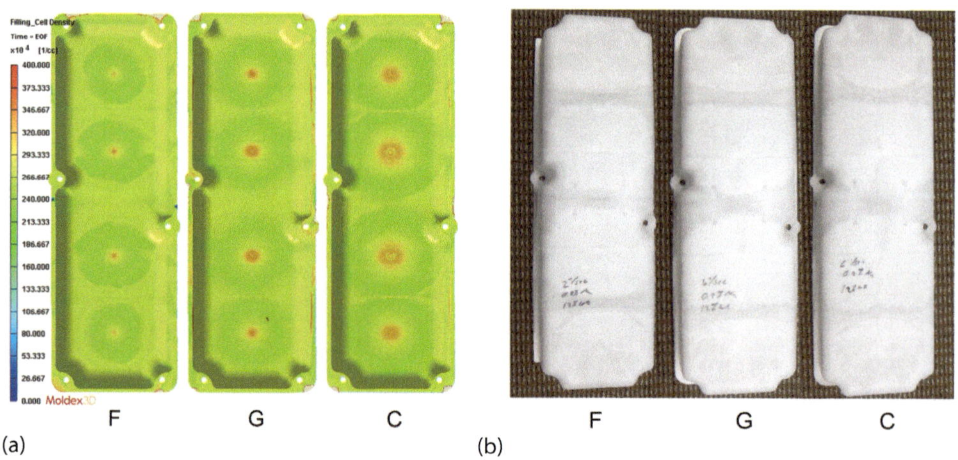

Figure 14.17 Results of samples F, G, and C: (a) simulated cell density and (b) actual appearance

Real and CAE results of part weight and weight reduction experiments and analyses of samples G and C are compared in Table 14.8 and illustrated in Figure 14.18. The cell size decreases slightly as the injection speed increases, which implies that a higher injection speed can avoid short shot. The injection speed has no effect on the cell density.

Table 14.8 Quantitative Comparison of Experimental and CAE Results of Samples G and C

Sample (injection speed)		G (4 in/s)	C (6 in/s)
Experiments	Part weight/weight reduction	713.5 g/12.14%	717 g/11.71%
CAE	Part weight/weight reduction	726 g/12.59%	728 g/12.35%
	Avg./std. cell size [mm]	42.65/7.29	40.69/6.13
	Avg./std. cell density [10^6/cm^3]	2.7/0.14	2.7/0.13

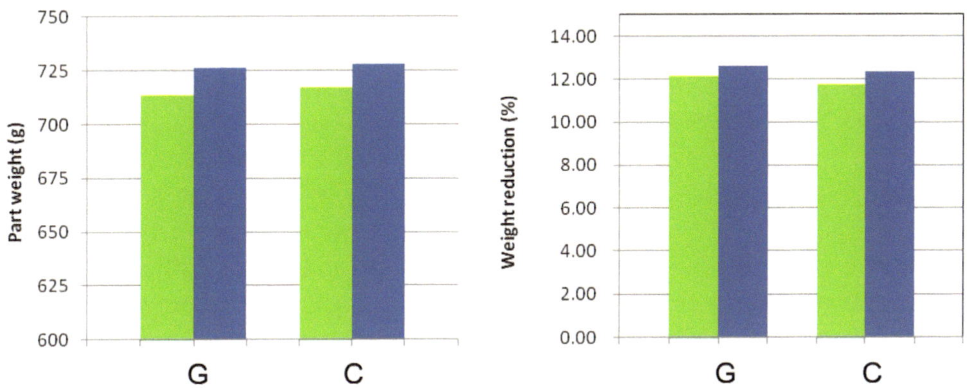

Figure 14.18 Part weight and weight reduction of samples G and C: green color represents real experiments and blue color for CAE results

14.2.2 CAE Verification of Microcellular Injection Molding: Case 2

1. Case Information

The other U-shaped CAE models were executed by Proplast, Inc. and Moldex3D [13]. In this case, the validation of sink marks, warpage, and the micro properties will be examined. As shown in Figure 14.19, the dimensions of the part are 200 × 220 × 15 mm^3. The thickness ranges from 1.0 to 11.3 mm and the average value is around 1.4 mm. There are several cross-linked ribs under the part with the height of 11.3 mm.

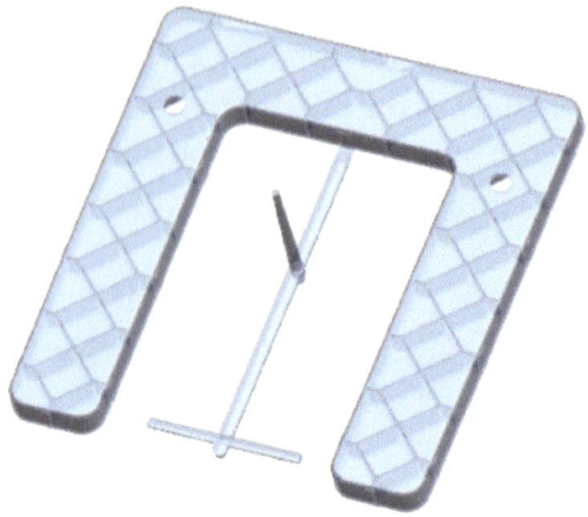

Figure 14.19 CAE model

2. Validation of Sink Marks

The sink-mark defect is heavily dependent on the rib design of the injection molding process. The main reason mentioned earlier is that heat dissipation at relatively thick ribs is slower and induces uneven shrinkage within the part. Therefore, it is common to employ relatively long packing to deal with this issue. However, that will increase either the cost of the material or the power of the injection machine. The other solution for preventing sink marks is foam injection molding, in which the internal bubble pressure plays an important role in generating a holding pressure that overcomes the uneven shrinkage. Below, we return to foam injection molding practice, using the model with a max. rib/wall thickness ratio of 2 : 1 and complex cross-linked geometry to compare with regular injection molding. As shown in Figure 14.20 and Figure 14.21, the sink mark is diminished by foam injection molding and fair agreement of the sink mark indicator between practice and simulation results can be seen.

(a) (b)

Figure 14.20 CAE results for sink marks: (a) regular injection molding and (b) foam injection molding

(a) (b)

Figure 14.21 Results for part appearance: (a) regular injection molding and (b) foam injection molding

3. Validation of Warpage

Generally, the U-shaped part amplifies warpage in the injection molding process. As is the case with shrinkage analysis, uneven stress inside the part leads to large deformation at the corner. The corner effect for both the compact regular part and the foaming part at the same time helped visualize the improvement in deformation that foam injection molding would achieve before the part is actually injected. The results are shown in Figure 14.22 and Table 14.9.

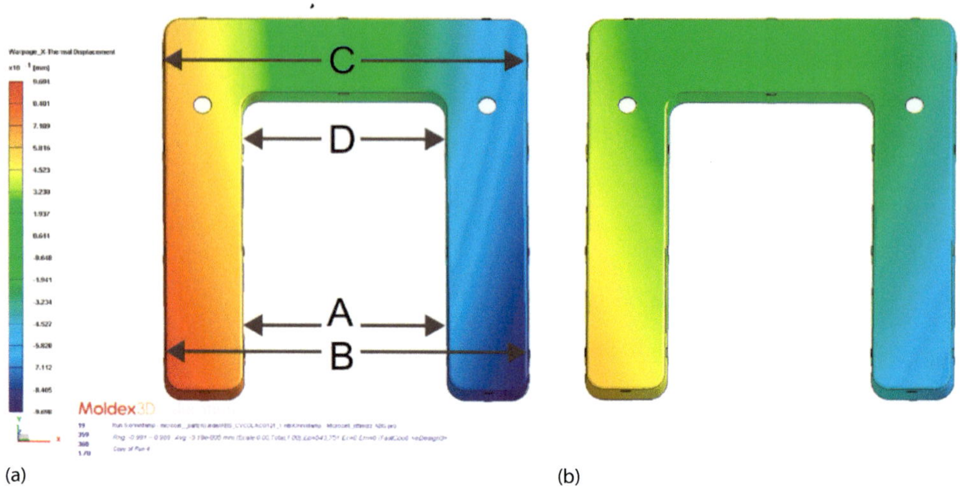

Figure 14.22 CAE results for warpage: (a) regular injection molding and (b) foam injection molding

Table 14.9 Comparison of Simulated and Real Part Deformation

	Nominal [mm]	Compact Molding [mm]		MuCell® [mm]		Variation [%]
		Simulation	Real part	Simulation	Real part	
A	110	108.76	109.17	109.22	109.16	−0.05
B	200	198.10	198.64	198.76	198.72	−0.02
C	200	198.61	198.92	199.18	198.86	−0.29
D	110	109.21	109.52	109.52	109.20	−0.29

4. Validation of Micro Properties

The cell size prediction is crucial for the foam injection molding process, as part quality and mechanical strength are extremely dependent on uniform cell distribution and adequate cell size. The correct prediction of micro properties along the flow direction is essential for ensuring the part design is reliable prior to production. Therefore, validation focused on three locations: 1. Gating region, 2. Middle of flow region, 3. End of flow region, respectively.

Location 1 – Gating region: The cells are more difficult to measure by SEM in this region. The small size of the bubbles or even the absence of bubbles is due to the high pressure near the gate location, where the pressure is higher than the saturation pressure of polymer/SFCs solution so the free energy is lower than the energy barrier of cell nucleation. The simulation result reveals this phenomenon and is validated by SEM, as shown in Figure 14.23 and Figure 14.24.

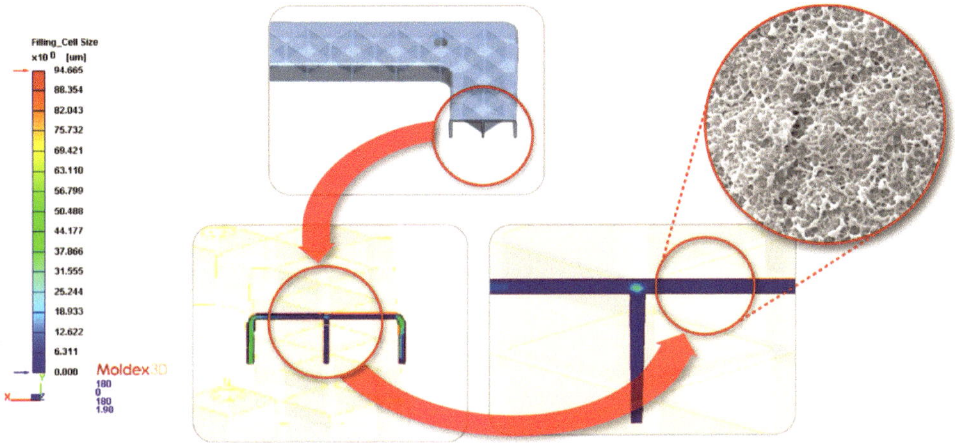

Figure 14.23 Cell size in the gating region: real part and CAE results

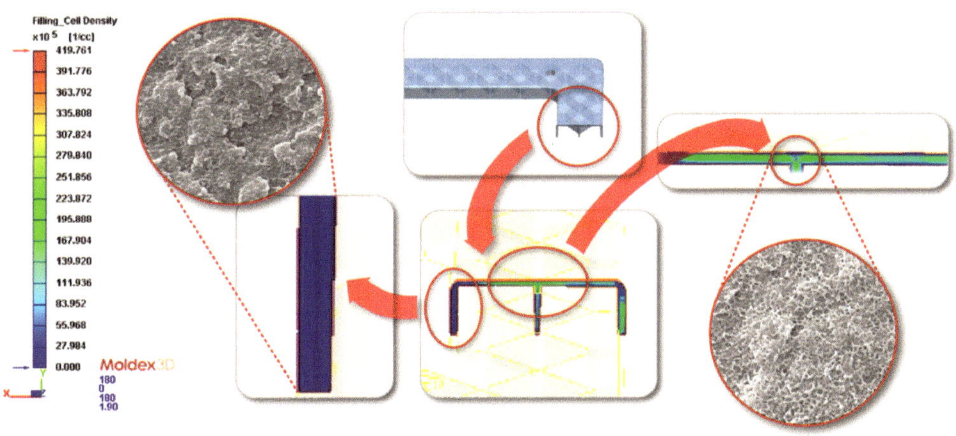

Figure 14.24 Cell density in the gating region: real part and CAE results

Location 2 - Middle of flow region: The cells grow as they go further from the gate; the size of these cells is therefore relatively larger than at the gating region. Both simulation and SEM revealed the same trend. The cells on the skin have less time to grow as compared to those at the core of the part, and so they are smaller. As bubble growth increases, a given space contains fewer cells, resulting in lower cell density. The results for cell size and cell density are shown in Figure 14.25 and Figure 14.26.

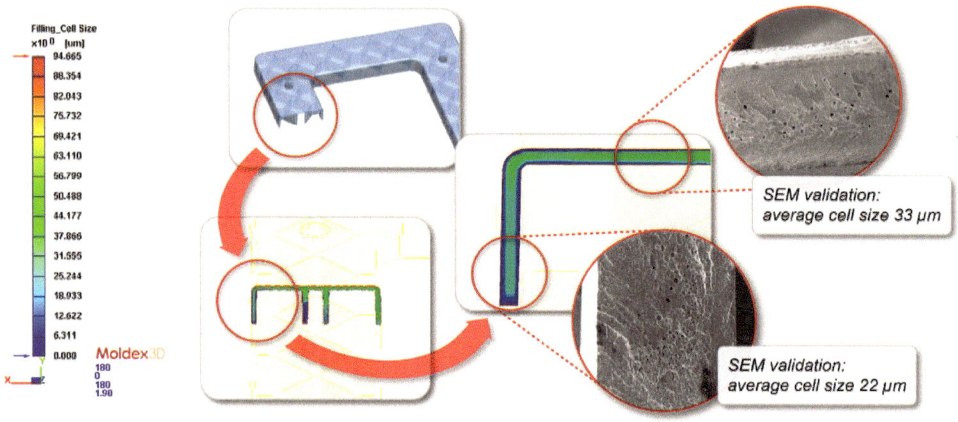

Figure 14.25 Cell size in the middle of the flow region: real part and CAE results

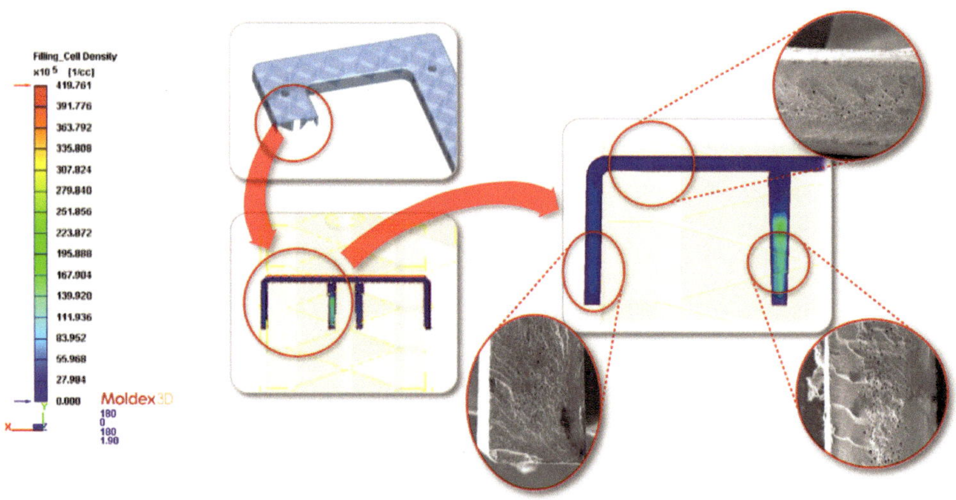

Figure 14.26 Cell density in the middle of the flow region: real part and CAE results

Location 3 - End of flow region: At the end of the flow region, the material has traveled the entire path and that allows the cell to grow; also, the pressure at the flow front is low enough to make no constraints on the growth of the bubbles. Therefore, the much-developed cell gives rise to a very low cell density, as revealed by SEM and shown in Figure 14.27 and Figure 14.28.

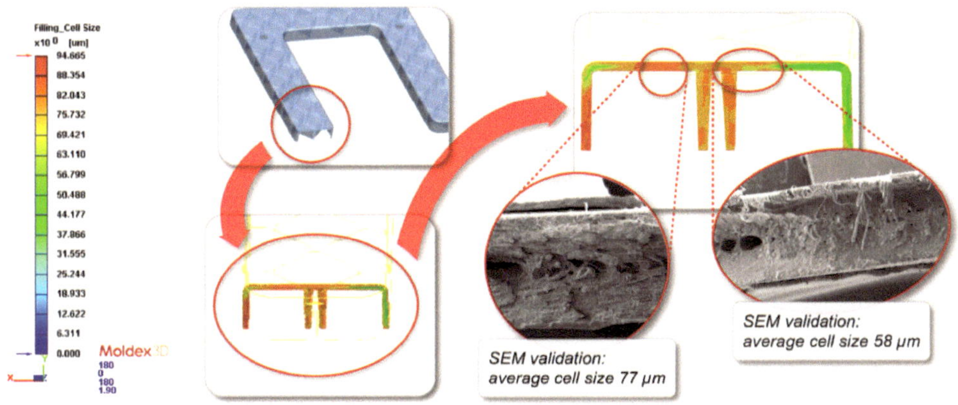

Figure 14.27 Cell size in the end of flow region: real part and CAE results

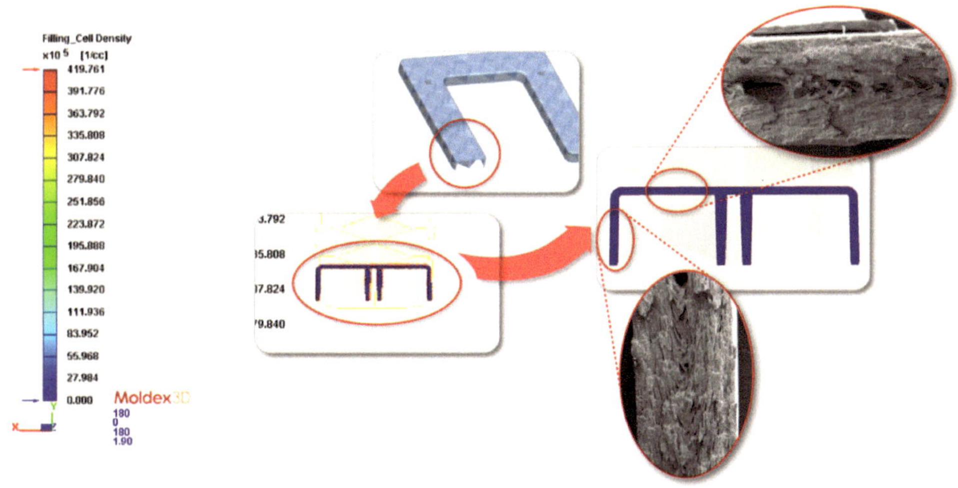

Figure 14.28 Cell density in the end of flow region: real part and CAE results

14.2.3 CAE Verification on Chemical Foaming Injection Molding

As previously mentioned, besides the microcellular foam injection process, there is another foaming process, characterized by the addition of blowing agents, which is called the chemical foaming process. This involves the addition of a blend of polymer and chemical blowing agent (CBA) via the hopper of a traditional injection molding machine. A notable feature of the chemical foaming process is that there is no need to modify the machine.

There are two categories of the CBAs: endothermic and exothermic compounds. The endothermic CBAs absorb heat energy in the heat decomposition reaction and mainly release carbon dioxide gas generated by the reaction. The general endothermic CBAs are sodium bicarbonate and zinc bicarbonate. These decompose over a wide temperature range, from 130 to 230 °C. Azodicarbonamide (AC) is an exothermic CBA which is generally decomposed in a temperature range from 205 to 215 °C with a generated gas amount of about 220 cm^3/g. It releases heat energy and mainly N_2 gas generated by the reaction. The decomposition temperature can also be reduced to as low as 150 °C by adding an activator. Because of its narrow decomposition temperature range, this CBA is easy to control.

The advantages of the chemical foaming process are similar to those of the microcellular foaming process, including weight reduction by 10 to 25%, sink removal, reduced warpage, reduced cycle times, elimination of surface imperfections, reduced clamping force, and so on. The main difference between chemical and microcellular foaming is the formation of larger bubbles by the chemical foaming process, which can be as large as 200 mm.

The application of CAE in chemical foaming is demonstrated in this chapter with a box model [14]. First, case information is provided and then the cell size distribution is analyzed by CAE.

1. Case Information

The box model is shown in Figure 14.29. The dimension of the part is 155 × 93.9 × 67.6 mm^3. The thickness ranges from 0.3 to 1.6 mm and the average is around 1.5 mm. The volume of the part is 66.6 cm^3. Polypropylene (PP) and azodicarbonamide are used as the matrix and CBA, respectively. The gas amount is 280 to 300 cm^3/g, and the composition of produced gas is 65% N_2, 32% CO, and 2% CO_2. The mold temperature is 60 °C and melt temperature 210 °C.

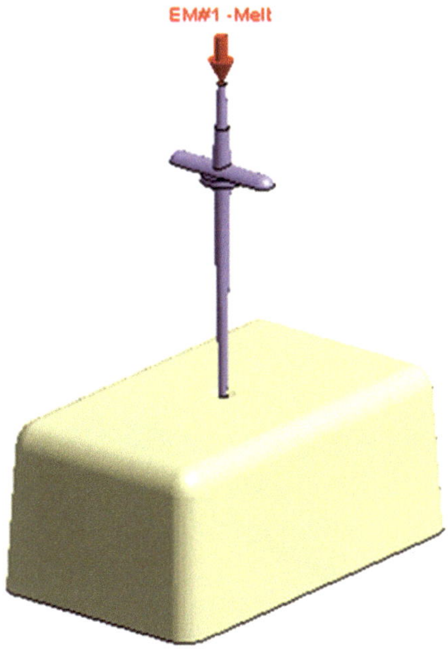

Figure 14.29 CAE model including the part and cold runner

2. Cell Size Distribution of Chemical Foaming Injection Molding

Figure 14.30 shows the cell size distribution CAE analysis results of the polypropylene box obtained by chemical foaming injection molding. The average cell size is 163.46 mm.

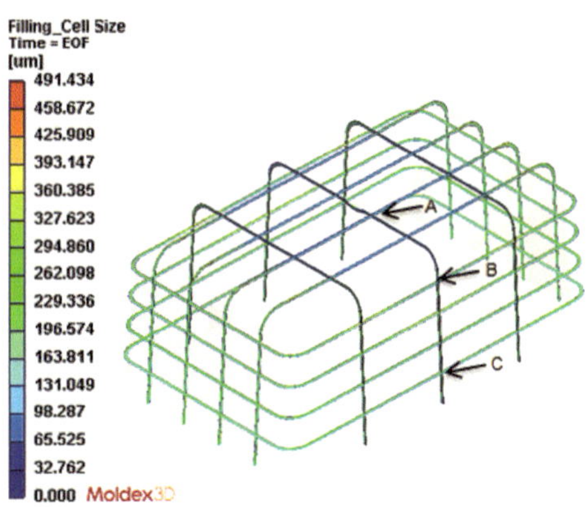

Figure 14.30 Cell size distribution by CAE of molded polypropylene box for 1 wt.% azodicarbonamide in various cross sections

Table 14.10 shows the cell sizes obtained by CAE analysis and scanning electron microscope (SEM) images at the positions A–C. The microcells formed have a larger size on average than those formed in microcellular injection molding for the same polymer. CAE results display a larger cell size in the flow end area. Both SEM images and simulation results also show the appearance of big cells with a cell size larger than 100 mm.

Table 14.10 Cell Size by CAE Analysis and SEM Images

Position	CAE (Legend scale: 0–491 mm)	SEM (Scale bar: 100 mm)
A	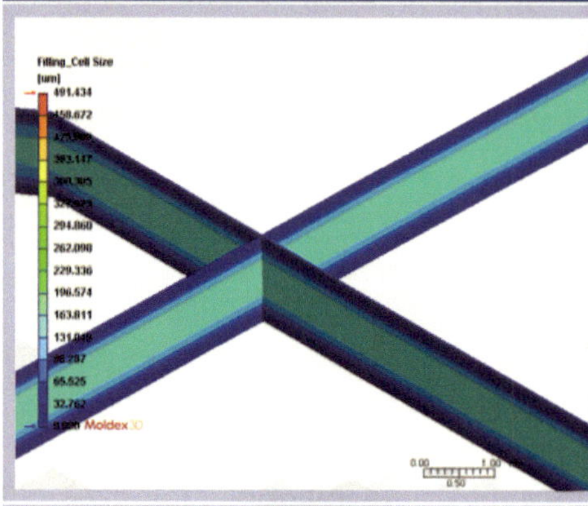	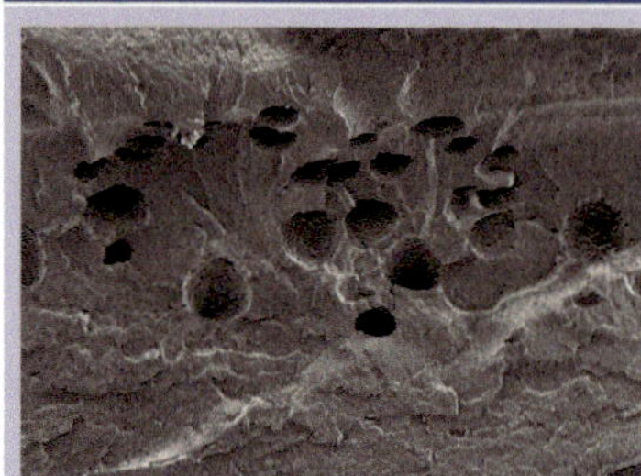
B	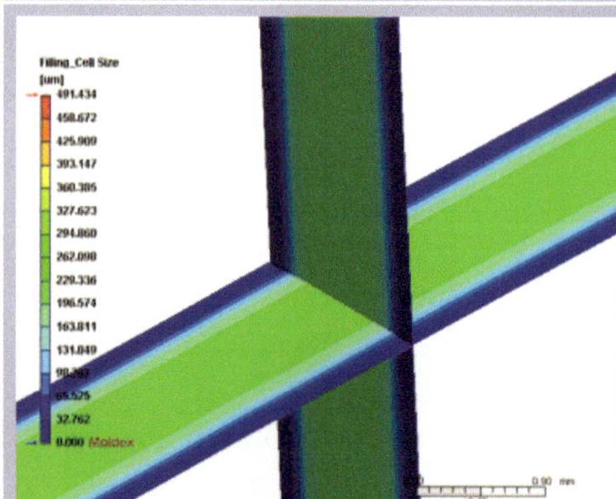	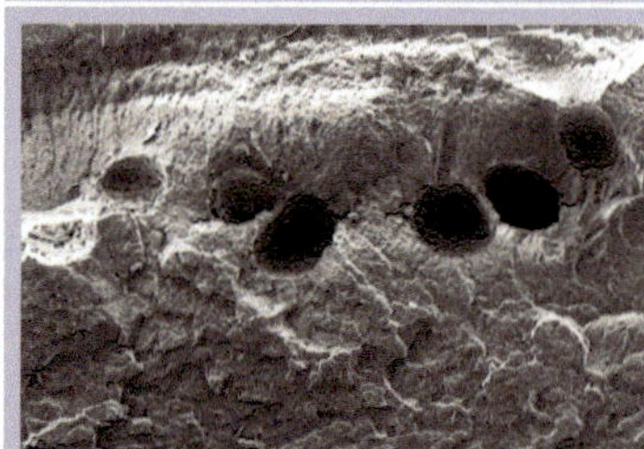
C	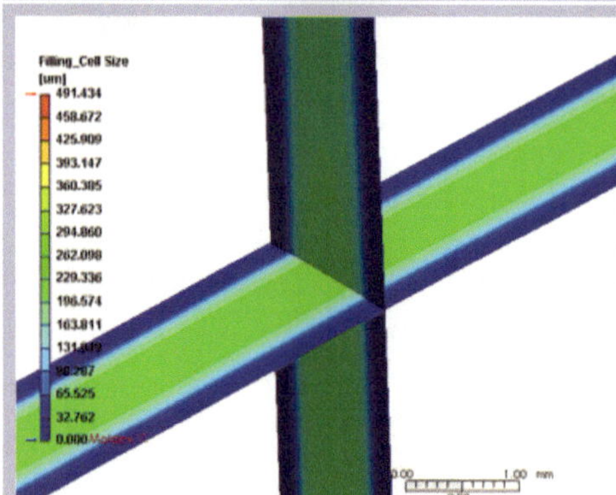	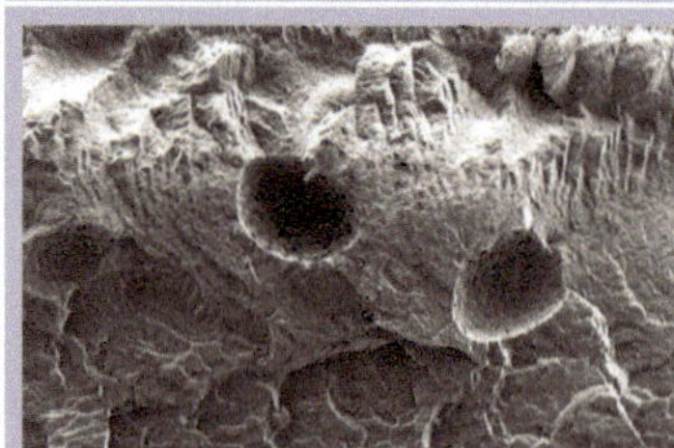

14.2.4 CAE Verification of Polyurethane Reactive Foaming Molding

1. Introduction

In order to alleviate the impact of global environmental pollution and the greenhouse effect, countries around the world continue to conduct research on lightweight technology, focusing particularly on foam (porous material). The principle here is that the density of resin (polymer matrix) will be reduced through foaming technology, which can basically lead to lighter weight and reduce the use of fossil raw materials. At present, the main polymer substrates are polyurethane (PU), polystyrene (PS), polyethylene (PE) and polypropylene (PP) foam materials. PU and PS account for about 90% of world market consumption of polymer foam materials.

Polyurethane resin (PU) is a widely used material, especially in the automotive, refrigeration, footwear, and medical industries. Polyurethane resin is mainly composed of polyol or polyether, which contains the OH group, and isocyanate. The molecule chain can be grown to form polymer chains by reaction and can polymerize to yield the cross-linked network structure. The basic process of polyurethane foam molding is mixing the polyol, isocyanate, and water and injecting into the mold cavity. In the injection stage, the cavity is usually not completely filled, and the remaining space is filled by foam expansion. In this process, the polyurethane releases the carbon dioxide gas formed in the chemical reaction, and the viscosity of the polyurethane also increases due to the crosslinking reaction. At the same time, the exothermic effect caused by the chemical reaction will increase the temperature of the mold, and further release the carbon dioxide into the polyurethane at the relatively high temperature. The foaming process is complete when the mold cavity is completely filled with polyurethane foam or the polyurethane is fully cured.

The production process and the final mechanical properties of foam are affected by the different combinations of formulas, such as polyols, water, isocyanate, additives, foaming agents, etc. That is to say, different combinations of chemical formulations will result in different foaming volumes, foaming speeds, curing rates, and exothermic heat as well. Furthermore, Figure 14.31 shows the process from the mixing of raw materials to foaming and curing of the foam in the foaming process, which can be divided into the four stages:

1. Mixing time: The time needed for the raw materials to be mixed.
2. Cream time: The time before the mixture begins to form a cloudy, creamy liquid, and expand.
3. Rise time: The time from when the creamy liquid starts to expand into a foam to when expansion is complete.
4. Set time: The time from when the foam expands to when the surface has solidified.

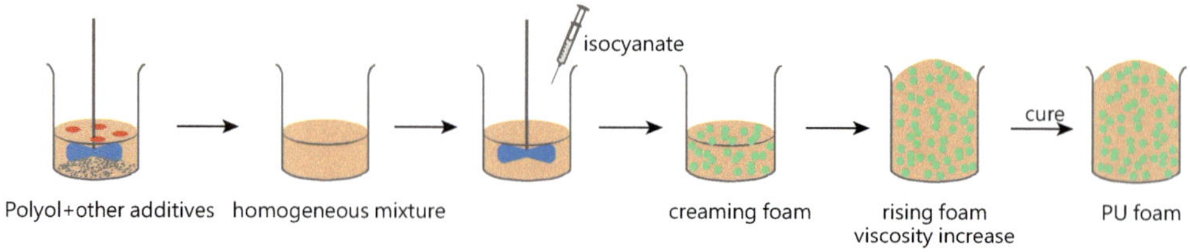

Figure 14.31 The PU foam process

The foaming process may cause cavity filling defects, such as trapped air bubbles and lack of material. The challenge of the polyurethane foam molding process is to fill the mold cavity with the exact doses of materials without a short shot happening. A short shot occurs when the cavity is filled with insufficient material, or insufficient foaming gas is used, or the polyurethane curing rate is too fast. However, large amounts of waste material occur when too much material is injected and too much gas is in the product. CAE simulation of the polyurethane foaming process makes it possible to more accurately predict the filling behavior of polyurethane and optimize the injected amounts of materials.

2. Material Characterization

Figure 14.32 shows a universal foam qualification system from FOAMAT which was used to thoroughly study the foaming characteristics of PU foam. We carried out an experimental investigation of the physical parameters during foam formation using FOAMAT 285 and PU material from the BASF ELASTOFOAM CI 4510/136/LD series. The PU raw material has the following specified characteristics at 23 °C: mixing ratio poly to isocyanate 100:63 w/w, water content 0.6 wt.%, cream time 9 s, rise time 32 s, and free rise density 172 g/L.

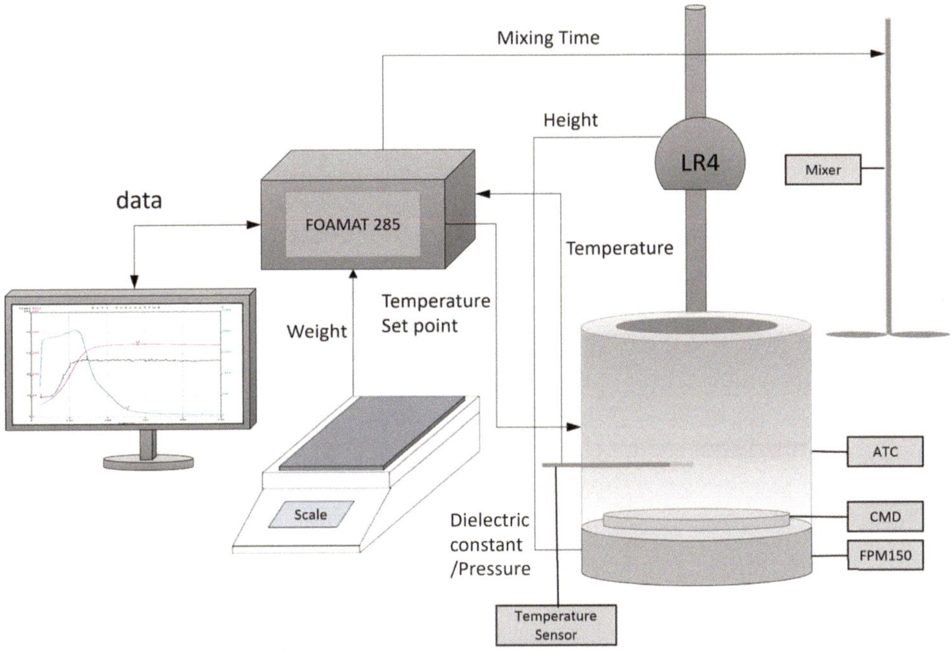

(a)

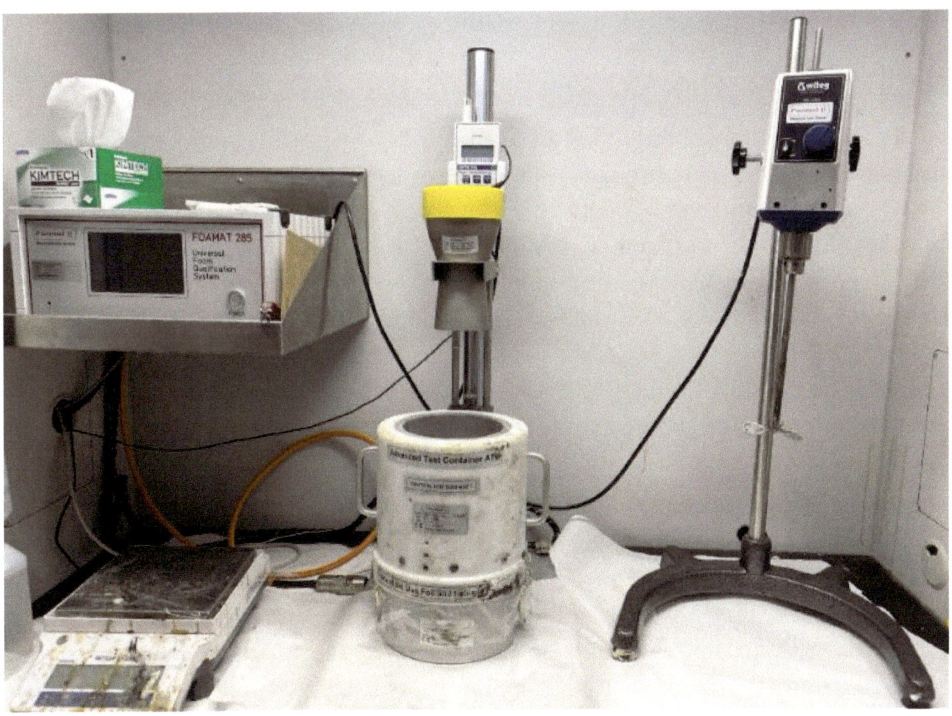

(b)

Figure 14.32 (a) Schematic of the FOAMAT foam qualification system and (b) the equipment

Given the cream time of 9 s, this PU material is extremely fast. The material starts to react and foam as soon as the two reagents are mixed, adding to measurement difficulty. The reagents were mixed in the specified amounts using a weighing balance, and the weight loss due to transfer from the container to the advanced temperature container (ATC) was carefully documented, so as to obtain the most accurate material parameters. The ATC provides a controlled temperature environment so that the effect of temperature on the reaction rate can be determined. The temperature dependence is crucial for simulation since, during actual production, elevated temperatures are used. The experiments were conducted at 30, 40, 50, and 70 °C. We observed that due to consistent temperatures, the measurement results of ATC are much more reproducible than those measured in non-thermostat test containers. The reaction temperature was also measured. The maximum core temperature was measured by placing the thermocouple in the lower third of the foam. The rise profile was recorded with a patented ultrasonic fan sensor, specifically designed for measuring the distance to the foam bun with high accuracy.

Dielectric polarization is a new measurement parameter that gives insight into the electrochemical processes occurring during foam formation. Dielectric polarization is essentially caused by chain-like molecules with a large dipole moment due to their polar ends (OH, NCO). Chain formation precedes the crosslinking reaction that ultimately suppresses all dipole mobility by curing. Hence, the dielectric polarization value can reveal the degree of curing during the course of the reaction, as shown in Figure 14.33.

The viscosity was measured using a parallel plate rheometer, model MCR502. The sample was first treated with a molecular sieve to remove water. The foaming material does not follow the Cox-Merz rule; therefore, we test its rheology using steady shear in parallel plate geometry. It was shown that shear viscosity was independent of shear rate at 0.01 s^{-1} and below. Furthermore, it was demonstrated that varying the gap between the plates (0.5 to 2 mm) showed that there was a small effect on the measured viscosity compared to the change in viscosity over time, and hence we did not attempt to correct the data for slip. Measurements were taken under 0.01 s^{-1}, with a gap of 1 mm, and at five temperatures from 30 to 50 °C in increments of 10 °C.

The importance of having separate rate equations describing these two competing foaming and curing reactions is highlighted in Figure 14.34, where the foaming reaction has a faster reaction rate than that of the curing reaction. The results are certainly consistent with the results reported by Sandia national laboratory [16]. In general, the foaming reaction is designed to have a faster reaction rate than the curing reaction for the obvious reason that if the curing reaction is too fast then the viscosity is too high for bubble formation.

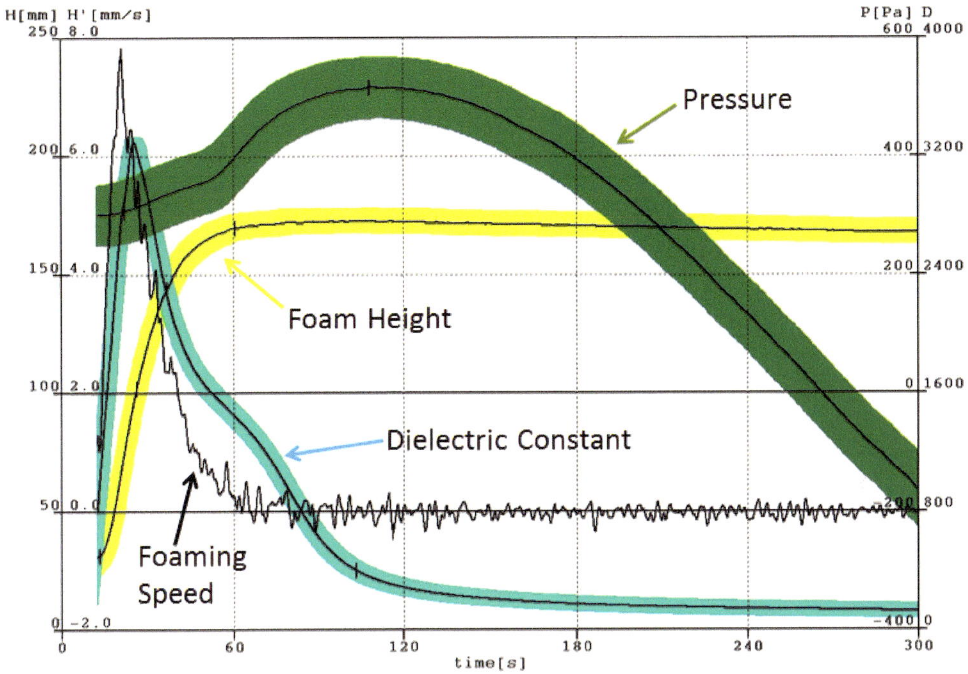

Figure 14.33 Typical experimental data for ELASTOFOAM CI 4510/136/LD series in respect of rise height, rise rate, rise pressure, and the simultaneously recorded dielectric polarization [15]

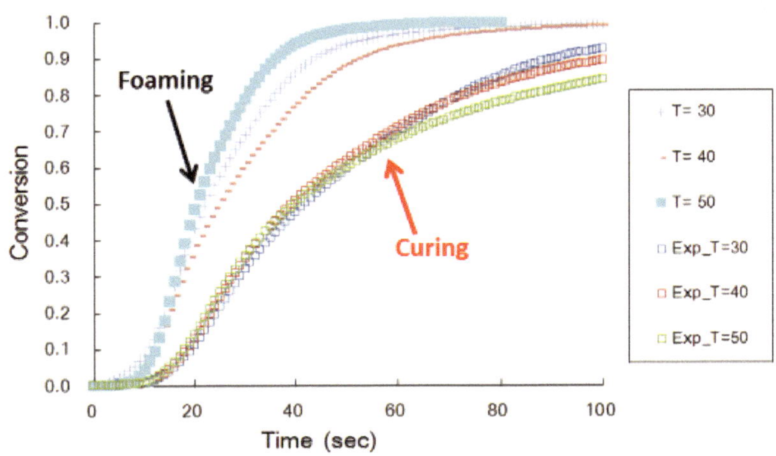

Figure 14.34 Comparison of foaming kinetics and curing kinetics at different temperatures [15]

3. Final Foam Height of Polyurethane Foam

Figure 14.35 to Figure 14.37 show the experimental validation of the foam height and temperature of polyurethane foam in ATC, in which the mold temperature and melt temperature were set to 30 °C. Figure 14.35 compares the simulation and experimental results of the final foam height in ATC.

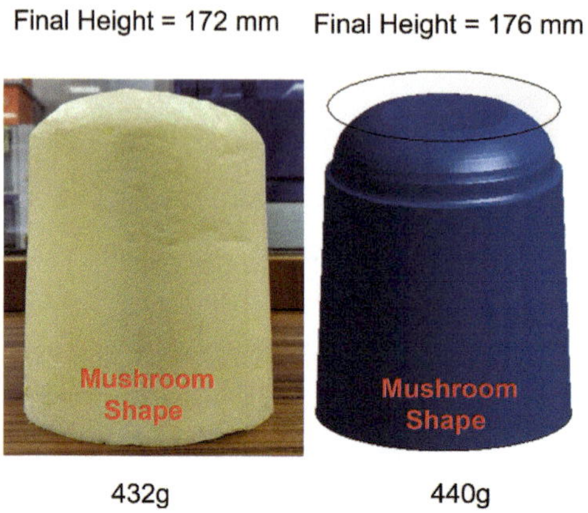

Figure 14.35 Comparison of the experimental (left) and simulation (right) results of the final foam height in the ATC [15]

Figure 14.36 shows the history curve of the foam height from simulation and the experimental results of the final foam height for comparison. In both cases the simulation results are consistent with the experimental results.

Figure 14.37 compares the experimental and simulation results for the temperature history curve. Since PU reactions are exothermic, the system temperature increases with time. The temperature simulation results are close to the experimental ones.

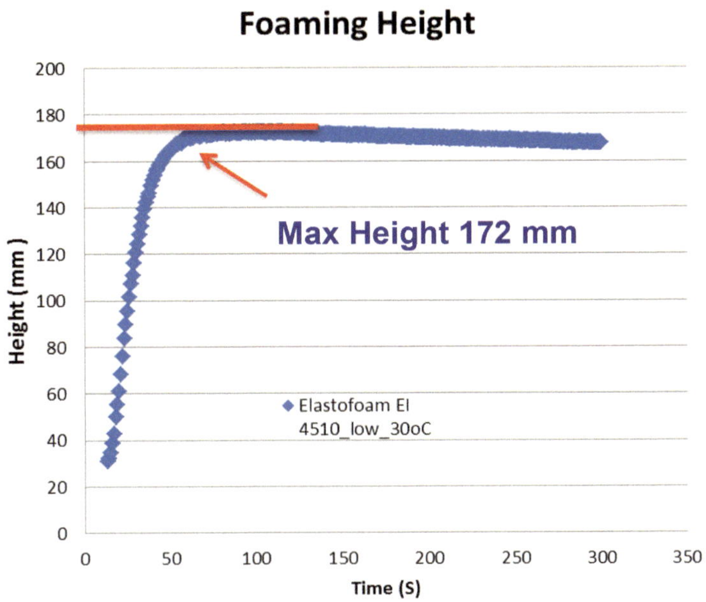

Figure 14.36 Comparison of simulated foam height as a function of time and maximal foam height in experiments [15]

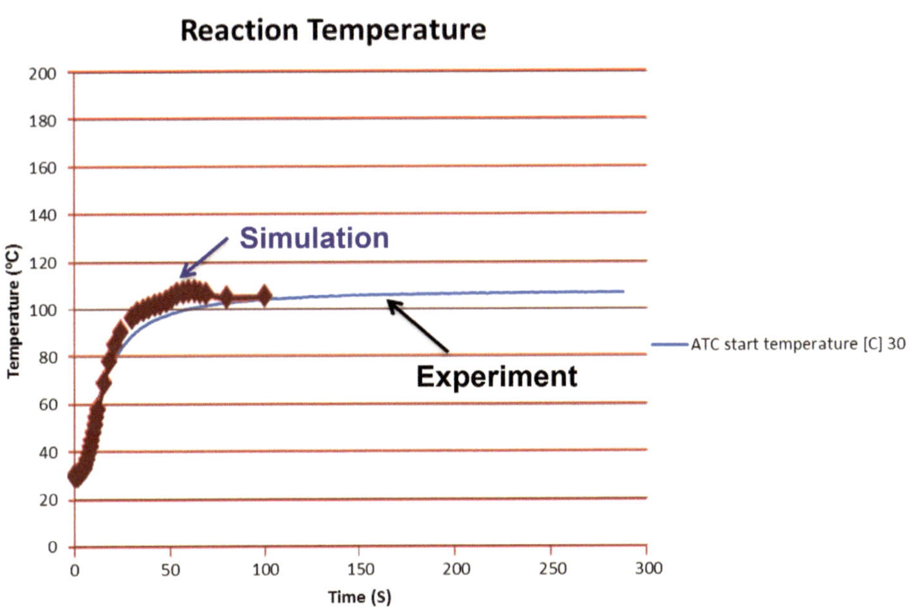

Figure 14.37 Comparison of the experimental (blue line) and simulation (red line) results for temperature as a function of time [15]

4. Applications in the Automotive Industry

Lastly, we input the measured PU material parameters into the Moldex3D simulation module, optimizing the related design and process parameters, such as heating rods design, mold temperature setting, PU injection volume, venting design and setting, and so on.

The part of interest is a steering wheel. The part volume is about 933 cm^3. In the process conditions, the partial fill volume percentage is 35% in 3.5 s. The mold temperature is 50 °C, and the inlet stream is at 25 °C.

Figure 14.38 shows the melt front of foamed polyurethane. First, the polyurethane melt will begin filling along the bottom of the cavity because of gravity. When the 35% of the cavity has been filled, CO_2 gas will be released by the polyurethane reaction, and the polymer viscosity will increase due to the gelation reaction. Meanwhile, the reaction heat of the exothermic reaction will be released to increase the temperature of the cavity. As a result, more and more CO_2 gases will be released to the polymer melt under a relatively high temperature. Eventually, polyurethane foam will completely fill the cavity.

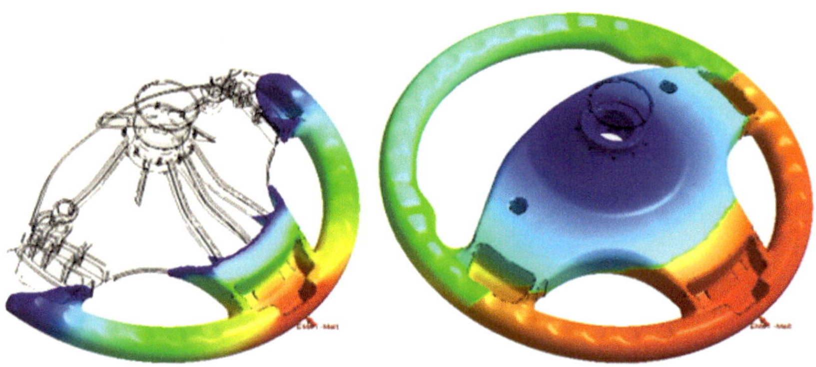

(a) (b)

Figure 14.38 Melt front at (a) the end of injection (35% filled) and (b) at the end of foaming [15]

Figure 14.39 shows the density distribution of foamed polyurethane. Density distribution reveals the expansion effect during the foaming process. When the foaming conversion is higher, the more CO_2 gas is produced. It is responsible for the larger volume of polyurethane foam and the lighter density.

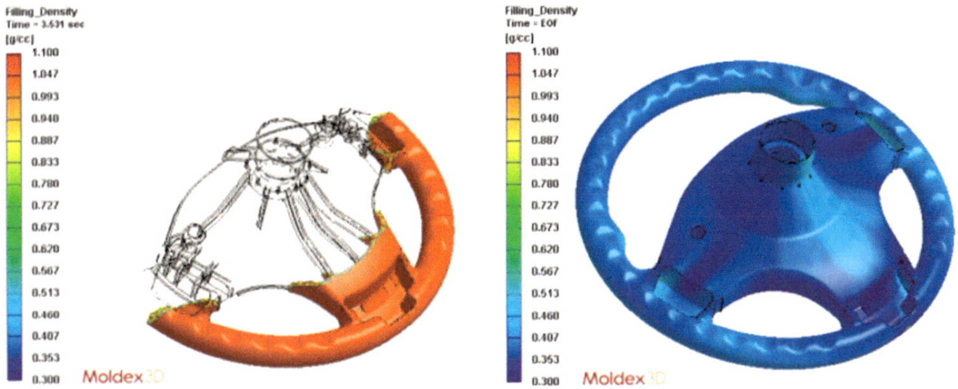

(a) (b)

Figure 14.39 Density distribution (a) at the end of injection (35% filled) and (b) at the end of foaming [15]

14.2.5 Summary

The cases described in Section 14.2.1 to Section 14.2.4 verified that CAE has good capability in the simulation of both microcellular, chemical foaming injection molding, and polyurethane reactive foaming processes. For microcellular injection molding, the effects of weight reduction, SCF dosage amount, and injection speed on the product appearance can be investigated and the correlations between the actual transparency and CAE cell density results are validated. The molding effects on cell density and cell size for microcellular injection molding are listed in Table 14.11. Therefore, CAE provides a very useful and powerful tool for innovative process development. Still many other molding condition effects in microcellular, chemical foaming injection molding, and polyurethane reactive foaming can be discussed.

Table 14.11 Molding Effects on CAE Results of Cell Density and Cell Size for Microcellular Injection Molding

	Cell Density	Cell Size
Injection speed	No clear effect	Decreases with speed
Weight reduction	No clear effect	Increases with reduction
Gas dosage amount	Increase with dosage	Decreases with dosage

Automotive products are the most popular applications of MuCell®. A famous instance is the instrument panel of Ford Motor Company, which won the Grand Award at the 41st SPE (Society of Plastics Engineers) Automotive Innovation Awards in 2011 [16]. *"By creating the instrument panel structure for the new Ford Escape in microcellular foam, weight is reduced more than 1 lb, mechanical properties are improved, molding cycle time is reduced 15%, and molding clamp tonnage is reduced 45%, saving an estimated $3 US per vehicle vs. solid injection molding,"* said MuCell® microcellular foam technology supplier Trexel, Inc.

14.3 CAE Case Study

Following the discussions of this chapter and the corresponding CAE verification cases, an exercise model, as illustrated in Figure 14.40, is provided in this section for readers to practice CAE application in foam injection molding optimization.

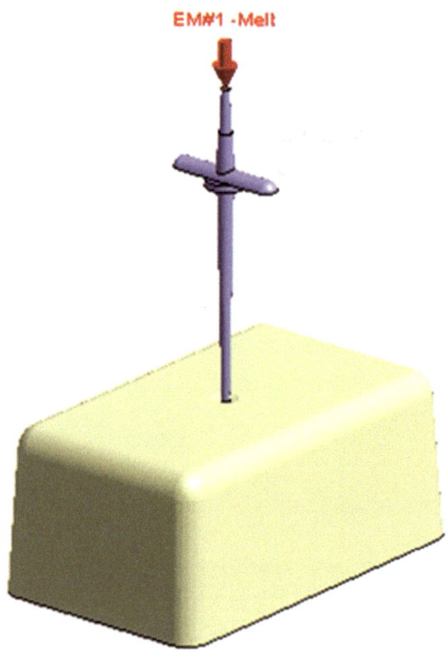

Figure 14.40 Case study sample

Please download the analysis files for CAE Case Study 14.3 from the following website:

https://moldex3d.box.com/s/zr6fvc1vlhbi4ocx111jwd3wmxt4ooif

Then open the .m3j or .mvj file in Moldex3D and answer the questions below according to the analysis results.

Questions:

1. Preanalysis of the model
   a) What is the maximum L/t of the case? Does short shot occur? Why?
   b) Where will be the filling ends?
   c) What will be the trend of warpage? Why?
2. Comparison of microcellular and solid (conventional, CIM) injection molding
   a) What is the default setting value of weight reduction?
   b) By changing from solid to microcellular injection molding:
      i. How much weight is reduced?
      ii. How much of the clamping force is reduced?
      iii. How much warp is reduced?
3. Comparison of microcellular injection molding process conditions
   a) In the foaming setting, select the option of "Volume percentage" for "Weight Reduction by", and describe how "Volume percentage" affects the part weight, clamping force, cell density, cell size, and warp.
   b) In the foaming setting, select the option of "Gas dosage amount" for "Initial Gas Concentration", and describe how "Gas dosage amount" affects the part weight, clamping force, cell density, cell size, and warp.
   c) Try to find the optimized settings for the following conditions to obtain minimal warp
      i. Weight reduction by volume percentage
      ii. Initial gas concentration (gas dosage amount)
      iii. Injection speed
      iv. Flow rate profile
   d) Which condition has the highest influence on warp? Why?

References

[1] Martini, J.E.; Waldman, F.A.; Suh, N.P.: "The Production and Analysis of Microcellular Thermoplastic Foams", in SPE-ANTEC Tech. Papers, 1982

[2] Kishbaugh, L.; Yang, A.; Shiu, L.; Chang, D.: "A Mould Filling Simulation and Validation Data for Microcellular Foaming", Blowing Agent and Foaming Processes, 2013

[3] Colton, J.S.; Suh, N.P.: Nucleation of Microcellular Foam: Theory and Practice, *Polymer Engineering & Science*, vol. 27, 1987, p. 500

[4] Xu, J.; Pierick, D.: Microcellular Foam Processing in Reciprocating-screw Injection Molding Machines, *Journal of Injection Molding Technology*, vol. 5, 2001, p. 152

[5] Suh, N.P., in: Stevenson, J.F. (ed.), Innovation in Polymer Processing: Molding, Hanser Gardner Publications, 1996, pp. 93–149

[6] Goodship, V.; Ogur, E.O.: *Polymer Processing with Supercritical Fluids*. Smithers Rapra Publishing, 2004

[7] Sato, Y.; Fujiwara, K.; Takikawa, T.; S. Takishima, S.; Masuoka, H.: Solubilities and Diffusion Coefficients of Carbon Dioxide and Nitrogen in Polypropylene, High-density Polyethylene, and Polystyrene under High Pressures and Temperatures, *Fluid Phase Equilibria*, vol. 162, 1999, p. 261

[8] Cha, S.W.; Suh, N.P.; Baldwin, D.F.; Park, C.B.: "Microcellular Thermoplastic Foamed with Supercritical Fluid". Patent US 5,158,986 A, 1992

[9] Xu, J., in: *Microcellular Injection Molding*, John Wiley & Sons, 2010, pp. 525–577

[10] Taki, K.: Experimental and Numerical Studies on the Effects of Pressure Release Rate on Number Density of Bubbles and Bubble Growth in a Polymeric Foaming Process, *Chemical Engineering Science*, vol. 63, 2008, p. 3643

[11] Han, C.D.; Yoo, H.J.: Studies on Structural Foam Processing. IV. Bubble Growth during Mold Filling, *Polymer Engineering & Science*, vol. 21, 1981, p. 518

[12] Blander, M.; Katz, J.L.: Bubble Nucleation in Liquids, *AIChE Journal*, vol. 21, 1975, p. 833

[13] Proplast: "Accurate Simulation Results on MuCell® Technology Enables Adoption of Moldex3D Solution", 2014 [Online]. *https://www.moldex3d.com/assets/2014/10/Customer-Success-Proplast.pdf*

[14] Chang, L.-Y.; Chang, Y.-J.; Hsu, C.-H.; Hwang, S.-S.; Yang, W.-H.: "Simulation and Experimental Charaterization of Foam Injection Molding", in FOAMS, 2014

[15] Wang, C.-W.; Chang, L.-Y.; Sun, S.-P.; Hsu, C.-C.; Hsu, C.-H.; Chang, R.-Y.: "Optimizing Process Condition of PU Chemical Foaming: Validation of Material Properties for Numerical Simulation", in SPE-ANTEC Tech. Papers, 2019

[16] Rao, R.R.; Mondy, L.A.; Noble, D.R.; Brunini, V.; Roberts, C.C.; Long, K.N.; Soehnel, M.M.; Celina, M.C.; Wyatt, N.B.; Thompson, K.R.; Tinsley, J.: "A Kinetic Approach to Modeling the Manufacture of High Density Structural Foam: Foaming and Polymerization", Sandia National Laboratories, 2015

[17] SPE Automotive Division: "The 41st-Annual SPE Automotive Innovation Awards Competition", 2011 [Online]. *http://www.speautomotive.com/Awards%20Modules/2011Awards/Home.html*

15 Powder Injection Molding

Huan-Chang (Ivor) Tseng

In this chapter, the powder injection molding (PIM) process is introduced. This normally involves two kinds of powder: metal powders and ceramic powders. Historically, PIM is a fairly young injection molding technique which can realize parts with a complex exterior that are otherwise highly difficult to produce, and thus significantly widens the application scope of powder metallurgy. Many hi-tech 3C products and medical appliances require PIM to manufacture metal parts with complex structures and proper strength, currently, and even in the future. However, challenges are encountered along the path of the PIM process and part development.

■ 15.1 Basics

PIM is a special molding technology developed in recent decades, which combines the features of powder metallurgy and conventional injection molding using two key technologies, injection molding and sintering. It also possesses the advantages of both technologies. Generally, PIM is categorized into ceramic injection molding (CIM) and metal injection molding (MIM) based on the powder material used. Due to its applicability and market scope, MIM has almost prevailed over CIM in industrial development. Common products produced by MIM and CIM such as cellphone, medical, mechanical, and knife components are shown in Figure 15.1.

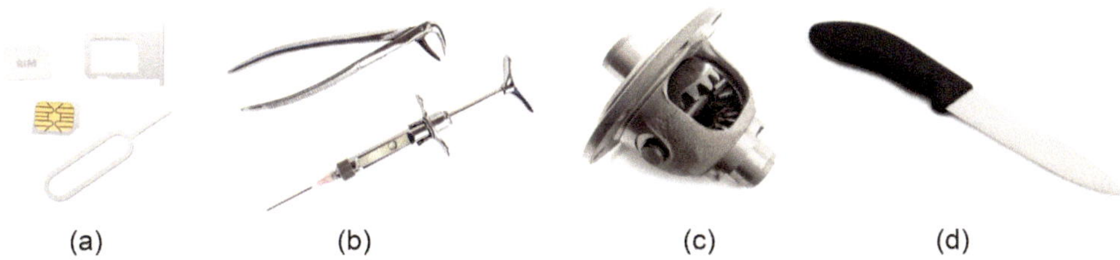

Figure 15.1 PIM products: (a) MIM cellphone components, (b) MIM medical components, (c) MIM mechanical components, and (d) CIM knife products

15.1.1 Process Principle

The PIM process includes five steps: (1) compounding and granulation, (2) injection molding, (3) debinding, (4) sintering, and (5) post-treatment.

1. Compounding and Granulation

As shown in Figure 15.2, to facilitate the flowing of powder during the injection process, the first step of the PIM process is to compound the raw powder material with the binder via a compound mixer according to certain proportions until the powder is evenly distributed in the binder to form an even mixture called feedstock. In addition, the mixture after compounding is often processed into particles. This process is called granulation, which should maintain feeding stability to avoid the bridging phenomenon when the feedstock enters the hopper.

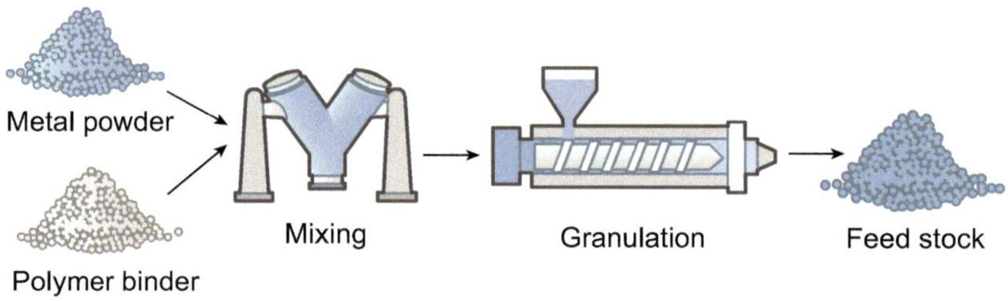

Figure 15.2 Compounding and granulation

2. Injection Molding

As shown in Figure 15.3, the injection molding step is the biggest difference between PIM and conventional powder metallurgy. After the feedstock has been poured into the injection machine and softened by the heating of the screw until the injection temperature is reached, it is then injected into the mold cavity for filling, and after packing, cooling, and demolding, a finished part is manufactured at last. This is called a green part.

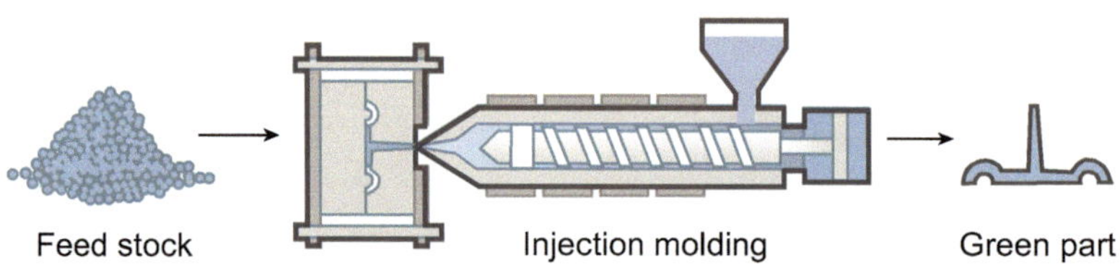

Figure 15.3 Injection molding

3. Debinding

To keep a good shape, the temperature should be increased with a mild gradient for removing the binder or debinding, which normally takes over 40 hours solely by heating. However, with a preprocess using solvent along with vacuum or a catalyst, the debinding can be accelerated and thus the production cycle time is reduced. The object processed with a two-stage debinding is called a "brown part". After debinding, the brown part becomes similar to a sponge, having many hollow channels containing air within.

4. Sintering

After debinding, a brown part is normally full of pores and with a theoretical density of 60%. Therefore, a sintering step is required to condense the finished part and give it appropriate strength and physical properties. The brown part is subjected to a linear shrinkage of 10–20% during the sintering and condensing process at high temperature. A neck-like structure would be induced from the high-temperature melting of powder particles, which is called "necking". As the space between particles shrinks, some closed or non-closed (open) pores are sometimes generated. The amount of closed and open pores greatly affects the final density of the finished part. Density is an important indicator for the adjustment of sintering conditions, and affects volumetric shrinkage of the finished part after sintering. The control of the sintering conditions to get an optimum finished part is a critical and difficult topic.

5. Post-Treatment

In addition, special demands such as polishing, electroplating, carbon doping, chemical grinding, and other post-treatments are sometimes required for the finished part after sintering. These can improve the precision and strength of the part, smoothen the surface, or remove some part defects, such as warpage, deformation, black lines, and other issues.

15.1.2 Advantages and Challenges

The size of the powder particles used in MIM and CIM differ a lot. Metal powders of submicron particle size are difficult to make and usually costly. Therefore, the metal powder used in MIM has particle sizes ranging from few mm to tens of mm, whereas CIM often uses ceramic powders with particle size smaller than 1 mm. Shape, dimension, and composition of the raw powder material have certain impacts on PIM, among which the more significant ones are as follows:

1. Particle Size

As shown in Figure 15.4, the smaller the particle size the higher the viscosity becomes due to the interaction between particles. Obviously, with the increase of shear rate, this effect becomes insignificant, showing a relatively weaker interaction between all kinds of particles at a high shear rate.

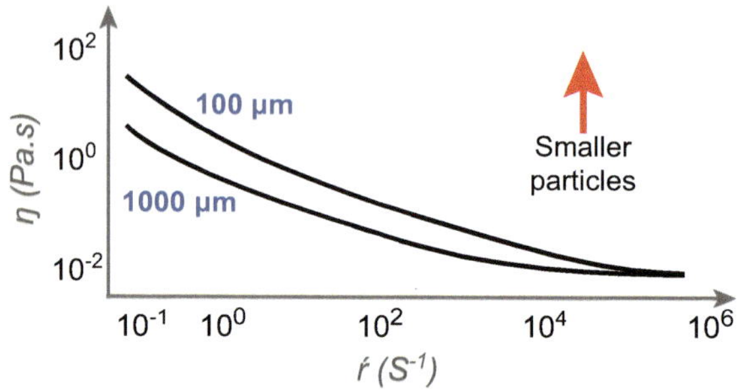

Figure 15.4 Effect of particle size on viscosity

Figure 15.5 shows the relationship between the volume fraction and the viscosity of the powder. Given a constant total volume of the powder, by reducing the particle size the viscosity would increase, as smaller particle size means more particles in number. Therefore, the chances and counts of interactions between particles also increase, and thus the viscosity of the sample increases as well. In contrast, increasing the particle size would lead to a decrease in viscosity.

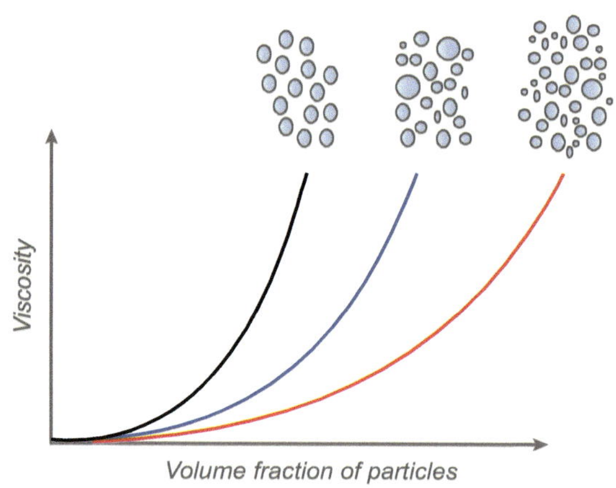

Figure 15.5 Effect of particle volume fraction on viscosity

2. Particle Size Distribution

As shown in Figure 15.6, the wider the distribution of the particle size the lower the viscosity becomes, because a wider distribution provides more free space for particles to move, meaning easier flowing and thus a lower viscosity. In contrast, tightening the particle size distribution can improve the stability of the system (higher viscosity). In summary of the above-mentioned two points, if large particles are mixed with small particles under the condition of same total volume, the viscosity would be lower than that of sole small particles or sole large particles. The minimum viscosity is close to that of sole large particles.

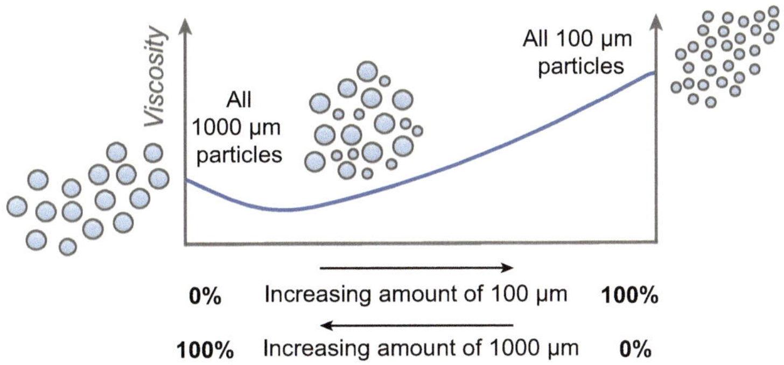

Figure 15.6 Effect of particle size distribution on viscosity

3. Shear-Induced Alignment

As shown in Figure 15.7, generally speaking, the shear viscosity at high shear rates is lower because slender particles align along the flow direction. At low shear rates the viscosity is higher due to randomly distributed particles.

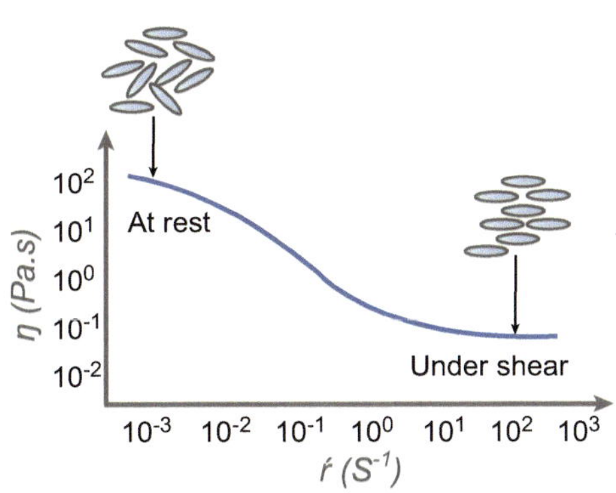

Figure 15.7 Shear-induced alignment and viscosity

15.1.3 Theory Models

Theory models of used in powder injection molding are described in this section.

1. Flow Governing Equation

The MIM feedstock is a concentrated suspension consisting of powder and binder, and assumed to behave like a generalized Newtonian fluid. A set of governing equations to describe the transient and non-isothermal fluid behaviors for channel flowing and mold filling are mentioned in Section 5.3.1.

2. Powder Concentration Equation

The suspension balance model developed by Miller and Morris [1] is considered for the flow of rigid spherical particles in a Newtonian fluid. The particle phase is approximated as a pseudo-continuum. The dominant interactions between the particles involve hydrodynamic, viscous, and non-Brownian forces with no external field except gravity.

Averaging over the particulate volume and dividing by the particle density yields the particle-phase conservation equation (Equation 15.1):

$$\frac{\partial \phi}{\partial t} + [\boldsymbol{u}] \cdot \nabla \phi = -\nabla \cdot \boldsymbol{j}_{\perp} \tag{15.1}$$

$$\boldsymbol{j}_{\perp} = \phi(U - [\boldsymbol{u}]) \tag{15.2}$$

where ϕ is the particle volume fraction, $[\mathbf{u}]$ the suspension average velocity, $\mathbf{j}_{\perp}$ the particle flux relative to the mean suspension motion, and U the local average velocity of the particulate phase.

Averaging the general momentum balance over the particle phase and restricting attention to low Reynolds numbers and neutrally buoyant, monodisperse spheres yields the relation for the particle flux (Equation 15.3).

$$\boldsymbol{j}_{\perp} = \frac{2a^2}{9\eta_f} f(\phi) \nabla \cdot \Sigma_p \tag{15.3}$$

where a is the particle radius, η_f is the viscosity of the suspending fluid, $f(\phi)$ is termed a sedimentation function, and Σ_p is the particle stress.

The sedimentation function (Equation 15.4) is described by Richardson and Zaki [2]:

$$f(\phi) = \left(1 - \phi / \phi_m\right)(1 - \phi)^{\alpha - 1} \tag{15.4}$$

where ϕ_m is the maximum packing fraction and α an index of particle friction.

The particle stress Σ_p contains both shear and normal stress portions (Equation 15.5).

$$\Sigma_p = 2\eta_f \eta_p(\phi)\boldsymbol{E} - \eta_f \eta_n(\phi)\dot{\gamma}\boldsymbol{Q} \tag{15.5}$$

where η_f is the suspension fluid viscosity, η_p the shear stress viscosity, η_n the normal stress viscosity, **E** the rate-of-strain tensor, $\dot{\gamma} = \sqrt{2\boldsymbol{E}:\boldsymbol{E}}$ the magnitude of rate of strain, and **Q** an anisotropic tensor.

The shear (Equation 15.6) and normal stress viscosity (Equation 15.7) are given in a dimensionless volume fraction of $\bar{\phi} = \phi / \phi_m$.

$$\eta_p(\phi) = 2.5\phi_m(1-\bar{\phi})^{-1} + K_s\bar{\phi}^2(1-\bar{\phi})^{-2} \tag{15.6}$$

$$\eta_n(\phi) = K_n\bar{\phi}^2(1-\bar{\phi})^{-2} \tag{15.7}$$

where K_s = 0.1 and K_n = 0.75 are factors fitted by experimental data of wide-gap Couette flow.

The anisotropic tensor is a constant tensor **Q**, which describes the anisotropy of the normal stresses (Equation 15.8).

$$\boldsymbol{Q} = \begin{bmatrix} \lambda_1 & 0 & 0 \\ 0 & \lambda_2 & 0 \\ 0 & 0 & \lambda_3 \end{bmatrix} \tag{15.8}$$

where λ_1, λ_2, and λ_3 are characteristic values for curvilinear flows.

3. Validation of Powder Concentration

As a validation [3], we use the suspension balance model to calculate the powder concentration and compare it with related experimental data. In a simple wide-gap Couette flow field, confined between two concentric cylinders, the experimental powder concentration of a volume fraction is measured by using nuclear magnetic resonance imaging for nearly monodisperse concentrated suspensions of spherical non-colloidal particles, as shown in Figure 15.8.

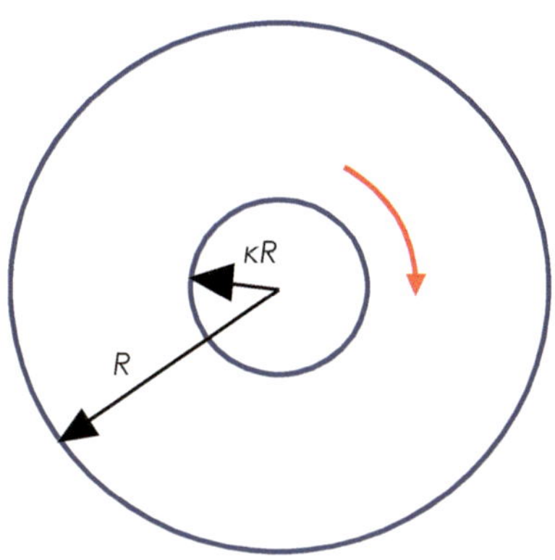

Figure 15.8 Illustration of Couette flow field

The experimental powder concentration data involve three average volume fractions (ϕ_{avg} = 0.45, 0.55, and 0.5) for poly(methyl methacrylate) (PMMA) spheres (average powder diameter of 675 mm) immersed in a Newtonian oil. In Couette flow, the stationary outer-cylinder radius is R (2.38 mm) and the rotating inner-cylinder radius is κR (0.64 mm) with an angular velocity. The length of the cylinder is 25 mm. The experimental details are available elsewhere [3].

We use the experimental data for comparison to our predictions with the maximum fraction of ϕ = 0.68. Figure 15.9 shows the reduced volume fraction distribution $\phi/\phi_{\max}$ through the reduced radius between the inner and outer cylinders. As a result, the agreement between the experimental data and the predictions is excellent. Clearly, this indicates that the suspension balance model is suitable for the prediction of the powder concentration.

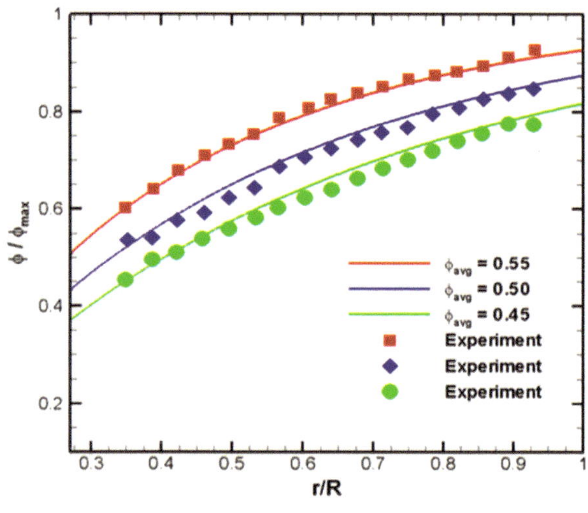

Figure 15.9 Dimensionless volume fraction against different average volume fractions

4. Application to Simple Flow Channels

The assessment of powder concentration has been validated. Here, we further use the suspension balance model for simple flow channels of the concentrated suspension fluid, including both plaque and contraction-expansion geometries, wherein the average volume fraction is 60 vol.% and the powder diameter 10 mm. The goal is to present the powder-binder phase separation during flowing.

The 16 mm × 16 mm × 1.5 mm plaque geometry is illustrated in Figure 15.10.

Figure 15.10 Plaque geometric illustration (16 × 16 × 1.5 mm³)

Figure 15.11 shows the powder concentration pattern in the plaque flow. The appearance of a powder concentration gradient is obvious. Near a wall, the concentration is relatively low, whereas the central concentration is high. For this reason, such a representation clearly indicates the powder-binder phase separation.

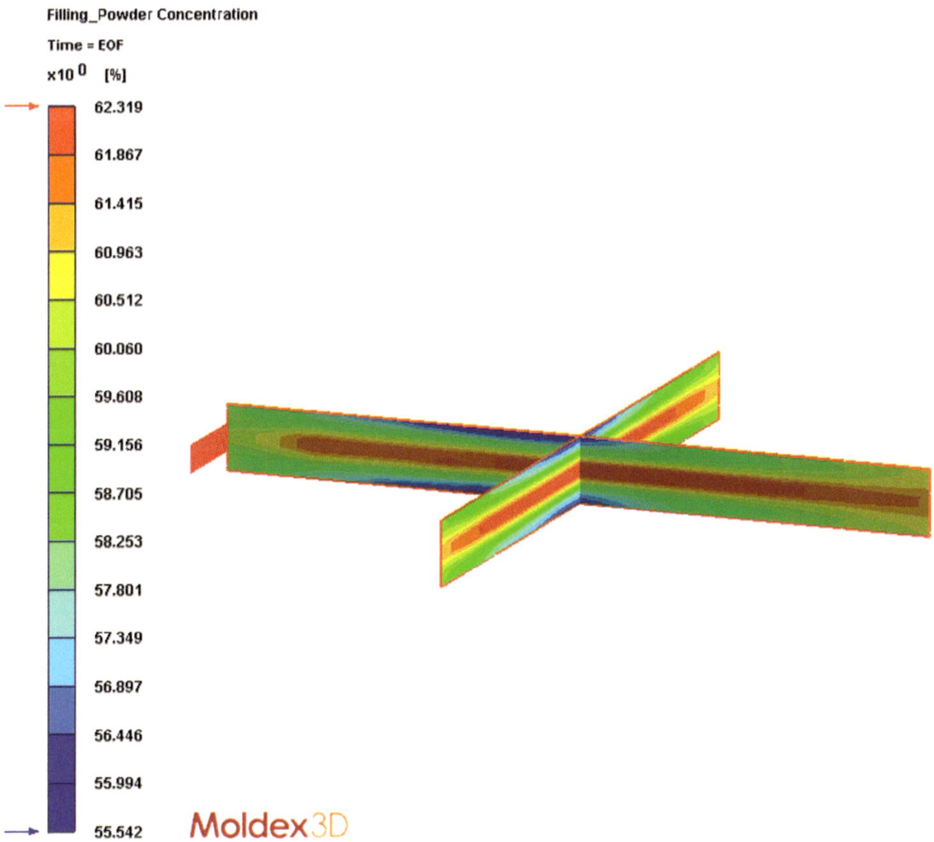

Figure 15.11 Powder concentration pattern for the plaque filling with an average volume fraction of 60 vol.% In addition, a schematic diagram of a 4:1:4 contraction-expansion geometry is given in Figure 15.12, wherein the contraction size is 1 mm and the expansion size 4 mm

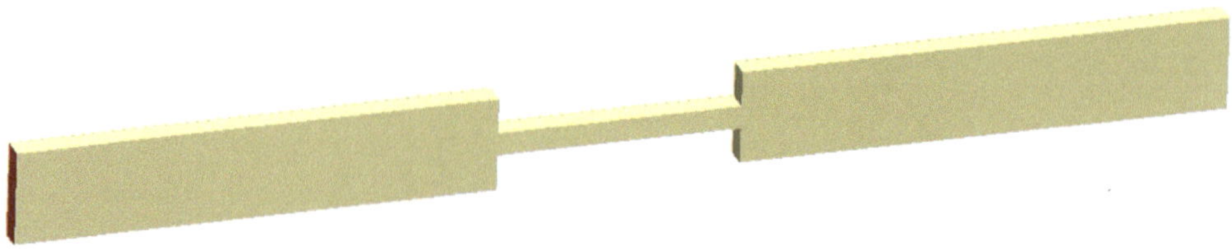

Figure 15.12 4:1:4 contraction-expansion geometric illustration with the biggest size of 4 mm and smallest size of 1 mm

Figure 15.13 shows the concentration profile. In the plaque, an obvious concentration gradient has developed. There are several interesting findings for the contraction-expansion flow. When the fluid flows into the contraction part, a maximum concentration obviously occurs in the proximity of the corner of the inlet expansion part. In particular, the contraction concentration pattern is very vivid from the low wall concentration to the high central concentration; such a dramatic profile indicates a powder-binder phase separation. In addition, the gradient is relatively weak through the expansion part so that the concentration profile is almost uniform.

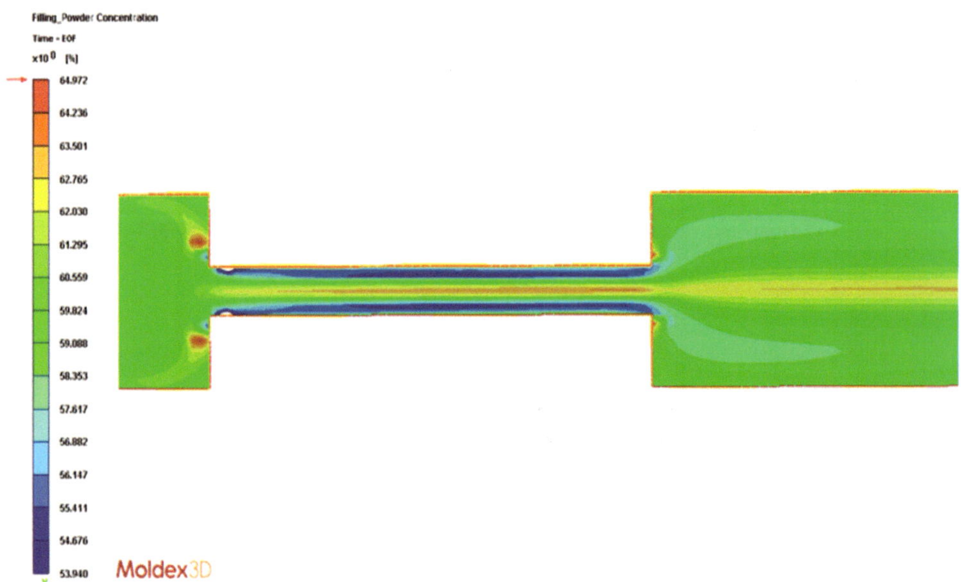

Figure 15.13 Powder concentration pattern for tensile-bar filling with an average volume fraction of 60 vol.%

■ 15.2 Practical Applications

How to resolve the black line issue in a finished part occurring in PIM by CAE?

When PIM is applied in industry, the black line is a common surface defect which is caused by the disengagement between powder and binder, which leads to a drop in powder concentration locally. Most black lines already occur when the feedstock is injected into the mold cavity, and sometimes cosmetic defects or cracks can be observed on green parts, which cannot be amended via sintering. Besides, for most manufacturers, the worst case is that such defects cannot be discovered until the sintering is completed, at which point much cost and time have already been spent.

Black lines denote regions with a lower powder concentration and a relatively higher concentration of binder. After the sintering process, such regions would generate cosmetic defects, poor mechanical properties, and other issues. As part models are becoming more complicated, the trial-and-error method used before is no longer cost-effective in solving the black line problem. To ensure good green part quality before the part is sintered, examination of the properties of the molding raw materials is especially important in the quality control of a molding process. Solutions for solving PIM issues can be derived via a complete CAE simulation, which can determine appropriate molding raw materials to secure the quality of the green part.

CAE Verification on an Electronic Device

The part model is shown in Figure 15.14. The part volume is about 4 cm^3. The MIM material used in this work consists of a polyoxymethylene (POM) binder and steel powder (Catamold 17-4 PH supplied by BASF). The average powder concentration is about 60 vol%. The mold temperature is held to approximately 100 °C and the inlet temperature to 200 °C. The filling and packing time are about 0.3 s and 2.3 s, respectively. Note that the green part mass is about 23 g and that of the sintered part about 21 g.

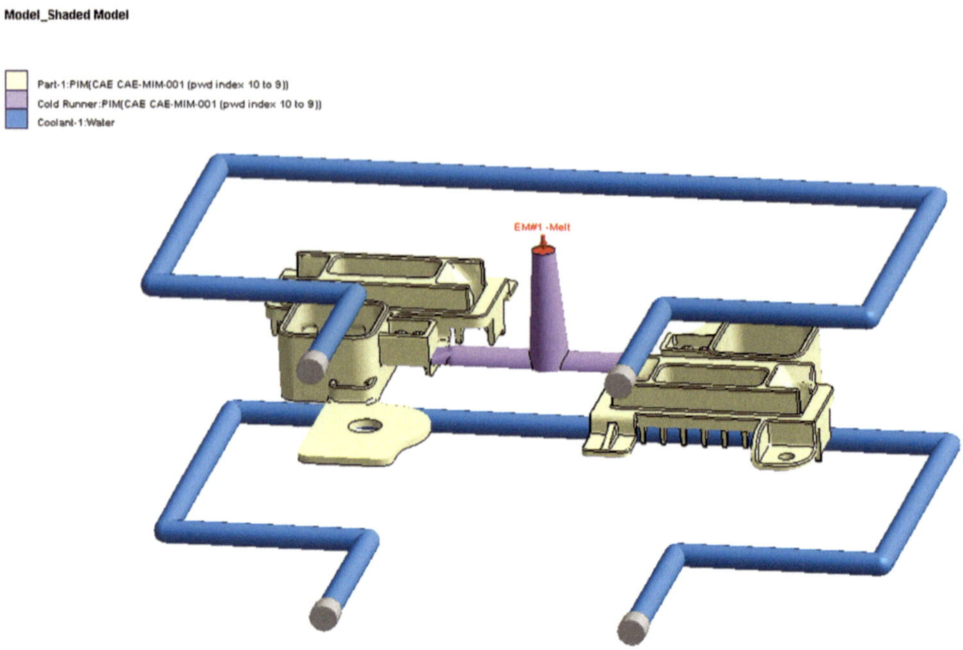

Figure 15.14 Geometry of the real MIM part

Figure 15.15 shows that the surface of the real sintered part exhibits some dark areas among the larger light areas. This is generally called a black line in MIM industrial fabrication or is known as powder-binder phase separation in the academic community.

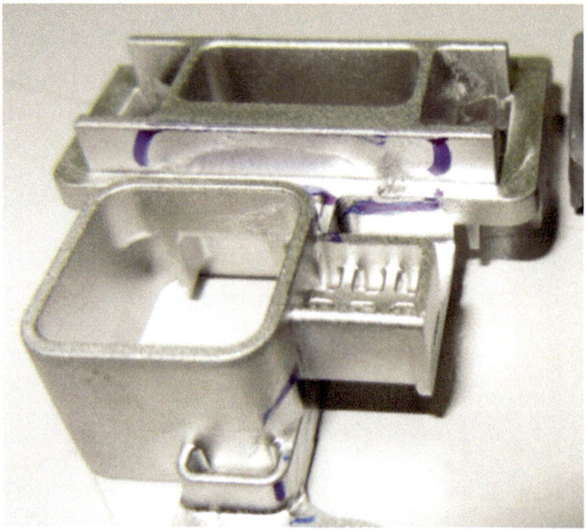

Figure 15.15 Black lines of the real sintered part are marked by blue lines

Furthermore, the surface powder concentration distribution is predicted by the Moldex3D PIM package, as shown in Figure 15.16. It is clear that the overall distribution is uniform, or is close to the average concentration (60 vol.%). Obviously, some areas of lower concentration can be seen. One can further observe a relationship between the black lines of the real part in Figure 15.15 and the powder concentration distribution shown in Figure 15.16. Surprisingly, the black lines are related to low concentration. Therefore, the black lines from the sintered part in the MIM process might indicate a high binder concentration or a low powder concentration for the green parts. This is helpful for the prediction of black lines, especially concerning surface defects of MIM fabrication.

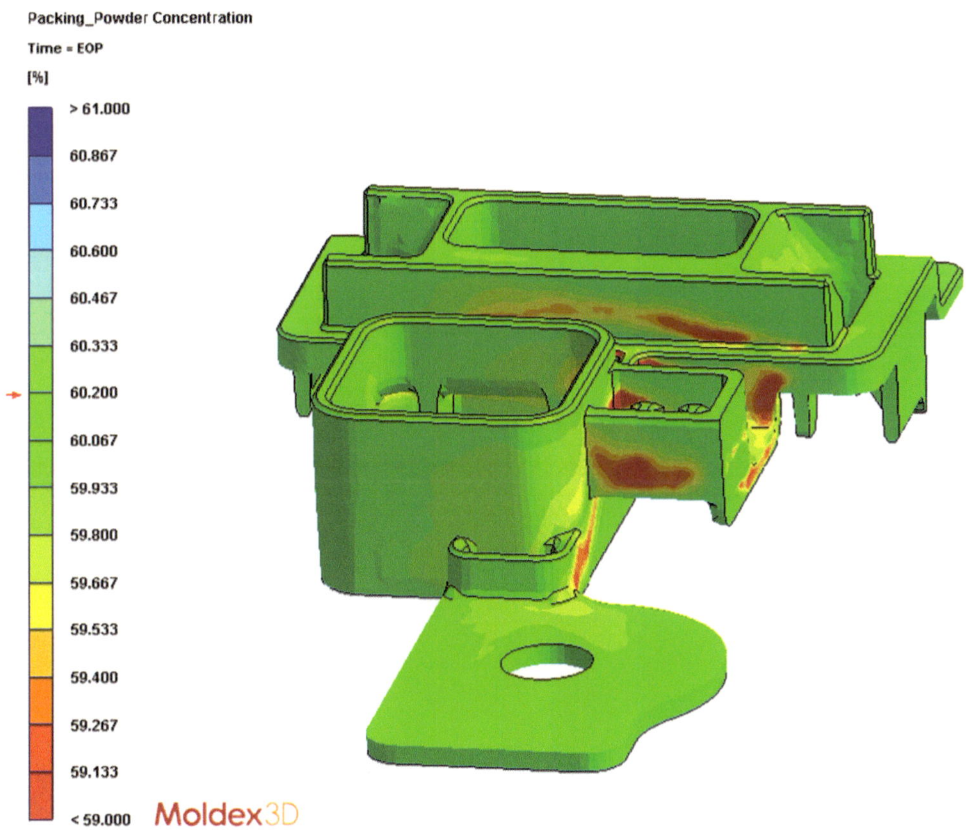

Figure 15.16 Powder concentration distribution prediction

In summary, this example of an actual industrial MIM part shows possible locations of black lines predicted by Moldex3D-PIM nearly identical to the actual locations of black line in the real part. Three important molding analysis results are summarized as follows:

1. Most of the surface has even powder concentration. The regions with lower concentration would possibly develop the black line issue.
2. The regions with excessive low concentration are the areas with medium to high viscosities, which is caused by large flow resistance.
3. The difference in viscosity distribution is one of the main reasons causing uneven powder concentration.
4. The black line issue in the part is solved by shortening the injection time.

15.3 CAE Case Study

Following the discussions of this chapter and the corresponding CAE verification cases, an exercise model, as illustrated in Figure 15.17, is provided in this section for readers to practice CAE application in powder injection molding optimization.

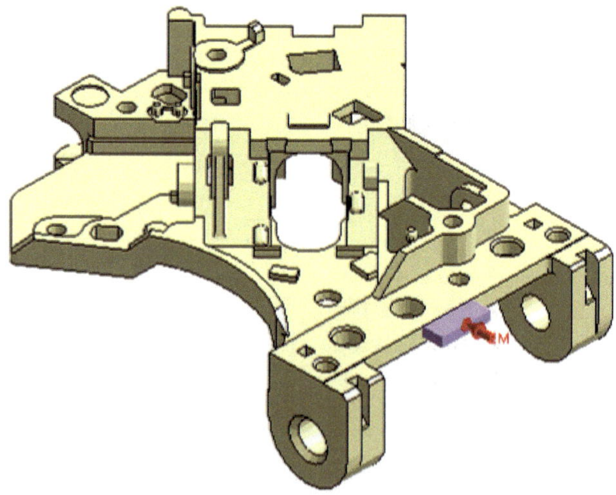

Figure 15.17 Case study sample

Please download the analysis files for CAE Case Study 15.3 from the following website:

https://moldex3d.box.com/s/zr6fvc1vlhbi4ocx11jwd3wmxt4ooif

Then open the .m3j or .mvj file in Moldex3D and answer the questions below according to the analysis results.

Questions:

1. Preanalysis of the model: Where will be the filling ends and weld lines?
2. Show analysis results under default conditions, including:
   a) melt front time,
   b) injection pressure history,
   c) shear rate, shear stress,
   d) powder concentration,
   e) and warpage.
3. How do analysis results change if adjusting:
   a) filling time,
   b) mold temperature,
   c) or melt temperature?

References

[1] Miller, R. M.; Morris, J. F.: Normal Stress-Driven Migration and Axial Development in Pressure-_Driven Flow of Concentrated Suspensions, *Journal of Non-Newtonian Fluid Mechanics*, 135, 2006, pp. 149–165

[2] Richardson, J.; Zaki, W.: Sedimentation and fluidization: Part I, *Transactions of the Institution of Chemical Engineers*, 32, 1954, pp. 35–47

[3] Phillips, R. J.; Armstrong, R. C.; Brown, R. A.; Graham, A. L.; Abbott, J. R.: A Constitutive Equation for Concentrated Suspensions That Accounts for Shear-Induced Particle Migration, *Physics of Fluids A: Fluid Dynamics*, 4, 1992, pp. 30–40

16 Resin Transfer Molding

Hsun (Fred) Yang, Yu-He (Zoe) Chen

Resin transfer molding (RTM) is a molding technology that offers great potential. It has been widely used in aerospace, transportation, shipbuilding, construction and other fields. It is one of the liquid composite molding (LCM) processes. Common liquid molding processes also include structural reaction injection molding (SRIM), vacuum assisted resin transfer molding (VARTM), wet pressing (Wet RTM), compression resin transfer molding (CRTM), and resin film infusion (RFI).

The RTM process is introduced in this chapter, where the basic principle, advantages, challenges, and theoretical models are described first. Then the real case of a wind turbine blade (Figure 16.1) demonstrates how to use CAE in assisting innovative process development. Finally, a case study provides the reader with some exercises to practice CAE settings and interpret the analysis results of RTM.

Figure 16.1 Wind turbine blade as example of a fiber-reinforced plastic product

16.1 Basics

Composite materials are multiphase solid materials that are composed of more than two components. The continuous phase is called the matrix. The other phase is the dispersion phase comprising reinforcement material. Polymer composite materials are divided into two types: thermosetting and thermoplastic. Normally, the mechanical strength of the reinforcement material is higher than that of the matrix, so the mechanical strength of the composite product is significantly greater than the matrix material alone.

The fiber-reinforced polymer (FRP) composite material consists of three fundamental components: fiber reinforcement, matrix, and fiber-matrix interface (Figure 16.2).

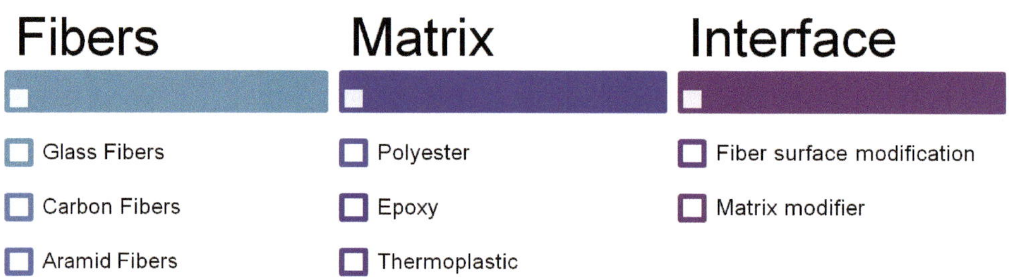

Figure 16.2 Fiber-reinforced plastic components

1. Fiber reinforcement:

   A fiber reinforcement is a reinforced material that gives FRP materials high strength and high elasticity, so that they do not bend or disrupt under stress. The fiber reinforcement is the main factor that determines the mechanical performance of a FRP material. The functions of fiber reinforcement on FRP materials can be summarized as the following five points: 1) sustain the main loading; 2) limit the extension of minor cracks; 3) improve material strength and stiffness; 4) improve anti-fatigue and anti-creep performance of the material; 5) extend the material lifetime and reliability. Due to various features of FRP materials, the design and application of anisotropic composites have better competitiveness and economic value.

3. Matrix (resin or plastics):

   The main function of the matrix is to transmit or disperse stress to each fiber and fix the fiber to the orientation required. The matrix can also protect fibers from being worn or corroded, and by integrating fibers into the matrix, the composite material is able to withstand breakage or warpage under stress. The type of matrix is the main factor that determines the highest operation temperature and the chemical properties of a fiber composite material.

The FRP material is the main factor that determines the mechanical properties of composite materials, which bears the majority of load, limits the extension of minor cracks, improves the material stiffness, and resists fatigue and creep. There are different materials and forms of fiber, among which the commonly used materials are glass fiber, carbon fiber, Kevlar fiber, boron fiber, and silicon carbide fiber, while in form they are divided into two types: continuous fiber and non-continuous fiber.

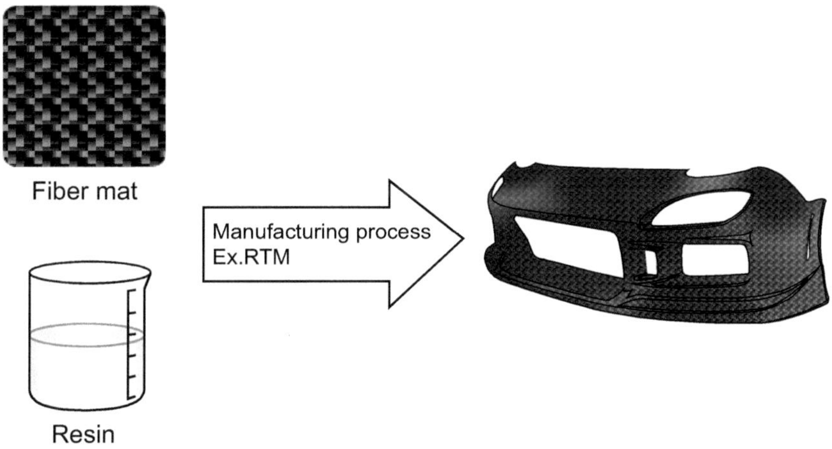

Figure 16.3 FRP product

The RTM process introduced in this chapter is mainly used for the production of polymer composite parts with continuous reinforcement fibers shown in Figure 16.3. The relationship between mechanical strength and fiber length is shown in Figure 16.4. When the fibers in a reinforcement material are longer, the mechanical strength of the part becomes higher. Compared to non-continuous fibers, the continuous fiber has better strength and performance. The process of making parts with continuous fibers is very different from the conventional non-continuous fiber process. The LCM process is often used in the production of fiber-reinforced polymer composite products with continuous fibers.

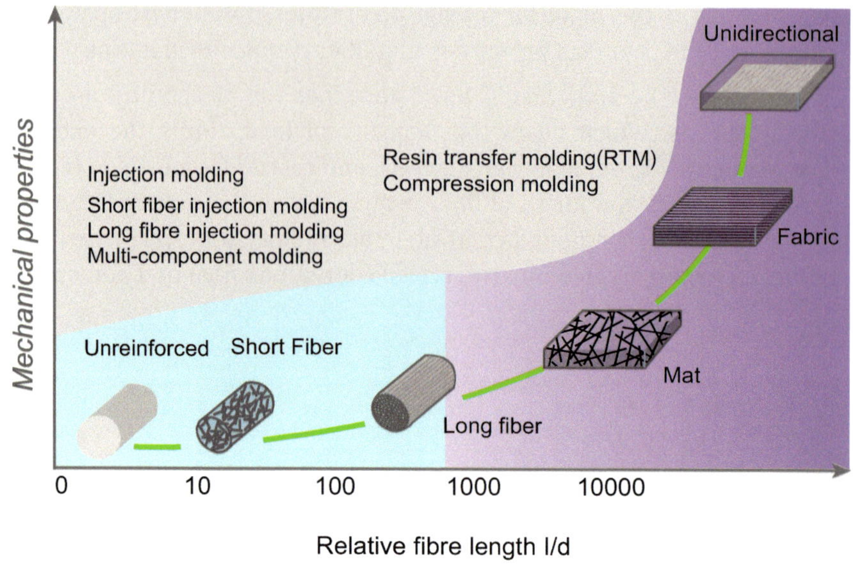

Figure 16.4 Mechanical strength of FRP products

Commercially Available Forms of Reinforcement

FRP parts can be subdivided into the categories shown in Figure 16.5 according to the type of matrix and reinforcement. Glass fiber (Figure 16.6) and carbon fiber (Figure 16.7) are the two most commonly used fiber types in the RTM process. Reinforcement materials can be distinguished by part appearance: glass fiber is white, but would turn transparent when being impregnated with resin, whereas woven carbon fiber is black and does not change much in color before and after impregnation. Normally, glass fiber fabric would be surface processed, so there is glass fiber sizing on the mat. It is necessary to wear gloves when handling fiberglass. A direct contact of glass fiber sizing could induce a stinging sensation or even cause allergy. Wear a lab coat and an N95 mask to avoid the dust generated by cutting. Figure 16.6(a) shows a glass fiber random mat consisting of non-continuous random fibers, whereas Figure 16.6(b) shows the parallel arrangement of fibers in a unidirectional fabric, which is formed by thin treads crossing between fibers. Figure 16.7 shows a carbon fiber woven. Carbon fiber woven is categorized according to the number of single threads in a bundle of fibers, and as K denotes a thousand, a 3 K woven means that there are 3000 single threads in a bundle.

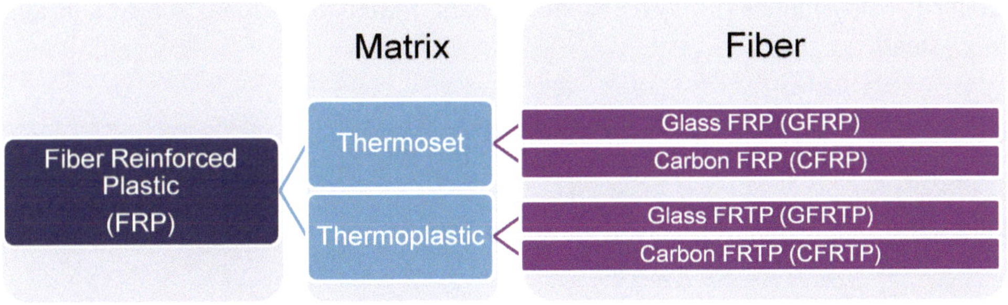

Figure 16.5 Kinds of FRPs

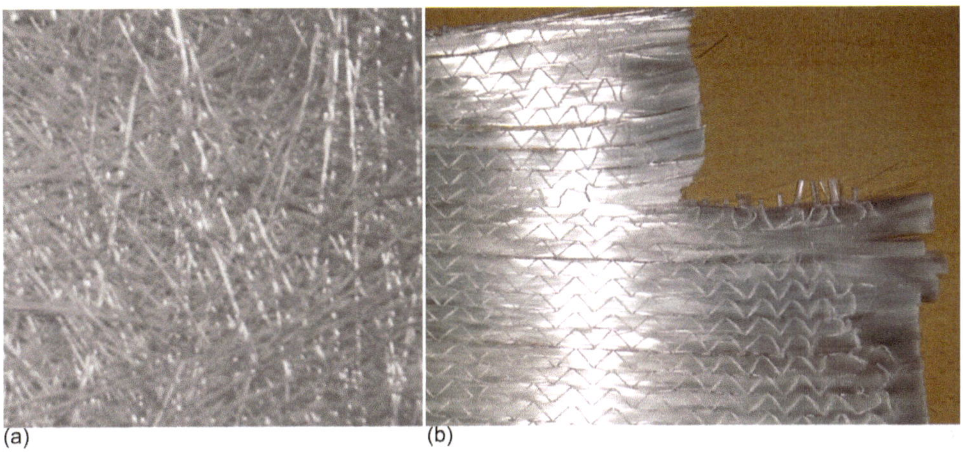

Figure 16.6 Glass fibers: (a) random mat and (b) unidirectional

Figure 16.7 Carbon fiber woven

Common Lay-up Structures

The sandwich structure as shown in Figure 16.8 is used in large parts such as yachts and wind turbine blades. Such a structure is composed of a face sheet on top and bottom, and the core material in the middle, which possesses excellent mechanical performance while being much lighter than traditional solid materials (e.g., steel).

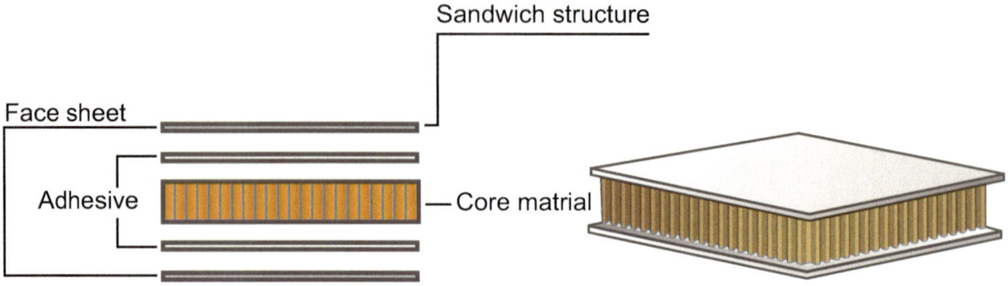

Figure 16.8 Sandwich structure

16.1.1 Process Principle

Generally, the simplest method of making a FRP part is by hand lay-up, as shown in Figure 16.9, which is to adhere the fibers layer by layer, followed by spreading the resin with a roller evenly on the fiber fabric. As a hand lay-up is done manually, a difference in part quality may occur due to different manufacturing conditions. When higher part precision is requested, such a manual operation would be substituted by the liquid composite molding process.

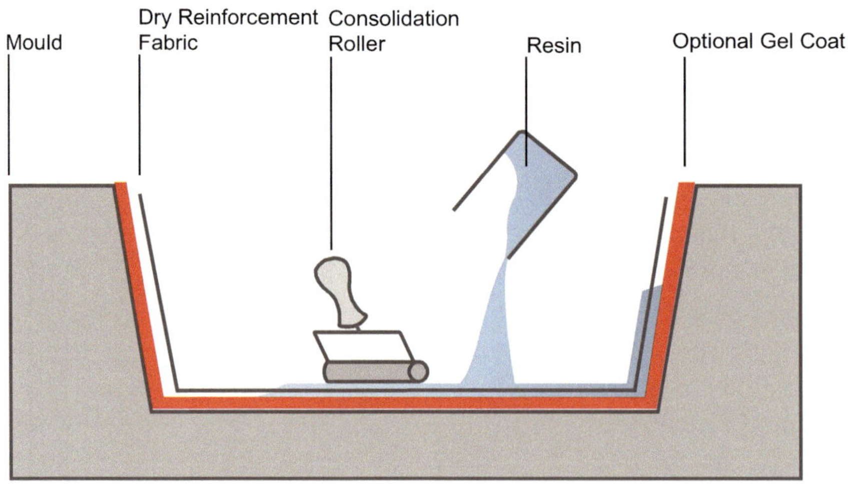

Figure 16.9 Hand lay-up

The RTM process, as shown in Figure 16.10, belongs to one of the branches of liquid composite molding processes. The main principle of a LCM process is to place a pre-molded reinforced part designed according to the performance and structural requirements, and inject resin into the mold cavity to impregnate the fiber. After being heated, cured, cooled, and demolded, a finished part is manufactured. Five identical steps are included in different LCM processes: (1) cut and place of pre-molded parts, (2) mold closing, (3) resin injection, (4) resin curing, and (5) demolding.

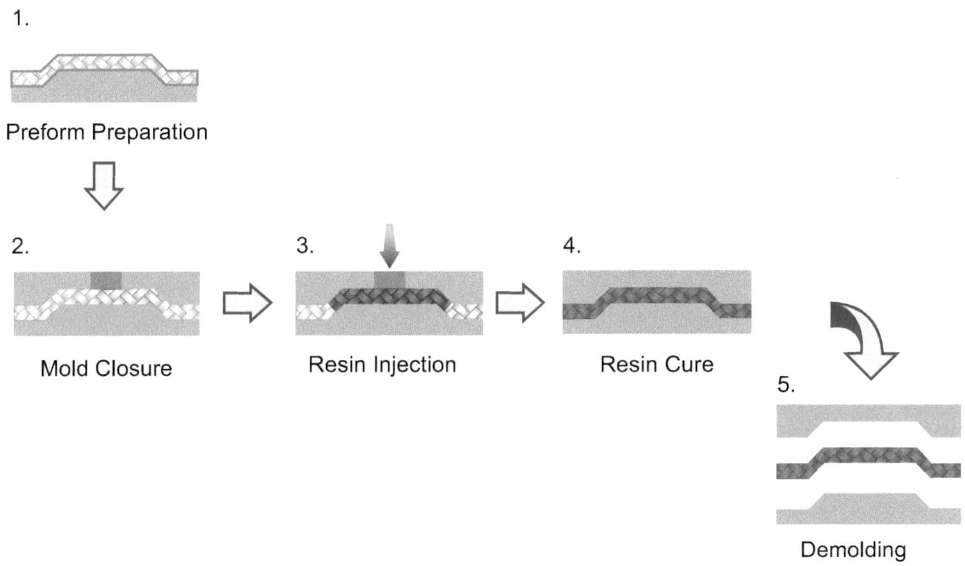

Figure 16.10 Flow chart of the RTM process

A characteristic feature of the RTM process is that a double-sided closed mold is used for manufacturing. The quality of the product is stable and high-precision parts can be produced with a smooth surface on both sides. This process is often adopted for the production of small-sized products.

The first step of RTM is to cut and place the pre-molded part. Preform cutting is shown in Figure 16.11. After coating a layer of demolding agent on the top and bottom mold, the preform is placed inside the mold. The second step is to remove the air from the mold. Normally, a RTM process uses thermosetting resins of low viscosity but a longer injection time. In the third step, the resin and hardener are mixed in proper proportions before being filled into the mold. Then the resin is injected into the mold by applying pressure or utilizing the inner and outer pressure difference of the mold until the mold is filled up completely, as shown in Figure 16.12. The fourth step is to heat up the mold to cure the resin, and after the mold has cooled down, the finished composite part can be retrieved after demolding.

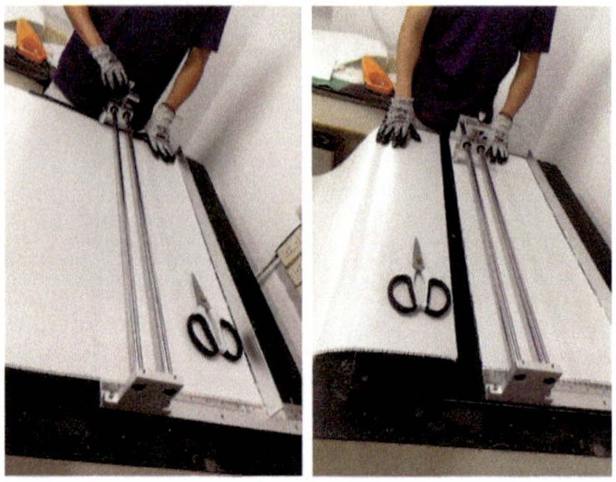

Figure 16.11 Preform preparation [1]

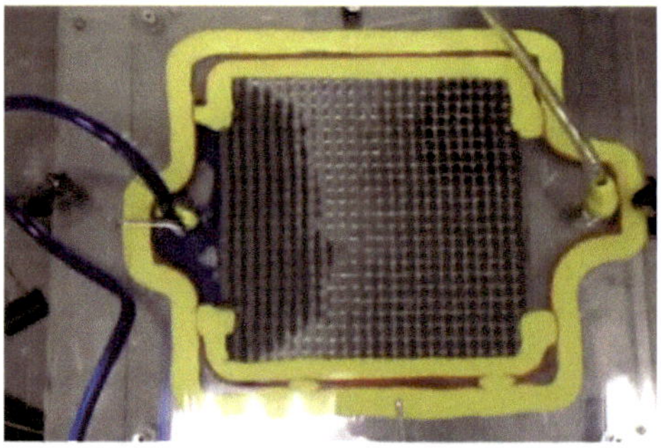

Figure 16.12 Mold filling

VARTM is chosen for the production of larger parts, such as yacht parts and large wind turbine blades. The replacement of the upper mold with a vacuum bag in a VARTM process is shown in Figure 16.13. As there is normally no requirement for high pressure or heating up to 93 °C [2], the manufacturing cost of the mold is far lower than that of the conventional RTM process. The most famous VARTM process is called SCRIMP (Seemann composites resin infusion molding process), which had been patented by William Seemann.

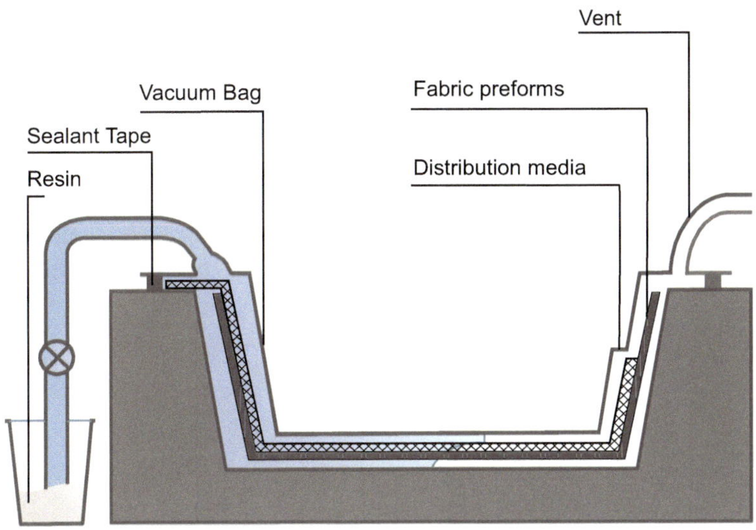

Figure 16.13 Vacuum-assisted resin transfer molding

Composite material components are often made using a variety of molding processes. For example, the continuous fiber part of the product is completed through hybrid molding technology, and then the discontinuous fiber part of the product is completed by injection molding or compression molding. Figure 16.14 shows the schematic of a hybrid molding part, which can be produced by thermal press molding with prepreg or RTM injection molding. The gray portion is the continuous fiber while the yellow portion is the ribs of discontinuous fiber processed by injection molding. The hybrid molding method enables a product of such design to be manufactured.

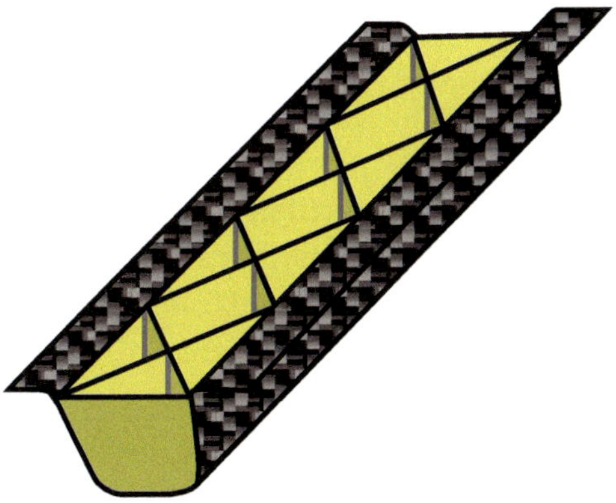

Figure 16.14 Schematic of hybrid molding product

16.1.2 Advantages and Challenges

Products manufactured by the RTM process, which are widely used in wind-turbine industry, yacht manufacturing (Figure 16.15), and automobile structural parts, have the following advantages: high dimensional precision, anti-corrosion, high mechanical strength, and smooth exterior surface. The advantages of the RTM process, compared to the traditional hand lay-up process, are weight reduction of the finished parts in the injection process, shorter process cycle time, and lower labor cost.

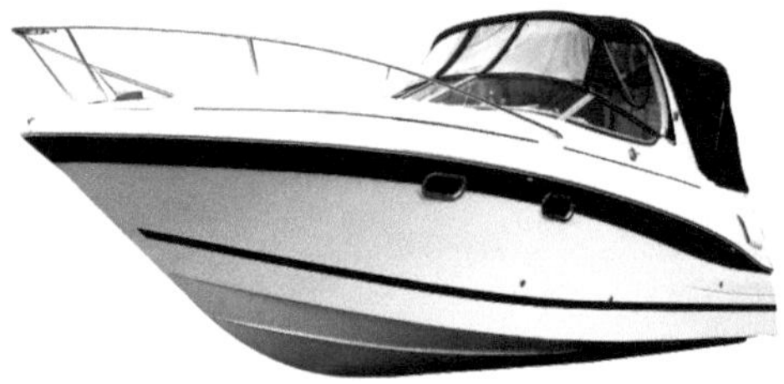

Figure 16.15 Parts of yacht are made by VARTM

The challenges that the RTM process faces are from various factors that affect the process, including the geometric structure of the mold, resin characteristics, material properties of the fiber preform, opening timing of inlets during the process, injection pressure, ambient temperature, and mold temperature. As the rheological properties of resins change with the reaction time and variation of resin temperature, the material properties differ a lot among different fiber preforms. To control the key factors in a RTM process, the material properties should be well-understood, which can be achieved by measuring the reactive and rheological properties of resin materials as well as the permeabilities of various woven preforms.

Common Defects of RTM Parts [3]

Defect parts with insufficient impregnation tend to be manufactured when the parts are molded under poor process conditions. The major two defects of insufficient impregnation of RTM parts are voids and dry spots. The existence of voids would decrease the impregnation of fibers, which worsens the adhesion at the interface between resin and fiber and results in lower mechanical strength in the composite part. Figure 16.16 shows a defect part suffering such a phenomenon; the right figure shows the exterior of the part, and the left shows the backside of the part. The issue can be verified without damaging the part via a nondestructive testing instrument. It can be seen that the red region in the left figure shows a difference in the test result when dry spots are present.

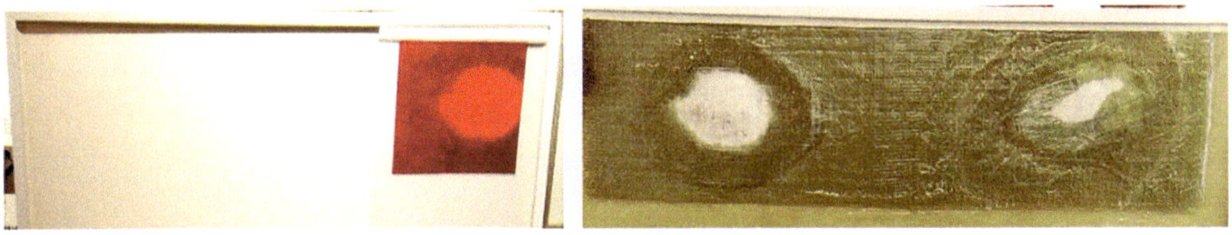

Figure 16.16 Dry spot [4]

There are three causes of void generation: 1) the curing time is shortened due to excessive heat release during the resin curing reaction, and thus the bubbles cannot be vented smoothly, 2) too much air is brought into the cavity during resin injection and cannot be vented during molding, and 3) the viscosity of the resin is too high so that the air cannot be vented during molding.

Dry spots tend to occur at the initial stage of part research and development when the process has not become stable yet. The causes of such issues include improper design of mold inlet and edge conditions of venting, improper placement of preforms, and edge effect of mold, which can be solved by: 1) setting more inlets and controlling the opening timing of the barrel as well as adjusting vacuum edge regions, 2) controlling the flow conditions of the resin during the part filling process, 3) adjusting the way the preform is placed, and 4) checking if the fiber fabric is properly paved at the edge regions after the preform is placed.

16.2 Theory Models

16.2.1 2.5D Analysis

An isothermal 2.5D analytic tool is used in conventional RTM simulation analysis, which is suitable for fabric preforms of single bulk material and simple geometry. As the simulation tool lacks one dimension, it requires two simplification measures for the analysis via a 2.5D tool to accommodate a 3D issue.

There are two major 2.5D analyses commonly used (Table 16.1):

Table 16.1 Methods of RTM Simulation in 2.5D

Simulation	Features	Disadvantages
Method 1	Flow in thickness direction	Numerous analyses required. Lack of regional interactions as a result
Method 2	Fully detailed flow behavior	Flow behavior prediction without thickness direction information

Method 1 uses multiple cross sections of geometry for analysis to simulate the flow behavior of the resin between lay-ups in each region. It is suitable for a simple flat plate geometry, but the downside is that the analysis must be performed multiple times and the overall result cannot be derived from the simulation result of each region.

In the other method, as shown in Figure 16.17, a multi-type prepreg is simplified into a single-property material to see the overall flow behavior of the resin. Such a method is applicable for a single material type, but the downside is that the influence on the flow behavior along the thickness direction cannot be considered, that is, when it comes to thick lay-ups or significant material property differences, the influence of thickness cannot be reflected.

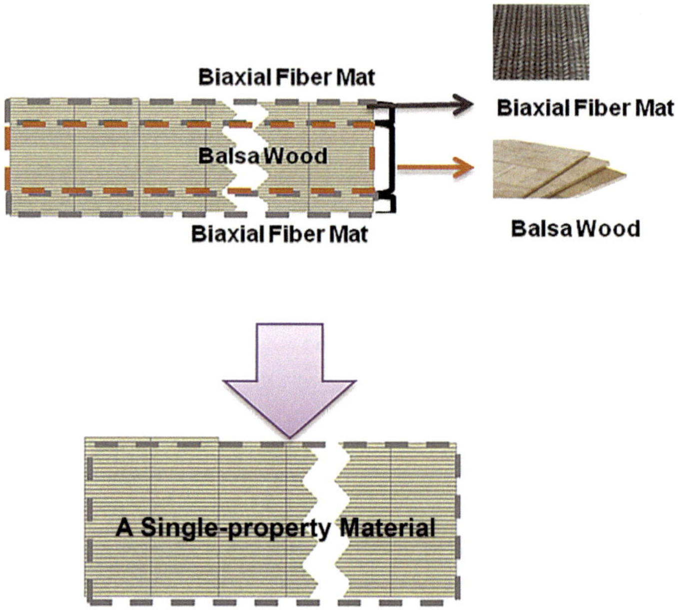

Figure 16.17 Multi-type prepreg is simplified into a single-property material [5]

The flow property of the fabric preform differs hugely between the thickness direction and the horizontal direction; therefore, the influence of the thickness direction on flow behavior cannot be neglected. Furthermore, as the properties of different fabric preforms vary a lot when lay-ups of different properties are molded together in a manufacturing process, and the curing reaction during the flow process of the resin is also influential, it is difficult to control the flow behavior of the resin using an isothermal viscosity model. A sandwich structure with complicated lay-ups and significantly different material properties is shown in Figure 16.18. The overall filling behavior of a resin (Figure 16.19) can only be simulated through non-isothermal real 3D simulation tools.

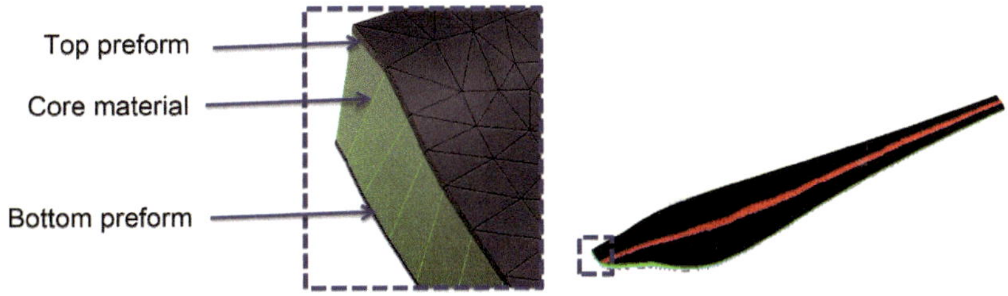

Figure 16.18 Sandwich structure

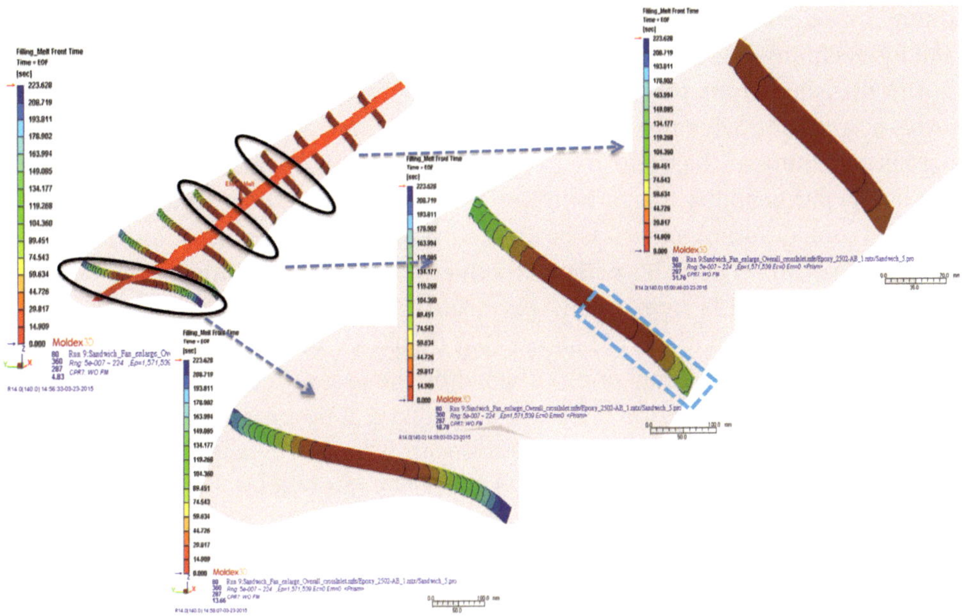

Figure 16.19 Slicing results of flow front time

16.2.2 3D Analysis

In contrast to injection molding simulation, which considers a fluid flowing in an air medium using Navier-Stokes equations as momentum governing equations, in RTM process simulation Darcy's law (Equation 16.1) is mostly used as the constitutive equation to model a fluid flowing through an anisotropic porous medium with the permeability tensor **K**, pressure gradient ∇P, fluid viscosity η, and Darcy velocity **u**.

$$\mathbf{u} = -\frac{1}{\eta}\mathbf{K} \cdot \nabla P \tag{16.1}$$

$$\mathbf{u} = \frac{Q}{\mathbf{A}} \tag{16.2}$$

$$\mathbf{v} = \frac{\mathbf{u}}{\phi} \tag{16.3}$$

Further in the above equations, Q is the volumetric flow rate, $\mathbf{A}$ is the cross-sectional area, $\mathbf{v}$ is the seepage velocity, ϕ is the porosity, and P is the pressure.

The definition of permeability is derived from hydrogeology, and it describes how difficult the interstitial flow of water is in soil. The French hydraulic engineer Darcy proved with an experiment in 1856 that when water exhibits laminar flow and by assuming it is an incompressible fluid, the Darcy velocity $\mathbf{u}$ is proportional to the pressure gradient induced by water. In Figure 16.20, the different water levels h_2 and h_1 provide the pressure drop for the fluid, and the permeability of the porous medium can be measured by recording the flow rate Q and viscosity of water. The permeability tensor $\mathbf{K}$ in Darcy's Law describes how difficult the flow of fluid between different materials is along each principal direction in an anisotropic porous material. When the principal value of the permeability tensor becomes larger, the material is more easily permeated. The Darcy velocity $\mathbf{u}$ can also be called Darcy flux Equation 16.2 shows that the Darcy velocity $\mathbf{u}$ in a 1D system is equal to the volumetric flow rate Q divided by the cross-sectional area $\mathbf{A}$. Because the porous medium contains the skeletal material, fluid can only pass the porous medium through the pores. The Darcy velocity $\mathbf{u}$ is not the actual velocity of the fluid $\mathbf{v}$ (seepage velocity) Equation 16.3 describes the relation between the Darcy velocity $\mathbf{u}$ and the actual velocity of the fluid $\mathbf{v}$.

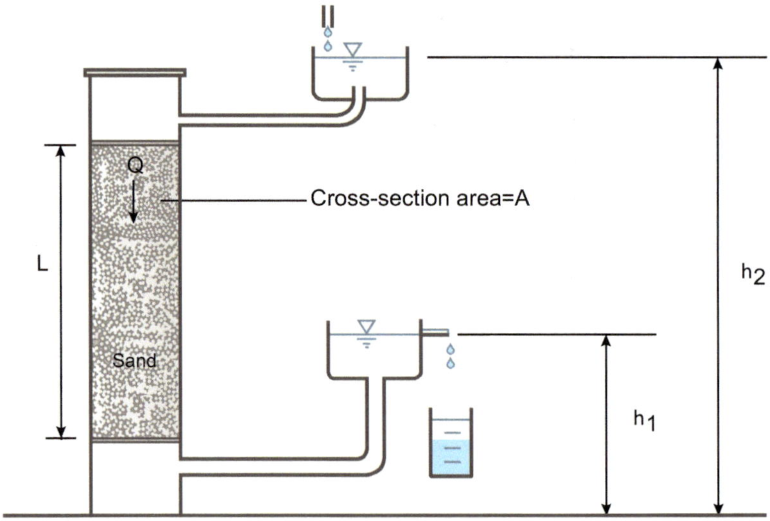

Figure 16.20 Schematic diagram of Darcy's law experiment [5]

The permeability tensor **K** can be written as:

$$\mathbf{K}=\begin{bmatrix} K_{xx} & K_{xy} & K_{xz} \\ K_{yx} & K_{yy} & K_{yz} \\ K_{zx} & K_{zy} & K_{zz} \end{bmatrix}=\begin{bmatrix} l_{11} & l_{12} & l_{13} \\ l_{21} & l_{22} & l_{23} \\ l_{31} & l_{32} & l_{33} \end{bmatrix}\begin{bmatrix} K_{11} & 0 & 0 \\ 0 & K_{22} & 0 \\ 0 & 0 & K_{33} \end{bmatrix}\begin{bmatrix} l_{11} & l_{21} & l_{31} \\ l_{12} & l_{22} & l_{32} \\ l_{13} & l_{23} & l_{33} \end{bmatrix} \tag{16.4}$$

where K_{ij} (i and j = x, y, or z) is the permeability tensor main value that represents how difficult the flow is along the main direction. A small value denotes a higher flow resistance along the main direction; l_{ij} denotes the direction cosine of the material coordinates and the individual components can be summarized as shown in Equation 16.5:

$$\begin{aligned} \mathbf{L}_1 &= (l_{11}, l_{12}, l_{13}) \\ \mathbf{L}_2 &= (l_{21}, l_{22}, l_{23}) \\ \mathbf{L}_3 &= (l_{31}, l_{32}, l_{33}) \end{aligned} \tag{16.5}$$

where $\mathbf{L}_1$, $\mathbf{L}_2$, and $\mathbf{L}_3$ denote the main directions of the permeability tensor. Usually $\mathbf{L}_1$ and $\mathbf{L}_2$ are set to the fiber directions or weaving directions of the preform, as shown in Figure 16.21, whereas the thickness direction of the preform is set to $\mathbf{L}_3$. A more detailed description of the permeability tensor **K** can be found in Chen Kailing's master thesis [5].

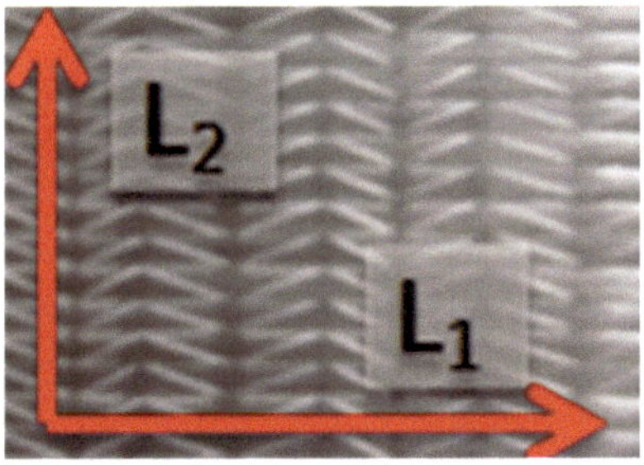

Figure 16.21 Preform and fiber main directions of permeability tensor

16.2.3 Measurement of Permeability

1. One-Dimensional Permeability

In a one-dimensional linear flow field, Equation 16.1 for Darcy's law can be rewritten as:

$$-\frac{K}{\mu}\left(\frac{dP}{dL}\right)=\phi\frac{dL}{dt} \tag{16.6}$$

For a filling system under constant pressure, the unsaturated permeability can be expressed by integrating Equation 16.6 to:

$$K = \frac{\phi \mu L^2}{2(-\Delta P)t} \tag{16.7}$$

where K is the permeability, ϕ is the porosity, μ is the viscosity, L is the filling distance, ΔP is the pressure difference, and t is the filling time.

2. Visualization Experiment

The parameters of the permeability tensor required for simulation can be measured and are obtained via a visualization experiment system. Such a system integrates a digital camera, an image processing software called LabVIEW, and a computer, and can measure the permeability of woven in the horizontal direction. Figure 16.22 shows all the instruments and equipment for this experiment. The overall experimental flow chart is shown in Figure 16.23. The camera sends the captured image to the computer, which can be converted into the data of flow front position via LabVIEW, and together with the information of pressure difference and time the value of permeability can be computed.

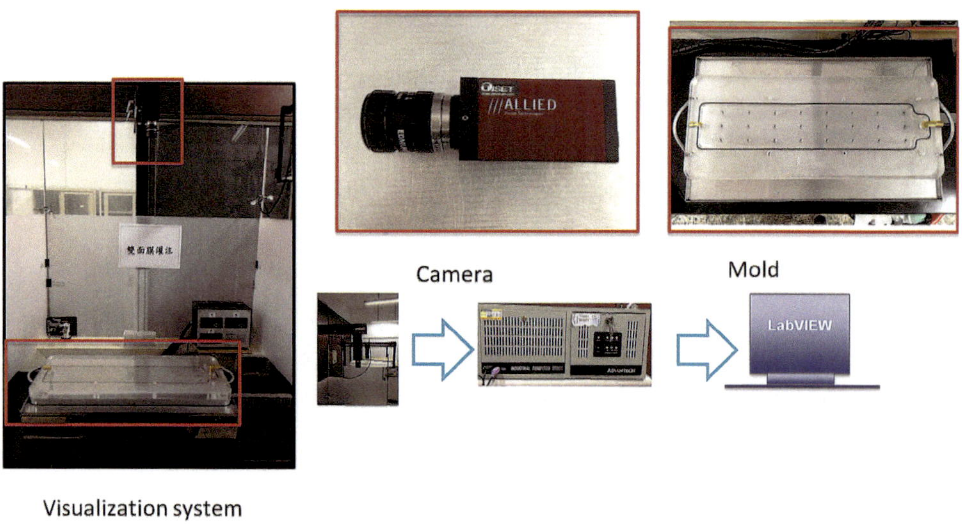

Figure 16.22 Equipment of visualization experiment [1]

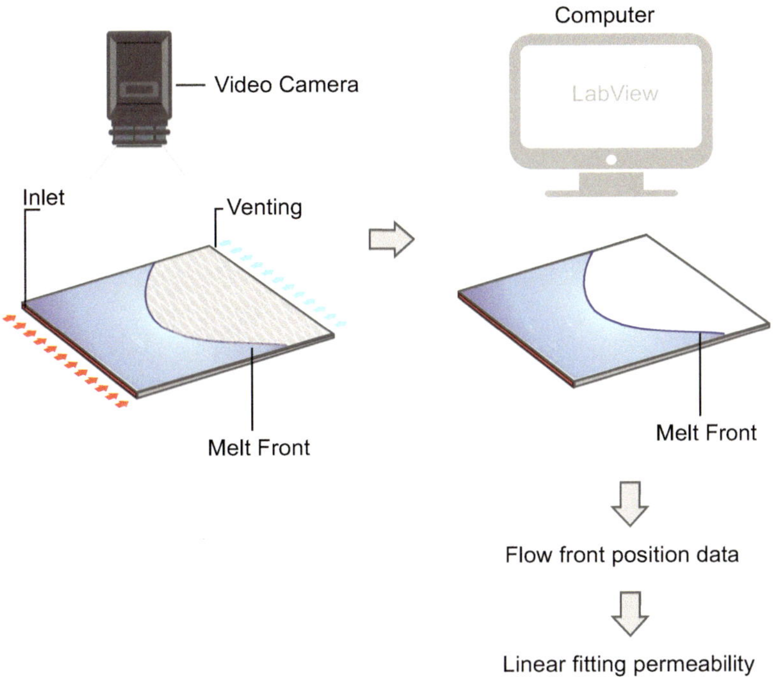

Figure 16.23 Flow chart of the experiment

3. EASYPERM Experiment

In addition, there are also professional instruments on the market that can measure the permeability, such as EASYPERM, which is used to measure the saturated permeability of the fluid flowing through the fiber mat at room temperature. This instrument is a pressure control system, which measures pressure changes at different positions by means of pressure sensors and records the weight of the electronic balance to calculate the flow rate. The permeability can be measured by collecting the above values. At the same time, experimental conditions such as the thickness of the fiber mat, the rotation angle of the fiber mat, and the importing pressure are controlled to obtain the experimental parameters under different process conditions. Moreover, the instrument can separately measure the permeability in the plane direction and in the thickness direction by changing the mold system and switching the direction of the fluid pipeline, as shown in Figure 16.24.

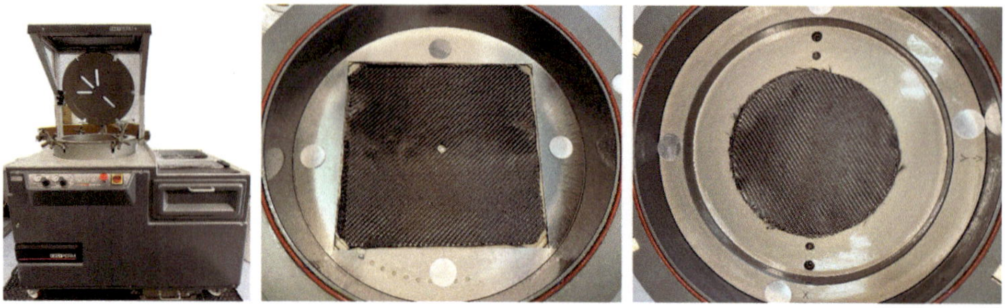
(a) (b) (c)

Figure 16.24 (a) EASYPERM Bench, (b) top view of the test area XY, and (c) top view of the test area Z

16.2.4 Porosity

The porosity ϕ denotes the ratio of the volume of the pores to the total volume of the woven, and the porosity of woven in a cavity is calculated by deducting the volume occupied by the woven, which affects the resin usage in the RTM molding process as well as the volume fraction of fiber in the part.

1. Measurement of Porosity

To obtain the porosity, first the thickness h of the n-layer woven is measured under reduced pressure as shown in Figure 16.25(a), followed by use of a density balance as shown in Figure 16.25(b) to measure the *body density* ρ_f of the woven:

$$\rho_f = \frac{W_a \times \rho_w}{W_a - W_w} \tag{16.8}$$

where ρ_w is the density of water, and W_a and W_w are the weights of the fiber in air and in water, respectively.

Then the porosity of the woven ϕ can be derived from Equation 16.9:

$$\phi = 1 - V_f = 1 - \frac{n\rho_s}{h\rho_f} \tag{16.9}$$

where V_f is the volume fraction of the fiber, n is the number of layers of the woven, h is the thickness of woven in vacuum, and ρ_s is the facial density of the woven.

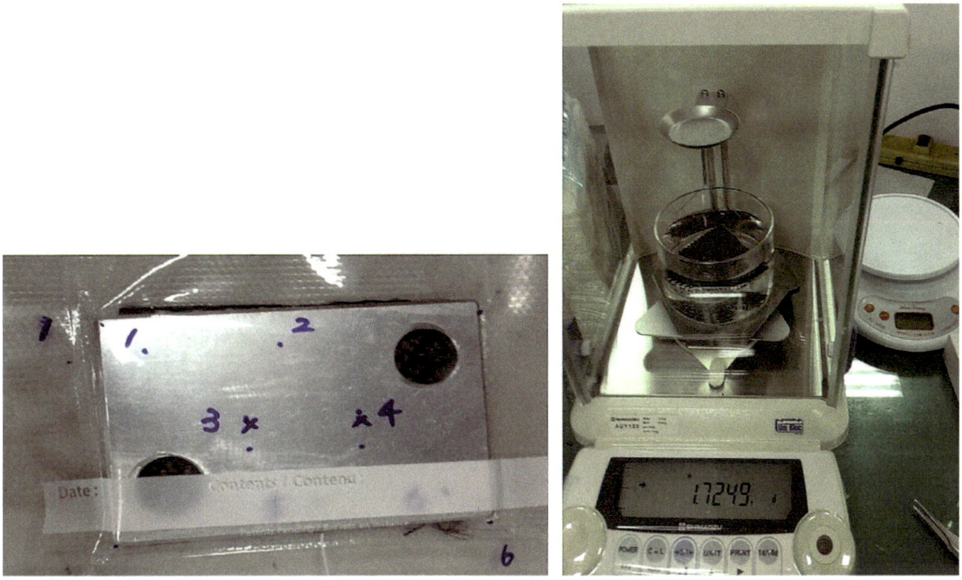

(a) (b)

Figure 16.25 (a) Fixture for thickness measurement and (b) density balance [1]

16.2.5 Measurement of Chemorheological Properties

RTM is a process utilizing polymerization reactions. The chemorheological properties of the composite material during the filling process can be described using (1) reaction kinetics and (2) the model of rheological viscosity.

1. Reaction Kinetics Equations

Chemical reaction kinetics can describe how the polymerization reaction rate varies with temperature and the amount of reaction. The reactivity of the resin can be properly described via a combined model:

$$\frac{d\alpha}{dt} = \left(k_1 + k_2\alpha^m\right)(1-\alpha)^n \tag{16.10}$$

$$k_1 = A_1 \exp\left(-\frac{E_1}{RT}\right)$$

$$k_2 = A_2 \exp\left(-\frac{E_2}{RT}\right)$$

where α is the conversion, k_1 and k_2 are the reaction rate constants, m and n are orders of reaction, A_1 and A_2 are correlation coefficients, and E_1 and E_2 are activation energies.

2. Rheological Viscosity Model

Because the curing reaction occurs during the flowing process of the resin, the flow behavior of a non-Newtonian fluid requires a kinetic model combined with the Castro-Macosko model [6] to predict the viscosity change of the resin under different conversions and shear rates.

$$\eta = \frac{\eta_0 \left(\frac{\alpha_g}{\alpha_g - \alpha} \right)^{C_1 + C_2 \alpha}}{1 + \left(\frac{\eta_0 \dot{\gamma}}{\tau^*} \right)^{1-n}} \tag{16.11}$$

$$\eta_0 = A \exp\left(\frac{T_b}{T} \right)$$

where η_0 is the zero-shear viscosity, α is the conversion, α_g is the gel point, n is the flow property index, and C_1 and C_2 are both constants.

16.2.6 Simulation Parameters

To approximate simulation results to actual manufacturing conditions, the material properties of the resin and the woven should be measured in advance, so as to obtain the parameters which are also required for the manufacturing process. The simulation parameters include:

1. Resin material parameters: kinetic parameters and rheological parameters
2. Material properties of fabric: in-plane and out-of-plane permeability, and porosity
3. Lay-up direction
4. Resin temperature
5. Mold temperature and room temperature
6. Arrangement of inlets and vacuum edges
7. Inlet pressure or flow rate
8. Opening and closing timing of inlet

16.3 Practical Applications

16.3.1 CAE Verification on Edge Effects

A case simulating the edge effect often seen in the molding process of double-sided molds should be examined. The flow of the resin is faster close to the edge of the mold during the filling process, and the main reason for such a phenomenon is that the flow resistance in this region is smaller because the fiber weaving at the edge of the mold is sparse compared to that in the middle area. For simulation, the edge regions should be set with a larger permeability to emulate an actual filling behavior when it comes to the setting of woven permeability parameters. Figure 16.26 shows the geometry of a plate model with 3 mm thickness used for the simulation of the edge effect of resin filling. A constant pressure is applied for molding, which is maintained at 1 atm. The viscosity of the resin is set to a constant 0.7 Pa · s, the permeability of woven is $K_{11} = K_{22} = K_{33} = 8.8 \times 10^{-10}$ m^2, and the porosity is 0.5.

Figure 16.27 shows the results of flow front changing with time considering the edge effect. Due to the constant pressure molding, the flow rate slows down with the increase of filling time.

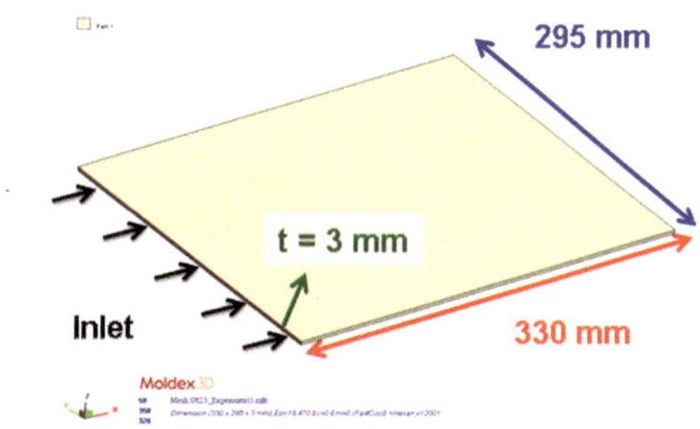

Figure 16.26 Model of plate

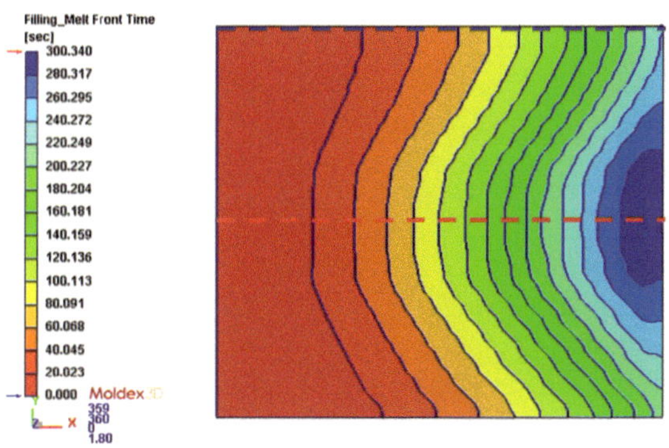

Figure 16.27 Flow front time of simulation with edge effect

Figure 16.28 and Figure 16.29 show the consistent simulation and experimental results of the flow front of the resin.

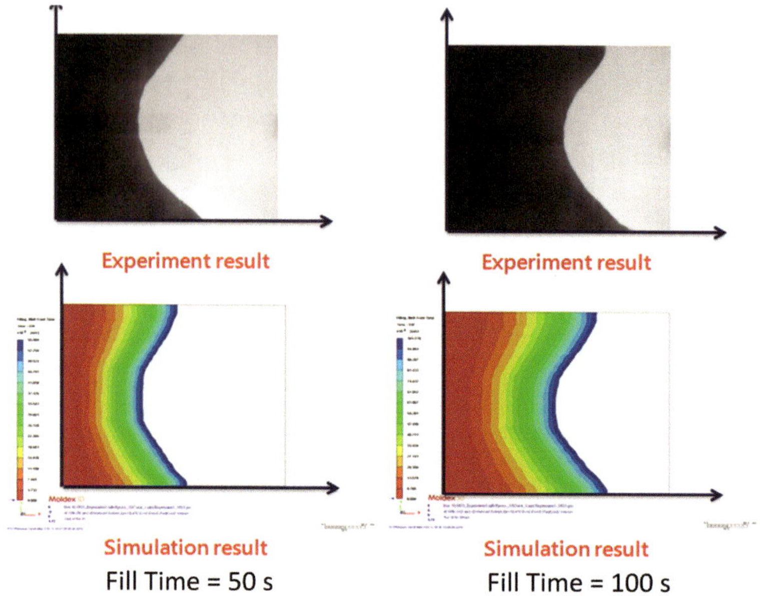

Figure 16.28 Comparison of experimental and CAE results at 50 and 100 s

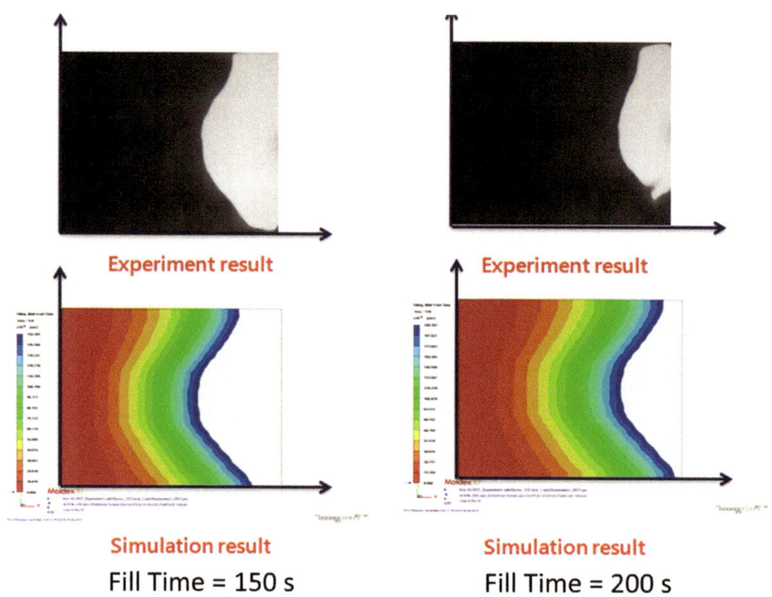

Figure 16.29 Comparison of experimental and CAE results at 150 and 200 s

The locations of the flow front at the edge and in the middle are recorded in Figure 16.30 for several time intervals (i.e., the middle and edge areas marked in red and blue color in Figure 16.27). The results clearly show that the edge effect leads to a faster flow at the edges and is the main factor for imbalanced flow, which should not be neglected.

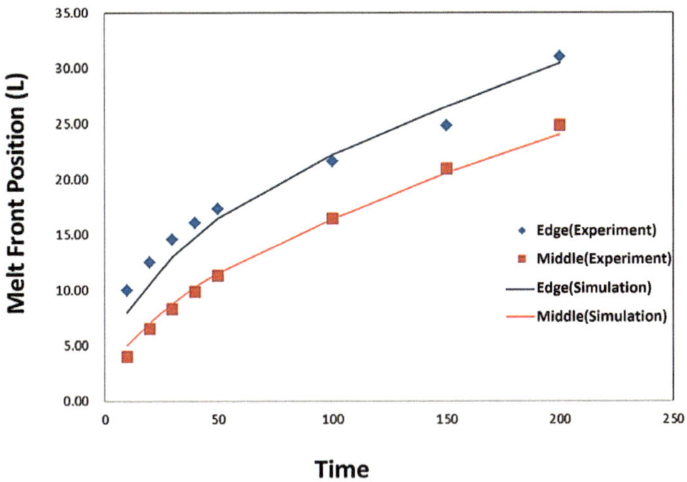

Figure 16.30 Comparison of experimental and CAE results at edge and middle zones

16.3.2 CAE Verification on Thickness-Direction Flow

The geometrical and dimensional model of the part used for simulation in this chapter is shown in Figure 16.31 and built based upon the case described in Chen Kailing's master thesis [5].

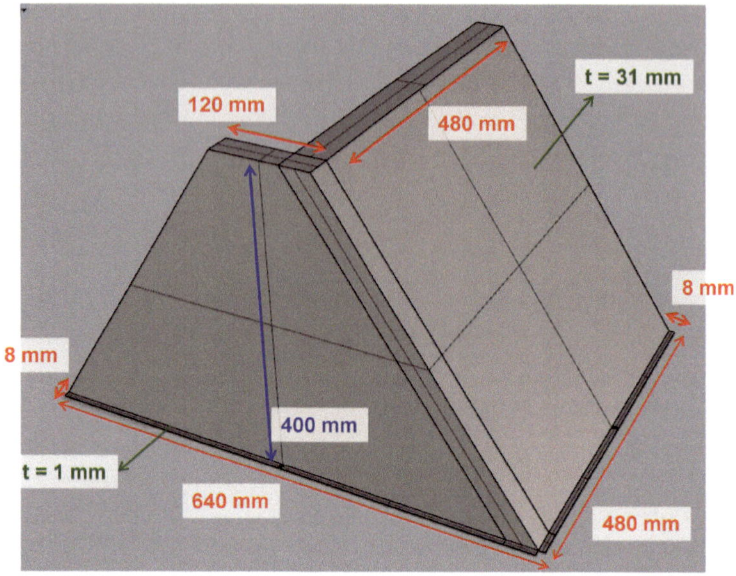

Figure 16.31 Model for CAE verification on thickness-direction flow

The parameters of the fiber fabric material are set at $K_{11} = K_{22} = K_{33} = 9.1829 \times 10^{-10}$ m^2 [4], and the parameters of the resin material are as shown in Table 16.2; these are measured from the resin SWANCOR 2502-A/B, which is commonly used in the manufacture of wind turbine blades.

Table 16.2 Material Properties of SWANCOR 2502-A/B Used in Equation 16.11

α_g	0.788
C_1	3.754
C_2	1.746
τ^* (Pa)	182.2
A (Pa · s)	3.035×10^{-7}
T_b (K)	4371.19
n	0.2

The 2.5D simulation shows results consistent with those of the experiment in the thesis. If the part is thin, a traditional 2.5D tool can be used to simulate and analyze the flow of the resin in the fiber fabric; however, the influence of thickness on the overall flowing conditions requires a 3D tool for analysis in the case of a complicated fiber fabric lay-up or geometry.

The influence of temperature, shear rate and processing time on the reaction of the thermosetting resin is simulated. Figure 16.32 (a) and Figure 16.32 (b) show the results for viscosity and conversion, respectively, of the resin material, which predict a variation of conversion and viscosity with increasing temperature under different heating rates.

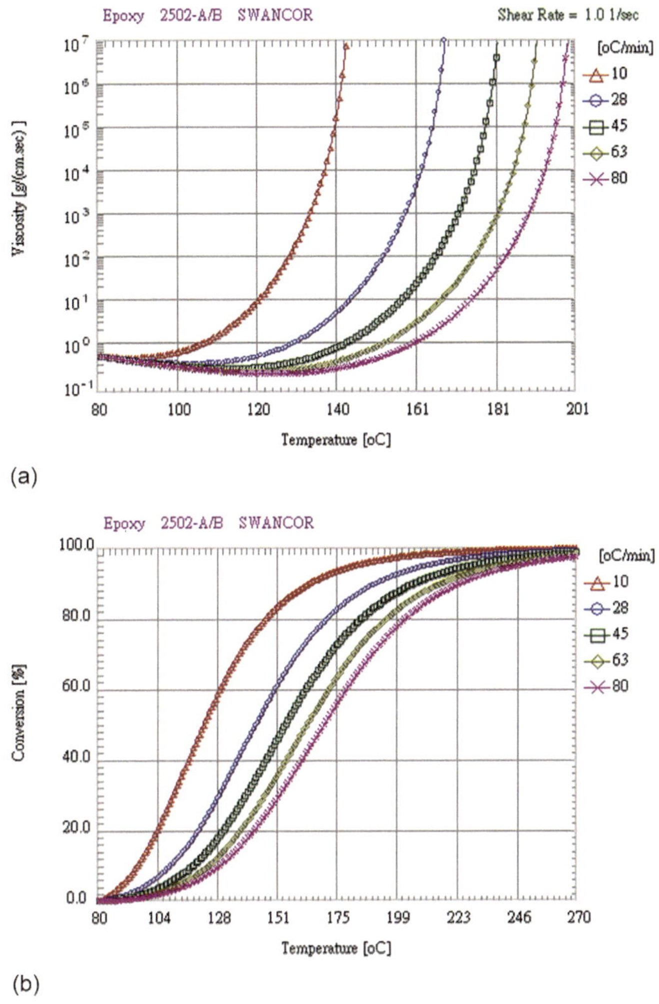

Figure 16.32 (a) Viscosity and (b) conversion curves of SWANCOR 2502-A/B under different temperature increasing rates

If we assume the woven is an isotropic material with the same flow resistance along the three main axes, the flow front simulation results can be obtained as shown in Figure 16.33(a) by setting the proportion of the permeability tensor main values to $K_{11} : K_{22} : K_{33}$ = 1:1:1 and the flow rate at a constant value of 6.35×10^{-6} m^3/s. As the resistance of the fiber fabric between the thickness direction and the horizontal direction varies significantly, it is difficult to display the difference of flow in both directions with such parameters. Figure 16.33(b) shows the flow behavior of the resin in the fiber fabric assuming that the resistance difference in horizontal direction and thickness direction reaches 100 times, that is, the proportion of the permeability tensor main values is $K_{11} : K_{22} : K_{33}$ = 100:100:1, which displays the difference of surface flow behavior of isotropic and anisotropic media.

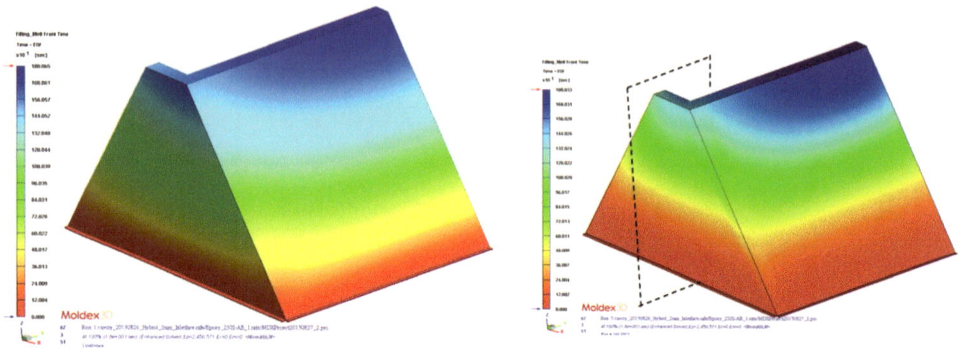

(a) (b)

Figure 16.33 Flow front time of resin with (a) isotropic permeability $K_{11}: K_{22} : K_{33}$ = 1:1:1 and (b) anisotropic permeability $K_{11} : K_{22} : K_{33}$ = 100:100:1

Figure 16.34(a) and Figure 16.34(b) are the flow front time isocontour results. Figure 16.34(a) demonstrates the resin infused into the isotropic media. Figure 16.34(b) demonstrates the status of resin flow along the geometric direction when the permeability of the thickness direction is far smaller than that of the plane direction. The results clearly show that the flow rate in the plane direction is faster than in the thickness direction. Due to the anisotropy effect, while the resin flows to the end of the mold on the top surface, the bottom surface of the mold is not completely impregnated, and if such a timing was taken as the filling complete time, the fiber fabric would not be impregnated completely.

Figure 16.35(a) and Figure 16.35(b) show the distribution of conversion and viscosity, respectively, when the filling is completed. They indicate a higher conversion and viscosity in the middle region, and if this is the cause of defects in the part during molding, the operating conditions of the process should be adjusted accordingly.

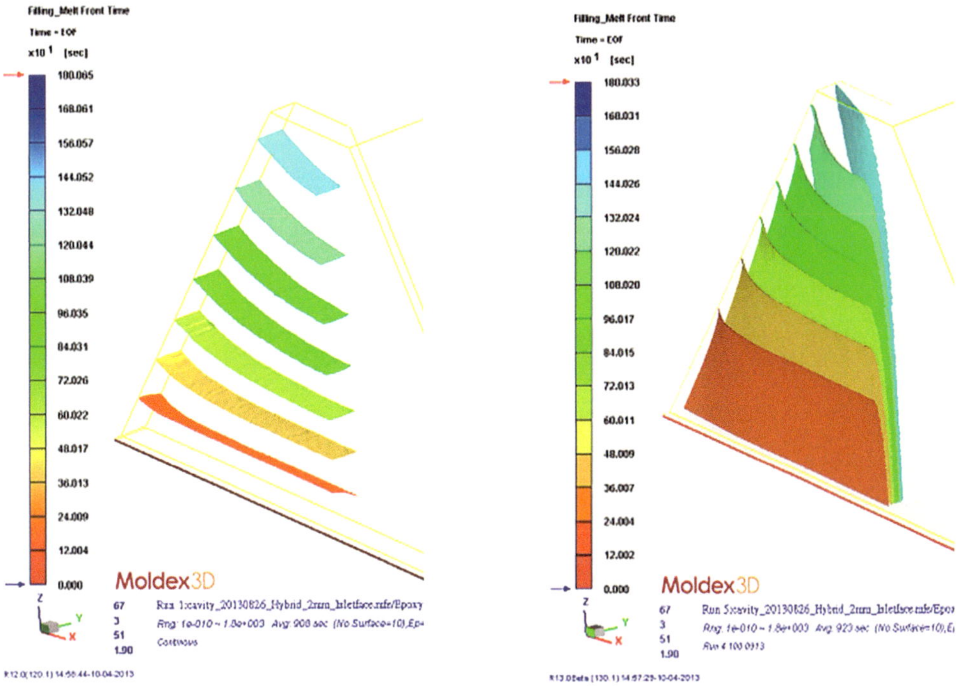

(a) (b)

Figure 16.34 Slicing results of flow front time at 200 s: (a) isotropic permeability $K_{11} : K_{22} : K_{33}$ = 1:1:1 and (b) anisotropic permeability $K_{11} : K_{22} : K_{33}$ = 100:100:1

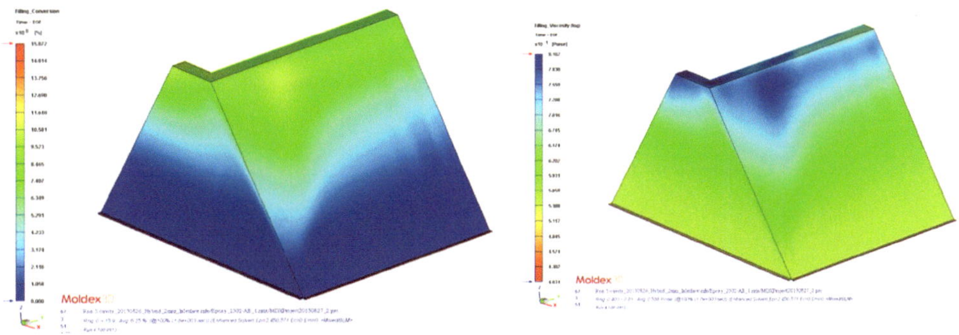

(a) (b)

Figure 16.35 (a) Conversion and (b) viscosity at end of filling

16.3.3 CAE Verification on a Wind Turbine Blade

The geometry of the outer skin of a 1-kW wind turbine blade (Figure 16.36) is shown in Figure 16.37, and Figure 16.38 shows a model built according to the geometry and molding design of this experiment.

Figure 16.36 1-kW wind turbine blade [4]

The inside of the mold would be put under vacuum during molding in the VARTM process, which creates a pressure difference to push the resin to flow. Because such a constant pressure feeding leads to a decreasing resin flow rate as the flowing distance increases, normally for manufacturing large parts by VARTM, multiple inlets would be designed for the overall process of the part, which can facilitate the filling process by reducing the distance between inlets and flow fronts. Also, the resin flow during the part filling process can be managed via opening and closing the inlets according to the flow front of the resin to avoid part defects. Sensor nodes should be configured for the simulation of such a control method, and by determining whether the flow front arrives at the sensor node or not the timing of opening and closing of each inlet can be controlled.

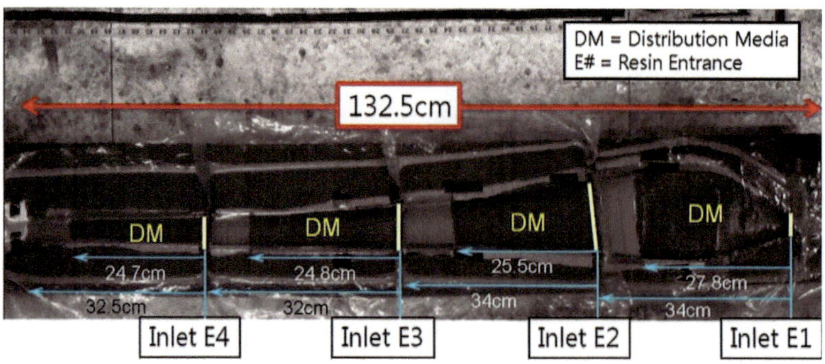

Figure 16.37 Geometry of the 1-kW wind turbine blade

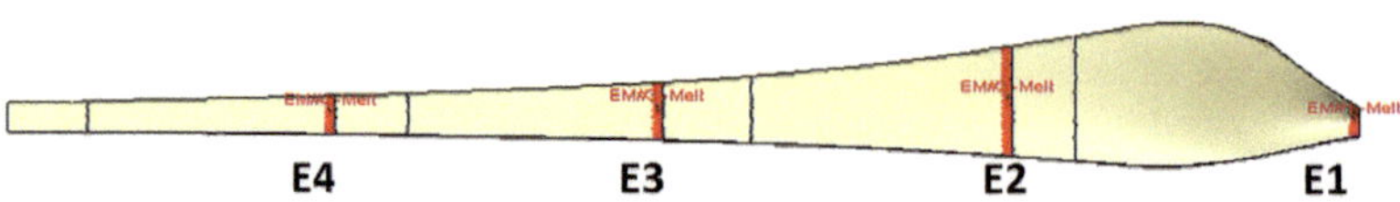

Figure 16.38 Model of the 1-kW wind turbine blade

The fiber preform lay-up uses an orthogonally woven carbon fiber fabric with a facial density of 200 g/m^2. Five layers are laid on the mold surface. The surface of the preform is divided into four regions, which are covered with distribution medium, peel plies, and inlets. The dark region is the distribution medium. The used resin is SWANCOR 2502-A/B, that is, the wind turbine blade is molded in epoxy resin. The molding experiment is conducted at normal pressure and temperature. The constant molding pressure is 1 atm. The opening time of the molding inlet E1 is at 0 seconds, and when the flow front reaches E2, E2 is then opened while E1 is closed. The opening times at E3 and E4 can be deduced in the same manner until the mold cavity is filled up by E4. The permeabilities of the materials of the preform used in this experiment, the carbon fabric and distribution medium, are listed in Table 16.3.

Table 16.3 Material Properties of the Carbon Fabric and the Distribution Medium

	Carbon Fabric	Distribution Medium
Thickness [mm]	0.9653	0.9263
K_{11} [m^2]	8.387×10^{-12}	1.071×10^{-9}
K_{22} [m^2]	8.387×10^{-12}	1.071×10^{-9}
K_{33} [m^2]	8.400×10^{-13}	1.071×10^{-9}
ϕ	0.399	0.491

Experimental and simulation results of flow front at two time intervals are compared in Figure 16.39.

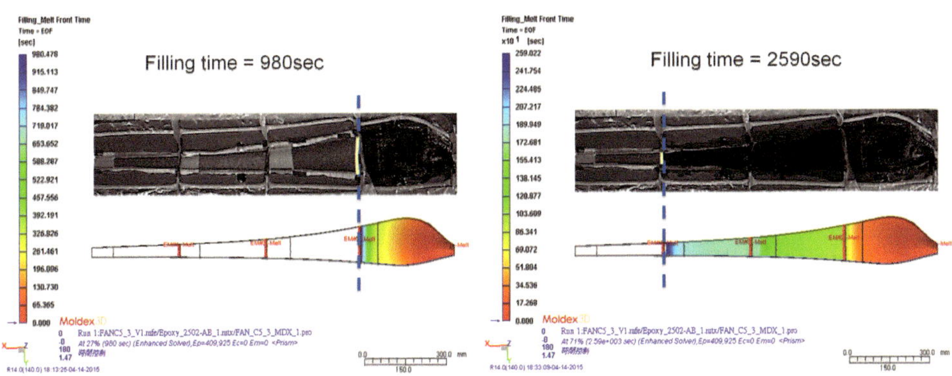

Figure 16.39 Experimental and simulation results

The experimental data captured by the visualization system along with the simulation results are plotted in Figure 16.40, showing the relationship between distance and time, which displays consistency between experiment and simulation.

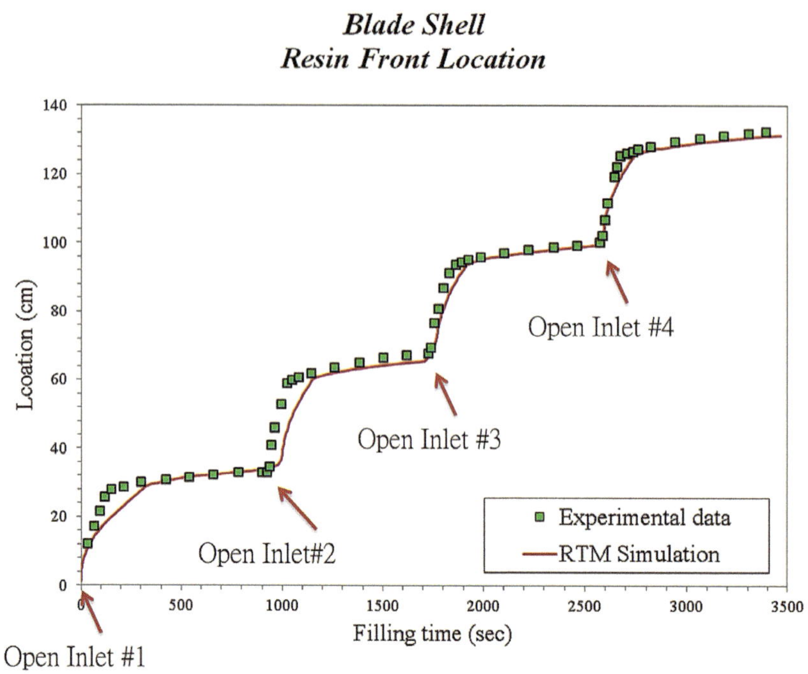

Figure 16.40 Flow front position at different time intervals during the RTM process

16.3.4 CAE Verification on Mat Effects

Usually many different lay-ups are used in the RTM process, and as the permeability of the fabric dominates the flow behavior, with many fabrics of multiple properties and different permeabilities in the mold cavity (e.g., the properties of core material and fabric in a sandwich structure vary significantly), the flow behavior can only be accurately predicted through simulation.

The following case is a constant-pressure molding experiment at 1 atm using a visualization permeability measurement system. The permeability of fabrics LT800, L900, and DBLT1900 of three different weaving methods were measured, and found to be 2.1×10^{-10} m^2, 5.93×10^{-11} m^2, and 4.19×10^{-11} m^2, respectively. Fabric LT800 has the largest permeability among the three fabrics, meaning a better resin flowability in comparison to the other two. By inputting the permeabilities into Moldex3D as parameters for analysis, the simulation results shown in Figure 16.41 are obtained; these are consistent with experimental data.

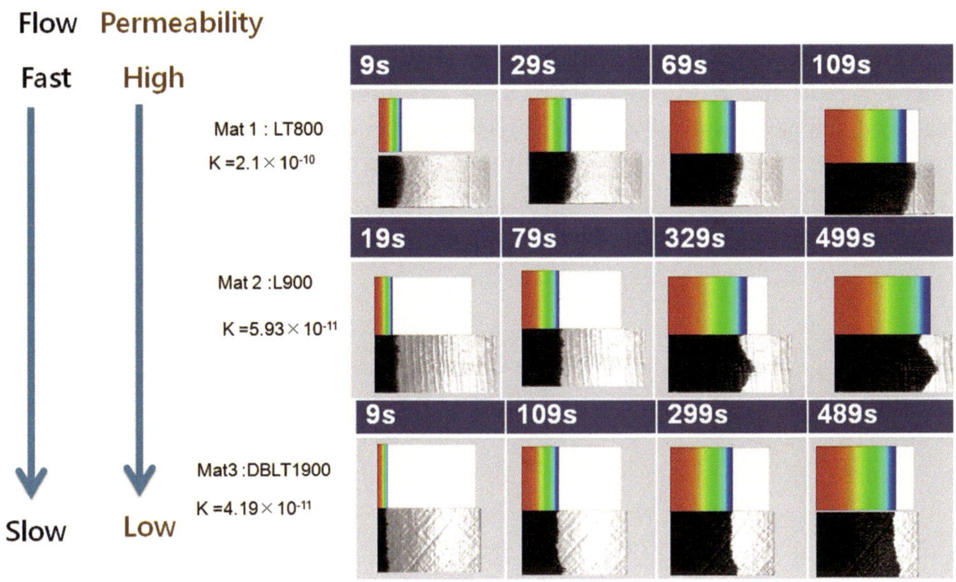

Figure 16.41 Simulation results and experimental photos

If the design of the fabric lay-ups is to be changed for a different strength of the part, the overall flow behavior would be significantly affected if the fabric substitution has a totally different permeability, as shown in Figure 16.42.

A material measurement and an analysis on the filling process can ensure that the process is not significantly influenced by the change of materials.

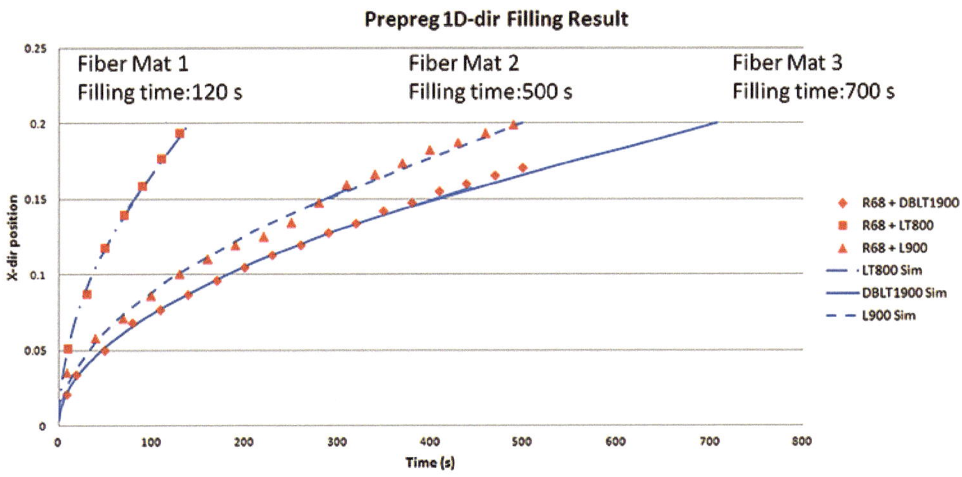

Figure 16.42 Flow front position at different time intervals: fabric effect

In the product design stage, the layer numbers, layup arrangement and the directions of the fabric laminate will be determined according to the product performance requirements. In the following case, the effect of fiber direction on the flow result is compared. This case of a Wet RTM process involves pressing the resin into the fabric through compression. Figure 16.43 is a schematic diagram of the fiber directions of the fabric in two cases, with the permeability in each direction of the material in the fabric being set as $K_{11} : K_{22} : K_{33} = 100{:}1{:}1$.

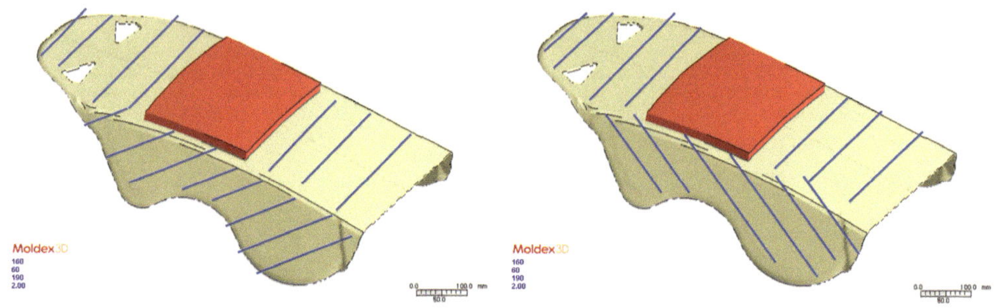

(a) (b)

Figure 16.43 Fiber directions for cases I and II

Figure 16.44 and Figure 16.45 are the results of melt front time in different fiber directions. From the results of the four filling ratios of 60%, 80%, 90%, and 100%, it can be seen that the difference in flow behavior is very obvious and that the resin flows along the fiber direction. The direction of fabric covering will also have a significant impact on the flow behavior of the resin; in particular, the flow of resin along the fiber direction is better than in the lateral and thickness directions for the uniaxial fiber mat. It is also obvious from the material properties that the permeability along the fiber direction is much larger than in other directions, so the fiber orientation setting in the simulation needs to be in accordance with the product design in order that the same result as the experiments may be achieved.

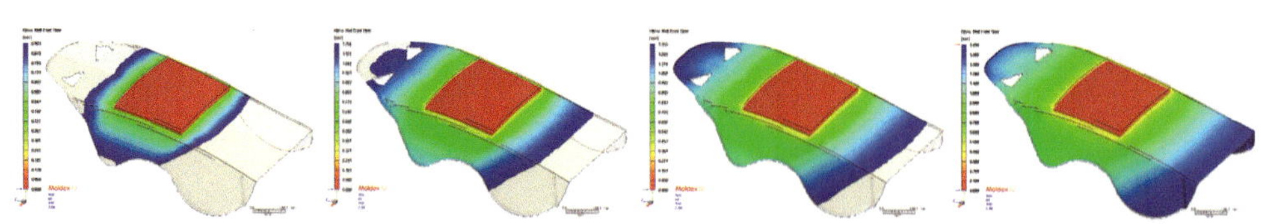

(a) (b) (c) (d)

Figure 16.44 Melt front time results at (a) 60%, (b) 80%, (c) 90%, and (d) 100% for case I

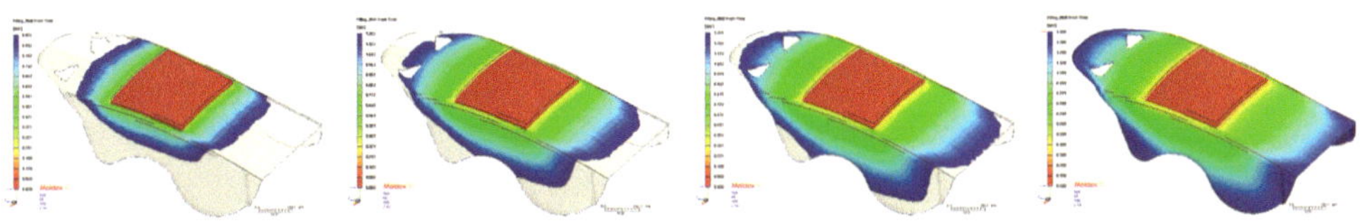

(a) (b) (c) (d)

Figure 16.45 Melt front time results at (a) 60%, (b) 80%, (c) 90%, and (d) 100% for case II

16.3.5 CAE Verification on Flybridge

In a yacht manufacturing process, large FRP yacht parts are often made by the SCRIMP method of VARTM. The following is a case of a flybridge manufacturing of a yacht in cooperation with Atech Composites Co., Ltd, who has expertise in yacht production.

(a) site 1 (b) site 2

Figure 16.46 Geometry of the flybridge components of site 1 and site 2

The molding of the flybridge is divided into two regions (site 1 and site 2) as shown in Figure 16.46(a) and Figure 16.46(b). Site 2 is used for the comparison of simulation and design change. Figure 16.47 shows the model geometry built according to site 2. The preform for molding uses glass fiber mats, and the resin material is an unsaturated polyester resin (UPR). The simulation should compare the influence on filling time between different numbers of inlets.

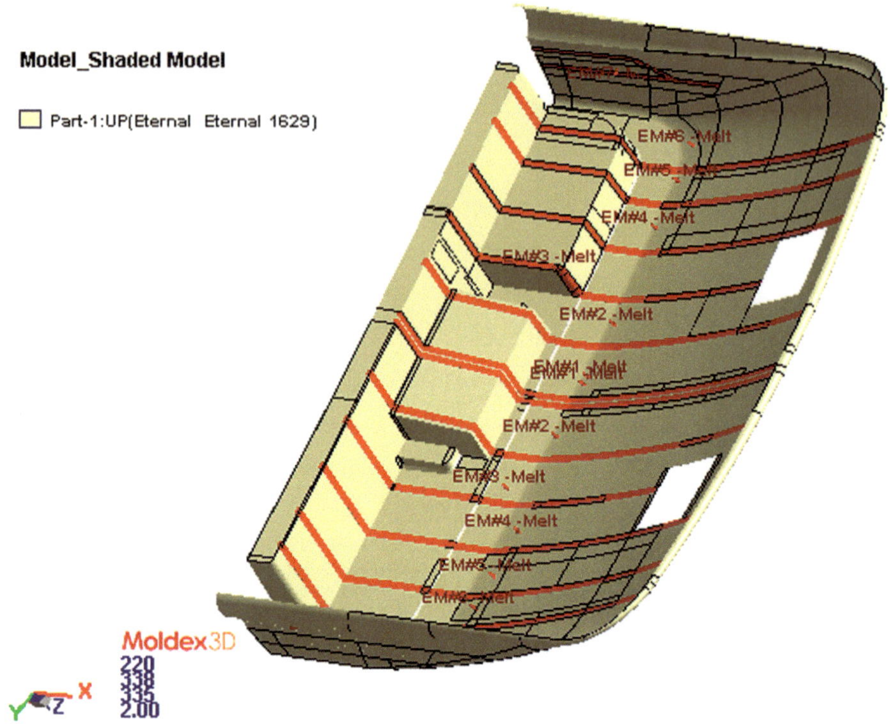

Figure 16.47 Model of flybridge component of site 2

By referring to the flybridge geometry, the mold cavity and the locations of inlets are created. The overall number of the solid mesh is around 3 million. The material feeding performed on site is by setting 14 inlets, which are sequentially opened from the center to the sides in a left-right symmetrical manner. As two inlets are controlled at a time, there are seven switching inlet groups (#1~#7) defined to control the opening timing of the inlet groups. The judgment for whether to open the inlets made on the molding site is to open the next inlets before the flow front arrives. To match the way of inlet opening control used on site, the simulation incorporates sensors or nodes near the next inlets, and the inlet is opened upon the arrival of the flow front. Such a setting for the control of inlet opening and closing is able to match the simulation result to the site operation behavior shows snapshots taken at different time intervals. The dark part denotes the resin impregnation region. The flow front simulation result obtained by incorporating site operation conditions shows a consistent flow trend with the actual manufacturing process.

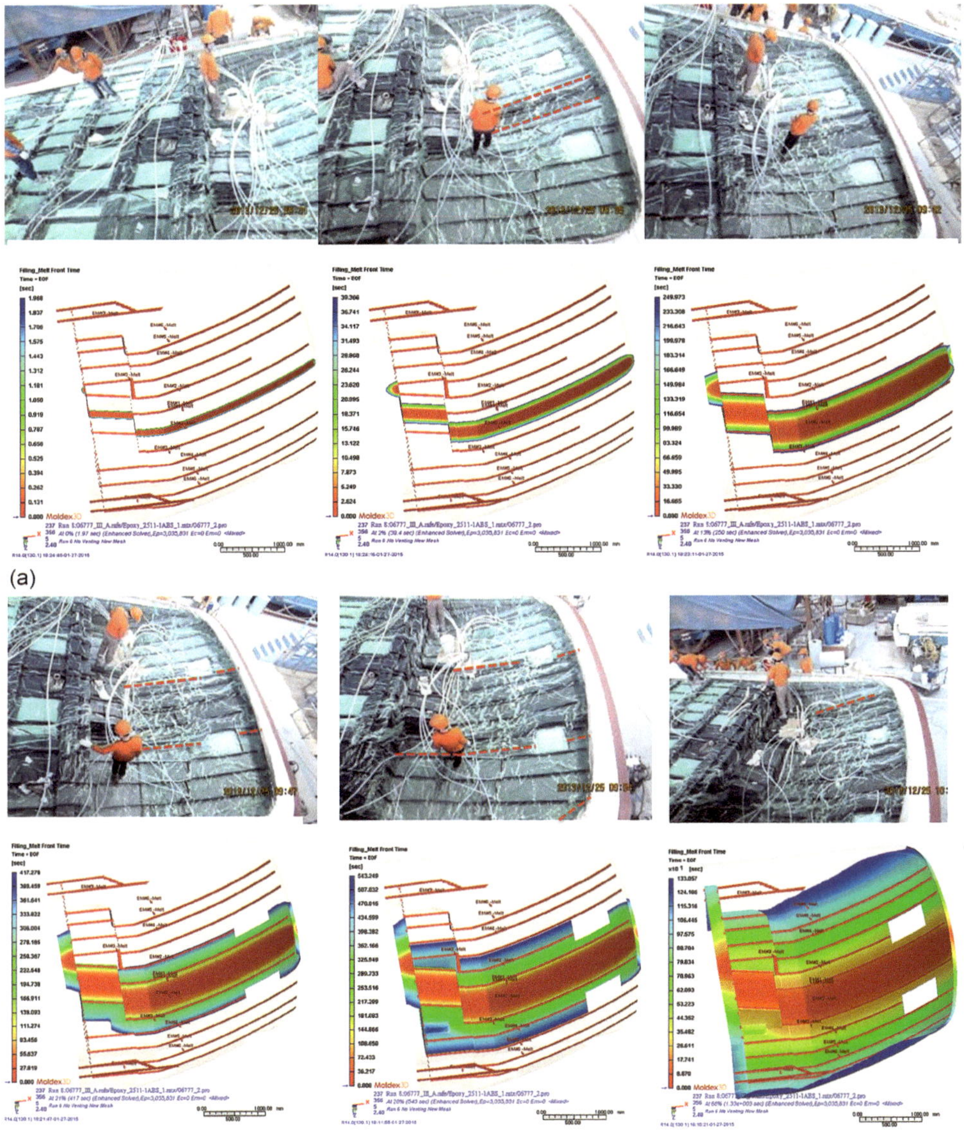

Figure 16.48 Snapshots and simulation results of the VARTM process

The inlet opening method of the site manufacturing design as process 1 should be compared to a changed design with a reduced number of inlets as process 2 to predict the influence on total molding time and resin conversion. The changed design reduces the original sequential opening of inlet groups #1~#7 (total 14 inlets) from the center to the sides to only opening inlet group #1, #3, #5, and #7 (total 8 inlets) for material feeding under the assumption of unchanged resin and preform material conditions. As the number of inlets is reduced, the usage of consumables is also decreased compared to process 1, and thus results in a longer flow distance of the resin. When molding is performed under constant pressure, as the flow distance gets longer, the flow speed of the resin gets slower and the conversion drastically increases, which can be seen from the difference of conversion distribution at the filling end as shown in Figure 16.49. As the resin viscosity increases drastically, the molding time surges from the original 1968 s to 9000 s as shown in Table 16.4. If such an inlet arrangement (reducing the number of inlets) were introduced, the mold cavity filling time would increase by four times, which greatly increases the overall process time. Through a simulation to predict the flow conditions before modifying the arrangement of inlets and other operation parameters, optimum manufacturing conditions can be found, by which the unnecessary waste generation during the manufacturing process can be reduced.

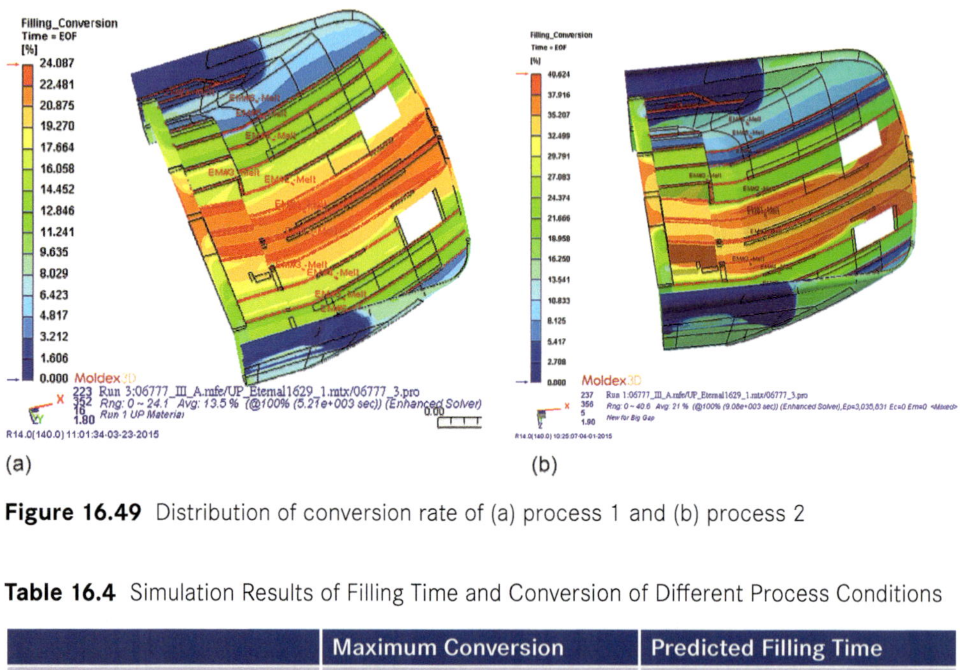

Figure 16.49 Distribution of conversion rate of (a) process 1 and (b) process 2

Table 16.4 Simulation Results of Filling Time and Conversion of Different Process Conditions

	Maximum Conversion	Predicted Filling Time
Process 1	24%	1968 s
Process 2	40%	9000 s

■ 16.4 CAE Case Study

Following the discussions of this chapter and the corresponding CAE verification case, an exercise model, as illustrated in Figure 16.50, is provided in this section for readers to practice CAE application in RTM.

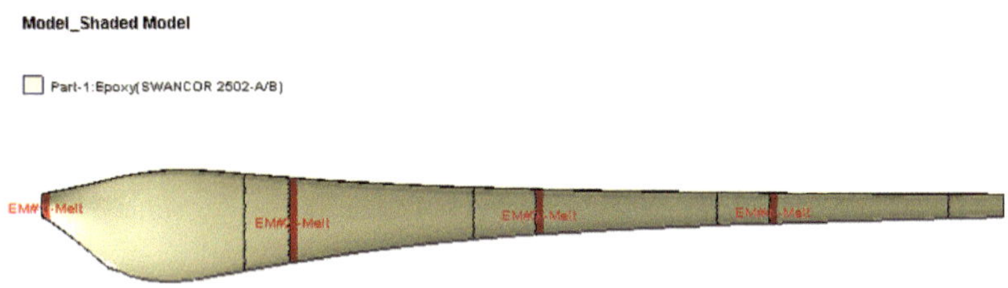

Figure 16.50 Case study sample

Please download the analysis files for CAE Case Study 16.4 from the following website:

https://moldex3d.box.com/s/zr6fvc1vlhbi4ocx11jwd3wmxt4ooif

Then open the .m3j or .mvj file in Moldex3D and answer the questions below according to the analysis results.

Questions:

1. According to the pre-analysis of the model
   a) What is the length of the wind turbine blade? How many inlets are there in this model? What is the sequential inlet control during the filling stage?
   b) How can the change of inlet of the process be determined?
2. According to the results under the default process conditions
   a) What is the resin filling behavior of each inlet?
   b) What is the flow front time distribution result in the end of filling/curing stage?
   c) What is the conversion distribution result in the end of filling/curing stage?

3. What is the result if the inlet pressure is changed to 0.2 MPa?
4. What is the result if the resin temperature is changed to 30 °C?
5. What is the result if the mold temperature is changed to 30 °C?
6. What is the difference if the curing pressure is changed to 0.1 MPa?

References

[1] Laboratory of Process Chemometrics and Systems Engineering, Department of Chemical Engineering, NTHU, Hsinchu, 2015

[2] *Composite Materials Handbook (CHM-17):* Volume 3: Polymer Matrix Composites, Warrendale: SAE International, 2012

[3] He, T. (ed.): *Lightweight Carbon Fiber Composite Technology* (Chinese Edition), Beijing: Science Press, 2015

[4] Atech Composites Co. Ltd., Kaohsiung, Taiwan, 2016

[5] Chen, K.-L.: "Study on the Application of VARTM Technique and Mold Flow Analysis to the Lamination of Sandwich Plates," Master Thesis, Department of Engineering Science and Ocean Engineering, National Taiwan University, Taiwan, 2005

[6] Castro, J. M.; Macosko, C. W.: Studies of mold filling and curing in the reaction injection molding process, *AIChE Journal,* 28, 1982, pp. 250–260

17 Integrated Circuit Packaging

Chih-Chung (Jim) Hsu, Chia-Peng (Victor) Sun,
Chen-An (Jennan) Wang, Yu-En (Joseph) Liang

17.1 Basics

As the size of an integrated circuit (IC) component made with wafers is small, a packaging is required to avoid damage by physical or chemical causes during manipulating processes and shipping. Furthermore, packaging also helps to establish the structure of an IC component, providing functions such as load bearing, protection, as well as electric signal arrangement and power transmission. A common packaging plastic product is shown in Figure 17.1.

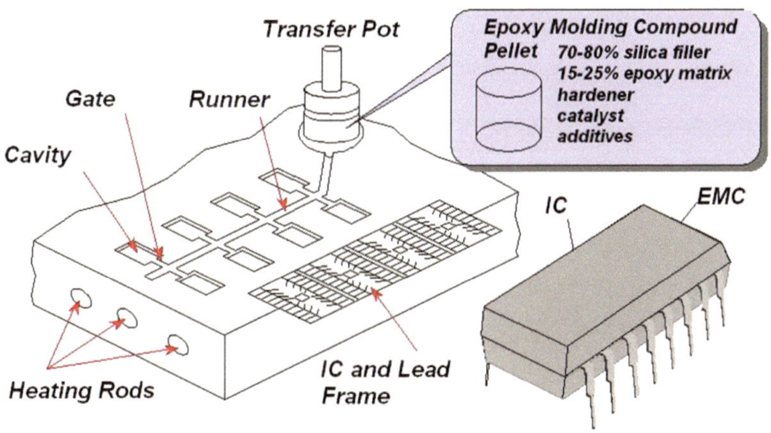

Figure 17.1 Components inside an IC packaging

Generally speaking, there are four major functions for IC packaging:

1. Protection by encapsulation
2. Electric circuit connection and signal transmission
3. Path for heat dissipation
4. Mechanical load bearing

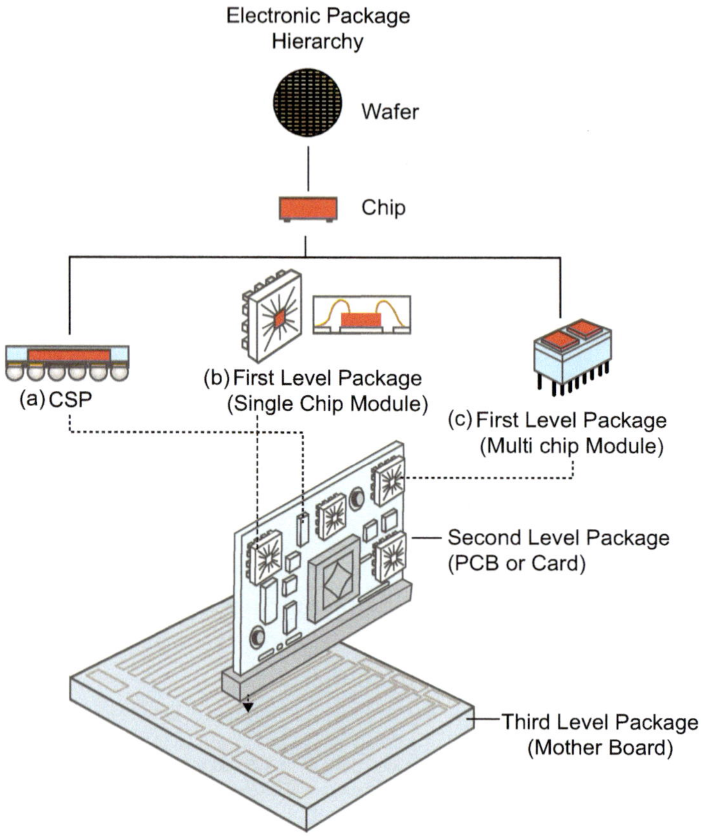

Figure 17.2 Category for IC packaging [1]

Due to the constant changing of IC packaging technology, here IC packaging is distinguished into several levels according to its system structure to systematically understand the difference between various packaging categories, which are depicted in Figure 17.2.

1. First Level Package: Chip Packaging

In chip packaging, the IC is connected to the outside electric circuit, which includes the combination of the chip with a substrate or a lead frame. Based on the connection method between chips and substrates, as the trend of reducing packaging size and increasing I/O pins develops, there are four types commonly seen: (1) wire bonding, (2) tape-automated bonding, (3) flip chip, and (4) wafer-level packaging.

Wire bonding is the earliest and most widely used technology nowadays. First, the chip is fixed on a lead frame, followed by using thin metal wires (made of gold or aluminum) to connect the electric circuits and lead frames on the chip to the pins, as shown in Figure 17.3. With the rise of other technologies in recent years, the market share of wire bonding technology has been gradually decreasing.

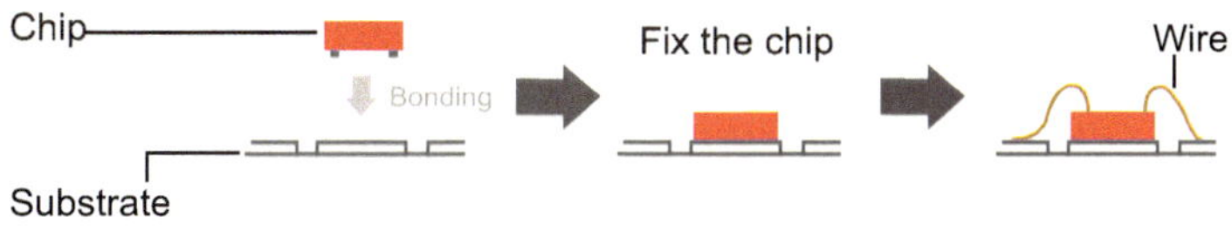

Figure 17.3 Wire bonding [2]

Tape-automated bonding was proposed by GE and first used in the 1960s. As shown in Figure 17.4, the chip is connected to the metal circuit on a polymer tape. The material used for the polymer tape is mainly polyimide and copper foils are mostly used as the metal layer on the tape. The small span between pins and the low thickness of tape-automated bonding make it suitable for IC products of light weight and small volume.

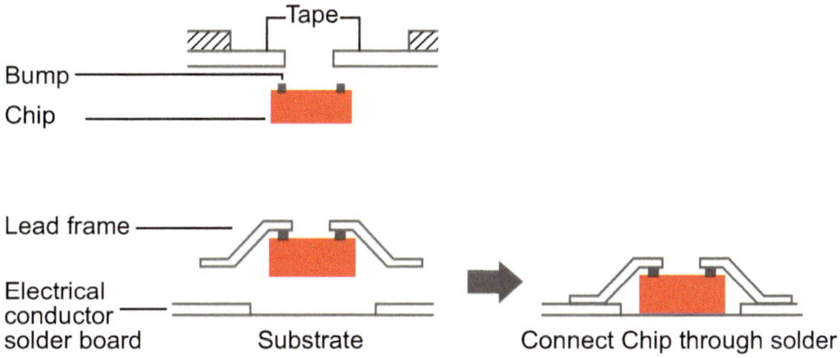

Figure 17.4 Tape-automated bonding [2]

Flip chip bonding was first developed by IBM in the 1960s. As shown in Figure 17.5, it creates solder bumps on the metal pad of the chip and contacts on the substrate corresponding to the solder bumps, followed by flipping the chip to align to all the contacts on the substrate for bonding. Flip chip bonding, compared to the traditional wire bonding package, has shorter connecting length, better electric properties, and higher density of input/output contact as well as the ability to reduce the IC size to increase the unit productivity of the wafer. Due to these advantages, it is the mainstream of current packaging processes on the market.

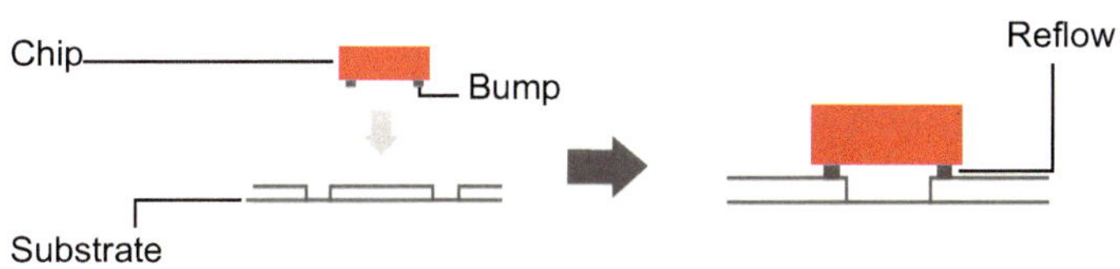

Figure 17.5 Flip chip bonding [2]

Wafer-level packaging is a process to encapsulate the whole wafer directly after manufacturing before the wafer is cut into single ICs; it requires no wire bonding or glue filling. As the IC remains the same size as the original chip after packaging, this process is also called wafer-level chip scale packaging (WLCSP), which has the advantages of integrating wafer manufacturing, packaging, and testing, simplifying the manufacturing process from wafer OEM (original equipment manufacturer) to product shipping. As shown in Figure 17.6, both fan-in and fan-out configurations are used to have a smaller packaging size and better electrical properties. Fan-in wafer-level packaging contains the interposer the same size as chip, while fan-out wafer-level packaging contains the interposer larger than chip to implement more I/O connections on the package. Fan-in wafer-level is suitable for products with fewer I/O connections. The concept of fan-out wafer-level packaging has become more popular in recent years because of the demands of large number I/O connection producs.

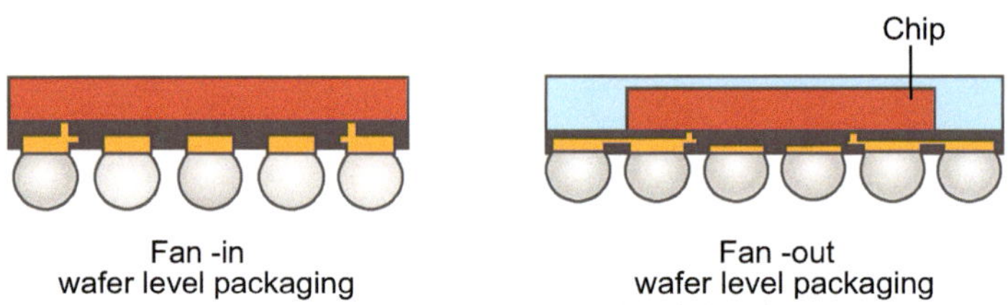

Figure 17.6 Fan-in and fan-out wafer-level packaging

2. Second Level Package: Chip-Substrate Packaging

In chip-substrate packaging, the major concept is to bond the finished electronic components on the first level with the electric circuit board. There are two ways of bonding: (1) pin through hole and (2) surface mounting:

In the pin through hole technology, metal pins of needle or thin plate shape are inserted into the socket or the electroplated through-hole on the circuit board and then soldered for fixation, as shown in Figure 17.7.

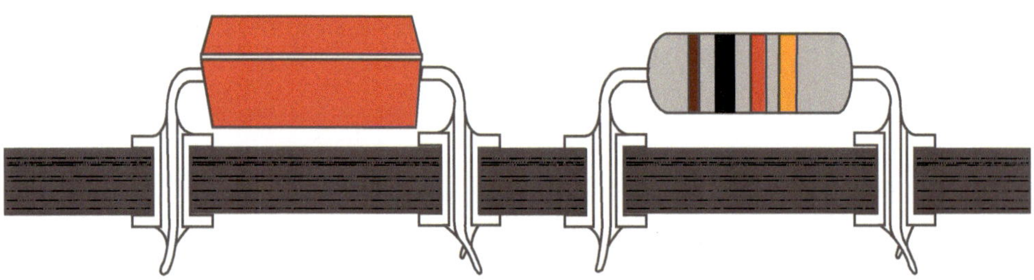

Figure 17.7 Pin through hole technology [3]

With the surface mounting technology, the component is temporally fixed to the solder pads on the circuit board by adhesive or solder paste, followed by wave soldering or reflow soldering techniques to finish bonding of the component and the circuit board, as shown in Figure 17.8.

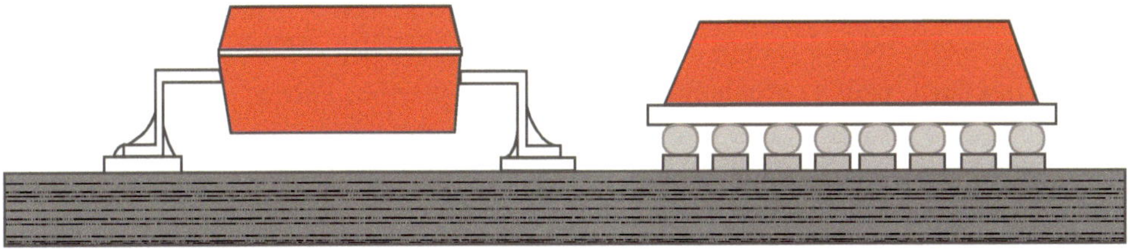

Figure 17.8 Surface mounting technology [3]

This level of packaging technology can be divided based on different types of application and variation as shown in Table 17.1 according to the above packaging configurations.

Table 17.1 Applications of Chip-Substrate Packaging

Technology	Applications	Derived Type
Pin through hole	Consumer electronics	Dual In-line Package (DIP)
Surface mounting	Memory Programmable logic ICLogic IC chip setLCD	Thin Small Outline Package (TSOP) Leaded/Leadless Chip Carrier (LCC) Quad Flat Package (QFP) Quad Flat No-leads (QF) Pin Grid Array (P) Land Grid Array (A) Ball Grid Array (BGA)

3. Third Level Package: Shelf Level Packaging

Such a packaging is mainly to assemble multiple circuit boards finished with second level package onto the motherboard to become a shelf level component.

4. Product Level Package: Frame Level Packaging

The main packaging integrates the shelf level components finished with third level packaging into a complete electronic product, also can be called as fourth level package.

The IC packaging technology then starts to enter a new stage of integrative packaging of multiple components and leaves the single component development behind.

This integrative approach has led to two major developments of relevant technology in the electronics industry: SoC (System-on-Chip) and SiP (System-in-a-Package). SoC and SiP are very similar: both integrate a system that consists of logic components, DRAM components, and even passive devices into a single unit. Starting from the design perspective, SoC highly integrates all the components needed for the system onto a single chip. SiP packaging has versatile configuration combinations based on different chip arrangement methods with different internal bonding technologies, which can also be customized or manufactured flexibly according to customer or product demands. Since 3D encapsulation can better meet the demands of initialization and higher performance, it has been adopted more and more in industry in recent years.

17.1.1 Process Principle

As first level packaging is mostly related to plastic packaging, the different processing conditions in various processes are discussed in this chapter. Because there is no absolute standard specification to be followed for completing a certain type of packaging, the packaging type is not used for the content differentiation, but instead the process techniques that can be used in encapsulation are taken as the object for description. Although based on the type of component packaging of the finished product, the above-described wire bonding, tape-automated bonding, and flip chip bonding techniques are covered here. According to the current major molding methods, three types of molding are described.

1. Transfer Molding

Transfer molding is the most common plastic packaging molding method nowadays. In a transfer molding process, lead frames or thin plate arrays are sent to the cavities of the mold manually or by automatic feeding, and then the mold is closed and applied with a high pressure for holding. The preheated (softened) plastic pellets of the packaging material are then pushed into the cavities via an injector. Once the thermosetting material fills up the cavities entirely, device encapsulation is then finished after baking, as shown in Figure 17.9.

In recent years, flip chip bonding has become very popular due to its high density, small volume, and high reliability. To overcome the filling resistance from the narrow gap at the solder bumps, application of transfer molding to flip chip bonding for molding requires a plastic material of higher flowability, which is usually used along with reduced pressure to assist molding. The process is generally called "molded underfill" in industry and illustrated in Figure 17.10.

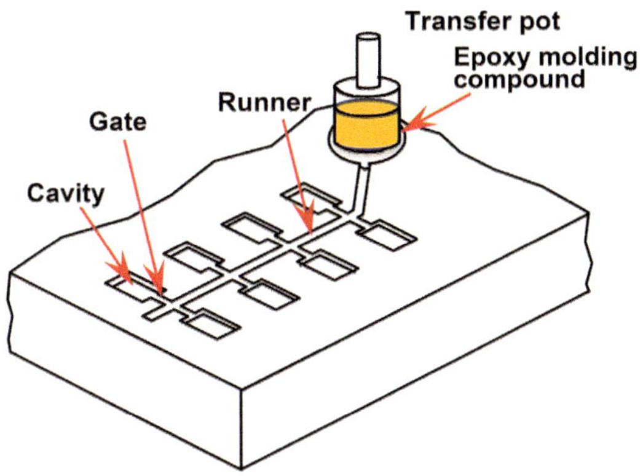

Figure 17.9 Transfer molding process

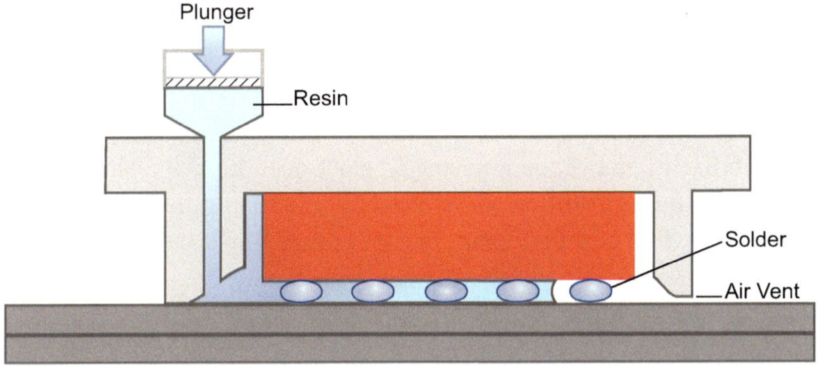

Figure 17.10 Molded underfill process

2. Compression Molding

Compression molding is a recently developed IC packaging technology, whose assembly method resembles the procedure of die attachment and its molding plastic can be in liquid or powder form. Firstly, the plastic is evenly placed or distributed on the substrate, which is then pressed by the chip coming from above to compress the plastic so it flows outward and thus filling is achieved. Next, the plastic is heated under pressure until it is completely cured. Figure 17.11 shows a compression molding process utilizing an anisotropic electroconductive resin.

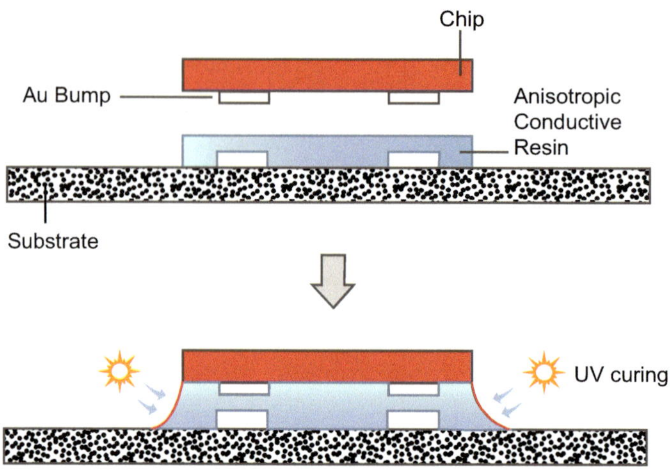

Figure 17.11 Compression molding process

Compression molding is now commonly used in wafer-level packaging because it offers the advantages of rapid molding and fewer mold constraints. The process concept is similar to that of the anisotropic electroconductive resin, with the only difference being the substitution of the wafer for the substrate. The plastic material is used in either pellet or liquid form. The process is illustrated in Figure 17.12.

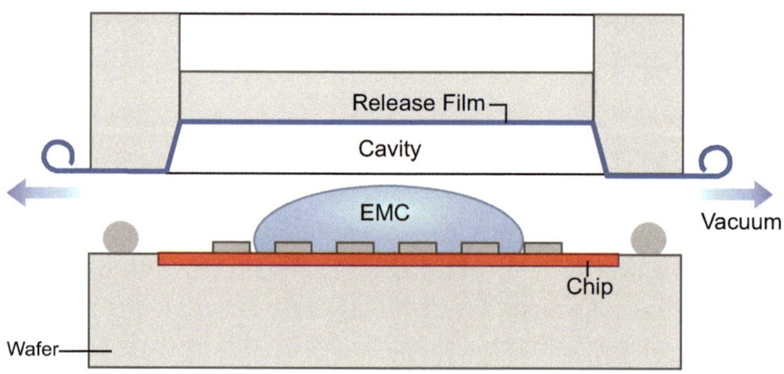

Figure 17.12 Compression molding process for wafer-level packaging

3. Capillary Underfill

In capillary underfill molding, as shown in Figure 17.13, the packaging material is injected directly on the edge of the chip, and via the capillary effect, the liquid packaging material keeps flowing until the whole bottom of the chip is covered completely. Generally speaking, apart from the glue volume control during the gluing process, normally the loader substrate is heated from the bottom to facilitate the capillary effect. This quickly lowers the viscosity of the packaging material upon injection. The packaging material then spreads faster to completely fill up the gap between the top and bottom sides.

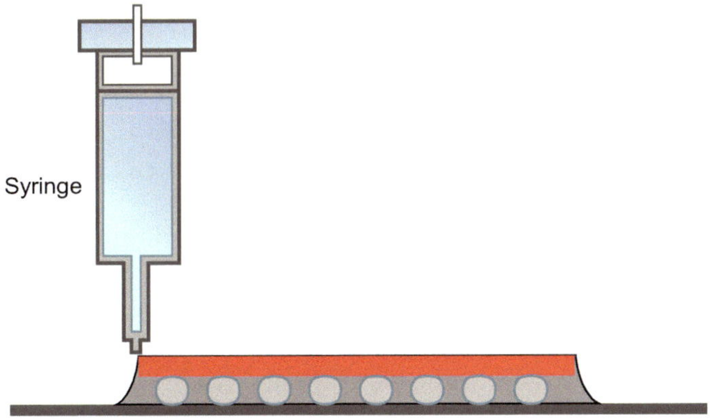

Figure 17.13 Capillary underfill molding process [4]

17.1.2 Advantages and Challenges

Regarding packaging types and the pin number growth for signal output, different bonding techniques are used in accordance with the packaging types, such as tape-automated bonding, ball grid array (BGA), flip chip, etc. The device occupation ratios of some packaging types are compared in Figure 17.14.

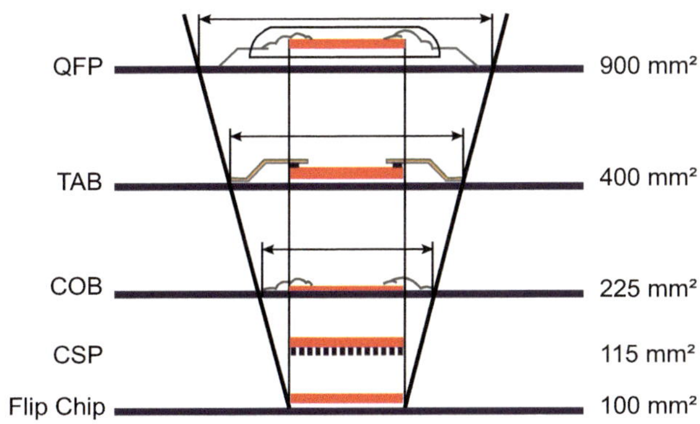

Figure 17.14 Comparison between different IC packages

Among all the above-described packing types, flip chip packaging has an extreme technological superiority. And along with the currently developing new packaging technologies such as wafer-level packaging and 3D stacking, both modified from flip chip packaging, these three types are generally called advanced packaging technology, which come with the following advantages:

1. Higher space utilization efficiency
2. No need for bonding wires and lead wires
3. Less heat generation and better heat dissipation effect
4. Saving of materials

For plastic packaging molding, a few critical challenges can be summarized:

1. Air trap:

   This is the most common issue occurring in the packaging process. Due to the difference in product device geometry, properties of packaging materials, and processing conditions, different flow resistances are generated in the plastic packaging process, which causes air traps and leads to insufficient filling.

2. Device shifting caused by flowing:

   This issue is mostly the result of an imbalance of flow distribution above and under the chip that causes shifting of the lead wires or lead frames inside the chip, or shifting of the chip in the case of a 3D stacking packaging type. Such a shifting behavior would further affect the follow-up flow behavior and cause the packaged device to deviate from its original designed location and thus results in the problems of short circuits or poor signals.

3. Warpage of molded finished products:

   Even if the molded product is free of air trap and device shifting, as the mechanical properties of each material in the package are different, the product would generate a deformation caused by stress release during the molding process (including the in-mold curing and post-mold process), which makes the follow-up reprocessing method become unpredictable and uncontrollable, and also increases the processing cost.

The advanced packaging technologies have more advantages than traditional IC packaging (QFP, CSP, etc.); however, compared with the traditional packaging technology, the advanced packaging manufacturing requires a more precise processing technology. The three issues of air trap, device shifting, and warpage mentioned above can make the parameters involved in the process become very complicated.

17.1.3 Theoretical Models

In a plastic packaging process, the plastic used for filling is a thermosetting material. The fundamental governing equations of a packaging process are that of a molding system: the continuous equation (Equation 5.1) describing the fluid field and temperature field, the momentum equation (Equation 5.2), and the energy equation (Equation 5.3) given in Chapter 5. The following assumptions are made:

1. The fluid is an incompressible and purely viscous fluid and no polymer elasticity effect is considered.
2. The influence of gravity is neglected.
3. As the preheating temperature of the plastic is different from the mold temperature and the material constantly releases reaction heat during the reaction process, the filling process does not take place at a constant temperature.
4. The heat capacity of the material, reaction heat, thermal conductivity, contact angle, and surface tension coefficient are all constants.
5. In the mold cavity, the boundary condition of a wall is a non-slip boundary condition, at which the fluid speed is a fixed value, that is, the boundary setting value.

Volume Fraction Transport Equation

$$\frac{\partial}{\partial t} f + \nabla \cdot f\mathbf{v} = 0 \tag{17.1}$$

where f is a volumetric fraction.

Normally, the volumetric fraction f in a controlled volume filled completely with melt is 1, and 0 for an empty controlled volume. The driving force comes from the force generated from the injector pushing the melt into the mold cavity. In general, the flow rate or pressure is used for the inlet boundary conditions. Because the mold face moves during a compression molding to push the plastic material into the mold cavity, moving boundary conditions should therefore be introduced into the governing equation above to take the moving force at the compression face into consideration. Hence, the mesh of the movement at the compression face is built into the analysis model as well, as shown in Figure 17.15.

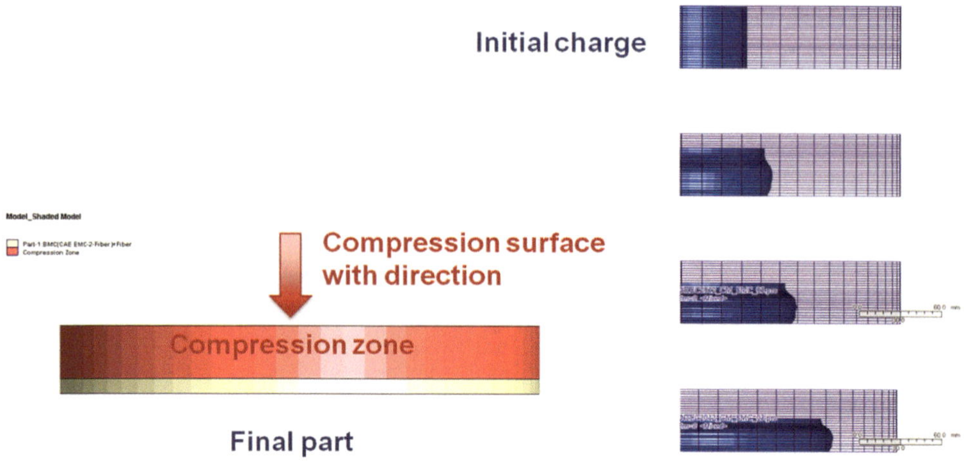

Figure 17.15 Moving mesh boundary assignment

In the context of capillary underfill, since the driving force for the filling process is surface tension, a key point is to introduce the surface tension and wall adhesion conditions. Regarding the description for surface tension, continuum surface force [6] is mostly used:

$$f_\sigma = -\sigma\kappa(\nabla f) \tag{17.2}$$

where σ is the surface tension coefficient, and κ is the free surface curvature.

Wall adhesion between the filler material and interface (the substrate) is described by the contact angle under steady state conditions. As Young's equation in Figure 17.16 shows, the contact angle represents the equilibrium conditions at the interface between gas, liquid and solid phase. Thus, the contact angle is an important property which keeps multiple phases in a steady state of equilibrium.

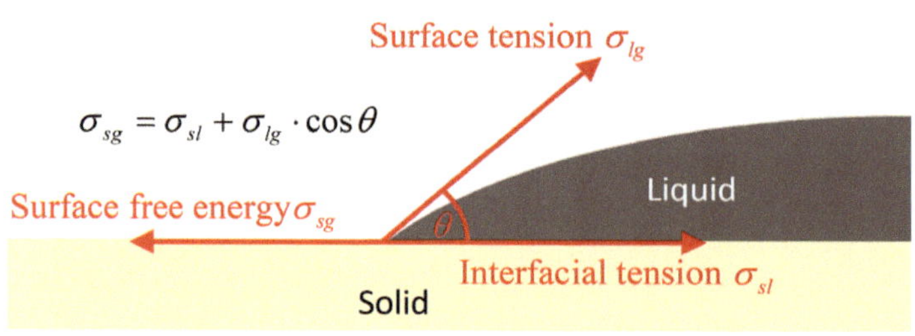

Figure 17.16 Young's equation

Reaction Kinetic Equation

The reaction kinetic equation adopted is the combined kinetic model (Equation 2.49) given in Chapter 2; the curing degree is referred to by the conversion a and is a function of time and temperature as:

$$\frac{D\alpha}{Dt}=\frac{\partial\alpha}{\partial t}+\mathbf{v}\cdot\nabla\alpha=\left(k_1+k_2\alpha^m\right)\cdot(1-\alpha)^n \tag{17.3}$$

where k_1, k_2, m, and n are constants of the material models. By utilizing the discrete form of the above equation, along with the plastic viscosity model and known boundary conditions, the plastic flow condition during the filling process can be calculated by solving it with the iteration method.

Drag Force Formula for Wire Sweep Analysis

A wire sweep analysis is used to predict the deformation of the wires during the encapsulation process. For the wire sweep analysis, to calculate the drag force exerted on the wires by the resin flow, the values of velocities and viscosity have to be determined from the mold filling simulation. After the drag force calculation is done, all the relevant data will be used for wire deformation calculation. Usually there are four different drag force models: Fruya's model, Nguyen's model, Lamb's model, and Takaisi's model. For example, Lamb's model is utilized to calculate the drag force as follow [6]:

$$F=\frac{C_D\rho U^2 d}{2} \tag{17.4}$$

where F is the drag force per unit length, ρ is the fluid density, U is the undisturbed upstream velocity, d is the wire diameter, and C_D is the drag coefficient, which can be written as:

$$C_D=\frac{8\pi}{\mathrm{Re}[2.002-\ln(\mathrm{Re})]} \tag{17.5}$$

where Re is the Reynolds number, which can be defined as:

$$\mathrm{Re}=\frac{\rho U d}{\eta} \tag{17.6}$$

The other popular model for drag force calculation is Takaisi's model. This method considers the effects of the cavity wall on the drag force. The calculation of the drag force on the wire is based on the following equation:

$$F=4\pi\eta\frac{U}{\ln\left(\frac{h}{d}\right)-0.9156+1.7243\left(\frac{h}{d}\right)^2} \tag{17.7}$$

where U is the undisturbed upstream velocity, d is the wire diameter, and h is the half-gap thickness of the mold cavity or the defined characteristic length.

Force Balance Equation for Paddle Shift Analysis

Paddle shift analysis is used to predict the deflection of the leadframe-pad and die during the encapsulation process. For paddle shift analysis, the melt pressure p during filling is first calculated. Moreover, it exerts a net lateral force on the paddle surface. Hence, the paddle shift can be obtained from the force balance:

$$\nabla\sigma + F = 0 \tag{17.8}$$

where σ is the stress and F is the body force from the melt pressure.

After filling analysis has been done, the force loading on the paddle exerted by this fluid and other relevant data are used for further paddle deformation calculation. Paddle deformation can then be investigated from the simulation results of Von Mises stress, and different stress and displacement components distributions.

For both the wire sweep and paddle shift analysis, we have to solve the force balance equation as Equation 17.8. Generally, geometric and material non-linear analyses are preferred for accuracy. We can use ANSYS or Abaqus for this kind of analysis.

Stress Equilibrium and Relaxation Equation for Post-Mold Cure Analysis

After the product is ejected from the mold, a free thermal and cure shrinkage occurs due to the temperature and conversion difference. The equilibrium equation representing the stress is expressed as follows:

$$\nabla\sigma = \mathbf{0} \tag{17.9}$$

And the relation between stress and strain is:

$$\sigma = \mathbf{C}\left(\varepsilon - \varepsilon_{Thermal} - \varepsilon_{Cure}\right) \tag{17.10}$$

with

$$\varepsilon = \frac{1}{2}\left(\nabla\mathbf{U} + \nabla\mathbf{U}^{\mathrm{T}}\right) \tag{17.11}$$

where σ is the stress, $\mathbf{C}$ is a fourth-order tensor and a function of the relaxation modulus $E(t, T, \mathrm{a})$, e is the strain tensor, and $\mathbf{U}$ is the displacement vector, respectively.

Thermal and cure induced strains can be expressed with the following equations:

$$\varepsilon_{Thhermal} = \alpha_{CLTE}\Delta T \tag{17.12}$$

where α_{CLTE} is the coefficient of the linear thermal expansion tensor.

VS(P,V,T,C) is the volume shrinkage of the polymer and can be calculated from the chemical volume shrinkage by the Pressure-Volume-Temperature-Cure (PVTC) model (see Section 2.3.3.3, Equation 2.41 of Tait-C model).

The PVTC model is a combination of the volume shrinkage data, measured by the PVTC Tester, and the differential scanning calorimeter (DSC) conversion data are used to formulate the specified volume of resin as below:

$$\varepsilon_{Cure} = \mathrm{VS}(P, V, T, C) \tag{17.13}$$

For the viscoelastic behavior, a master function for the relaxation modulus is described by using the Generalized Maxwell model (also known as the Prony series). The shift factor $a = a_T a_C$ is considered with $E(t)$ to have the temperature and curing degree shift; a_T is the temperature shift factor, and a_C is the curing shift factor.

$$\log a_C = a_1 C^n + a_2 C^{(n-1)} + \ldots a_n \tag{17.14}$$

■ 17.2 Practical Applications

Due to the complexity of multiple material components, the precise development needs higher simulation accuracy to control chip, lead frame, and wire density conditions. It is a challenge to prevent issues such as damage or corrosion due to incomplete filling, void, wire sweep, die shift, warpage, etc., in IC packaging processes. The simulation of advanced packaging deals mostly with (1) void prediction during molding, (2) fluid-structure interaction of components during molding (both wire sweep and paddle shift (3) in-mold/post-mold warpage prediction, and (4) the effect of dispensing control and creeping behaviors on the underfill process.

17.2.1 CAE Verification on Void Prediction

In this case [8], the molded underfill (MUF) process of a flip-chip multi-chip module is investigated. The rheological and cure kinetic parameters of a commercial MUF compound are acquired using a parallel plate rheometer and dynamic differential scanning calorimeter (DSC) analysis. The experimental results show severe void at the underfill area for different type gate locations. The packaging structure of the flip chip multi-chip module is shown in Figure 17.17. The underfill gap between substrate and die is only 60 mm. The total mold height is 0.27 mm. The z-directional mesh of the underfill area is divided by 3 and the die area is divided by 3. The mold clearance between the top die surface and the mold chase is divided by 3 meshes. The total z-directional number of the mesh is 9.

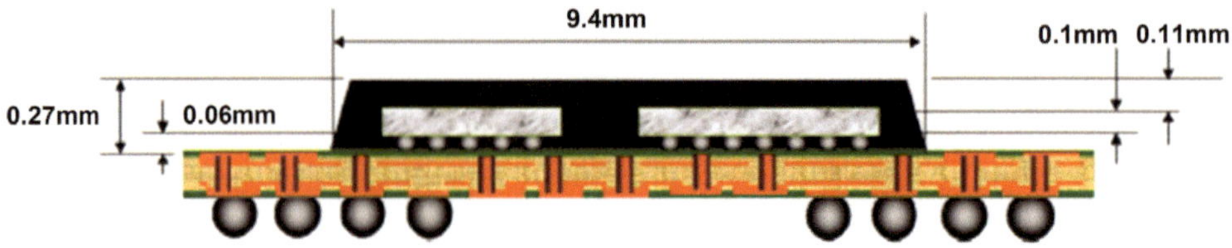

Figure 17.17 Flip chip multi-chip module structure [7]

Five different kinds of gate types are used in the study [7]. Figure 17.18 shows the pin-type gate with top left gate location. The total number of finite element meshes used for the full 3D model for current studies is about 600,000–800,000, depending on the gate type. The transfer time for the mold filling process is 7.4 s with optimum ram speed profile control obtained from the mold process.

Figure 17.18 Flip chip multi-chip module: top left pin type gate [7]

Figure 17.19 shows the simulated melt front advancement contour results for the top left pin gate type with the substrate side view. From the melt front advancement, we can see that the void trapping mechanism of a large die is almost similar to that of a small die. However, because of the increased x-y dimension of a large die, the flow separation and void trapping are much more severe compared to those in a small die. Figure 17.19 shows the predicted void trapping at the final filling stage where the MUF filled volume is about 99.8%.

After the MUF process is finished, the samples are analyzed by SAT (Scanning Acoustic Tomography) imaging. We can observe the entrapped voids in the underfill area of the large die for both the top-left and the bottom-right pin gate type cases as shown in Figure 17.20. The locations of the simulated and experimental voids entrappes are almost identical.

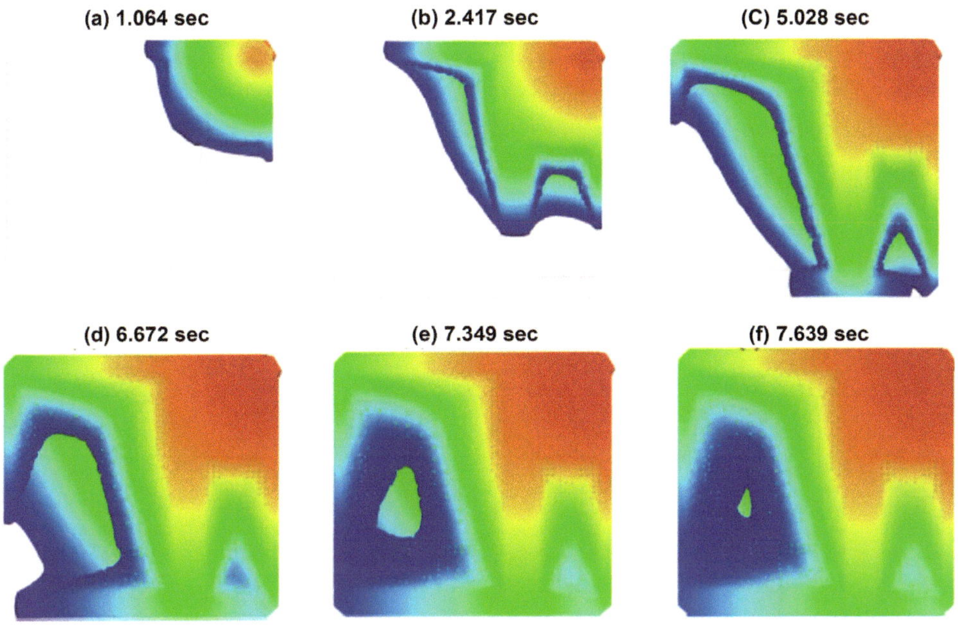

Figure 17.19 Predicted MUF melt front advancement contours for the top left pin gate type for a specified transfer time [7]

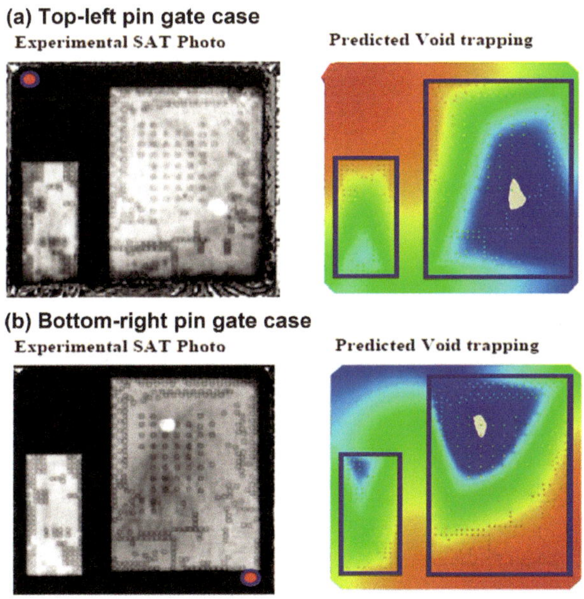

Figure 17.20 Experimental and simulation results for void defects [7]

As mentioned above, the major void trapping mechanism is the flow speed difference between the underfill and other areas. Even if trapping occurs, the trapping size is getting smaller and smaller when the bulk flow in the mold cavity is not stopped. Based on these flow characteristics, an optimized design provides the room for the MUF flow to drag out the melt fronts of the void trapping area through mold vent (Figure 17.21).

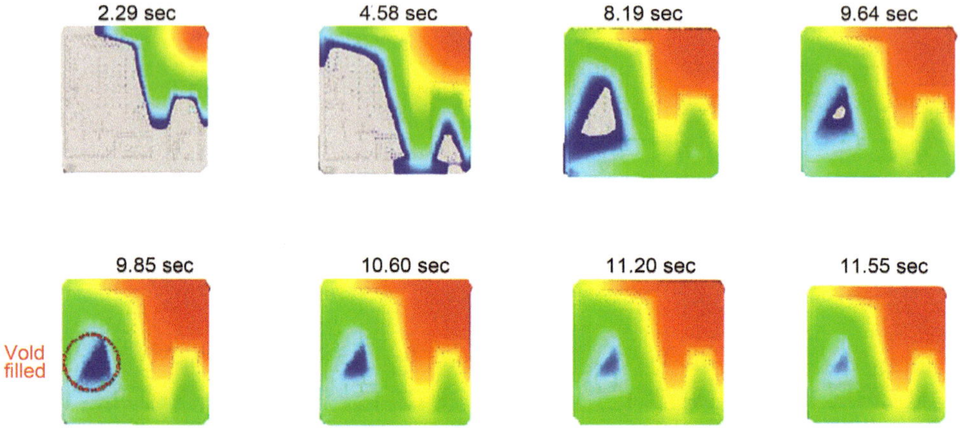

Figure 17.21 MUF filling analysis: melt front advancement of optimized mold design for top-left gate (substrate side view) [7]

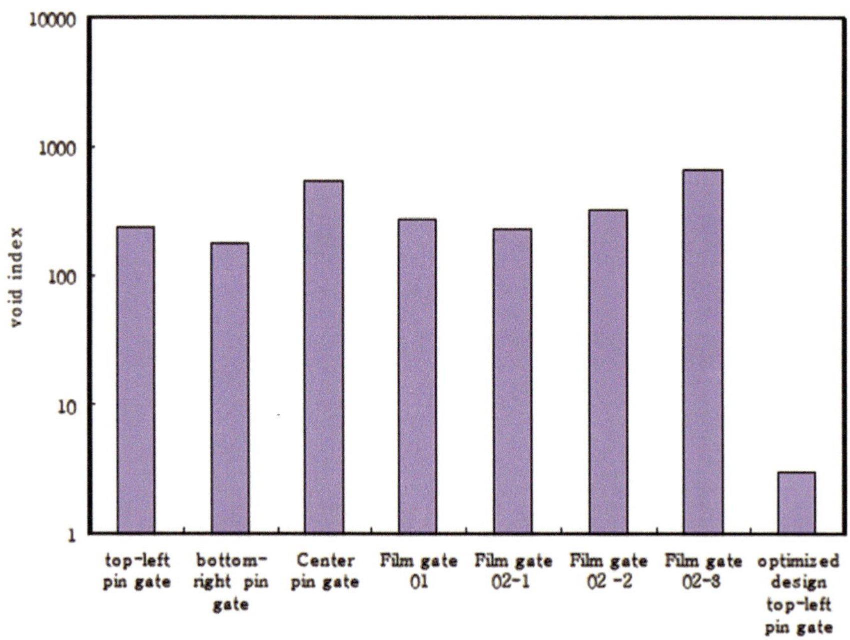

Figure 17.22 Number of void-trapping elements (void index) for various gate types and process conditions [7]

Figure 17.22 shows the void index which is defined as the number of elements in a void-trapping area. The simulated results of void index values for a model case are compared. The film gate 1 represents the top-edge type film gate case and film gate 2 is the right edge film gate where 2-1, 2-2, and 2-3 are different transfer time conditions with 4.0, 7.4, and 12.0 s, respectively. For the same film gate type 2 with different transfer time, as the transfer time is increased from 4.0 to 12.0 s, the void trapping index increases linearly. However, the optimized mold design case shows the lowest void index value. From the results, we can see that the optimized mold design very effectively eliminates void defects in the underfill area in a MUF process.

17.2.2 Fluid-Structure Interactions: Wire Sweep Analysis

In this case [8] , a full three-dimensional rheokinetic simulation of the mold filling behavior of the three-die stack CSP 294LD of 4 × 4 mold array with four different mold gate types is performed with experimental validation by checking wire sweeping dependencies for each gate type.

Figure 17.23 shows the meshes for the different kinds of gate types used in this study. The center gate is the widest gate and the half gate is 1/2 of the center gate. The quarter and corner gates are 1/4 and 1/8 of the center gate, respectively. The total number of finite element meshes used for full 3D models for current studies is about 700,000–900,000, depending on the gate type. The transfer time for the mold filling process is 7.4 s (pot and runner filling time of 3.4 s and cavity filling time of 4.0 s) with optimum ram speed profile control obtained from the molding process.

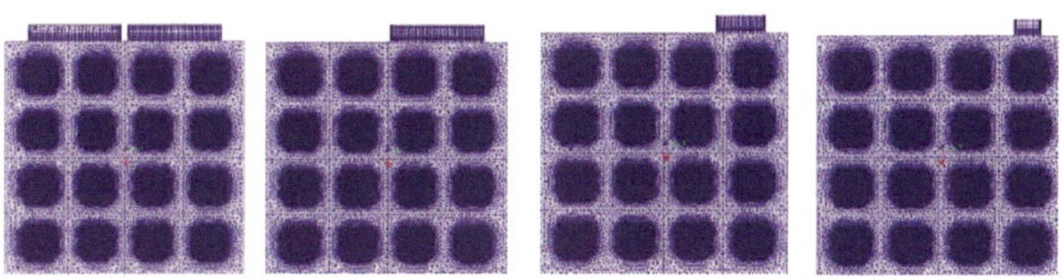

Figure 17.23 Finite element mesh generation for four layouts of gates: (a) center, (b) half, (c) quarter, and (d) corner [8]

Figure 17.24 shows the actual wire sweeping results of the 294LD three-die stack case. An index called wire sweep index (WSI) is a commonly used parameter to determine the degree of wire deformation, which is calculated from making the largest deformation normal to the wire divided by the projected length of the wire. The wire sweeping results show that, as expected, wire sweeping of the corner gate type is more severe near to the gate and the final filling area of the packages. From the simulation results, the center gate is supposed to have less positional dependencies and maintains a low wire sweeping value for the whole area. The actual experimental results reveal that the wire sweeping value for the center gate is less dependent on its position and the maximum wire sweeping values for the gate area, center area, and final filling zone are 3–4%.

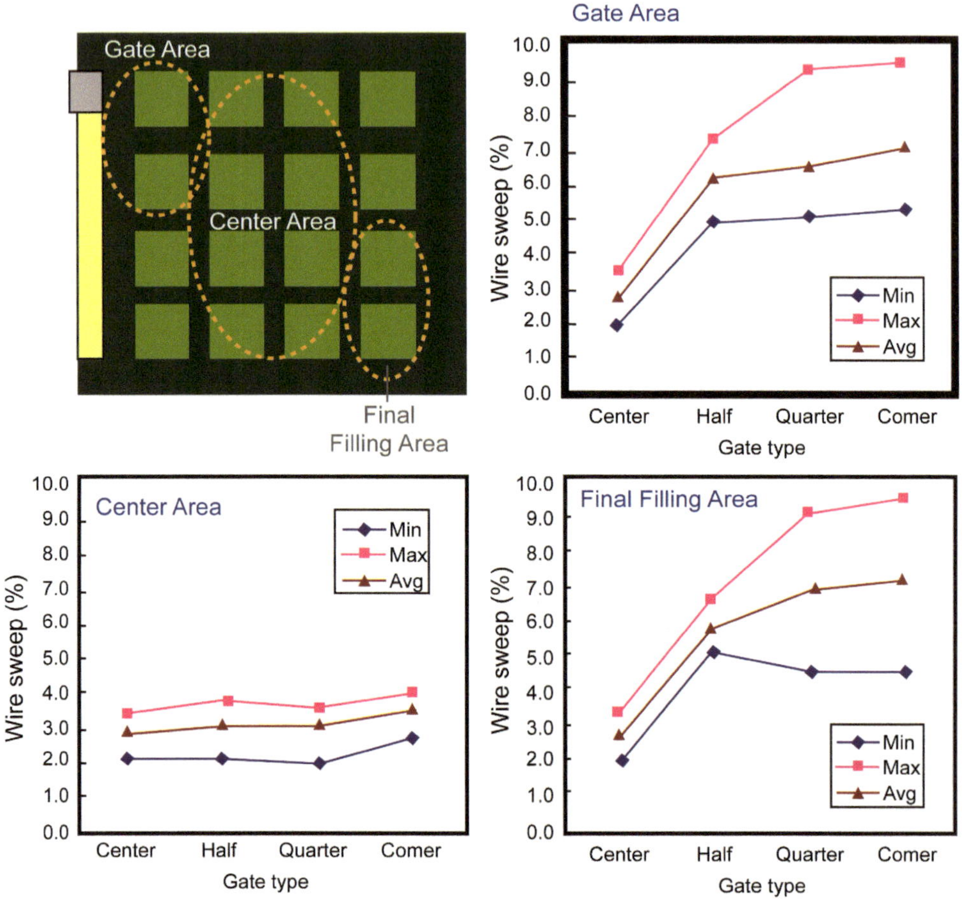

Figure 17.24 Comparison of experimental minimum, average and maximum wire sweeping results for gate types of Figure 17.23 and their positional dependencies in mold array packaging [8]

Figure 17.25 shows a comparison between experimental and simulated results for several selected units from mold array. The experimental case study and the mold filling simulation show acceptable match on the wire sweep related prediction. In Figure 17.25(a) and Figure 17.25(c), resin flows from the left side to the right side, which show that the wires perpendicular to the flow direction have larger deformation. The same phenomenon is shown in Figure 17.25(b) and Figure 17.25(d).

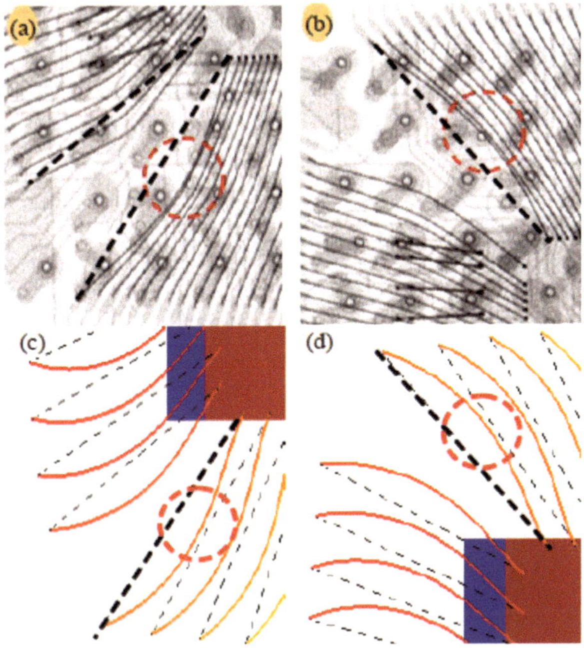

Figure 17.25 X-ray inspection pictures and simulation results for the wire short occurred region near the gate area of corner gate type [8]

The full 3D simulation methodologies enable the prediction of microscale mold filling behavior in multi-die stacking and other more complicated packaging structures such as SiP and package-in-package (PiP) for future application.

17.2.3 Fluid-Structure Interactions: Paddle Shift and Chip Deformation Analysis

The fluid-structural interaction experimental results given in reference 10] are used in this example to validate the two-way fluid-structure interaction modeling results of single- and stacked-chip packages. The scaled-up package for both models is located at the cavity and the dimension is the same as the reference. The geometry of both experimental models has been depicted including cavity, chip, solid ball, and substrate. The models of single- and stacked-chip packages for simulation are shown in (a) and (b), respectively, and illustrate that the fluid flows to the cavity through the entrance at the left lower corner of a cavity and separates to two flow streams, upper and lower, for a single-chip model. In a stacked-chip model, the fluid separates to upper, middle, and lower streams to flow around the chips.

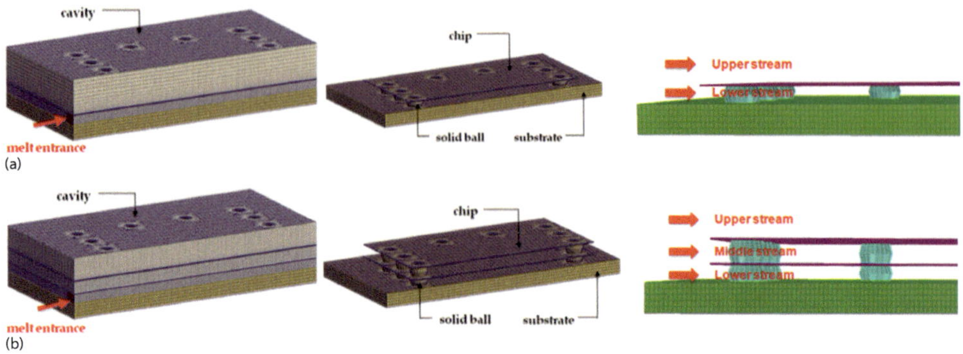

Figure 17.26 Models of (a) single-chip and (b) stacked-chip packages

The test fluid is a Newtonian fluid with a constant viscosity of 4 Pa·s. The mechanical properties of chip, solid ball, and substrate are tabulated in Table 17.2 and the material properties of the fluid are listed in Table 17.3.

Table 17.2 Mechanical Properties of Chip, Solid Ball, and Substrate

Parameters	Chip	Solid Ball	Substrate	Fluid
Elastic modulus *E* [GPa]	1.571	71.7	68.98	548
Poisson ratio	0.33	0.33	0.33	0.33
Solid density *ρ* [kg/m³]	4454	2850	2711	937

Table 17.3 Material Properties of Fluid

Density [kg/m³]	937
Heat capacity [J/(kg·K)]	1460
Thermal conductivity [W/(m·K)]	0.182
Viscosity [Pa·s]	4

During the injection molding process of this system, a non-uniform force loading around the chip results in chip deformation, and a pervasive problem especially in fluid-structure interaction occurs. The details will be addressed in what follows. To model the fluid-structural interaction effect accurately, the filling procedure is divided into two time steps. The node displacements on the chip are simulated by a previous structure analysis at each time. Simultaneously, the solid meshes of a cavity are deformed to maintain the mesh quality as good as possible after each chip deformation as shown in Figure 17.27.

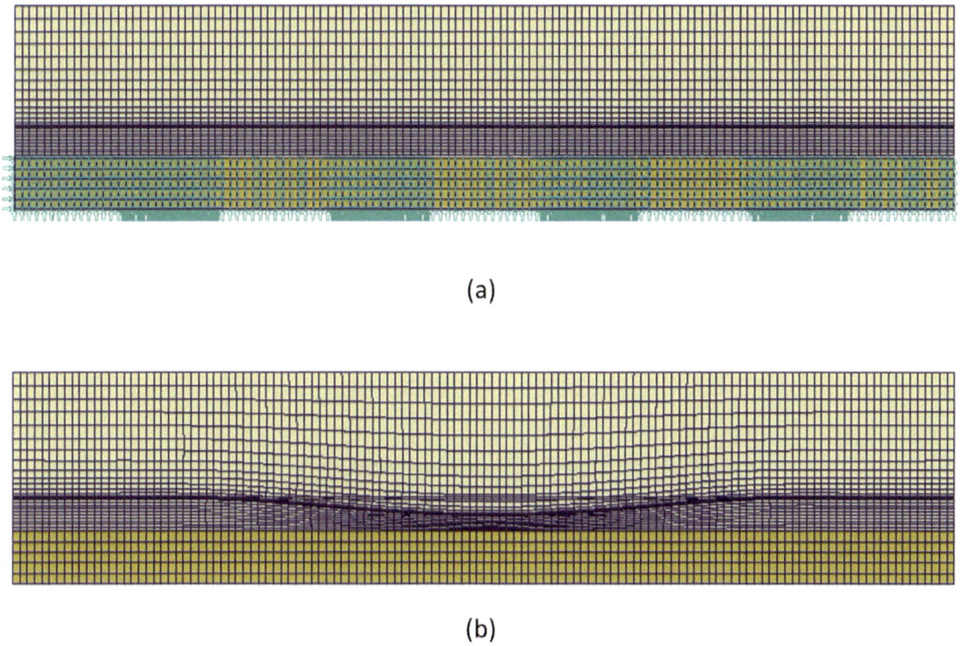

Figure 17.27 Simultaneous deformation of elements for chips and cavity during encapsulation: (a) at initial time step, and (b) at the end of filling [9]

The comparison of the predicted deflection of a chip with the experimental deformation shape is depicted in Figure 17.28. The flow front advancement of the model nearly corresponds to that of an experiment in shape and flow behavior. The deformation of the chip by simulation (1.718 mm) agrees with that by experiment (1.72–1.74 mm) at the middle region of the chip where there is no support by the solid balls. Therefore, two-way fluid-structure interaction modeling can provide substantial results to predict the flow front and the deformation of chip or paddle.

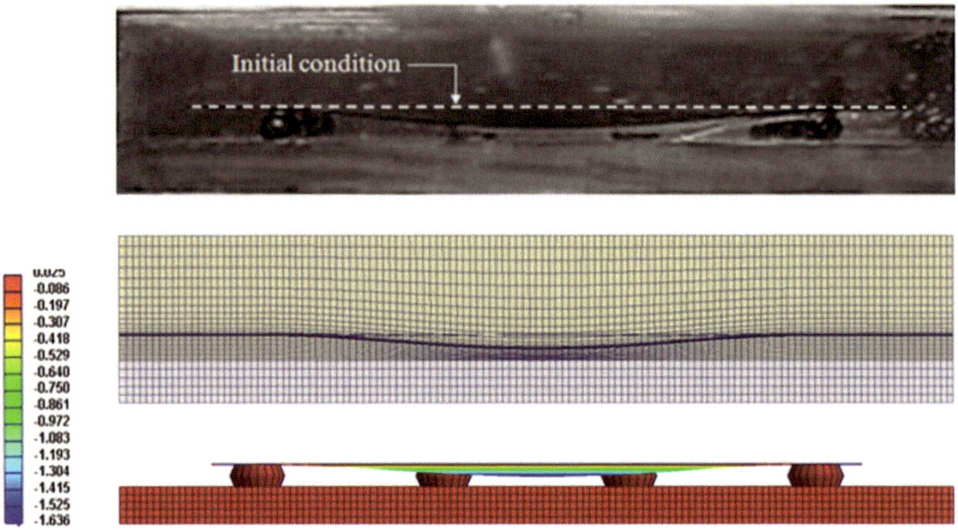

Figure 17.28 Deformation of chip in experiment and simulation [9]

From experimental investigations, flow profiles and deformations of the chip at the middle region for various filling times are shown in Figure 17.29. At 5% filling volume, the fluid enters and reaches the edge of a chip, which demonstrates the initial condition without deformation in the chip. The fluid continues to flow toward the center of the chip and the contour of tiny deflection of the chip develops slightly around 30% filling volume; meanwhile, a race-track effect and cross flow occur during the filling stage because of the unbalanced flow resulting from the discrepancy in the two free spaces at the upper and lower stream region. The fluid flows faster at the upper space region than at the lower region of the chip.

The comparison of experimental with modeling results for the stacked-chip package is tabulated in Figure 17.30. The fluid starts to reach the chip at nearly 15% filling volume and separates then into three flow paths, upper, middle, and lower stream, with balanced flow behavior at 20% filling volume. The flow behavior does not lead to the deformation of both chips. However, a tiny deformation develops at nearly 50% filling volume has good agreement between simulation and experiment.

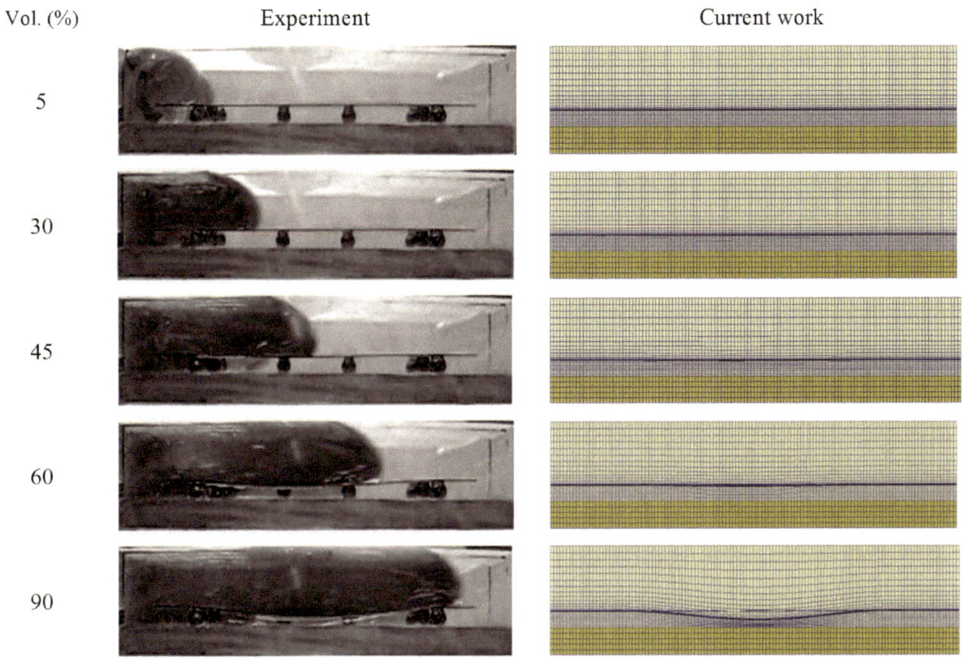

Figure 17.29 Entire fluid-structure interaction process record and modeling deformation results of chip with various filling volume (%) [9]

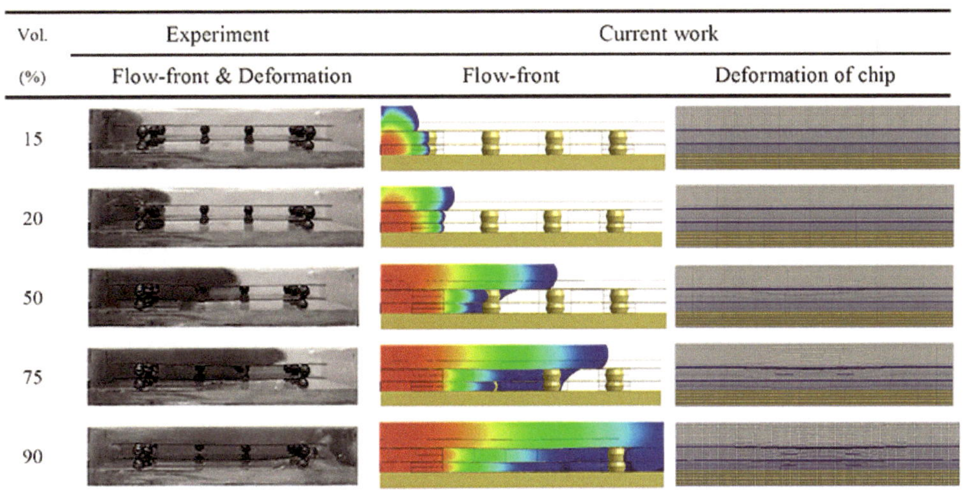

Figure 17.30 Comparison of flow profiles and chip deformation of experiment with simulation for stacked-chip at various stages [9]

As paddle shift is also a flow-induced phenomenon, paddle shift analysis can be realized as well with the current two-way fluid-structure interaction framework. The change of paddle deformation with time can be examined. Side, front, and back views of the 3D geometry model used for paddle shift analysis are shown in Figure 17.31. The predicted results could be validated by experiments. However, in cases with a long slender paddle, such as the two-cavity model shown in Figure 17.31, paddle contact with each other leading to short may be a severe problem during encapsulation. The material properties of the fluid are tabulated in reference [0].

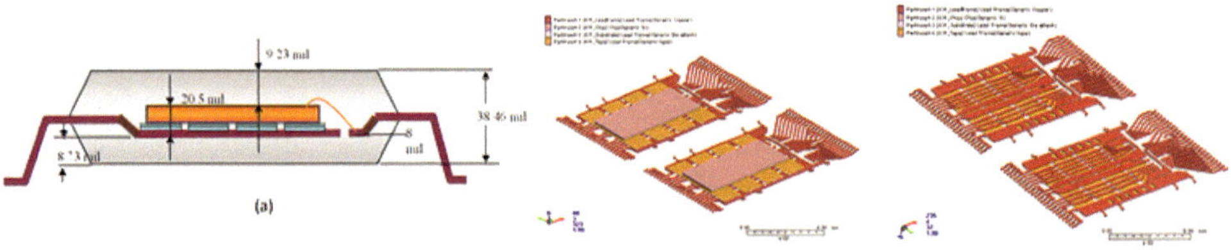

Figure 17.31 Models for paddle shift: (a) side view, (b) front view, and (c) back view [9]]

shows a comparison of the filling pattern between experimental and simulation results at distinct filling times, and the results show a close agreement. Also, from the simulation result shown in, we can see that paddles with larger deflection occur at positions where melt surges into a cavity from the entrance and melt merges at the end of filling. However, the large paddle deflection caused by the large lateral force exerted by the flow overwhelms the paddle deflection suffered from the pressure drop within a cavity. Therefore, paddles used to support chips can be analyzed individually.

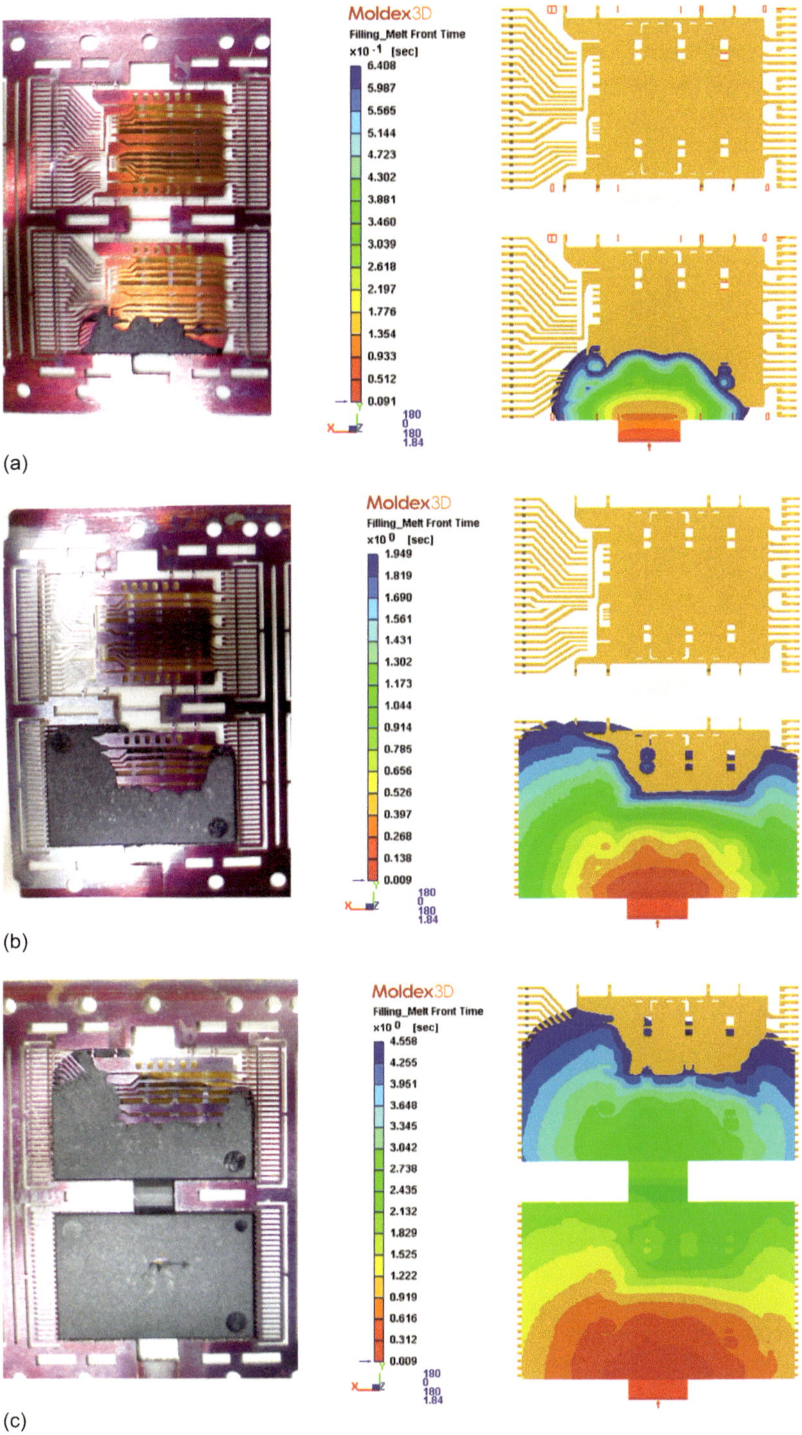

Figure 17.32 Comparison of filling pattern between experimental and simulation results at distinct filling times: (a) 0.66 s, (b) 1.962 s, and (c) 4.544 s [9]

A comparison of the deformed paddle shape at the end of filling between simulation and experiment is shown in Figure 17.33, and displays a coincidence in the position where the deformed paddle is exposed outside the package.

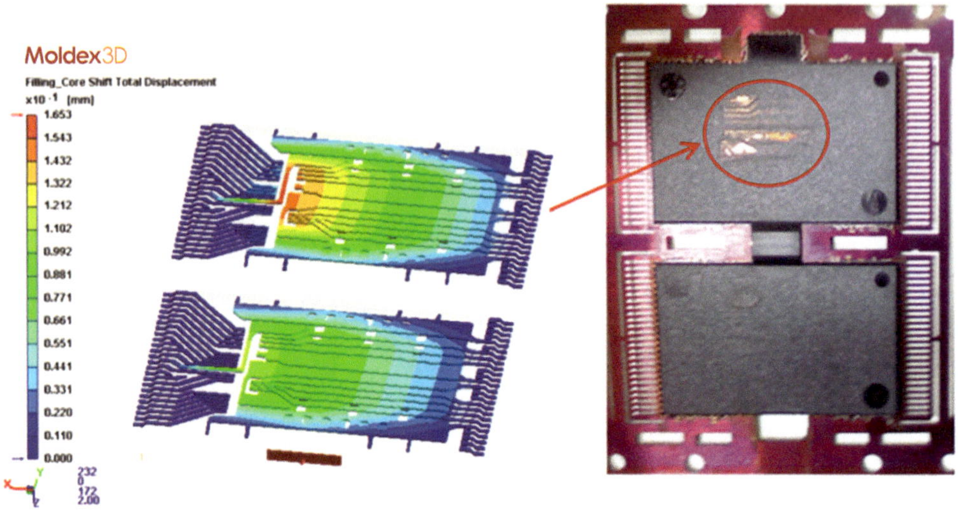

Figure 17.33 Comparison of deformed shape at end of filling between simulation and experimental results [9]

The side view of a deformed paddle shown in Figure 17.34 shows that the predicted tendency of the paddle shift direction matches the experimental trend. Therefore, from the results of filling and paddle shift analyses, the feasibility of the two-way fluid-structure interaction method to analyze mold designs for microchip encapsulation can be demonstrated.

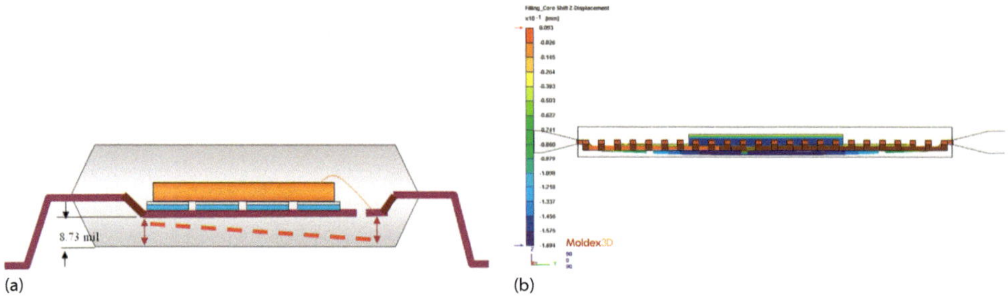

Figure 17.34 Side view of paddle shift phenomenon: (a) illustration of paddle shift within package and (b) predicted paddle shift result [9]

In summary, a two-way fluid-structure interaction simulation approach has been implemented for single- and stacked-chip package models and paddle analysis using a two-way fluid-structure interaction simulation framework. Furthermore, a novel approach has been introduced for both models for further investigation. The obviously observed deformation of chips in single-chip packages during experiment is well simulated with the current two-way fluid-structure interaction algorithm. A race-track effect causes unbalanced flow profiles at the upper and lower streams through the single chip. In addition, for the stacked-chip package, the fluid flows separately in three flow paths to the upper, middle, and lower regions along the chips. Owing to nearly the same gap height between the chips, the filling behavior in the three regions is closer than that at the single-chip package. Therefore, the more balanced the flow front advancement the less unstable the pressure exerted onto the surface of chips.

17.2.4 Warpage Prediction for a Bilaminate

To verify the reliability of the proposed method in this case study, we refer to experimental results in the literature [10] using the bilaminate geometry. The specimen is made by epoxy molding compound (EMC)/Si or EMC/Cu. The EMC used is Hitachi CEL9200 and its chemorheological parameters can be found in the literature [10]. The specimen size is 40 × 10 × 1.7 mm^3 with a thickness of 1.7 mm. The dimensions of the part with EMC thickness of 1.55 mm and Cu thickness of 0.15 mm ars shown in Figure 17.35.

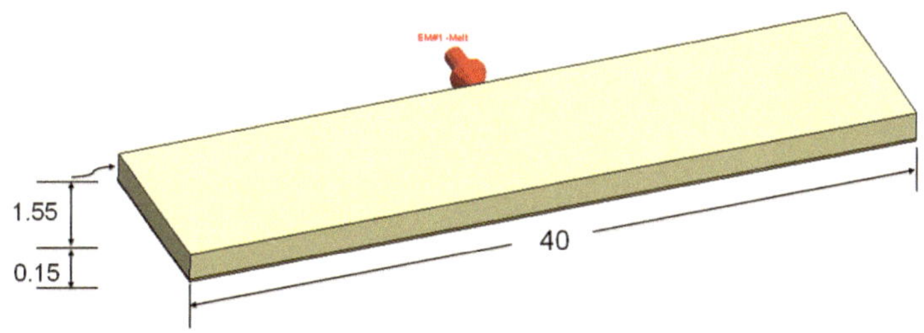

Figure 17.35 Schematic diagram of the bilaminate (unit: mm)

In the actual process, the resin is preheated at the beginning and the resin is pushed by the plunger into the cavity for mold filling. Once the cavity is fully filled, the part is kept for the curing stage for 80 s with the mold temperature at 175 °C. The finished product is placed at room temperature (25 °C) for cooling and put into the oven for post-mold cure for 6 hours. Then a second cooling process follows at room temperature, as shown in Figure 17.36. The warpage values are measured in the first and second cooling stage to compare the difference between in-mold and post-mold cure.

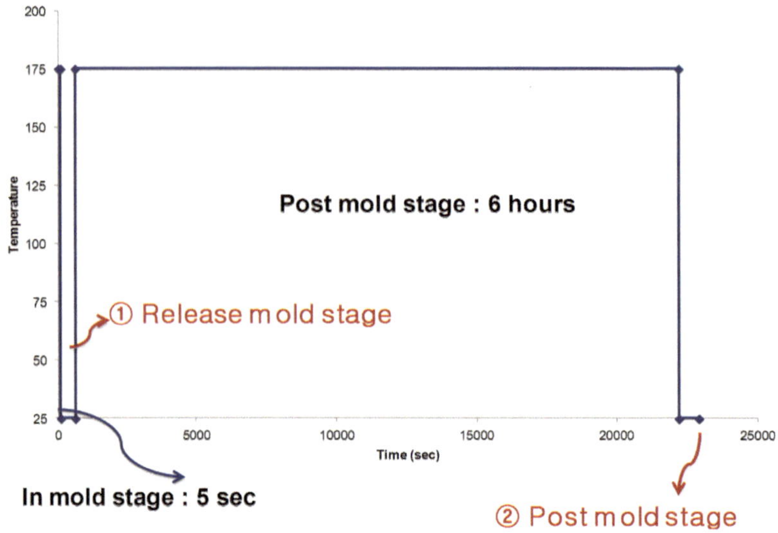

Figure 17.36 Process conditions for post-mold cure analysis

Based on the experimental material measurement data, we can refit the parameters of the PVTC model (see Section 2.3.3.3, Tait-C model) and viscoelastic constitutive behavior. Table 17.4 and Table 17.5 list all the parameters used in the simulation and additional simulation parameters can be found in reference [10].

Table 17.4 Parameters of PVTC Model

Parameters	b_1 [cm^3/g]	b_2 [cm^3/g K]	b_3 [dyne/cm^2]	b_4 [1/K]	ζ
Values	0.529	1.12×10^{-5}	3.0×10^{9}	2.0×10^{-6}	0.008

Table 17.5 Mechanical Properties for Different Components

Material	Poisson's Ratio	CTE [ppm/°C]	Young's Modulus [GPa]
EMC	0.25	7 (below 120 °C) 33 (above 120 °C)	-
Cu	0.34	17.1	115
Si	0.26	3	170

Firstly, from simulation results, the Z-displacement is 0.309 mm for EMC/Cu and the final Z-displacement is 0.625 mm for EMC/Si. The final experimental part warpage in the Z-direction (Z-displacement) is 0.32 mm for EMC/Cu and 0.65 mm for EMC/Si, as shown in Figure 17.37. Clearly, both of the simulation results are quite close to the literature.

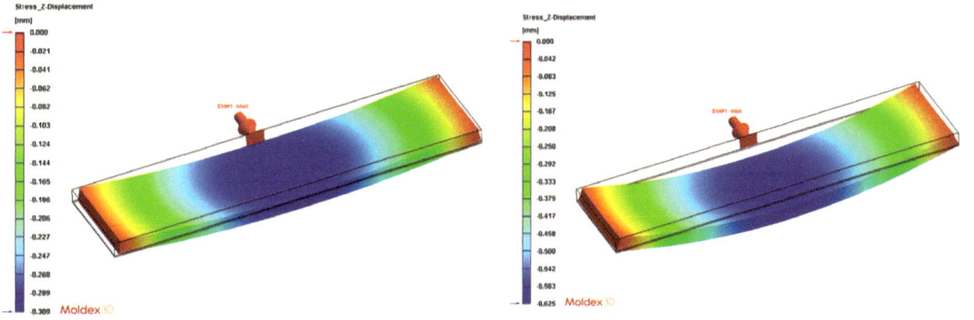

Figure 17.37 Simulation results of final warpage for (a) EMC/Cu and (b) EMC/Si

We can further investigate the warpage values induced by thermal and cure effects, as shown in Figure 17.38. It is found that the main difference between EMC/Cu and EMC/Si is due to thermal-induced warpage. The EMC/Si case has the larger deformation, which is caused by the larger mismatch of coefficient of thermal expansion (CTE), as shown in Table 17.5. The final warpage amount can be divided into contributions from the release-mold stage and post-cure stage as shown in Figure 17.39.

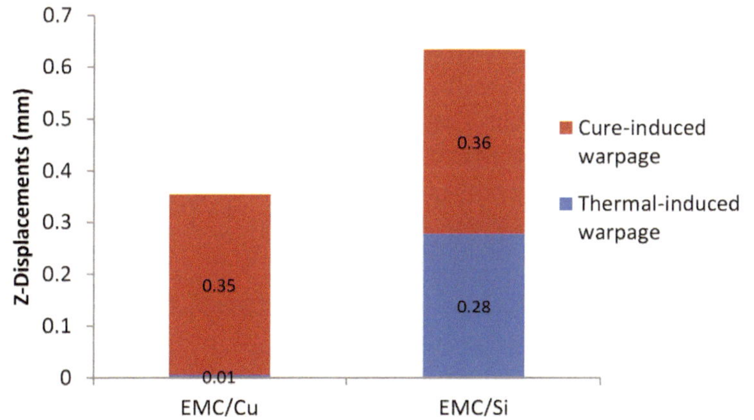

Figure 17.38 Warpage contributions of different bilaminates in release-mold stage [10]

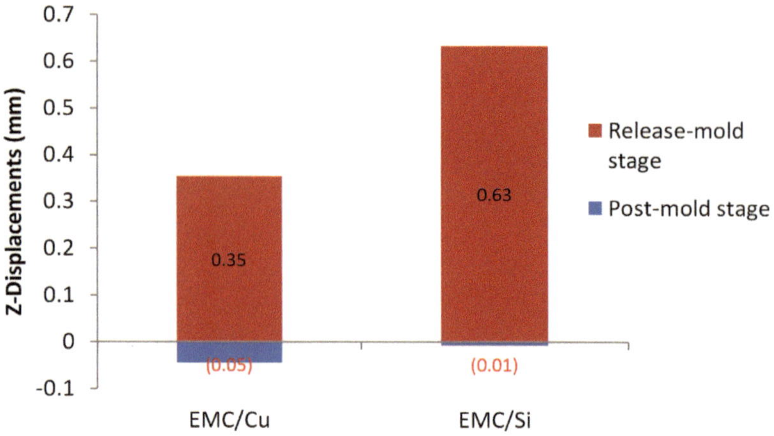

Figure 17.39 Warpage contributions of different bilaminates [10]

For the EMC/Si case, before the part is moved to the oven, the major influence on warpage of the part is due to the volumetric shrinkage caused by the curing reaction which can be analyzed using Moldex3D simulation. The Z-displacement during the pre-oven stage is 0.633 mm. Furthermore, after putting the cured part into the oven for post-mold cure, the main factor on warpage becomes the relaxation of stress which might be due to the viscoelasticity and the post-curing process inside the encapsulated unit. At this stage, the warpage behavior is solely caused by the release of internal stress. The Z-displacement from this stage is only −0.008 mm. Therefore, the final Z-displacement of 0.625 mm is the sum of those two results. Obviously, the post-cure process can release stress and hence reduce warpage defects of the final products.

17.2.5 Warpage Prediction for a Bi-Material Component Model

The viscoelastic (VE) effect and PVTC effect were characterized for a bi-material component model [11]. The model was an arbitrary one with length and width 10 mm. The 200-µm-thick mold compound was applied to the 100-µm-thick silicon. The mesh element size was 5 × 5 and the whole model contained 1600 mesh elements, as shown in Figure 17.40. The anchor plane was set to ensure stress analysis convergence.

The stress-free temperature was 175 °C for this model. The model experienced warpage due to ambient temperature change. Figure 17.41 shows the ambient temperature profile, which suddenly dropped from 175 to 25 °C at the beginning of analysis. The ambient temperature was kept at 25 °C for 100 s, and then surged to and remained at 260 °C until 200 s in total had elapsed. The heat transfer coefficient was 20 W/m^2/K around all external faces.

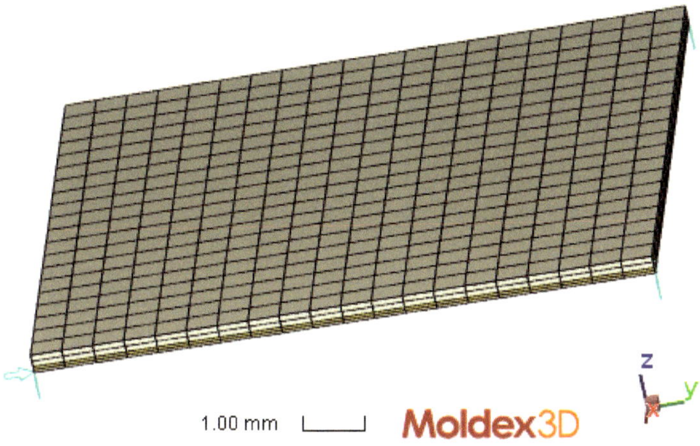

Figure 17.40 Geometry of the bi-material component model

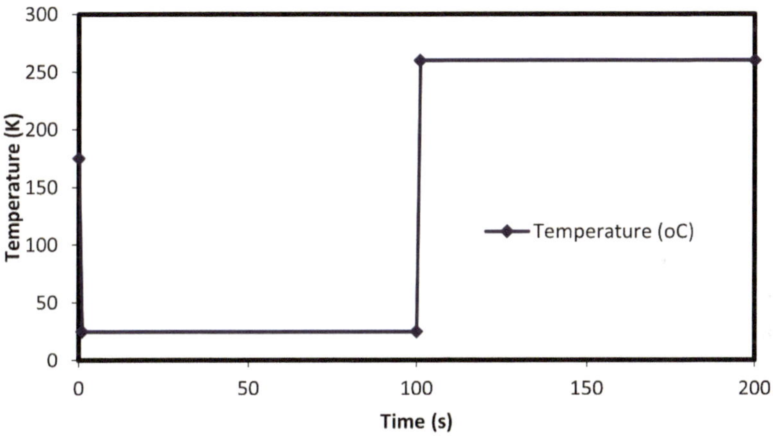

Figure 17.41 Bi-material component model ambient temperature profile

Moldex3D computed the heat transfer and performed a structural analysis to obtain warpage results. The warpage results for different material models are shown in Figure 17.42. The silicon material properties were set to linear-elastic while the molding compound material properties were simulated individually by linear-elastic, viscoelastic with time-temperature shift factor (VE-TTS), and VE-TTS with chemical shrinkage.

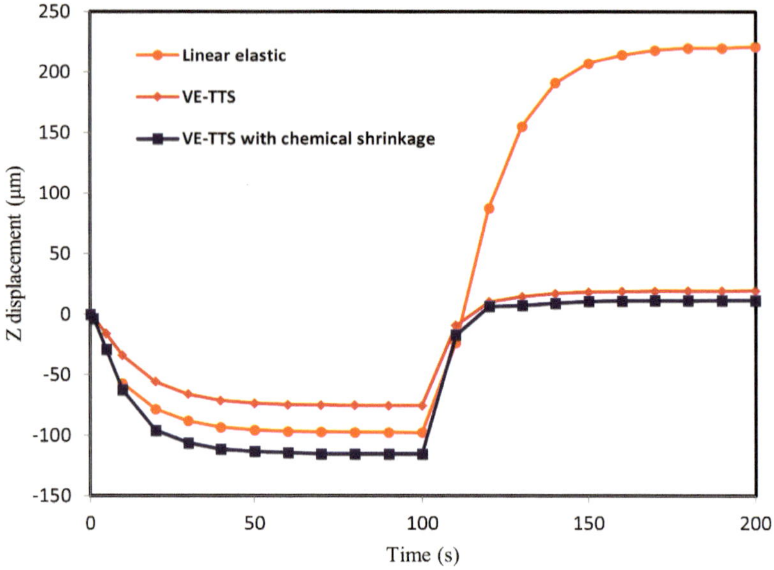

Figure 17.42 Warpage results for the bi-material component model

Warpage magnitude is indicated by the Z-displacement of the model's top center point. In the linear-elastic cases, the model warped into a concave shape with a magnitude of 98 µm, due to high contraction of the mold, as shown in Figure 17.43. In contrast, at 260 °C, the warpage shape was convex, with a magnitude of 220 µm, due to high expansion of the mold. For the VE-TTS case, the Z-displacement was 76 µm at 25 °C. The warpage difference was caused by thermal strain relaxation. At 260 °C, the warpage magnitude decreased to 19 µm, due to the change in material properties that was induced by TTS. In the case of chemical shrinkage, the Z-displacement magnitude decreased to −115 µm, which was due to a higher CTE of the uncured mold material, leading to thermal shrinkage.

The conversion rate rose to 100% as the ambient temperature surged to 260 °C, as shown in Figure 17.44. Conversion led to additional chemical shrinkage of the mold, and so the Z-displacement dropped to 11.7 µm. This example shows that the VE-TTS and chemical shrinkage effect cannot be neglected when package warpage is being simulated.

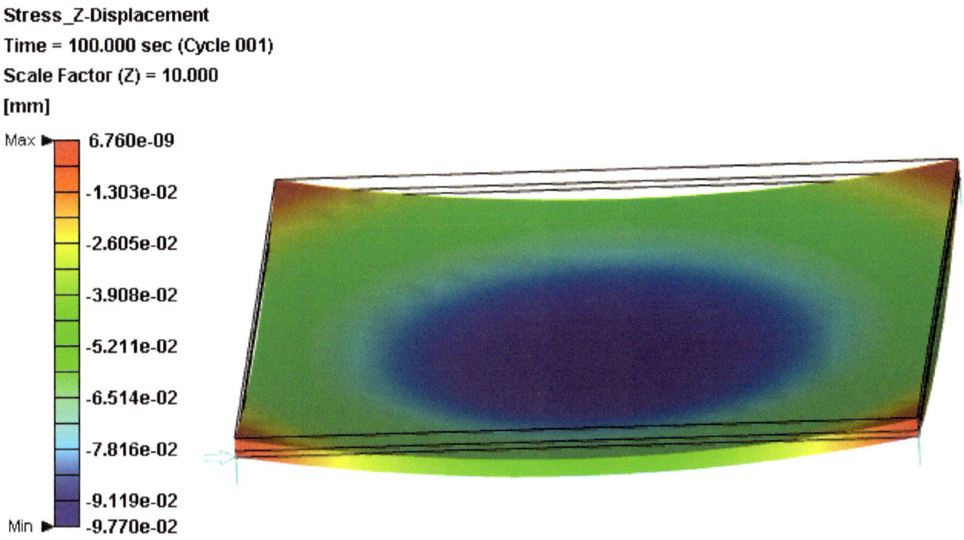

Figure 17.43 Warpage results for bi-material component model for linear-elastic at 100 s

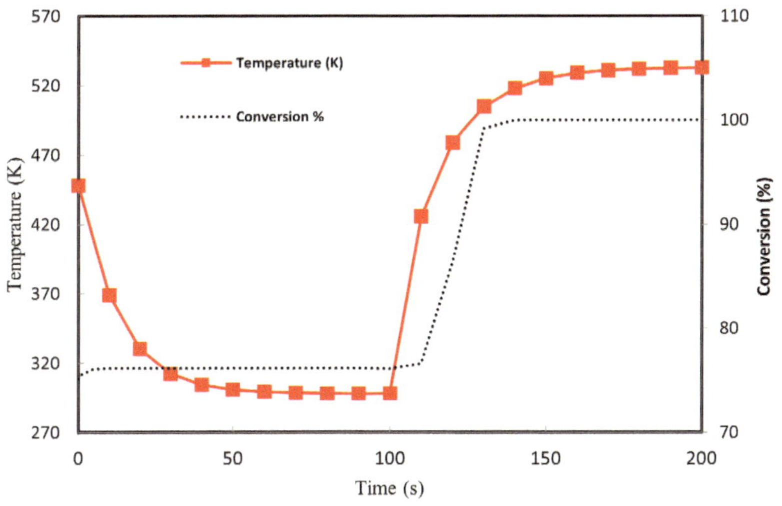

Figure 17.44 Bi-material component model – average temperature and degree of conversion

17.2.6 Warpage of Bi-Material Strip

The other bi-material model comprised epoxy and copper strip [12]. The copper strip measured 240 × 74 × 0.35 mm and two strips of mold compound measuring 115.5 × 66 × 0.255 mm were transferred to it. The strip geometry is shown in Figure 17.45. The stress-free temperature was 175 °C. The anchor plane boundary condition was set so as to ensure stress analysis convergence.

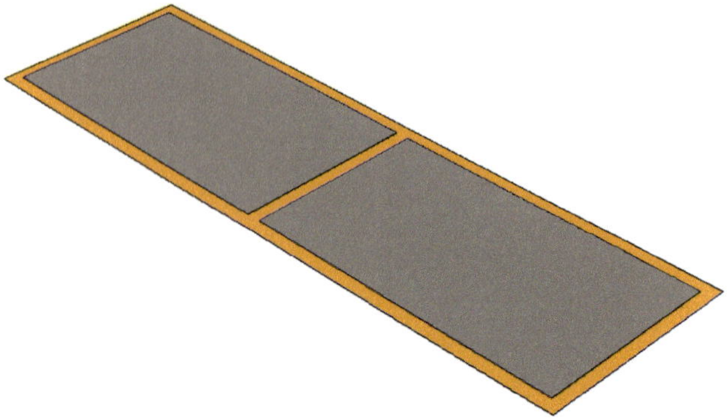

Figure 17.45 Bi-Material Strip Geometry

Two ambient temperature profiles were simulated for this model. Profile A featured cooling and a 260 °C reflow, as shown in Figure 17.46(a), while profile B featured an additional post-mold cure process, as shown in Figure 17.46(b).

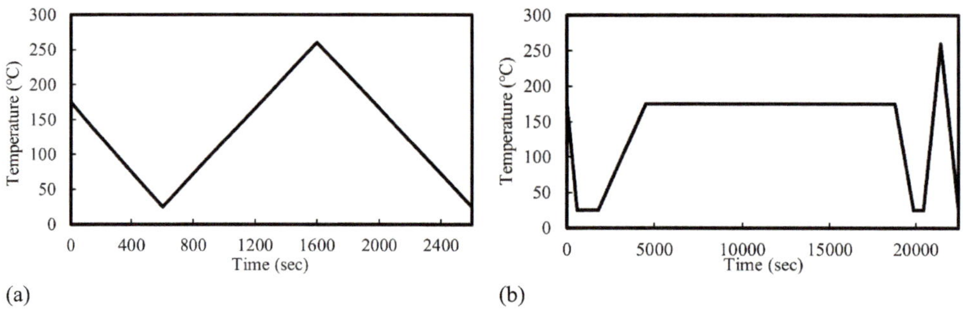

Figure 17.46 Ambient temperature profiles of (a) A and (b) B

To observe the effect of mold-curing shrinkage, the difference in mold curing time was simulated to get the different initial degrees of curing. Four kinds of initial degrees of curing were simulated for profile A, as shown in Figure 17.47. The initial degree of curing surged rapidly at the beginning of the analysis because the temperature was high enough to induce curing. However, curing stopped after about 200 s because the temperature was too low to induce it. The curves for the degree of curing all reached 100% after 1300 s, when post-mold curing started.

Top center node Z-displacement was recorded so as to describe the warpage. The Z-displacement of profile A showed that, for the fully cured case, the strip turned concave between 69 and 175 °C, and maximum Z-displacement existed at room temperature, as shown in Figure 17.48. The trend showed that the Z-displacement was shifting downward as the initial degree of curing decreased. The 0 s cured case had a different Z-displacement trend between 250 and 1000 s. As 18% of epoxy remained uncured during that period, the material's thermal strain properties were different.

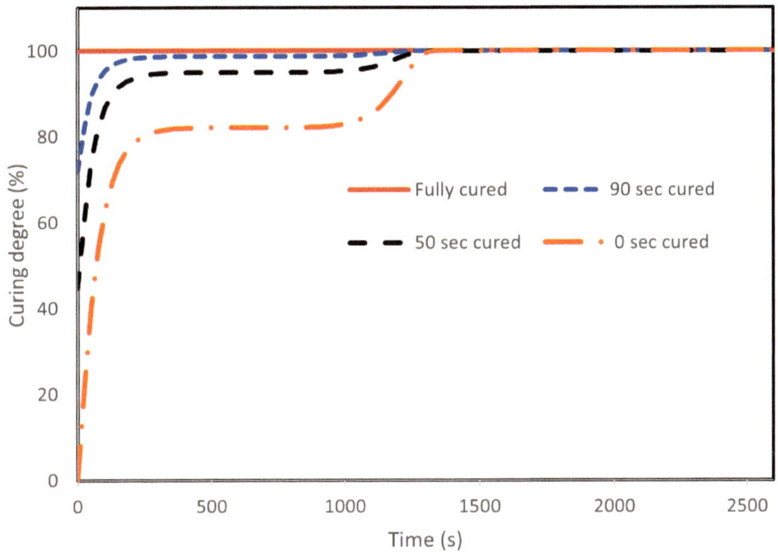

Figure 17.47 Degree of curing versus time for profile A

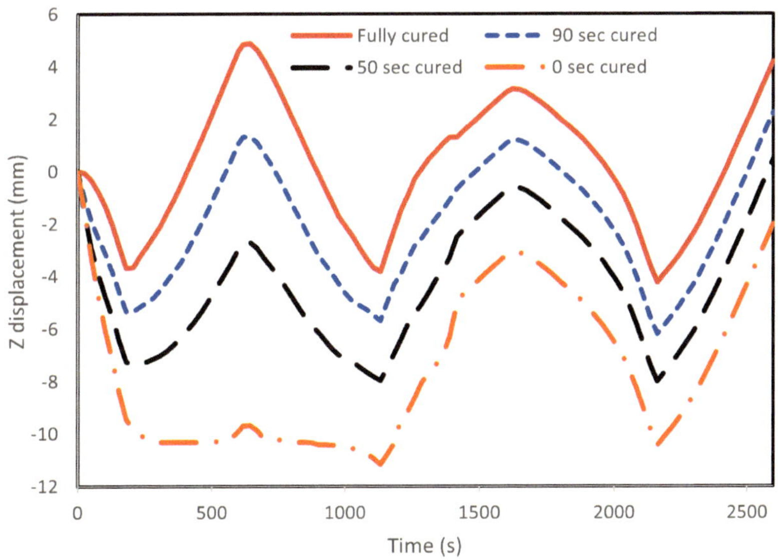

Figure 17.48 Warpage versus time for profile A

Figure 17.49 shows the Z-displacement of profile B for two kinds of initial curing ratio. We can also see the Z-displacement shift downward as the initial degree of curing is lower. Furthermore, the Z-displacement at the beginning of reflow was larger for profile B than for profile A, which implies that the warpage is related to the material's viscoelastic properties. In summary, this case illustrates that curing exerts a significant effect on warpage.

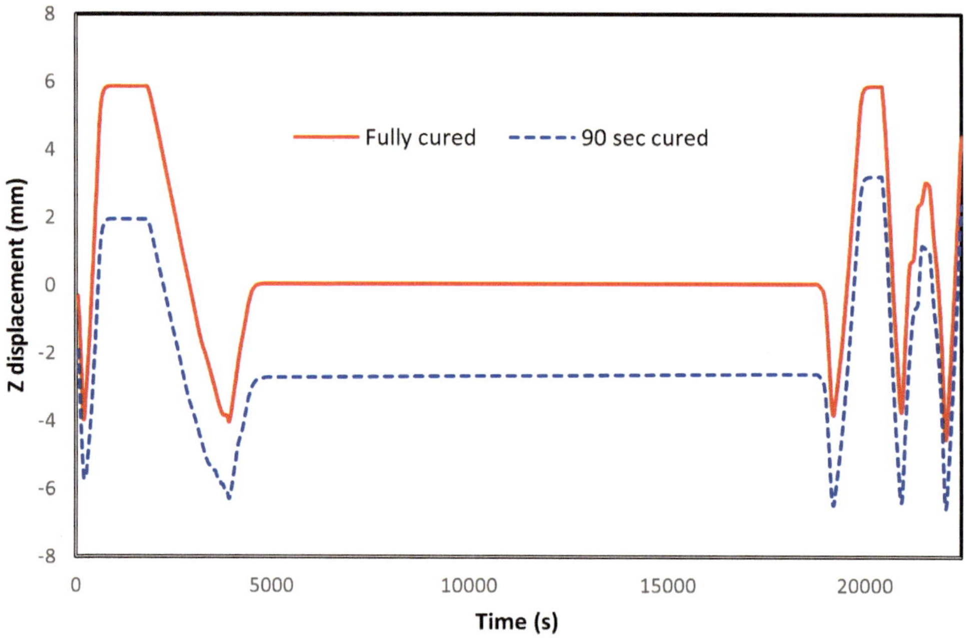

Figure 17.49 Z-displacement versus time for profile B

17.2.7 The Effect of Dispensing Control and Creeping Behavior on the Underfill Process

Due to the nature of thermoset plastics, the underfill material cannot be reprocessed after encapsulation. Moreover, trial-and-error cost at the development stage is high because both the capillary-driven flow and the curing process are time-consuming work. As a means of reducing development costs, CAE simulation is an efficient tool for discovering the best solution for design control that provides good agreement with requests.

Underfill materials are costly, hence mass control is also an important issue. Besides the dispensing area, underfill can also flow into the bump zone from the edge of the chip, due to overflow caused by creeping. Therefore, overflow simulation not only helps with understanding mass control of underfill, but also with the analysis of void locations caused by the melt front. As shown in Figure 17.50, after dispensing, the melt flow is predominantly driven by capillary forces, gravity, and fluid viscosity. Thus, the amount of material needs to allow for filling by capillary forces, edge flow due to creeping and extension flow by spreading forces when the melt sinks on the substrate. Therefore, any analysis aimed at comprehensively studying the use of material through CAE simulation should consider all three kinds of flow.

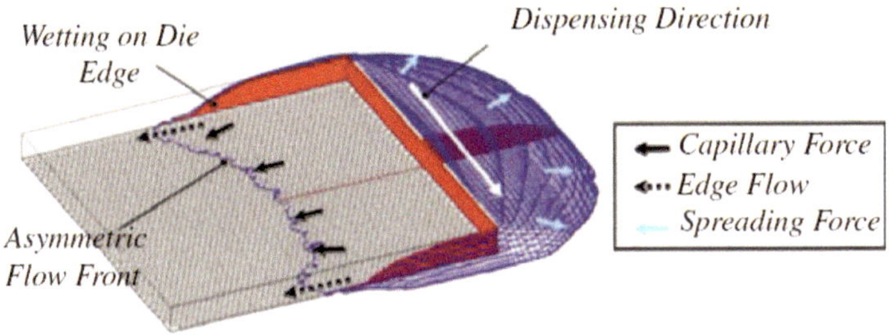

Figure 17.50 Chart of the relations existing in melt flow [13]

The simulation and experimental results in Figure 17.51 show that the dispensing and creeping flow area must be considered in the simulation. Because the effects of the dispensing and creeping flow area are neglected in traditional simulation, the simulation result shows the lack of the underfill from edge flow. More qualified simulation results are obtained with the Moldex3D 2020 Capillary Underfill Module, which provides comprehensive analysis.

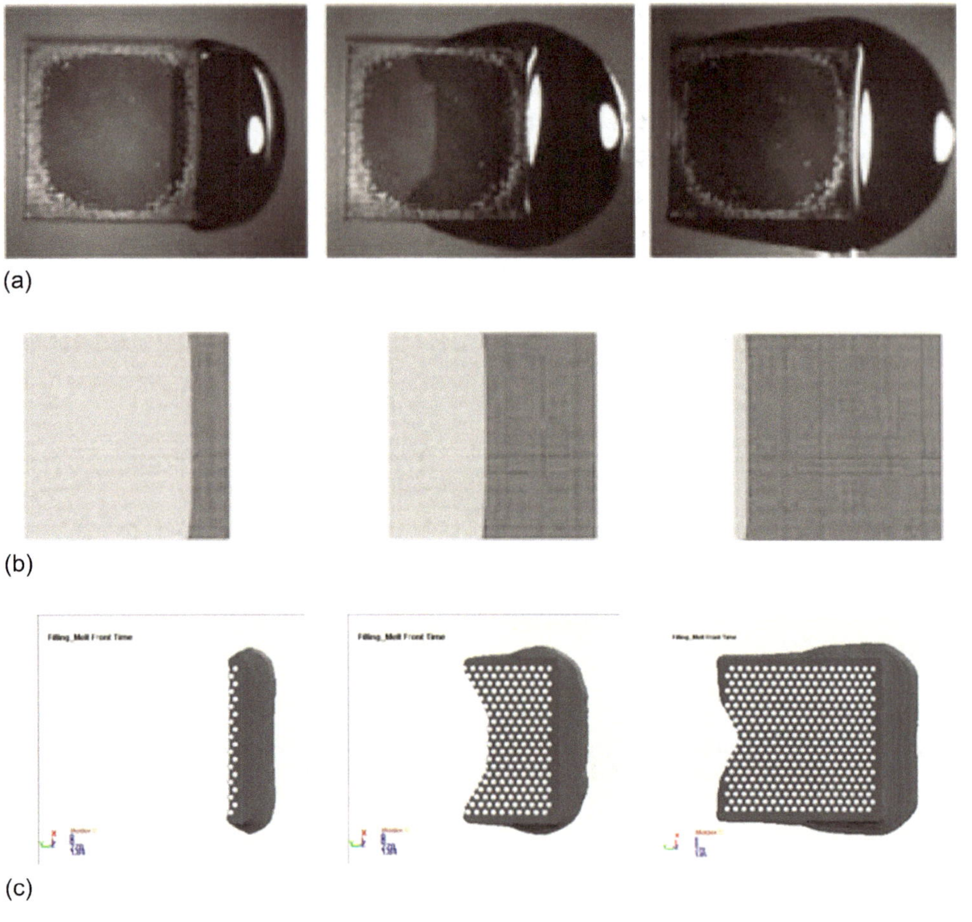

Figure 17.51 Creep-up behavior of (a) experimental results [13], (b) traditional simulation results [13], and (c) Moldex3D 2020 simulation results

This module utilizes the 3D model with flip chip product details, including the bump distribution and dies, as shown in Figure 17.52.

The Dispensing setting supports multiple paths, including the dispensing amount for every path, the start time and the speed of the dispenser movement, as shown in Figure 17.53.

The Material Parameter setting supports the contact angle between the filling material and other materials, as shown by the capillarity settings in Figure 17.54, permitting those polymer behaviors which are affected by environmental factors to be simulated.

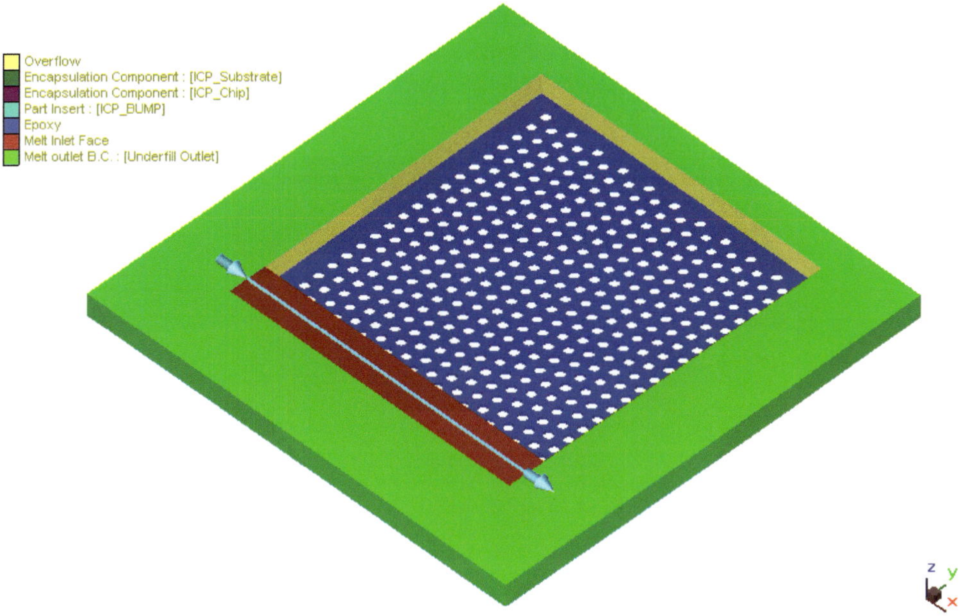

Figure 17.52 The 3D model

Dispensing passes

Color	Pass	Start time (s)	Velocity (mm/s)	Weight (mg)	Dispensing Times	Delay time (s)
	Pass 1	0.00	10.00	10.600	1	0.000

Maximum filling time: 20 (s)

Dispenser size (diameter): 1.2 (mm)

Figure 17.53 The Dispensing settings

Surface Tension Properties

Coefficient : 30 dyne/cm

Contact angle between :

Interface	Degree
Mold wall	30
Part insert-1 [ICP_BUMP]	30
Part insert-2 [ICP_Substrate]	30
Part insert-3 [ICP_Chip]	30

Figure 17.54 The Capillarity settings

Moldex3D simulation includes the dispensing process during flow, underfill of the bump area and flow outside of the dies (creep-up and extension flow), as shown in Figure 17.55.

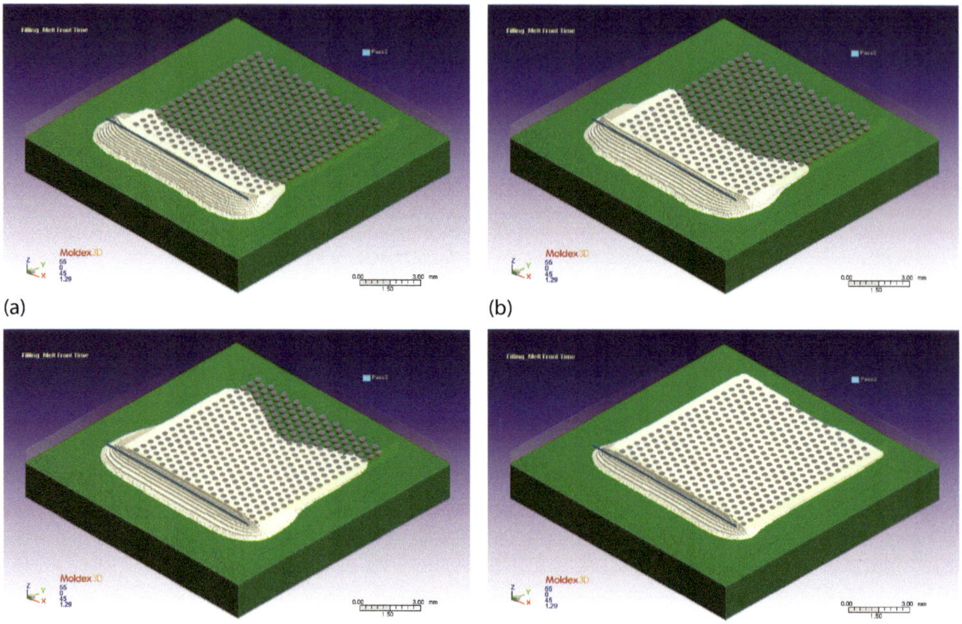

(a) (b) (c) (d)

Figure 17.55 Melt front time results at (a) 20%, (b) 40%, (c) 75%, and (d) 97%

Figure 17.56 shows the changes in melt sink behavior along with the underfill process in the dispensing area. From the simulation results, we can see the necessity to comprehensively consider the physical behaviors in the scientific simulation.

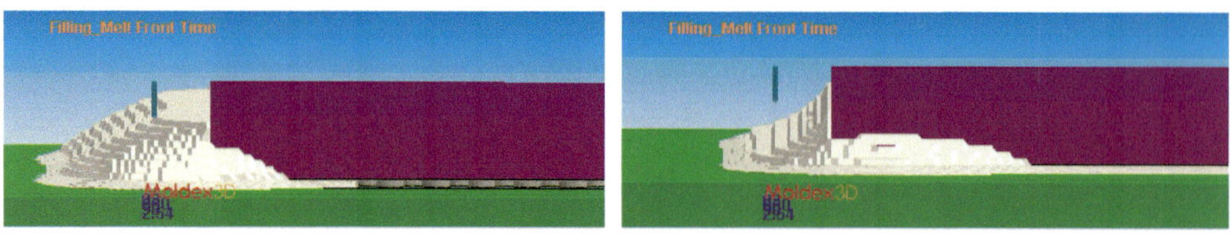

(a) (b)

Figure 17.56 Schematic diagrams of changes in the dispensing area at (a) 15% and (b) 90%

In summary, Moldex3D can help to understand the effects of dispensing control and creeping behavior on the underfill process at the design stage. Thus, the goal of achieving the desired design can be achieved through mass control of underfill and avoidance of costly trial and error.

17.3 CAE Case Study

Following the discussions of this chapter and the corresponding CAE verification case, an exercise model, as illustrated in Figure 17.57, is provided in this section for readers to practice CAE application in IC packaging.

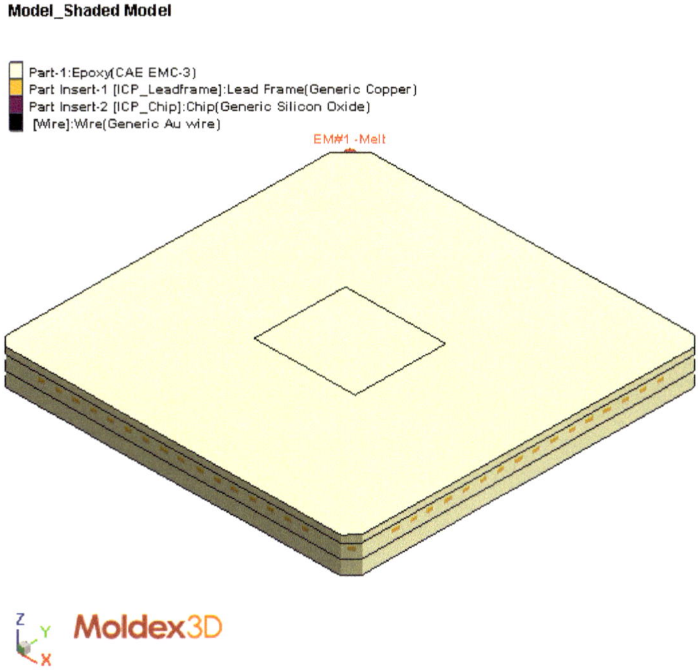

Figure 17.57 Case study sample

Please download the analysis files for CAE Case Study 17.3 from the following website:

https://moldex3d.box.com/s/zr6fvc1vlhbi4ocx111jwd3wmxt4ooif

Then open the .m3j or .mvj file in Moldex3D and answer the questions below according to the analysis results.

Questions:

1. According to Run 1
   a) Look at the model and show the layout of a lead frame, chip, and wire (hint: open model manager toolbar).
   b) Where will the model exhibit the worst wire sweep index value and what are the relations between wire sweep and filling/curing analysis?
   c) When is the maximum displacement of paddle during filling analysis and what are the relations between paddle shift value and filling/curing analysis?
2. According to Run 1
   a) What will happen if we change the process conditions such as:
      i. transfer time,
      ii. curing pressure,
      iii. curing time,
      iv. resin temperature,
      v. mold temperature?
   b) Summarize the effects of different process conditions on filling/curing analysis for:
      i. sprue pressure,
      ii. conversion,
      iii. temperature.
   c) Summarize the effects of different process conditions on wire sweep analysis for:
      i. drag force,
      ii. wire sweep index,
      iii. wire crossover.
   d) Summarize the effects of different process conditions on paddle shift analysis for:
      i. total displacement,
      ii. Von Mises stress.
   e) Suggest the optimal process conditions for a small wire index and a small paddle shift value.

References

[1] Lau, H.; Wong, C.; Prince, J.: *Electronic Packaging: Design, Materials, Process, and Reliability*, McGraw-Hill Professional, 1998

[2] Tian, M.-B.: *Electronic Packaging Technology for Semiconductor Device*, Wu-Nan Book Inc., 2005

[3] Chen, X.-W.: *Electronic Packaging Technology and Materials*, Gau Lih Books Co., 2004

[4] Zhang, Z.; Wong, C. P.: Recent Advances in Flip-Chip Underfill: Materials, Process, and Reliability, *IEEE Transactions on Advanced Packaging*, vol. 27, 2004, p. 515

[5] Brackbill, J.; Kothe, D.; Zemach, C.: A Continuum Method for Modeling Surface Tension, *Journal of Computational Physics*, vol. 100, 1992, p. 335

[6] Han, S.; Wang, K. K.: A Study on Wire Sweep in Encapsulation of Semiconductor Chips Using Simulated Experiments, *Journal of Electronic Packaging*, vol. 117, 1995, p. 178

[7] Lee, M. W.; Jung, W. K.; Sohn, E. S.; Lee, J. Y.; Hwang, C. H.; Lee, C. H.: "A Study on the Rheological Characterization and Flow Modeling of Molded Underfill (MUF) for Optimized Void Elimination Design", in 58th Electronic Components and Technology Conference, 2008

[8] Lee, M. W.; Khim, J. Y.; Yoo, M.; Chung, J. Y.; Lee, C. H.: "Rheological Characterization and Full 3D Mold Flow Simulation in Multi-die Stack CSP of Chip Array Packaging", in 56th Electronic Components and Technology Conference, 2006

[9] Hsu, C.-C.; Wang, T.-C.; Chen, Y.-C.; Y.-K. Lin, Y.-K.: "A Novel Method to Predict Fluid/Structure Interaction in IC Packaging", in 64th Electronic Components and Technology Conference, 2014

[10] Hsu, C.-C.;Vallury, S.; Lin, K.; Yang, A.: "A Study on Warpage Behavior of EMC in Post-Mold Cure Stage using Moldex3D", in 12th International Conference and Exhibition on Device Packaging, 2016

[11] Loh, W.; Ong, K.; Hsu, C.-C.; Wang, J.; Kulterman, R.; Fu, H.: "Molded Electronic Package Warpage Predictive Modelling Methodologies", in International Conference on Electronics Packaging, Toki Messe, Niigata, 2019

[12] Wang, J.; Eu, O.; Weng, W.; Hsu, C.-C.; Loh, W.; Kulterman, R.; Fu, H.: "Predictive Modelling Methodologies for Bi-material Strip Warpage", in 15th International Microsystems, Packaging, Assembly and Circuits Technology Conference, 2020

[13] Moon, S.-W.; Li, Z.; Gokhale, S.; Wang, J.: 3-D Numerical Simulation and Validation of Underfill Flow of Flip-Chips, *IEEE Transactions on Components, Packaging and Manufacturing Technology*, vol. 1, 2011, p. 1517

Index